METHODS IN MOLECULAR BIOLOGY

Series editor
John M. Walker
School of Life and Medical Sciences
University of Hertfordshire
Hatfield, Hertfordshire, AL10 9AB, UK

For further volumes:
http://www.springer.com/series/7651

Bacterial Protein Secretion Systems

Methods and Protocols

Editors

Laure Journet and Eric Cascales

Institut de Microbiologie de la Méditerranée, Aix-Marseille Univ - CNRS, Marseille, France

Humana Press

Editors
Laure Journet
Institut de Microbiologie de la Méditerranée
Aix-Marseille Univ - CNRS
Marseille, France

Eric Cascales
Institut de Microbiologie de la Méditerranée
Aix-Marseille Univ - CNRS
Marseille, France

ISSN 1064-3745 ISSN 1940-6029 (electronic)
Methods in Molecular Biology
ISBN 978-1-4939-7031-5 ISBN 978-1-4939-7033-9 (eBook)
DOI 10.1007/978-1-4939-7033-9

Library of Congress Control Number: 2017944103

Printed on acid-free paper

This Humana Press imprint is published by Springer Nature
The registered company is Springer Science+Business Media LLC
The registered company address is: 233 Spring Street, New York, NY 10013, U.S.A.

Preface

In their ecological niches, bacteria are in contact with other prokaryotic and eukaryotic cells. Bacteria therefore evolved mechanisms to communicate and collaborate with these cells. They also developed belligerent behaviours to eliminate competitors and to infect eukaryotic host cells. These aggressive actions are mediated by effector toxins with specific activities which will ultimately cause target cell lysis or re-routing of metabolic or trafficking pathways in the host.

The proper delivery of bacterial effectors into the milieu or directly into target cells is assured by dedicated structures called secretion systems. Up to now, nine secretion systems have been described in bacteria, and additional systems allow the exposition of toxins or adhesins at the cell surface or at the extremity of a pilus structure. These multi-protein trans-envelope complexes differ in composition, mechanism of assembly, and mode of recruitment and transport of toxins. However, studying these macromolecular complexes requires common techniques, ranging from the bioinformatic identification of machine components and effectors to methods of defining interactions amongst the different subunits and the development of reporters to follow effector translocation in vivo. Finally, state-of-the-art techniques have made significant progress in the analysis of these large complexes directly in the bacterial cell or from purified samples.

The purpose of this book on bacterial secretion systems is to provide protocols that cover the broad arsenal of techniques used to study a secretion system from A to Z: identifying and localising the different subunits, defining interactions within subunits, monitoring conformational changes, purifying and imaging large complexes, defining the assembly pathway by fluorescence microscopy and the role of energy during assembly or secretion, and identifying secreted effectors as well as using reporters to follow effector transport. Most of these techniques are not restricted to the study of secretion systems but are also of specific interest for any researcher interested in multi-protein complexes of the bacterial cell envelope.

The book starts with a chapter describing a recently developed software program aimed at identifying gene clusters encoding secretion systems within bacterial genomes. Then six chapters describe methods for defining the subcellular localisation of the different subunits of a multi-protein system: prediction programs, fractionation, cell surface exposition and isopycnic density gradients to partition inner and outer membranes. Three techniques to determine the topology of membrane proteins (substituted cysteine accessibility, protease accessibility and reporter fusions) are then detailed. This is followed by eleven chapters covering genetic, cytologic and biochemical methods used to study protein–protein and protein–peptidoglycan interactions. Two chapters are dedicated to methods for unveiling the role of energy and conformational changes, and then six chapters describe techniques for purifying and imaging large complexes. Finally, methods to identify effectors as well as the reporter fusions used to validate effector secretion and translocation are presented.

Marseille, France *Laure Journet*
Marseille, France *Eric Cascales*

Contents

Contributors

SOPHIE S. ABBY • *Microbial Evolutionary Genomics, Institut Pasteur, Paris, France; CNRS, UMR3525, Paris, France; Université Grenoble Alpes, Laboratoire Techniques de l'Ingénierie Médicale et de la Complexité - Informatique, Mathématiques et Applications, Grenoble (TIMC-IMAG), F-38000 Grenoble, France; Centre National de la Recherche Scientifique (CNRS), TIMC-IMAG, Grenoble, France*

JULIANA ALCOFORADO DINIZ • *Division of Molecular Microbiology, School of Life Sciences, University of Dundee, Dundee, UK*

JULIE ALLOMBERT • *CIRI, Centre International de Recherche en Infectiologie, Inserm, U1111, Université Claude Bernard Lyon 1, CNRS, UMR5308, École Normale Supérieure de Lyon, Univ Lyon, Villeurbanne, France*

KRISHNAMOHAN ATMAKURI • *Translational Health Science and Technology Institute, NCR Biotech Science Cluster, Faridabad, India*

INA ATTRÉE • *Bacterial Pathogenesis and Cellular Responses, Centre National pour la Recherche Scientifique (CNRS), University Grenoble Alpes, INSERM, Biosciences and Biotechnology Institut, CEA Grenoble, Grenoble, France*

SARAH BIGOT • *Molecular Microbiology and Structural Biochemistry, CNRS UMR 5086, Université Lyon 1, Institut de Biologie et Chimie des Protéines, Lyon, France*

MIKHAIL BOGDANOV • *Department of Biochemistry and Molecular Biology, University of Texas Health Science Center at Houston, McGovern Medical School, Houston, TX, USA*

EMMANUELLE BOUVERET • *Laboratoire d'Ingénierie des Systèmes Macromoléculaires, UMR7255, Institut de Microbiologie de la Méditerranée, Aix-Marseille University—CNRS, Marseille, France*

NIENKE BUDDELMEIJER • *Institut Pasteur, Biology and Genetics of the Bacterial Cell Wall Unit, Inserm Group Avenir, Paris, France*

MARIA GUILLERMINA CASABONA • *Bacterial Pathogenesis and Cellular Responses, Centre National pour la Recherche Scientifique (CNRS), University Grenoble Alpes, INSERM, Biosciences and Biotechnology Institut, CEA Grenoble, Grenoble, France; Division of Molecular Microbiology, School of Life Sciences, University of Dundee, Dundee, Scotland, UK*

ERIC CASCALES • *Laboratoire d'Ingénierie des Systèmes Macromoléculaires, (LISM, UMR7255), Institut de Microbiologie de la Méditerranée (IMM), Aix-Marseille Université—Centre National de la Recherche Scientifique (CNRS), Marseille, France*

SUMA CHAKRAVARTHY • *Plant Pathology and Plant-Microbe Biology Section, School of Integrative Plant Science, Cornell University, Ithaca, NY, USA*

XAVIER CHARPENTIER • *CIRI, Centre International de Recherche en Infectiologie, Inserm, U1111, Université Claude Bernard Lyon 1, CNRS, UMR5308, École Normale Supérieure de Lyon, Univ Lyon, Villeurbanne, France*

NUNO CHARRO • *UCIBIO—REQUIMTE, Faculdade de Ciências e Tecnologia, Universidade NOVA de Lisboa (FCT NOVA), Caparica, Portugal*

TIAGO R.D. COSTA • *Institute for Structural and Molecular Biology, School of Biological Sciences, Birkbeck College, London, UK*

SARAH J. COULTHURST • *Division of Molecular Microbiology, School of Life Sciences, University of Dundee, Dundee, UK*

MARILYNE DAVI • *Unité de Biochimie des Interactions Macromoléculaires, Département de Biologie Structurale et Chimie, Institut Pasteur, CNRS, UMR 3528, Paris, France*

ANDREAS DIEPOLD • *Department of Biochemistry, University of Oxford, Oxford, UK; Department of Ecophysiology, Max Planck Institute for Terrestrial Microbiology, Karl-von-Frisch-Str., Marburg, Germany*

TOBIAS DIETSCHE • *Section of Cellular and Molecular Microbiology, Interfaculty Institute of Microbiology and Infection Medicine (IMIT), University of Tübingen, Tübingen, Germany*

BADREDDINE DOUZI • *Laboratoire d'Ingénierie des Systèmes Macromoléculaires (LISM, UMR 7255), Institut de Microbiologie de la Méditerranée (IMM), Aix-Marseille Université—Centre National de la Recherche Scientifique (CNRS), Marseille, France*

DENIS DUCHÉ • *Laboratoire d'Ingénierie des Systèmes Macromoléculaires (LISM, UMR 7255), Institut de Microbiologie de la Méditerranée (IMM), Aix-Marseille Université—Centre National de la Recherche Scientifique (CNRS), Marseille, France*

RHYS A. DUNSTAN • *Department of Microbiology and Infection and Immunity Program, Biomedicine Discovery Institute, Monash University, Melbourne, VIC, Australia*

ERIC DURAND • *Laboratoire d'Ingénierie des Systèmes Macromoléculaires (LISM, UMR7255), Institut de Microbiologie de la Méditerranée (IMM), Aix-Marseille Univ and CNRS, Marseille, France*

MARC ERHARDT • *Helmholtz Centre for Infection Research, Braunschweig, Germany*

MARIO F. FELDMAN • *Department of Molecular Microbiology, Washington University School of Medicine, Saint Louis, MO, USA*

NICOLAS FLAUGNATTI • *Laboratoire d'Ingénierie des Systèmes Macromoléculaires, (LISM, UMR7255), Institut de Microbiologie de la Méditerranée, Aix-Marseille Univ—CNRS, Marseille, France*

KATRINA T. FOREST • *Department of Bacteriology and Biophysics Program, University of Wisconsin-Madison, Madison, WI, USA*

IRINA S. FRANCO • *UCIBIO—REQUIMTE, Faculdade de Ciências e Tecnologia, Universidade NOVA de Lisboa (FCT NOVA), Caparica, Portugal*

EMILIE GAULIARD • *Unité de Biochimie des Interactions Macromoléculaires, Département de Biologie Structurale et Chimie, Institut Pasteur, CNRS, UMR 3528, Paris, France; Université Paris Diderot, Sorbonne Paris Cité, Cellule Pasteur, Paris, France*

ALBA KATIRIA GONZALEZ RIVERA • *Department of Bacteriology and Biophysics Program, University of Wisconsin-Madison, Madison, WI, USA*

DIDIER GRUNWALD • *Bacterial Pathogenesis and Cellular Responses, Centre National pour la Recherche Scientifique (CNRS), University Grenoble Alpes, INSERM, Biosciences and Biotechnology Institut, CEA Grenoble, Grenoble, France*

BIRGIT HABENSTEIN • *Institute of Chemistry & Biology of Membranes & Nanoobjects (UMR5248 CBMN), CNRS, University of Bordeaux, Institut Européen de Chimie et Biologie, Pessac, France*

IAIN D. HAY • *Department of Microbiology and Infection and Immunity Program, Biomedicine Discovery Institute, Monash University, Melbourne, VIC, Australia*

BIRTE HOLLMANN • *Division of Molecular Microbiology, School of Life Sciences, University of Dundee, Dundee, UK*

Laetitia Houot • *Laboratoire d'Ingénierie des Systèmes Macromoléculaires (LISM, UMR 7255), Institut de Microbiologie de la Méditerranée (IMM), Aix-Marseille Université—Centre National de la Recherche Scientifique (CNRS), Marseille, France*

S. Peter Howard • *Department of Microbiology and Immunology, College of Medicine, University of Saskatchewan, Saskatoon, SK, Canada*

Bethany Huot • *Department of Energy, Plant Research Laboratory, Michigan State University, East Lansing, MI, USA*

Raffaele Ieva • *Laboratoire de Microbiologie et de Génétique Moléculaires, Centre de Biologie Intégrative (CBI), Université de Toulouse, CNRS, UPS, Toulouse, France*

Athanasios Ignatiou • *Institute for Structural and Molecular Biology, School of Biological Sciences, Birkbeck College, London, UK*

Laure Journet • *Laboratoire d'Ingénierie des Systèmes Macromoléculaires, (LISM, UMR7255), Institut de Microbiologie de la Méditerranée, Aix-Marseille Univ—CNRS, Marseille, France*

Gouzel Karimova • *Unité de Biochimie des Interactions Macromoléculaires, Département de Biologie Structurale et Chimie, Institut Pasteur, CNRS, UMR 3528, Paris, France*

Anna Konovalova • *Lewis Thomas Laboratory, Department of Molecular Biology, Princeton University, Princeton, NJ, USA*

Brian H. Kvitko • *Department of Plant Pathology, University of Georgia, Athens, GA, USA*

Daniel Ladant • *Unité de Biochimie des Interactions Macromoléculaires, Département de Biologie Structurale et Chimie, Institut Pasteur, CNRS, UMR 3528, Paris, France*

Erh-Min Lai • *Institute of Plant and Microbial Biology, Academia Sinica, Taipei, Taiwan*

Ray A. Larsen • *Department of Biological Sciences, Bowling Green State University, Bowling Green, OH, USA*

Gang Li • *Department of Microbiology and Immunology, College of Medicine, University of Saskatchewan, Saskatoon, SK, Canada*

Jer-Sheng Lin • *Institute of Plant and Microbial Biology, Academia Sinica, Taipei, Taiwan*

Trevor Lithgow • *Department of Microbiology, and Infection and Immunity Program, Biomedicine Discovery Institute, Monash University, Melbourne, VIC, Australia*

Roland Lloubes • *Laboratoire d'Ingénierie des Systèmes Macromoléculaires, (LISM, UMR7255), Institut de Microbiologie de la Méditerranée, Aix-Marseille Univ—CNRS, Marseille, France*

Laureen Logger • *Laboratoire d'Ingénierie des Systèmes Macromoléculaires, UMR7255, Institut de Microbiologie de la Méditerranée (IMM), Aix-Marseille Univ and CNRS, Marseille, France*

Antoine Loquet • *Institute of Chemistry & Biology of Membranes & Nanoobjects (UMR5248 CBMN), CNRS, University of Bordeaux, Institut Européen de Chimie et Biologie, Pessac, France*

Arthur Louche • *Molecular Microbiology and Structural Biochemistry, CNRS UMR 5086, Université Lyon 1, Institut de Biologie et Chimie des Protéines, Lyon, France*

Pek Man Ly • *Department of Molecular Microbiology, Washington University School of Medicine, Saint Louis, MO, USA*

João M. Medeiros • *Department of Biology, ETH Zürich, Institute of Molecular Biology and Biophysics, Zürich, Switzerland*

Julia V. Monjarás Feria • *Section of Cellular and Molecular Microbiology, Interfaculty Institute of Microbiology and Infection Medicine (IMIT), University of Tübingen, Tübingen, Germany*

Luís Jaime Mota • *UCIBIO—REQUIMTE, Faculdade de Ciências e Tecnologia, Universidade Nova de Lisboa (FCT NOVA), Caparica, Portugal*

Henrik Nielsen • *Technical University of Denmark, Lyngby, Denmark*

Elena V. Orlova • *Institute for Structural and Molecular Biology, School of Biological Sciences, Birkbeck College, London, UK*

Scot P. Ouellette • *Division of Basic Biomedical Sciences, Sanford School of Medicine, University of South Dakota, Vermillion, SD, USA*

Sara V. Pais • *UCIBIO—REQUIMTE, Faculdade de Ciências e Tecnologia, Universidade Nova de Lisboa (FCT NOVA), Caparica, Portugal*

Melissa Petiti • *Laboratoire d'Ingénierie des Systèmes Macromoléculaires, (LISM, UMR7255), Institut de Microbiologie de la Méditerranée, Aix-Marseille Univ—CNRS, Marseille, France*

Martin Pilhofer • *Department of Biology, ETH Zürich, Institute of Molecular Biology and Biophysics, Zürich, Switzerland*

Mylène Robert-Genthon • *Bacterial Pathogenesis and Cellular Responses, Centre National pour la Recherche Scientifique (CNRS), University Grenoble Alpes, INSERM, Biosciences and Biotechnology Institut, CEA Grenoble, Grenoble, France*

Eduardo P.C. Rocha • *Microbial Evolutionary Genomics, Institut Pasteur, Paris, France; CNRS, UMR3525, Paris, France*

Suzana P. Salcedo • *Molecular Microbiology and Structural Biochemistry, CNRS UMR 5086, Université Lyon 1, Institut de Biologie et Chimie des Protéines, Lyon, France*

Yoann G. Santin • *Laboratoire d'Ingénierie des Systèmes Macromoléculaires, (LISM, UMR7255), Institut de Microbiologie de la Méditerranée, Aix-Marseille University—CNRS, Marseille, France*

Mehari Tesfazgi Mebrhatu • *Section of Cellular and Molecular Microbiology, Interfaculty Institute of Microbiology and Infection Medicine (IMIT), University of Tübingen, Tübingen, Germany*

Claudia E. Torres-Vargas • *Section of Cellular and Molecular Microbiology, Interfaculty Institute of Microbiology and Infection Medicine (IMIT), University of Tübingen, Tübingen, Germany*

Julie P.M. Viala • *Laboratoire d'Ingénierie des Systèmes Macromoléculaires, UMR7255, Institut de Microbiologie de la Méditerranée (IMM), Aix-Marseille Université—Centre National de la Recherche Scientifique (CNRS), Marseille, France*

Anne Vianney • *CIRI, Centre International de Recherche en Infectiologie, Inserm, U1111, Université Claude Bernard Lyon 1, CNRS, UMR5308, École Normale Supérieure de Lyon, Univ Lyon, Villeurbanne, France*

Maxence S. Vincent • *Laboratoire d'Ingénierie des Systèmes Macromoléculaires, (LISM, UMR7255), Institut de Microbiologie de la Méditerranée (IMM), Aix-Marseille Université—Centre National de la Recherche Scientifique (CNRS), Marseille, France*

Samuel Wagner • *Section of Cellular and Molecular Microbiology, Interfaculty Institute of Microbiology and Infection Medicine (IMIT), University of Tübingen, Tübingen, Germany; German Center for Infection Research (DZIF), Tübingen, Germany*

Brent S. Weber • *Department of Molecular Microbiology, Washington University School of Medicine, Saint Louis, MO, USA*

GREGOR L. WEISS • *Department of Biology, ETH Zürich, Institute of Molecular Biology and Biophysics, Zürich, Switzerland*
SUSANN ZILKENAT • *Section of Cellular and Molecular Microbiology, Interfaculty Institute of Microbiology and Infection Medicine (IMIT), University of Tübingen, Tübingen, Germany*
ABDELRAHIM ZOUED • *Laboratoire d'Ingénierie des Systèmes Macromoléculaires, UMR7255, Institut de Microbiologie de la Méditerranée, Aix-Marseille Univ—CNRS, Marseille, France; Division of Infectious Diseases and Harvard Medical School, Department of Microbiology and Immunobiology, Howard Hughes Medical Institute, Brigham and Women's Hospital, Boston, MA, USA*

Chapter 1

Identification of Protein Secretion Systems in Bacterial Genomes Using MacSyFinder

Sophie S. Abby and Eduardo P.C. Rocha

Abstract

Protein secretion systems are complex molecular machineries that translocate proteins through the outer membrane, and sometimes through multiple other barriers. They have evolved by co-option of components from other envelope-associated cellular machineries, making them sometimes difficult to identify and discriminate. Here, we describe how to identify protein secretion systems in bacterial genomes using MacSyFinder. This flexible computational tool uses the knowledge stemming from experimental studies to identify homologous systems in genome data. It can be used with a set of predefined models—"TXSScan"— to identify all major secretion systems of diderm bacteria (i.e., with inner and with LPS-containing outer membranes). For this, it identifies and clusters colocalized components of secretion systems using sequence similarity searches with hidden Markov model protein profiles. Finally, it checks whether the genetic content and organization of clusters satisfy the constraints of the model. TXSScan models can be customized to search for variants of known systems. The models can also be built from scratch to identify novel systems. In this chapter, we describe a complete pipeline of analysis, including the identification of a reference set of experimentally studied systems, the identification of components and the construction of their protein profiles, the definition of the models, their optimization, and, finally, their use as tools to search genomic data.

Key words Comparative genomics, Genome annotation, Bioinformatics detection, Macromolecular systems, Bioinformatic modeling

1 Introduction

Bacteria produce proteins to interact with other individuals, prokaryotes or eukaryotes, to effect changes in their local environment, or to take up resources. Many of the proteins involved in these processes need to be secreted to the outside of the cell. Bacteria with an LPS-containing outer membrane (henceforth called diderms) face a formidable challenge in secreting these proteins because they must transport them through the inner membrane, the cell wall, the outer membrane, and eventually other barriers such as the bacterial capsule and membranes of other cells. The complexity of these molecular processes and the key roles of

Laure Journet and Eric Cascales (eds.), *Bacterial Protein Secretion Systems: Methods and Protocols*, Methods in Molecular Biology, vol. 1615, DOI 10.1007/978-1-4939-7033-9_1, © Springer Science+Business Media LLC 2017

protein secretion systems in bacterial ecology and virulence have spurred much interest in their study (for recent reviews *see* refs. [1–8]). There are nine well-known protein secretion systems in diderms (numbered from T1SS to T9SS), but others probably remain undiscovered [9].

Few computational tools exist to identify and characterize protein secretion systems in bacterial genomes (for a list *see* Table 1 of [9]). Their development becomes urgent in light of the availability of many thousands of genomes and the ease with which new ones can be sequenced. These tools should be able to identify components of the protein secretion systems and to assess whether they are sufficient at defining an instance of a given system. When the components are highly conserved proteins, they can be identified with high sensitivity. The identification of fast-evolving components might be more complicated because of poor sequence conservation. Additionally, some components may not be strictly necessary for a functional system, and it may be difficult to know which of the two factors explains their absence from an instance of a system. Under these conditions, it is useful for bioinformatics purposes to split the components of secretion systems into those that should be present in the instance ("mandatory") and those that may be absent ("accessory"). The former correspond to highly conserved, easily identifiable components, the latter to components that may be lacking in systems either because they are missing or because they were not detected. This nomenclature presumes nothing about the biological role of the accessory components: they may be biologically essential but unidentifiable by sequence similarity searches. The goal of this classification is to describe the system in a way that facilitates its identification in genomes. We will use it throughout this text.

The evolution of secretion systems has involved the co-option of many components from other molecular machineries. These components have sometimes been co-opted in their turn for other cellular machineries. As a result, many components of protein secretion systems have homologs in other systems [10, 11]. This increases the risk of misidentifications. For example, the T3SS and the flagellum are evolutionarily related, and several of their core components belong to homologous families [10]. In this specific case, discrimination between the two systems is facilitated by the existence of flagellum-associated mandatory proteins that are always absent from T3SS (e.g., FlgB), and vice versa (e.g., secretin). These components can be qualified as "forbidden" in the other system to prevent misidentifications.

The discrimination between protein secretion systems and other molecular systems can also be improved by the analysis of the genetic context. For example, T3SS components are usually encoded in a single locus, and this facilitates their discrimination from the components of the flagellum [12]. Another example is provided by the three mandatory components of T1SS, all of which have homologs in other systems [13], even if none of the other systems actually includes all three of them [14]. Importantly, the abc and mfp components are systematically encoded in the same locus in T1SS (and only in this system). Hence, three types of information facilitate the unambiguous detection of secretion systems: the identification of pertinent and forbidden components, the completeness of the set of components, and their genetic organization. This information can be put together in a model of the system that can be used by a computer program called MacSyFinder. This program searches genomes for instances satisfying the characteristics described in the model [15].

Sometimes good computational models of protein secretion systems are not available. The creation of novel (or better) models requires the identification of the relevant components and their genetic organization. Most secretion systems are studied on a small number of bacteria, where they are sometimes remarkably well characterized in terms of their components, their genetic regulation, their structure, and sometimes their assembly pathways. In contrast, the other instances of the systems can be very poorly characterized. The challenge posed to the researcher interested in identifying novel instances of a given type of system is thus to produce models with pertinent descriptions of the current knowledge of the system. This is difficult because the number of components and their organization may vary widely. For example, the T4SS locus of *Legionella pneumophila* is encoded by more than twice the number of genes of the *vir* T4SS of *Agrobacterium tumefaciens* [16, 17]. Most T4SS are encoded in a locus, but there are intracellular pathogens where they are encoded in several distant loci [18]. The key point in the production of novel models is thus the identification of traits that are conserved and can be most useful to identify a certain type of system.

The production of models involves generalizing knowledge obtained from specific examples. These models are quantitative representations of the composition and organization of known systems. When they work, they dramatically facilitate the identification of homologous systems. When they fail, they highlight gaps in our understanding of systems, which often raises interesting biological questions.

This chapter shows how to use MacSyFinder to identify protein secretion systems with the predefined models of TXSScan (*see* Fig. 1, Subsection 3.3). These models define the components of the system, the minimum number of mandatory and accessory components (quorum), and their genetic organization. They have been validated and shown to perform well: they identify the vast majority of known systems [9]. They have recently been used to identify over 10,000 systems in bacteria (available in MacSyDB/TXSSdb; *see* Table 1). Yet these models may be inadequate in

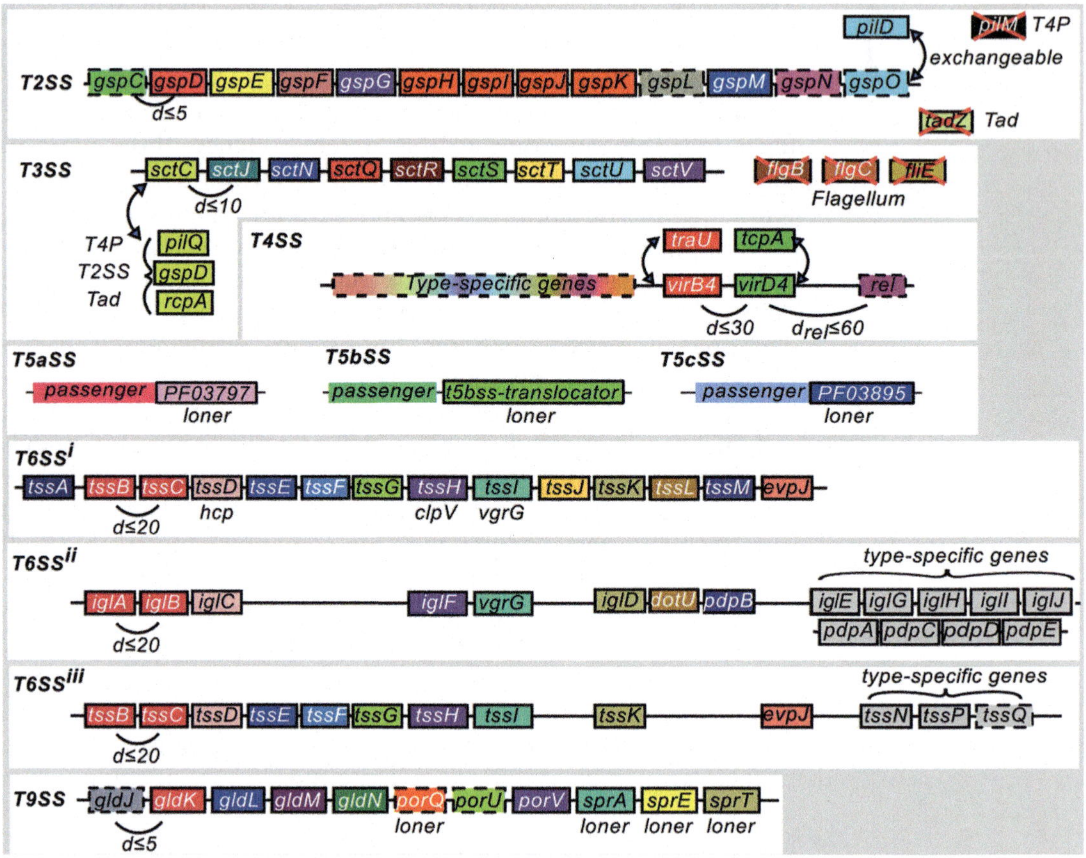

Fig. 1 Models of protein secretion systems available in TXSScan. The T1SS model is displayed in Fig. 2. *Each box* represents a component and its status in the system's model: "mandatory" (*plain*), "accessory" (*dashed*), or "forbidden" (*red cross*). Within a system panel: the families of homologous proteins are represented in columns and colored identically. When applicable, the system of origin is indicated next to the *box(es)*. The colocalization parameter of the system is indicated (*d*). When this parameter is specific to a gene (e.g., the relaxase in T4SS), it is indicated by a *subscript* (e.g., d_{rel}). Additional features specific to a component can be found in Table 2. *Curved double-headed arrows* indicate exchangeable components. Figures and legends are freely reproduced with modification from reference [9] [as specified by the Creative Commons Attribution (CC BY) license version 4.0]

certain situations. We show how they can be modified or built from scratch to identify novel variants of a secretion system. The models can then be easily shared.

MacSyFinder uses hidden Markov model (HMM) protein profiles specified in a model to search for components of a system in a file of protein sequences. It collects all clusters of colocalized components in the genome and checks whether they are valid relative to the specifications of the model. MacSyFinder then outputs the results of the protein profile searches and the information about the identified secretion systems. In the next section, we indicate

Table 1
Useful online resources to design models for the detection of secretion systems

Resource [reference]	Type	Commands
NCBI/Blast [32]	Sequence database indexing Sequence similarity search	makeblastdb blastp
Silix [25]	Sequence clustering	silix
MAFFT [21]	Multiple sequence alignment	mafft
Muscle [22]	Multiple sequence alignment	muscle
Seaview [23]	Sequence alignment, edition, and phylogeny	–
Jalview [24]	Sequence alignment, edition, and analysis	–
HMMER [20]	Build HMM profiles and use them for sequence similarity searches	hmmbuild hmmsearch
PFAM [34]	Database of HMM protein profiles	–
TIGRFAM [35]	Database of HMM protein profiles	–
InterProScan [36]	Identify conserved domains using several resources	–
MacSyFinder [15]	Macromolecular system detection from models of systems	macsyfinder
MacSyView [15]	Visualization of MacSyFinder's results in a web browser	–
TXSScan [9]	MacSyFinder-based tools to detect T1SS-T9SS and related appendages (T4P, Tad, bacterial flagellum)	macsyfinder
TXSSdb [9]	Database of secretion systems detected with TXSScan in 1,528 genomes of diderm bacteria	–

the data and software required to use and design models for MacSyFinder. In the last section, we describe how to define models and protein profiles to identify protein secretion systems.

2 Materials

2.1 Sequence Data

MacSyFinder analyzes protein sequences to identify protein secretion systems. The protein sequences should all be stored in a single file in Fasta format. This file represents one of several types of information that must be specified by the "--db-type" option:

- When the protein sequences are from diverse sources (e.g., peptides from a metagenome), the file type is "*unordered*." In this case, the program simply outputs the results of the identification of the systems' components using the protein profiles.

- When the proteins are from a single genome but the relative order of the corresponding genes in the genome is not known, the file type is "unordered_replicon." In this case, the program can identify components and check whether the quorum of components is respected. It cannot, however, check the genetic organization of the model.
- When proteins are from a single genome and are ordered following the position of genes in the genome, then the file type is "ordered_replicon." The "ordered_replicon" mode allows the use of all available criteria (quorum of components and genetic organization) to identify instances of the system and as such is the most powerful mode. It should be used whenever possible to create, test, and validate a new model and the corresponding protein profiles. It will constitute the focus of this protocol.
- The "gembase" type is similar to the "ordered_replicon" but has special identifiers that allow the analysis of multiple "ordered" genomes in a single step (*see* MacSyFinder documentation).

2.2 Predefined Models Available in TXSScan

TXSScan is a set of predefined MacSyFinder models to detect the best-studied protein secretion systems in diderms (T1SS, T2SS, T3SS, T4SS, T5SS, T6SS, T9SS, and related appendages). These models are used as examples throughout the following sections. The files of TXSScan (HMM profiles and model files) should be placed at a recognizable location (*see* Subsection 3.3.1) to be used locally with the standalone program MacSyFinder. One can add to these directory profiles built by the user or retrieved from public databases. MacSyFinder can also be used online with the TXSScan models (*see* Table 1). Currently, only the standalone version allows the modification of TXSScan models and the introduction of novel protein profiles.

2.3 Software

Table 1 displays resources of interest for this protocol. To run MacSyFinder, one needs to install the NCBI/BLAST tools (in particular, makeblastdb version 2.8 or higher or formatdb), HMMER, and MacSyFinder [15, 19, 20]. The latter requires a Python interpreter (version 2.7) that must be installed beforehand. See MacSyFinder's online documentation for more details [15].

To build novel HMM protein profiles, one also needs a program to make multiple sequence alignments (e.g., MAFFT or Muscle [21, 22]), an alignment editor (e.g., Seaview or Jalview [23, 24]), and a program to cluster proteins by sequence similarity (e.g., Silix or MCL [25, 26]). The program MacSyView can be used to visualize the results of MacSyFinder.

3 Methods

The procedure to design a novel model follows a number of steps (*see* Fig. 2). Briefly, it starts with the identification of the relevant components, and their genetic organization, from the reference data set of experimentally validated instances of the system. The HMM protein profiles for each component can be built from the reference data set or retrieved from public databases. These three types of information (component list, genetic architecture, and profiles) can then be used to formulate the model. Finally, the model is used to analyze an independent data set of experimentally validated systems. At the end of this process, one should be able to use the model to identify novel instances of the system and know the sensitivity of the procedure.

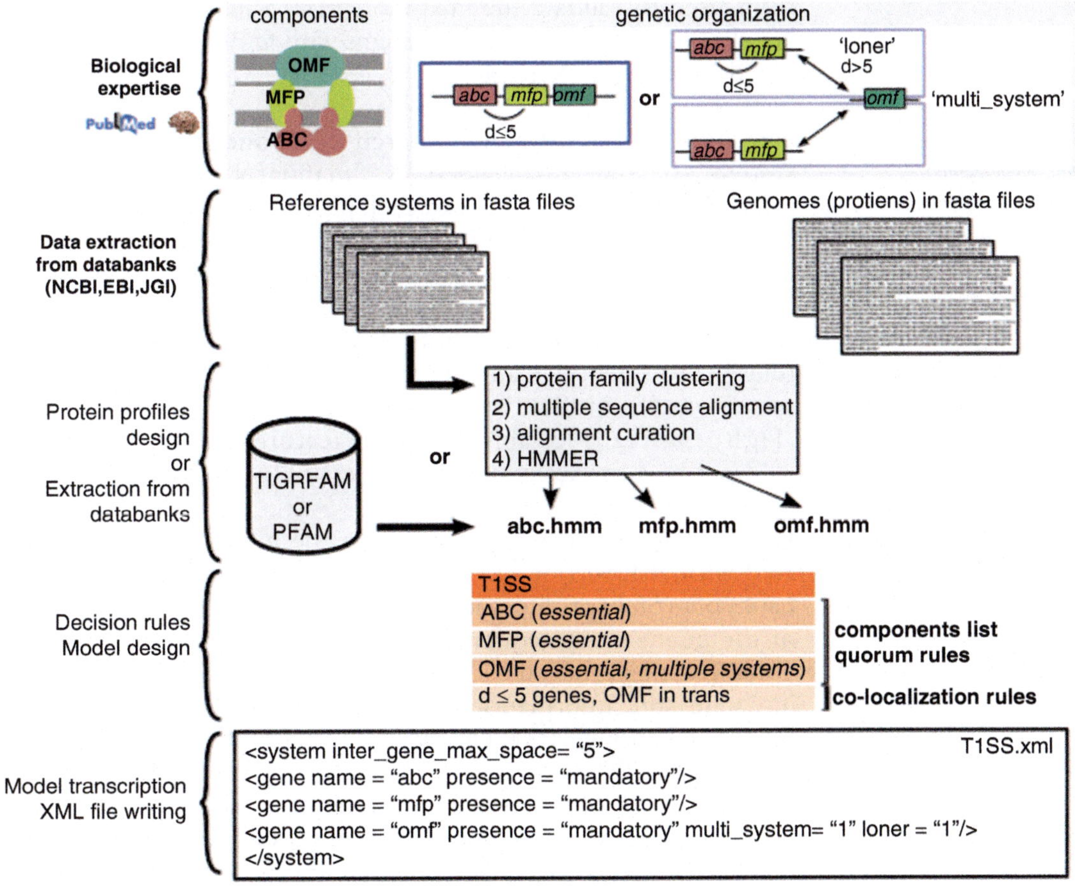

Fig. 2 Overview of the protocol

3.1 Compilation of Available Information on the System

Gathering the current biological knowledge of the system, especially of its experimentally studied instances, may not be simple. The genes and proteins of experimentally studied systems are often named differently; sequence similarity searches may thus be necessary to establish which components in a system are homologous to those of other systems. Occasionally, gene fusions and fissions further complicate the identification of families of homologs. Once the relationships of homology between known instances of the system are well established, its key components can be inventoried and qualified (as mandatory, accessory, and forbidden). One can then characterize the system in terms of its genetic organization and quorum (i.e., how many components are required for a system, *see* Subsection 3.1.4). The set of experimentally validated protein secretion systems should be split into two independent data sets (*see* **Note 1**). The *reference* data set is used to characterize the protein secretion systems, build the models, and construct the protein profiles (if required; *see* **Notes 2** and **3**). The *validation* data set is used to validate the final model (*see* Subsection 3.5). There are alternative procedures for systems with few experimentally validated systems (*see* **Note 4**).

3.1.1 Identification and Classification of Secretion System Components

The most frequently encountered components of a secretion system are typically classed as "mandatory." They may be inferred from the frequency of each protein family in the *reference* data set or retrieved from the literature (*see* Subsections 3.1.3 and 3.1.4) (*see* **Note 2**). The other components are classed as "accessory." When it is necessary to distinguish one system from other systems with many homologous components, it may be relevant to introduce and class some genes specific to the other systems as "forbidden," as for the T2SS (e.g., PilM) or the T3SS (FlgB) (*see* Fig. 1). Table 2 lists the features and terms available to describe a model.

3.1.2 Extraction of HMM Protein Profiles from Databanks

Each component (called a "gene" in the model) must be associated with at least one HMM protein profile. Several public databanks of protein profiles can be queried by keyword (e.g., the name of the gene) or by sequence for a match to the component (*see* Table 1). InterProScan is a good starting point for this task since it integrates the information on protein profiles from many databanks, including PANTHER, PFAM, SUPERFAMILY, and TIGRFAM [27].

At the end of this procedure, one often has many profiles for some components and none for others. One can pick the best matching profile for each component (*see* **Note 1**). However, sometimes there is no profile that is clearly better than the others. This occurs when profiles match only certain subfamilies of components, for example, because they are very divergent in sequence. In these cases, it is possible to specify multiple profiles (*see* **Note 5**).

Table 2
Keywords to specify models

Keyword (Command-line override)	Required? (Default)	Level(s)	Description
system	Yes	Upper (first)	Keyword needed to start definition of system model
inter_gene_max_space (--inter-gene-max-space)	Yes	system, gene	"colocalization" parameter. Defines maximum number of genes between two consecutive components of system for a cluster of components to be created
min_genes_required (--min-genes-required)	No (number of mandatory genes or min_mandatory_genes_required if declared)	system	Minimum number of mandatory and accessory components
min_mandatory_genes_required (--min-mandatory-genes-required)	No (number of mandatory genes, or min_mandatory_genes_required if declared)	system	Minimum number of mandatory components
multi_loci (--multi-loci)	No (false, i.e., "0")		Whether system can be encoded on several main loci
gene	Yes	Second	Component of system
multi_system	No (false, i.e., "0")	gene	Whether gene can participate in several instances of a system
name	Yes	gene	Name of gene to be used in report files, and base name of corresponding HMM protein profile

(continued)

Table 2
(continued)

Keyword (Command-line override)	Required? (Default)	Level(s)	Description
loner	No (false, i.e., "0")	gene	Whether gene can be found encoded outside of system's main loci, i.e., an exception to colocalization rule defined at system level
exchangeable	No	gene	Whether gene can be replaced by another gene in list of components (quorum); the gene can then be exchanged with one of its "homologs" or "analogs"
analogs	No	gene	List of "analogs" to gene
homologs	No	gene	List of "homologs" to gene
system_ref	No	gene	Specifies original system of gene when it was described in another system's model

For more details, *see* http://macsyfinder.readthedocs.org/en/latest/system_definition.html#the-xml-hierarchy

The user must build a novel protein profile in one of three typical situations: (1) when none is available in the databases, (2) when one profile can replace a large set of very specific profiles, or (3) when one wishes to build profiles that are specific to a system because existing profiles also match components of other types of systems. The new profile should be built using the sequences of the *reference* data set (*see* also **Notes 1** and **3**).

3.1.3 Establishing the Model of the Genetic Architecture

By default, MacSyFinder searches for clusters of colocalized genes encoding components of the system ("ordered" data sets) (*see* Subsection 2.1). The maximum distance allowed between consecutive components ("inter_gene_max_space") can be the same for all components or be specific to a particular component. The distance is measured in genes, i.e., a distance of three means that there can be up to three genes between two consecutive components of the system. This distance can be inferred from the *reference* data set or from the secretion systems identified in bacterial genomes (*see* **Note 6**). Additionally, one may define some genes as "loners," in which case the model allows them to be encoded outside the clusters of colocalized genes (Subsection 3.2.1).

The overall genetic architecture of a system can be specified through the "multi_loci" attribute. The default value (False or "0") indicates that the system is encoded in a single locus (except for the loner components), whereas the alternative ("1") authorizes the existence of multiple clusters for an instance of the system. For example, this option is useful for describing the type IV pili or T9SS, since they are often encoded in several loci [9].

3.1.4 Defining the Quorum of Components

The quorum of the model is the minimum number of components required to validate an instance of the system. It is defined by two parameters: the minimum number of mandatory components ("min_mandatory_genes_required") and the minimum number of mandatory and accessory components ("min_genes_required"). The values of both parameters are set by default to the number of mandatory components, and in this case the accessory components do not actually count in the quorum. To make them count, one must specify higher values for the second than for the first criterion. It is particularly important to use prior knowledge to assess the relevance of a locus lacking a mandatory gene.

The quorum can be optimized using the information from the secretion systems identified in bacterial genomes (*see* **Note 7**). Changes in the values of the quorum affect the sensitivity and specificity of the method. Low values authorize the validation of systems with fewer components that are less similar to the reference data set, thereby increasing the number of detected systems. This might come at the cost of misidentifications or the validation of nonfunctional systems. Running the models with higher values of the quorum results in the identification of instances that are more similar to the reference data set and thus more likely to be true. On the other hand, this results in the identification of fewer instances and may exclude those that are too distinct from those in the *reference* data set.

3.2 Model Formulation

After gathering the available information on the protein secretion system, one must write down the model in the relatively simple XML format predefined for MacSyFinder. A list of the keywords of this hierarchical XML grammar is presented in Table 2 and in MacSyFinder's online documentation.

3.2.1 Defining the Model in an XML Text File: Example 1—T1SS

We illustrate the formulation of a simple model with the example of the type I secretion system. This system has three mandatory components (Fig. 2): an ABC transporter ("abc"), a membrane fusion protein ("mfp"), and an outer membrane porin ("omf"). The latter can be colocalized with the two other components (less than six genes apart, "inter_gene_max_space" set to 5), or encoded apart in the genome (loner attribute set to True: "1"). In addition, omf can be involved in several occurrences of the system ("multi-system" set to True ("1")). The three mandatory components are required to form a full system. The quorum is therefore not specified; it is left at its default value. The file with the model should be named after the system ("T1SS.xml"). Its content is displayed in Fig. 2.

3.2.2 Defining the Model in an XML Text File: Example 2—T9SS

Larger systems tend to require more complex models. This is well exemplified by the model to identify T9SS (file "T9SS.xml" [9]).

This model, in line with the existing literature [28–31], states that T9SS consists of several mandatory and accessory components encoded in multiple loci (multi_loci = "1"). The quorum allows one mandatory and several accessory components to be missing. Some components form clusters, whereas others are defined as loners (Fig. 1). The SprA component was matched by several profiles of the PFAM and TIGRFAM databanks; we used the "exchangeable" and "homologs" attributes to include them all (*see* **Note 5**).

3.3 Running MacSyFinder

3.3.1 Organizing the Data

HMM protein profiles should be in individual files named after the corresponding component and placed in the same directory. The path to this directory must be provided with the option "-p". All profiles must have the same file name extension (".hmm" by default, though it can be changed with the option "--profile-suffix"). For the T1SS model mentioned earlier, the program requires three HMM files in a "profiles" directory: "abc.hmm," "mfp.hmm," and "omf.hmm".

The path to the directory within the current working directory containing the XML model file must also be given in the input (option "-d," named "definitions," for example).

3.3.2 Identification of Secretion Systems

In this section, we provide an example of the procedure to identify instances of T1SS in a protein file of the type "ordered_replicon" ("CP000521_proteins.fasta"). The command line to launch the program is as follows:

```
macsyfinder -p profiles -d definitions --db-
type ordered_replicon --replicon-topology
circular --sequence-db CP000521_proteins.
fasta T1SS
```

In our experience, the identification of multiple unrelated secretion systems is best done independently, i.e., in successive runs for the different models. This is especially important when the systems to be searched for have many homologous components. Yet, when systems are unrelated, one can identify them at the same time. For example, to identify both T1SS and T9SS, one should type:

```
macsyfinder -p profiles -d definitions --db-
type ordered_replicon --replicon-topology
circular --sequence-db CP000521_proteins.
fasta T1SS T9SS
```

3.4 Output Files, and Visualization of Results with MacSyView

3.4.1 Finding Relevant Information in MacSyFinder's Output Files

The results of the detection are printed to files in simple text format that can be read by any word-processing application. They are stored in the MacSyFinder's output directory (specified using the "-o" option, or named automatically). They include configuration and log files, along with the results of MacSyFinder and HMMER. The description of the output files is detailed in MacSyFinder's documentation: http:// macsyfinder.readthedocs.org/en/latest/outputs.html.

Some output files are particularly useful for improving the design of a new model. The standard output ("macsyfinder.out") shows how the tool processed the instances of the different components to validate the systems. Together with the files storing raw or filtered HMMER hits (in the "hmmer_results" directory), it gives an overview of the detection process, from the detection of components to the validation of the clusters. For automated downstream analyses of MacSyFinder's results, the files "macsyfinder.report" and "macsyfinder.summary" are of particular interest. The former contains the list of components identified in each instance with information on the quality of the HMMER matches. The latter describes the content of each instance of a secretion system and can be loaded as a dictionary object using a Python script. It can be particularly useful in the optimization steps (*see* **Notes 6** and 7).

3.4.2 Visualization of Results with MacSyView

MacSyView allows the user to visualize the results of MacSyFinder in a web browser. It reads the "results.macsyfinder.json" output file. Clicking on a system opens a window with three panels showing the number of components per category, the genomic context, and the statistics of the detection of each component (Fig. 3).

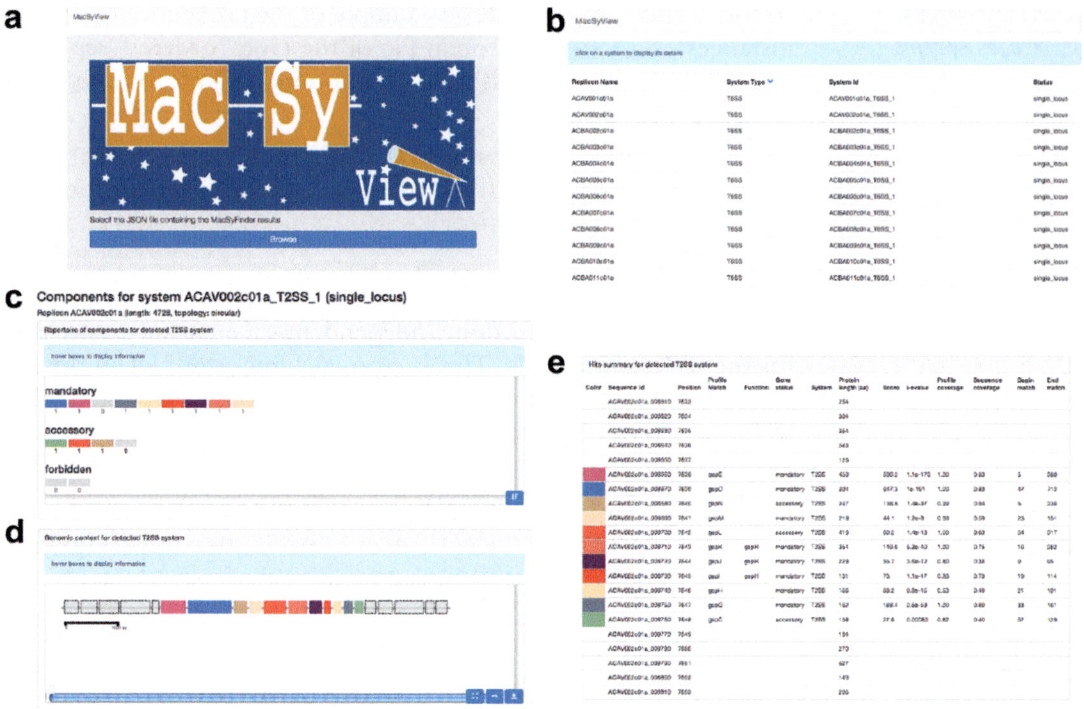

Fig. 3 Screenshots of MacSyView application. (**a**) The application reads the "results.macsyfinder.json" file from MacSyFinder's results directory. (**b** and **c**) The selection from a list of detected systems leads to a page with a description of the instance in relation to the model. (**d**) The graphical representation of the instance in its genomic context can be downloaded. (**e**) A table details the identified components (e.g., name, length) along with their HMMER scores and other statistics

3.5 Optimization and Validation

3.5.1 Optimization

The initial models should be very general to assess the diversity of the instances of the system. They can then be optimized iteratively depending on the results obtained in the analysis of the *reference* data set (and external data). The output files discussed in Subsection 3.4.2 can be used to identify the parts of the model that must be improved. The specific case of model optimization for poorly studied systems is described in **Notes 6** and **7**.

3.5.2 Validation

The final model must be tested on an independent *validation* data set (*see* **Note 1**). This procedure allows the quantification of the sensitivity of the model, i.e., to know how well it identifies a secretion system and its different components. It is more difficult to assess its specificity, i.e., its ability to separate true from false instances, since usually there is no reliable information on invalid protein secretion systems.

At the end of the validation process, it might seem tempting (or necessary) to correct the initial model to identify a larger fraction of the instances of the validation data set. It must be borne in mind that this violates the hypothesis of independence between the model and the validation data set. Hence, if the model is changed to fit the *validation* data set, then it can no longer be validated with the same data set. One can build an additional independent validation data set to circumvent this difficulty.

4 Notes

1. Constructing a reference or validation set of secretion systems
   The diversity of the protein secretion systems in the *reference* and *validation* data sets should encompass their diversity in the genomes where they are going to be searched for. Otherwise, the model will miss the instances that are very divergent, in sequence or genetic organization. One can increase the diversity of these sets by sampling instances from all previously described subtypes of a system (e.g., SPI1, SPI2, Hrp1, Hrp2, Ysc for T3SS [12]). When this information is unavailable, one can use all the known systems to build phylogenetic trees of previously identified key components and then sample representatives of every major clade to build the reference data set.
2. Building protein families for the reference secretion systems
   We propose the following procedure to build and analyze families of homologous proteins:
   (a) Store the protein sequences of the reference set of secretion systems in a multi-fasta file (e.g., "reference_systems.fasta"). Index the file using the makeblastdb command:
   ```
   makeblastdb -dbtype prot -in reference_systems.fasta
   ```
   (b) Run Blastp for each pair of proteins and generate a tabulated output. Pick a statistical threshold of significance for the hits (here E-value $<10^{-6}$):
   ```
   blastp -query reference_systems.fasta -db reference_systems.fasta -evalue 0.000001 -outfmt 6 -out blastall_reference_systems.out
   ```
   Alternatively, this can be done using other sequence similarity search tools, such as psiblast or jackhmmer.
   (c) Identify the protein families present in the data set by clustering the blastp results with Silix.
   ```
   silix reference_systems.fasta blastall_reference_systems.out -f FAM > reference_systems.fnodes
   ```
   (d) Save the different families obtained in separate fasta files:
   ```
   silix-split reference_systems.fasta reference_systems.fnodes
   ```

(e) Annotate/label the families. If different components are clustered together, they must be disambiguated. One can split families by clustering the proteins with parameters enforcing higher similarity or alignment coverage. A phylogenetic analysis may also show how to separate two protein families that cluster together (not covered here). In some cases, clearly separate protein families (and, thus, protein profiles) are unobtainable. The best solution is to list them all as a single component in the model and specify a quorum of one (*see* **Note 5**).

3. Designing HMM protein profiles from families of homologous proteins
Follow these steps to build HMM profiles for the protein families (e.g., those obtained following **Note 2**):

(a) Store the sequences of each protein family in a different text file in Fasta format. Name the file in accordance with its content, e.g., "abc.fasta" for the ABC-transporter protein of T1SS.

(b) Make a multiple-sequence alignment for each family using MAFFT [21] (or other analogous program):

```
mafft abc.fasta > abc.aln-mafft.fasta
```

(c) Check each multiple alignment (e.g., using Seaview [23]). Trim its extremities if they are poorly conserved. Do not remove columns inside the region of the alignment that is going to be used to construct the profile. Save the edited alignment in a new file, e.g., "abc.mafft-edit.fasta."

(d) Run hmmbuild from the HMMER package [20] on the edited alignment to create the HMM protein profile:

```
hmmbuild --informat afa --amino abc.hmm
abc.mafft-edit.fasta
```

4. Analyzing poorly characterized systems
The typical model-building procedure cannot be used to model systems lacking enough experimentally validated instances to make a reference data set. In this case, one must use the few known instances to collect homologs from the genome databases by sequence similarity searches (e.g., using blast [32], *see* **Note 2**). The hits can be used to build protein families and HMM profiles. A careful analysis of the patterns of co-occurrence of the components usually highlights those that should be specified in the model because they are sufficiently conserved.

It is usually better to build simple models with weak constraints in terms of quorum and colocalization when the reference data sets are small. For example, every component could be set as "loner" and the quorum set to one or two. The model can then be optimized iteratively (*see* Subsection 3.5.1).

Naturally, one should be cautious when drawing conclusions from studies using models that could not be validated with an independent data set of experimentally validated systems.

5. Identifying a component using multiple protein profiles

A single protein profile may not be sufficient to identify all instances of a given component. In such cases, one can associate several protein profiles to a single component in the quorum of the system. These genes should be qualified as "exchangeable," and "homologs" or "analogs." For example, T3SS has one of three different subfamilies of secretins (T3SS_sctC, T2SS_gspD, Tad_rcpA, depending on the T3SS subtype, [12]). To detect them correctly, one can define it in the T3SS model as follows:

```
<gene name="T3SS_sctC" presence="mandatory"
exchangeable ="1">
<homologs>
   <gene name="T2SS_gspD" system_ref="T2SS"/>
   <gene name="Tad_rcpA" system_ref="T4P"/>
</homologs>
</gene>
```

With this model, MacSyFinder will identify a secretin when it finds a hit for any of these three protein profiles. The quorum will increase by one when there is one or multiple hits to any of these profiles. The keyword "system_ref" indicates that the two genes T2SS_gspD and Tad_rcpA are defined in the models of T2SS and Tad, respectively. When the other models are accessible to MacSyFinder, the components do not need to be redefined in the "homologs" or "analogs" section of T3SS. When the other models are unavailable, then the components must be described in the system's model. We show an example of the latter case concerning the SprA gene of T9SS in Subsection 3.2.2.

6. Optimizing the colocalization criterion

The procedure for optimizing the colocalization criterion starts by setting it to a value higher than expected (but not too high; otherwise, several occurrences of the system could be agglomerated). This produces large clusters that are expected to contain all relevant colocalized genes. This parameter may be subsequently refined by plotting the distribution of the maximum distance found between two consecutive components in the clusters of the *reference* data set. To illustrate, we searched for "single-locus" T6SS in a set of 1,528 bacterial genomes using a minimum colocalization distance of 20 genes [9]:

```
macsyfinder -p profiles -d definitions --db-
type gembase --sequence-db bacterial_genom-
es_proteins.fasta -o macsyfinder_opt_coloc_
T6SS --inter-gene-max-space T6SS 20 T6SS
```

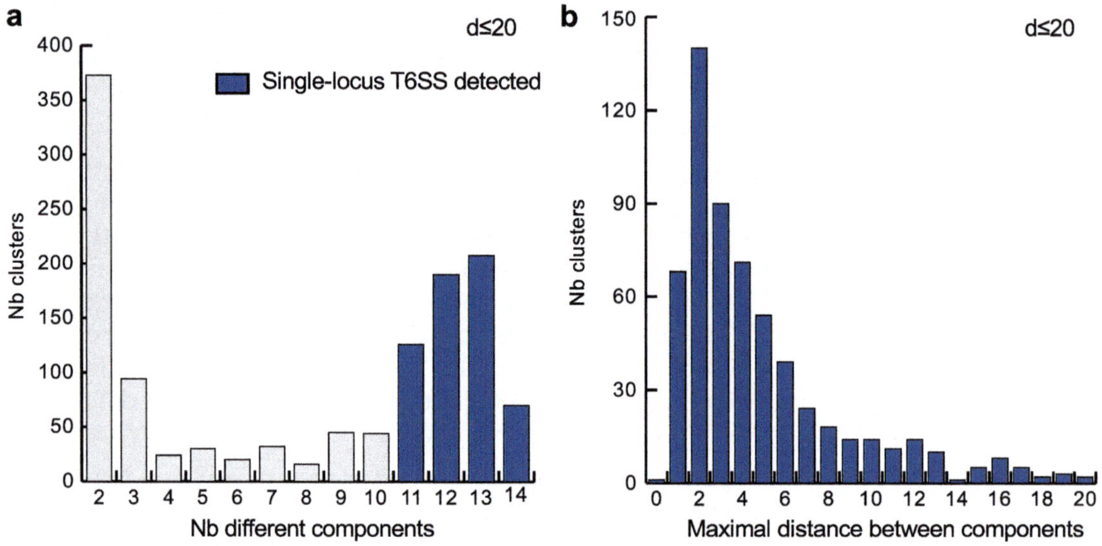

Fig. 4 Optimization of the quorum and colocalization parameters for T6SS^i. (**a**) Distribution of number of different components of T6SS^i colocalized ($d \leq 20$). (**b**) Distribution of maximum distance between consecutive components in T6SS^i detected with $d \leq 20$. The minimum number of components required for a T6SS^i was set to 11 in the T6SS^i model, which corresponds to the start of the second peak in the distribution represented in panel (**a**). T6SS^i detected as full systems are colored in *blue*. Figures and legends are freely reproduced with modification from reference [9] (as specified by the Creative Commons Attribution (CC BY) license version 4.0)

The distribution of distances actually observed in genomes suggests that a smaller value (e.g., 14) is enough to identify most of the relevant clusters (Fig. 4b).

7. Optimizing the quorum criterion
   The quorum can be optimized in multiple ways, depending on the distribution of the secretion systems across genomes.

   If the system is encoded at a single locus (but may be found in several copies per replicon), the quorum can be optimized by studying the distribution of the number of different components detected in each cluster. For this purpose, one can use a model with a very relaxed quorum criterion (e.g., set to "1") and draw the distribution of the number of components found in each cluster (with at least one component) (Fig. 4a). This can be done directly in the command line, for example, using the T6SS model available in TXSScan:

```
macsyfinder -p profiles -d definitions --db-type gembase --sequence-db bacterial_genomes_proteins.fasta -o macsyfinder_opt_quorum_T6SS --min-genes-required T6SS 1 --min-mandatory-genes-required T6SS 1 T6SS
```

   The number of (accessory and mandatory) components in each cluster is indicated in the "macsyfinder.summary" output file in the "macsyfinder_opt_quorum_T6SS" folder, in the eighth column, "Nb_Ref_Genes_detected_NR." In this specific example, it shows many clusters with more than 11 components (Fig. 4a). This is in line with the presence of more than 13 core components in most T6SS [33]. A final model with a quorum set to 11 would thus be able to identify novel instances of the T6SS accurately [9].

When the system is typically encoded in a single copy per genome scattered in several loci ("multi_loci"), the analysis of the clusters is less informative, especially if there are other systems in the genome with homologs to these components. Nevertheless, it can be complemented with the information on the number of components per replicon. This analysis should use low stringency colocalization and quorum parameters, for example, all genes can be set to "loner" and the quorum to a small value. We illustrate this by searching to optimize the T9SS model (*see* Fig. 1 and Subsection 3.2.2):

(a) Copy the "T9SS.xml" file as "T9SS_loner.xml."

(b) Add "loner = '1'" to the definition line of each gene.

(c) Alter the values of the parameters "min_genes_required" and "min_mandatory_genes_required," either by using the command line (as in the following command line) or by modifying the XML file in the system definition line ("min_mandatory_genes_required = '1′ min_genes_required = '1′").

(d) Run MacSyFinder on the modified version of the model:

```
macsyfinder -p profiles -d definitions
--db-type gembase --sequence-db bacte-
rial_genomes_proteins.fasta -o macsyfind-
er_opt_quorum_T9SS --min-genes-required
T9SS_loner 1 --min-mandatory-genes-re-
quired T9SS_loner 1 T9SS_loner
```

The distribution of the number of (accessory and mandatory) components shows many replicons with more than seven components. The model using a quorum of seven is able to identify novel instances of T9SS accurately [9].

References

1. Kanonenberg K, Schwarz CK, Schmitt L (2013) Type I secretion systems—a story of appendices. Res Microbiol 164(6):596–604
2. Campos M, Cisneros DA, Nivaskumar M, Francetic O (2013) The type II secretion system—a dynamic fiber assembly nanomachine. Res Microbiol 164(6):545–555
3. Korotkov KV, Sandkvist M, Hol WG (2012) The type II secretion system: biogenesis, molecular architecture and mechanism. Nat Rev Microbiol 10(5):336–351
4. Galan JE, Lara-Tejero M, Marlovits TC, Wagner S (2014) Bacterial type III secretion systems: specialized nanomachines for protein delivery into target cells. Annu Rev Microbiol 68:415–438
5. Alvarez-Martinez CE, Christie PJ (2009) Biological diversity of prokaryotic type IV

secretion systems. Microbiol Mol Biol Rev 73:775–808

6. van Ulsen P, Rahman S, Jong WS, Daleke-Schermerhorn MH, Luirink J (2014) Type V secretion: from biogenesis to biotechnology. Biochim Biophys Acta 1843(8):1592–1611
7. Zoued A, Brunet YR, Durand E, Aschtgen MS, Logger L, Douzi B, Journet L, Cambillau C, Cascales E (2014) Architecture and assembly of the Type VI secretion system. Biochim Biophys Acta 1843(8):1664–1673
8. McBride MJ, Nakane D (2015) Flavobacterium gliding motility and the type IX secretion system. Curr Opin Microbiol 28:72–77
9. Abby SS, Cury J, Guglielmini J, Néron B, Touchon M, Rocha EPC (2016) Identification of protein secretion systems in bacterial genomes. Sci Rep 6:23080
10. Ginocchio CC, Olmsted SB, Wells CL, Galan JE (1994) Contact with epithelial cells induces the formation of surface appendages on Salmonella typhimurium. Cell 76(4):717–724
11. Peabody CR, Chung YJ, Yen MR, Vidal-Ingigliardi D, Pugsley AP, Saier MH Jr (2003) Type II protein secretion and its relationship to bacterial type IV pili and archaeal flagella. Microbiology 149(Pt 11):3051–3072
12. Abby SS, Rocha EP (2012) The non-flagellar type III secretion system evolved from the bacterial flagellum and diversified into host-cell adapted systems. PLoS Genet 8(9):e1002983
13. Holland IB, Schmitt L, Young J (2005) Type 1 protein secretion in bacteria, the ABC-transporter dependent pathway. Mol Membr Biol 22(1–2):29–39
14. Paulsen IT, Park JH, Choi PS, Saier MH Jr (1997) A family of gram-negative bacterial outer membrane factors that function in the export of proteins, carbohydrates, drugs and heavy metals from gram-negative bacteria. FEMS Microbiol Lett 156(1):1–8
15. Abby SS, Neron B, Menager H, Touchon M, Rocha EP (2014) MacSyFinder: a program to mine genomes for molecular systems with an application to CRISPR-Cas systems. PLoS One 9(10):e110726
16. Christie PJ (2004) Type IV secretion: the Agrobacterium VirB/D4 and related conjugation systems. Biochim Biophys Acta 1694(1–3):219–234
17. Franco IS, Shuman HA, Charpentier X (2009) The perplexing functions and surprising origins of Legionella pneumophila type IV secretion effectors. Cell Microbiol 11(10):1435–1443
18. Gillespie JJ, Brayton KA, Williams KP, Diaz MA, Brown WC, Azad AF, Sobral BW (2010) Phylogenomics reveals a diverse Rickettsiales type IV secretion system. Infect Immun 78(5):1809–1823
19. Camacho C, Coulouris G, Avagyan V, Ma N, Papadopoulos J, Bealer K, Madden TL (2009) BLAST+: architecture and applications. BMC Bioinformatics 10:421
20. Eddy SR (2011) Accelerated profile HMM searches. PLoS Comput Biol 7(10):e1002195
21. Katoh K, Toh H (2010) Parallelization of the MAFFT multiple sequence alignment program. Bioinformatics 26(15):1899–1900
22. Edgar RC (2004) MUSCLE: multiple sequence alignment with high accuracy and high throughput. Nucleic Acids Res 32:1792–1797
23. Gouy M, Guindon S, Gascuel O (2010) SeaView version 4: a multiplatform graphical user interface for sequence alignment and phylogenetic tree building. Mol Biol Evol 27(2):221–224
24. Waterhouse AM, Procter JB, Martin DM, Clamp M, Barton GJ (2009) Jalview Version 2—a multiple sequence alignment editor and analysis workbench. Bioinformatics 25(9):1189–1191
25. Miele V, Penel S, Duret L (2011) Ultra-fast sequence clustering from similarity networks with SiLiX. BMC Bioinformatics 12:116
26. Enright AJ, Van Dongen S, Ouzounis CA (2002) An efficient algorithm for large-scale detection of protein families. Nucleic Acids Res 30(7):1575–1584
27. Mitchell A, Chang HY, Daugherty L, Fraser M, Hunter S, Lopez R, McAnulla C, McMenamin C, Nuka G, Pesseat S, Sangrador-Vegas A, Scheremetjew M, Rato C, Yong SY, Bateman A, Punta M, Attwood TK, Sigrist CJ, Redaschi N, Rivoire C, Xenarios I, Kahn D, Guyot D, Bork P, Letunic I, Gough J, Oates M, Haft D, Huang H, Natale DA, Wu CH, Orengo C, Sillitoe I, Mi H, Thomas PD, Finn RD (2015) The InterPro protein families database: the classification resource after 15 years. Nucleic Acids Res 43(Database issue):D213–D221
28. Shrivastava A, Johnston JJ, van Baaren JM, McBride MJ (2013) Flavobacterium johnsoniae GldK, GldL, GldM, and SprA are required for secretion of the cell surface gliding motility adhesins SprB and RemA. J Bacteriol 195(14):3201–3212
29. McBride MJ, Zhu Y (2013) Gliding motility and Por secretion system genes are widespread among members of the phylum Bacteroidetes. J Bacteriol 195(2):270–278
30. Zhu Y, McBride MJ (2014) Deletion of the Cytophaga hutchinsonii type IX secretion system gene sprP results in defects in gliding motility and cellulose utilization. Appl Microbiol Biotechnol 98(2):763–775

31. Kharade SS, McBride MJ (2015) Flavobacterium johnsoniae PorV is required for secretion of a subset of proteins targeted to the type IX secretion system. J Bacteriol 197(1):147–158
32. Altschul SF, Gish W, Miller W, Myers EW, Lipman DJ (1990) Basic local alignment search tool. J Mol Biol 215:403–410
33. Boyer F, Fichant G, Berthod J, Vandenbrouck Y, Attree I (2009) Dissecting the bacterial type VI secretion system by a genome wide in silico analysis: what can be learned from available microbial genomic resources? BMC Genomics 10:104
34. Finn RD, Tate J, Mistry J, Coggill PC, Sammut SJ, Hotz HR, Ceric G, Forslund K, Eddy SR, Sonnhammer EL, Bateman A (2008) The Pfam protein families database. Nucleic Acids Res 36(Database issue):D281–D288
35. Haft DH, Selengut JD, Richter RA, Harkins D, Basu MK, Beck E (2013) TIGRFAMs and Genome Properties in 2013. Nucleic Acids Res 41(Database issue):D387–D395
36. Quevillon E, Silventoinen V, Pillai S, Harte N, Mulder N, Apweiler R, Lopez R (2005) InterProScan: protein domains identifier. Nucleic Acids Res 33(Web Server issue):W116–W120

Chapter 2

Protein Sorting Prediction

Henrik Nielsen

Abstract

Many computational methods are available for predicting protein sorting in bacteria. When comparing them, it is important to know that they can be grouped into three fundamentally different approaches: signal-based, global-property-based and homology-based prediction. In this chapter, the strengths and drawbacks of each of these approaches is described through many examples of methods that predict secretion, integration into membranes, or subcellular locations in general. The aim of this chapter is to provide a user-level introduction to the field with a minimum of computational theory.

Key words Protein sorting, Subcellular location, Secretion, Transmembrane proteins, Prediction, Machine learning

1 Introduction

Protein sorting prediction—in other words, inferring the subcellular location (SCL) of proteins from their amino acid sequences—has a long history in bioinformatics. The first attempts at predicting the best known sorting signals, the transmembrane α-helix (TMH) and the secretory signal peptide (SP), were published in 1982–1983, long before bioinformatics was even established as a field [1, 2]. Since then, a plethora of methods for predicting sorting signals and SCL have been published, and it can be a daunting task to select the most relevant and reliable methods for analyzing a set of sequences.

Of course, the development of algorithms and the growth in available training data have led to an increase in the predictive performance of the available methods. In 2005, some of the authors of the PSORTb method for predicting SCL in bacteria [3] even concluded that "on average, recent high-precision computational methods such as PSORTb now have a lower error rate than laboratory methods" [4]. This conclusion should be taken with a grain of salt; first, it applies only to high-throughput laboratory methods; second, it should be remembered that computational methods will

Laure Journet and Eric Cascales (eds.), *Bacterial Protein Secretion Systems: Methods and Protocols*, Methods in Molecular Biology, vol. 1615, DOI 10.1007/978-1-4939-7033-9_2, © Springer Science+Business Media LLC 2017

never be better than the data used to train them. Nevertheless, the authors had a point regarding the experimental sources of error that can easily render a high-throughput experiment less reliable than a well-trained computational method.

It can be difficult, however, to decide what to believe when the authors of every computational method tend to describe their performance as being superior to all others. There are different ways of defining the problem, different ways of measuring the performance, and different prerequisites used for prediction. The aim of this chapter is not to provide a definitive answer to which method is best for which problem—such a checklist would quickly become outdated—but instead to give the reader a toolbox for critically evaluating bioinformatics algorithms. This will involve a number of examples of computational methods selected for their relevance for bacteria, with the main focus being on Gram-negative bacteria. A similar chapter with focus exclusively on Gram-positive bacteria has been published elsewhere [5]. In general, prediction methods will be mentioned only if they either provide publicly available web servers or have strong historical relevance.

2 Three Approaches to Prediction

It is crucial to understand that there are basically three different approaches to protein sorting prediction. The first approach is recognition of the actual sorting signals. The aforementioned early methods for TMH and SP recognition [1, 2] were examples of this. A number of more recent examples will be given in Subheadings 5 and 7.

The second approach is prediction based on global properties of the proteins, for example, their amino acid composition. This approach was first used to discriminate between intracellular and extracellular proteins in both prokaryotic and eukaryotic proteins in 1994 [6]. It has been shown that the main part of the differences in amino acid composition between intracellular and extracellular proteins resides in surface-exposed amino acids, which makes sense since the surfaces should be adapted to the varying physicochemical environments of the different SCLs [7]. This analysis was done for eukaryotic proteins only, but it would be fair to assume that the observation holds true also for bacterial proteins.

Two early SCL prediction methods that used only the amino acid composition were NNPSL in 1998 [8] and SubLoc in 2001 [9] (*see* Table 1), based on artificial neural networks (ANNs) and support vector machines (SVMs), respectively (*see* Subheading 3). They were limited in their applicability because their data set did not include any membrane proteins, and they did not distinguish between Gram-positive and Gram-negative bacteria.

Table 1
Web addresses of servers reviewed in this chapter

Name	Website address	References
SubLoc	http://www.bioinfo.tsinghua.edu.cn/SubLoc/	[9]
PROSITE	http://prosite.expasy.org/prosite.html	[28]
Pfam	http://pfam.xfam.org/	[29]
TIGRFAMs	http://www.jcvi.org/cgi-bin/tigrfams/index.cgi	[30]
InterPro	http://www.ebi.ac.uk/interpro/	[31]
SignalP	http://www.cbs.dtu.dk/services/SignalP/	[48–51]
PrediSi	http://www.predisi.de/	[56]
SOSUIsignal	http://harrier.nagahama-i-bio.ac.jp/sosui/sosuisignal/sosuisignal_submit.html	[57]
Signal-BLAST	http://sigpep.services.came.sbg.ac.at/signalblast.html	[58]
LipoP	http://www.cbs.dtu.dk/services/LipoP/	[60]
SPEPlip	http://gpcr.biocomp.unibo.it/cgi/predictors/spep/pred_spepcgi.cgi	[62]
PRED-LIPO	http://bioinformatics.biol.uoa.gr/PRED-LIPO/	[63]
PROSITE profile PROKAR_LIPOPROTEIN	http://prosite.expasy.org/PS51257 and http://prosite.expasy.org/PDOC00013	
TatFind	http://signalfind.org/tatfind.html	[65]
TatP	http://www.cbs.dtu.dk/services/TatP/	[66]
PRED-TAT	http://www.compgen.org/tools/PRED-TAT/	[67]
PROSITE profile TAT	http://prosite.expasy.org/PS51318 and http://prosite.expasy.org/PDOC51318	
Pfam profile TAT_signal	http://pfam.xfam.org/family/PF10518	

(continued)

Table 1
(continued)

Name	Website address	References
TIGRFAMs profile TAT_signal_seq	http://www.jcvi.org/cgi-bin/tigrfams/HmmReportPage.cgi?acc=TIGR01409	
SecretomeP 2.0	http://www.cbs.dtu.dk/services/SecretomeP/	[70]
SecretP 2.0	http://cic.scu.edu.cn/bioinformatics/secretPV2/	[71]
SecretP 2.1	http://cic.scu.edu.cn/bioinformatics/secretPV2_1/	[72]
T4EffPred	http://bioinfo.tmmu.edu.cn/T4EffPred/	[78]
T4SEpre	Web server at http://biocomputer.bio.cuhk.edu.hk/T4DB/T4SEpre.php, downloadable version at http://biocomputer.bio.cuhk.edu.hk/softwares/T4SEpre/	[79]
SIEVE	http://www.sysbep.org/sieve/	[82]
EffectiveT3	http://www.effectors.org/	[83]
Löwer and Schneider's method	http://gecco.org.chemie.uni-frankfurt.de/T3SS_prediction/T3SS_prediction.html	[84]
BPBAac	http://biocomputer.bio.cuhk.edu.hk/T3DB/BPBAac.php	[85]
T3_MM	http://biocomputer.bio.cuhk.edu.hk/T3DB/T3_MM.php	[86]
BEAN	http://systbio.cau.edu.cn/bean/	[87, 88]
pEffect	http://services.bromberglab.org/peffect/	[89]
TMHMM	http://www.cbs.dtu.dk/services/TMHMM/	[95]
HMMTOP	http://www.enzim.hu/hmmtop/	[96]
Phobius & PolyPhobius	http://phobius.sbc.su.se/	[101, 108]
Philius	http://www.yeastrc.org/philius/	[102]
MEMSAT3 and MEMSAT-SVM	Available through the PSIPRED Protein Sequence Analysis Workbench, http://bioinf.cs.ucl.ac.uk/psipred/	[103, 104]

OCTOPUS and SPOCTOPUS	http://octopus.cbr.su.se/	[105, 106]
SCAMPI	http://scampi.cbr.su.se/	[109]
BPROMPT	http://www.ddg-pharmfac.net/bprompt/BPROMPT/BPROMPT.html	[111]
TOPCONS	http://topcons.cbr.su.se/ or http://topcons.net/	[112, 113]
TOPCONS-single	http://single.topcons.net/	[114]
PRED-TMBB	http://biophysics.biol.uoa.gr/PRED-TMBB/	[117, 118]
ProfTMB	https://rostlab.org/owiki/index.php/Proftmb	[119, 120]
B2TMPRED	http://gpcr.biocomp.unibo.it/cgi/predictors/outer/pred_outercgi.cgi	[122]
TBBpred	http://www.imtech.res.in/raghava/tbbpred/	[123]
ConBBPRED	http://bioinformatics.biol.uoa.gr/ConBBPRED/input.jsp	[121]
BOCTOPUS	http://boctopus.bioinfo.se/	[124, 125]
BOMP	http://services.cbu.uib.no/tools/bomp	[126]
HHomp	http://toolkit.tuebingen.mpg.de/hhomp	[127]
BetAware	http://betaware.biocomp.unibo.it/BetAware/	[128, 129]
transFold	http://bioinformatics.bc.edu/clotelab/transFold/	[130, 131]
TMBpro	http://tmbpro.ics.uci.edu/	[132]
PSORTb	http://www.psort.org/psortb/	[3, 42, 134]
Proteome Analyst	http://pa.wishartlab.com/pa/pa/ Note: the website requires login, but registration is free.	[15, 33]
Gneg-PLoc	http://www.csbio.sjtu.edu.cn/bioinf/Gneg/	[19]
Gpos-PLoc	http://www.csbio.sjtu.edu.cn/bioinf/Gpos/	[20]

(continued)

Table 1 (continued)

Name	Website address	References
Gneg-mPLoc	http://www.csbio.sjtu.edu.cn/bioinf/Gneg-multi/	[21]
Gpos-mPLoc	http://www.csbio.sjtu.edu.cn/bioinf/Gpos-multi/	[22]
iLoc-Gneg	http://www.jci-bioinfo.cn/iLoc-Gneg	[23]
iLoc-Gpos	http://www.jci-bioinfo.cn/iLoc-Gpos	[24]
PSLpred	http://www.imtech.res.in/raghava/pslpred/	[137]
LocTree3	https://rostlab.org/services/loctree3/	[138]
CELLO	http://cello.life.nctu.edu.tw/	[13]
SOSUI-GramN	http://harrier.nagahama-i-bio.ac.jp/sosui/sosuigramn/sosuigramn_submit.html	[140]
MetaLocGramN	http://iimcb.genesilico.pl/MetaLocGramN/	[135]

Using only the amino acid composition for prediction, of course, means throwing away all sequence information, including possible signatures of actual sorting signals. One way to retain some of this information while still keeping a fixed number of parameters is to count the occurrences of amino acid pairs, either adjacent or separated by a small distance. Nakashima and Nishikawa in 1994 [6] thus found that including the composition of amino acid pairs with a separation distance of up to five positions improved predictive performance.

The third approach is prediction by sequence homology. When trying to predict functional aspects of an unknown protein, the standard procedure is to do a BLAST search [10] and then infer such aspects from the functional annotations of the found homologs. Therefore, the intuitive expectation is that such a procedure will also work for SCL—in other words, that a protein tends to stay in the same compartment in the course of evolution. Indeed, a significant part of the so-called subcellular location annotations of bacteria in Swiss-Prot (the manually annotated part of UniProt [11]), are found with "sequence similarity" as the evidence (more than five times as many as the corresponding annotations with experimental evidence).

However, it is not trivial to determine how similar a pair of proteins must be in order to draw an inference about SCL. Nair and Rost [12], working with eukaryotic proteins only, concluded that more than 70% identical residues in a pairwise BLAST search are needed to correctly infer SCL for 90% of the query proteins. On the other hand, the authors of the CELLO method for both eukaryotes and bacteria [13] found that SCL prediction by a simple BLAST search was better than a machine learning method above a pairwise identity cutoff as low as 30%.

The simplest possible homology-based prediction is the direct transfer of annotation from the best BLAST hit, i.e., the query protein is used to search a database of proteins with experimentally known SCLs, and then the SCL of the best hit is assigned to the query. However, more advanced approaches to homology-based prediction are also possible, using indirect means to infer SCL from the annotation of homologs that do not necessarily have experimentally known SCLs. This annotation could be derived from keywords or functional descriptions [14, 15], titles or abstracts of literature references [16, 17], or Gene Ontology terms [18–24].

In this context, it should be mentioned that many signal-based and global-property-based methods use a BLAST search to build a profile of related sequences to enhance the prediction. This does not make these methods homology-based since they do not use the annotations of the found hits.

In addition to the three approaches described here, it is of course possible to construct hybrids of them. Most of the multicategory methods described in Subheading 8 are of the hybrid type.

When comparing methods based on sorting signals, global properties, or homology, it is important to realize that each approach has its strengths and weaknesses. Homology-based methods, or hybrid methods containing homology-based components, often present the best measured performances; but the performance depends critically on the source of the query protein. Organisms that have been subjected to intense research will naturally tend to have more high-quality annotations, so proteins from those and their close relatives will find more close homologs with richer annotations from which to make predictions, while predictions for less-well-studied organisms will suffer from a lack of annotations of close homologs. This is typically not taken into consideration when reporting on the predictive performances of such methods. Signal-based and global property methods should be expected to be less sensitive to the source of the query protein, unless the signals in the training data are very organism specific.

There are two advantages to using global-property-based or homology-based methods. First, they can be used also for those compartments where the actual sorting signals are not known, or are too poorly characterized to support a prediction method. Second, they may work for sequences that are fragments from which the actual sorting signal may be missing or for amino acid sequences derived from genomic or metagenomic sequence where the start codon of the protein has not been correctly predicted, thus obscuring any N-terminal sorting signals. On the downside, global-property-based or homology-based methods do not provide the same degree of insight into the information processing in the cell since they ignore which parts of the sequence are actually important for sorting. Another drawback is that such methods will not be able to distinguish between very closely related proteins that differ in the presence or absence of a sorting signal, and they will not be able to predict the effects of small mutations that destroy or create a sorting signal.

3 Algorithms for Prediction

A rich variety of computational algorithms have been used in the prediction of SCL from amino acid sequences. Common to all of them is that they take a number of sequence-derived inputs and produce an output that can be the presence or absence of a sorting signal (for signal-based predictors) or an assignment of the protein into one of a number of possible SCL classes (for multicategory predictors). For Gram-negative bacteria, the number of SCL classes is most often defined as five (cytoplasm, inner membrane, periplasm, outer membrane, and extracellular), while for Gram-positive bacteria, the corresponding number is four (cytoplasm, membrane, cell wall, and extracellular).

Some algorithms, such as sequence alignment and hidden Markov models (HMMs), are naturally designed to work with sequences, while others, such as ANNs and SVMs, take only a fixed number of input values. When working with the latter category, one can either input the sequence as a series of overlapping windows of fixed length (typical for signal-based predictors) or extract a fixed number of features from the sequence (typical for global property predictors).

Numerical prediction algorithms can roughly be divided into two groups, statistical and machine learning, although it can sometimes be a matter of definition where to draw the distinction. Both classes of methods have a number of free parameters that must be estimated from the data, but while the parameters in statistical methods can be calculated directly, machine learning methods depend on an iterative optimization process where parameters are gradually changed until the classification error has reached a minimum.

The simplest sequence pattern recognition method is the *consensus sequence* or *regular expression*, for example, "RR.[FGAVML][LITMVF]" for SPs following the twin-arginine translocation (Tat) pathway. It is interpreted like this: there should be two consecutive arginines, followed by any amino acid, then one amino acid from the FGAVML group, and then one from the LITMVF group. It is easy to check whether such a pattern is present in a sequence, but it is also a very crude method because it defines absolute requirements for certain amino acids at certain positions and only provides yes or no answers. The preceding pattern, for example, ignores the fact that not all amino acids from the FGAVML and LITMVF groups are equally probable at positions 4 and 5 (cf. heights of individual letters in positions 17 and 18 in Fig. 1).

An alternative to the consensus sequence or regular expression is the *position-weight matrix* (PWM) [25], a statistical window-based

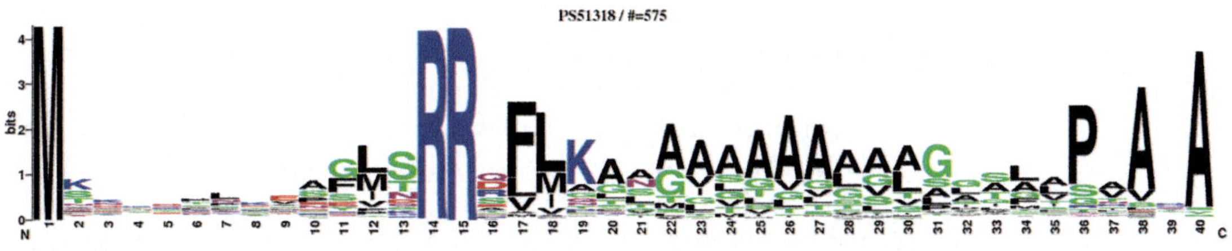

Fig. 1 Sequence logo of Tat (twin-arginine translocation) signal peptides from both Gram-positive and Gram-negative bacteria, aligned to the PROSITE profile PS51318/TAT. The height of each stack of letters corresponds to the information (conservation) at that position, while the height of each individual letter is proportional to the fraction of that amino acid at that position. Note that individual sequences may be shorter or longer than 40 amino acids; in the logo, they have been stretched or shortened to fit the model. Picture from PROSITE [28] made with WebLogo [146]

method that is very useful for characterizing and predicting short sequence motifs. The procedure for constructing a PWM involves using a set of examples of the motif of interest (the training set) to estimate the probability of each amino acid at each position and then convert those probabilities into weights. The score for a new sequence window can then be calculated by looking up the weights for each amino acid in each position in the window and adding them up. In this way, the weight matrix can give a *quantitative* answer to how well a sequence window fits the pattern.

A graphical counterpart to the PWM is the *sequence logo* [26], where each position is summarized by a stack of letters. The height of each stack is equivalent to the information content—a measure of the conservation—of that position, while the height of each letter is proportional to the probability of the corresponding amino acid at that position. Figure 1 presents an example of a sequence logo.

A straightforward extension of the PWM is the *sequence profile*, which allows for insertions and deletions in the sequence and therefore can model motifs of variable length. It is possible to formulate a profile in probabilistic terms, in which case it becomes a profile HMM [27]. An HMM is a machine learning algorithm, since the probabilities it contains are found by an iterative optimization process from a set of training data. After training, new sequences can be evaluated in terms of the probability that the sequence was generated by the model (a process known as HMM decoding). It should be emphasized that not all HMMs are profile HMMs—any grammar that can be described as a diagram of connected states can be modeled as an HMM. For example, a cyclic HMM can describe a repeating pattern, and a branched HMM can describe a choice between alternative patterns. Examples of cyclic HMMs are given in Subheading 7.

Several publicly available databases specialize in creating and storing profiles for protein families or domains. These include PROSITE [28] (*see* Table 1), which contains both regular expression patterns and PWM-like profiles, and Pfam [29] (*see* Table 1) and TIGRFAMs [30] (*see* Table 1), which are both databases of profile HMMs. InterPro [31] (*see* Table 1) is a special case since it does not create its own profiles but rather collects profiles from a number of contributing databases, including PROSITE, Pfam, and TIGRFAMs. Most profiles in these databases are evolutionarily related families or domains, but there are also instances of functional motifs that are similar owing to common selection pressure rather than common descent. Among these are a few protein sorting motifs, which can be used as prediction tools; examples will be given in Subheading 5.

Among the methods that use a fixed set of numbers as input, the simplest possible one is the Naïve Bayes classifier, which assumes that all the input variables are independent. It can show surprisingly good performance also in cases where the assumption of independence is known to be violated [32], and it is sometimes preferred over more advanced machine learning methods because it offers the opportunity to explain exactly which input variables were important for each prediction [33, 34]. Two very popular machine learning algorithms that are widely used in biological sequence analysis are the ANN [35] and the SVM [36]. This is not the place to go into technical details about the details of ANNs and SVMs, but they both have large numbers of parameters that must be estimated via a set of training data, and they are both potentially able to model situations where there are correlations between input features.

4 Performance of Prediction Methods

After having trained a statistical or machine learning method, it is crucial to test its predictive performance on another data set. This is a very important point: it is not enough that a trained method can reproduce its input examples exactly—in fact, it is not even interesting, since a database can do the same. What is interesting is whether a model can *generalize* from the examples in the training set and produce useful output for sequences it has not "seen" before.

There is often a certain degree of trade-off between training set and test set performance. If a model reproduces its training examples in too much detail, it uses its parameters to fit not only the common pattern in the data but also the individual noise in each data point. When this happens, the performance on the test set declines, and the model is said to be *overfitted*—colloquially speaking, it cannot see the forest for the trees.

Avoiding overfitting can be tricky; it may involve limiting the number of free parameters in the model, adding some regularizing terms to the parameters, or—especially in the case of ANNs—terminating the training early. In some cases, this is done using the test set performance as a criterion for choosing the optimal number of free parameters or the best point to stop the training; but in fact this is cheating, since the test set in such a procedure has been part of the training process. Instead, three data sets should be used: a training set, a validation set for optimizing the model architecture and training process, and a true test set (also known as evaluation set) for measuring the performance.

Instead of using a fixed part of the data as the test set, performance evaluation is often done by *cross-validation*, where the data set is divided into a number of folds, and each fold is in turn used as a test set while the others are used as the training set. The final performance is then calculated as an average of the test set performances. The number of folds can vary; most often, five- or tenfold cross-validation is used, but some authors prefer N-fold cross-validation, where N is the number of data points—in other words, just one example at a time is held out, while the training is performed on all other examples. This is also known as leave-one-out cross-validation or jackknife test.

The necessity for splitting the data into training and test is not particular to bioinformatics; it applies to all prediction tasks. However, bioinformatics has an added complication: sequences are related by descent. If there are sequences in the test set that are closely related to sequences in the training set, the measured performance is arguably not a true test performance. This can be taken into account by reducing homology in the data set before splitting it into folds (homology reduction) or by ensuring that no pairs of sequences that are too closely related end up in different folds (homology partitioning). Two widely used algorithms for homology reduction were published early in the history of bioinformatics [37].

There are diverging views concerning exactly how closely related two sequences should be allowed to be in order to be separated into different folds. Some authors arbitrarily set a rather high cutoff, for example, 80% or 90% identity in a pairwise alignment [8, 38]. One approach to a nonarbitrary definition is to identify a cutoff above which the problem could be better solved by alignment than by machine learning [39, 40]. Another approach is to use a cutoff in the alignment score above which there is a statistical significance of homology [41]. These approaches tend to result in much lower cutoff values, typically corresponding to ≈25% identity in long alignments [39]. When comparing reported performances of different methods, it is important to take into account which type of homology reduction or partitioning was used (if any).

When reporting performances of prediction methods, a variety of measures may be used, potentially confusing the untrained reader. The conceptually simplest performance measure, the fraction or percentage of correct answers (also known as *accuracy*), can be misleading if the classes are not the same size. As an example, consider a data set with 99 negative examples for each positive example. If a prediction method consistently returns an answer of no, it will be correct 99% of the time, even though the supposed prediction is completely noninformative. Instead, a number of alternative measures are often used. When discriminating between two classes, the most important performance measures can be defined in terms of the numbers of true positives (TPs), true negatives (TNs), false positives (FPs) (type I errors or overpredictions), and false negatives (FNs) (type II errors or misses):

- Sensitivity (also known as recall or the TP rate—how many positive examples are found?):

$$Sn = \frac{TP}{TP + FN};$$

- Specificity (also known as the TN rate—how many negative examples are found?):

$$Sp = \frac{TN}{TN + FP};$$

- Precision (also known as positive predictive value—how many positive predictions are true?):

$$Pr = \frac{TP}{TP + FP};$$

- Matthews correlation coefficient (MCC)—a measure that takes values between −1 and 1, where 1 is a perfect prediction, 0 is a random guess or noninformative prediction, and −1 is a prediction that is consistently wrong:

$$MCC = \frac{TP \times TN - FP \times FN}{\sqrt{(TP + FP)(TP + FN)(TN + FP)(TN + FN)}}.$$

It should be mentioned that the term *specificity* is not unequivocal; it has sometimes been used to denote what is here referred to as precision (e.g., in ref. 42).

Whenever a prediction method gives a quantitative output, there is a trade-off between sensitivity and specificity, controlled by the threshold (also known as cutoff) above which a prediction is considered positive. Lowering the threshold reduces the number of FNs, thereby increasing the sensitivity, but it also increases the number of FPs, thereby reducing the specificity (and the precision). It is possible to plot the sensitivity as a function of the FP rate (1 minus the specificity) for varying threshold values; such a plot is known as a receiver operating characteristic (ROC) curve (*see* Fig. 2). The area under the ROC curve (usually referred to as AUC or AROC) can be used as a threshold-independent performance measure; it will be 1 for a perfect prediction, 0.5 for random guesses, and 0 for a consistently wrong prediction.

When predicting more than two classes, for example, a number of SCLs, the maximum information about the prediction is provided by the so-called *confusion matrix*: a table showing, for each observed class, how many examples were predicted to belong to each class. This can be used to see not only how well each class was predicted, but also which classes were particularly difficult to distinguish. From the confusion matrix, the sensitivity, specificity, precision, and MCC can be calculated for each class. There are also measures that summarize a whole confusion matrix in one number, such as the Gorodkin correlation coefficient, which is a generalization of the MCC to more than two classes, or the normalized mutual information coefficient [43, 44]. In practice, these are rarely calculated, and the percentage of correct answers is often used instead, despite the shortcomings of this measure.

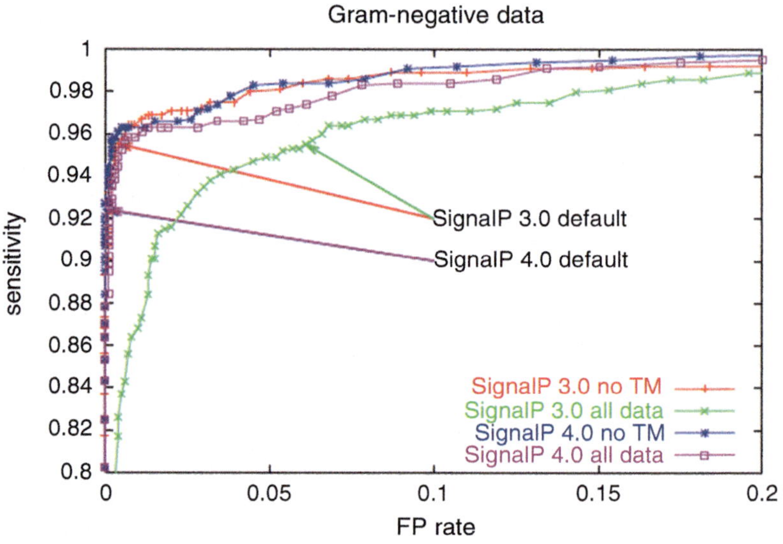

Fig. 2 ROC curve showing performance of SignalP versions 3 and 4 as sensitivity versus FP rate. "No TM" means performance when the negative set did not contain transmembrane segments; "all data" means that sequences with transmembrane segments were included in the negative data. Observe that although SignalP 4 by default has a lower sensitivity than SignalP 3, this is only a question of cutoff; the curve for SignalP 4 when using "all data" is consistently closer to the upper left corner, showing that SignalP 4 is a better method. Note that the sensitivity and FP rate values depicted here are not cross-validation performances but measured by applying the finished method to the whole data set

5 Recognition of Signal Peptides

The secretory SP is among the earliest prediction targets for bioinformatics algorithms. The oldest SP prediction methods used a simple PWM for the SP cleavage site, first with a reduced alphabet [2] and later with weights for all amino acids [45]. Another very early SP prediction method used two simple sequence-derived features, peak hydrophobicity and length of the uncharged region, to discriminate SPs, but it did not predict the cleavage site [46].

SPs are present in all domains of life, but it was early discovered that there are differences between broadly defined systematic groups [47]. SPs of Gram-positive bacteria are longer than those of Gram-negative bacteria, which in turn are longer than those of eukaryotes. A sequence logo of SPs from Gram-negative bacteria is shown in Fig. 3.

In 1997, the SP predictor SignalP (*see* Table 1) was among the first to use ANNs to predict sorting signals [48]. Later, in versions 2 and 3, an HMM was added to the method [49, 50], while version 4 is again purely ANN-based [51]. SignalP is among the most cited prediction servers in bioinformatics, and it has performed well in comparative studies [52–55].

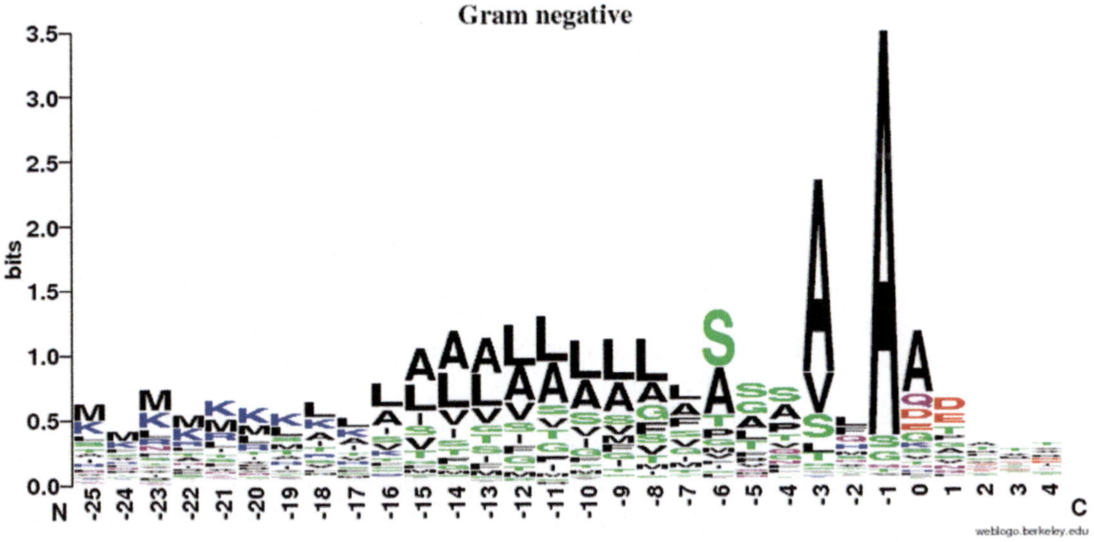

Fig. 3 Sequence logo of signal peptides from Gram-negative bacteria, aligned after their cleavage site (between positions −1 and 0). The visible features are the cleavage site specifying residues in −3 and −1 (strong preference for alanine), the hydrophobic region that approximately stretches from −16 to −7, and a preference for the positively charged lysine in the N-terminal region. Note that no stretching or shortening of the sequences has been performed; they have simply been aligned by the cleavage site; this is why no completely conserved methionine is seen on the *left-hand* side. Picture made with WebLogo [146]

Other SP prediction methods worth mentioning are the PWM-based PrediSi [56] (*see* Table 1) and SOSUIsignal (*see* Table 1), which is based on amino acid propensities in regions [57]. There is also Signal-BLAST [58] (*see* Table 1), which, quite unusually for a sorting signal prediction method, is homology-based. It uses BLAST [10] with some customized settings to search a reference set of SPs and non-SPs and returns the class of the best hit as its prediction. In addition, some TMH prediction programs also offer SP prediction; *see* Subheading 7 for details.

The performance of SP prediction in Gram-negative bacteria is fairly high, with SignalP 4.0 reporting an MCC of 0.85 in distinguishing between SPs and non-SPs and a cleavage site precision of 71%. Note that these performances are cross-validation performances on a strictly homology-reduced data set, so they reflect the performance you would expect if you submitted sequences that were completely unrelated to any in the SignalP 4.0 data set. The performance measured by applying the finished method (where the outputs of the different data set partitions are averaged) to the whole data set is considerably higher, with an MCC of 0.96. Note that SignalP 4.0, despite a higher MCC, has a lower sensitivity than SignalP 3.0; the cutoff had simply been placed at a higher value in order to maximize the MCC. In the slightly modified SignalP 4.1, there is an option to select a cutoff that reproduces the sensitivity of SignalP 3.0. Of course, this comes at a price of a higher FP rate, but it is still lower than that of SignalP 3.0, as can be seen from the ROC curves in Fig. 2.

It should be stressed that the presence of an SP does not necessarily mean that the protein is secreted. First, it may be periplasmic or integrated into the outer membrane; second, there may be downstream TMHs keeping the protein integrated in the cytoplasmic membrane. It has been reported that cleavable SPs are rarely found in bacterial cytoplasmic membrane proteins [59], but a quick search in UniProt [11] revealed that they are not that rare after all, so a prediction of SPs should always be combined with a search for TMHs (*see* Subheading 7) before drawing conclusions about the SCL.

SignalP and the other SP predictors mentioned so far only predict classical SPs, translocated by the Sec system and cleaved by type I signal peptidases. For lipoproteins cleaved by lipoprotein signal peptidase, there are other prediction methods. LipoP [60] (*see* Table 1) is an HMM-based method (although an ANN was also trained during the development of the method). Even though LipoP has been trained on sequences from Gram-negative bacteria only, both the original paper and a later study [61] reported that it demonstrated good performance on sequences from Gram-positive bacteria also. Other methods include the ANN-based SPEPlip [62] (*see* Table 1), which has separate options for Gram-negative and Gram-positive bacteria, and the HMM-based PRED-LIPO [63] (*see* Table 1), which is specific to Gram-positive bacteria. In addition to these methods, there is also a profile in PROSITE [28], dedicated to lipoproteins from both Gram-negative and Gram-positive bacteria, called PROKAR_LIPOPROTEIN (*see* Table 1). A sequence logo of lipoprotein SPs aligned to this model is shown in Fig. 4. It should be noted that LipoP, SPEPlip, and PRED-LIPO are all able to predict classical SPs as well, differentiating between the two types of SP. Additionally, LipoP differentiates between SPs and N-terminal TMHs.

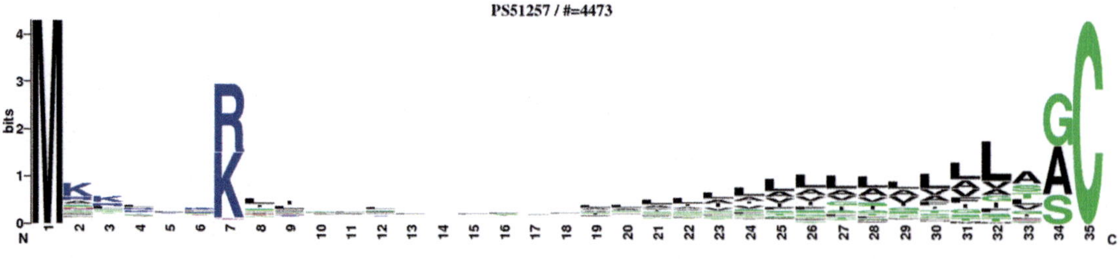

Fig. 4 Sequence logo of lipoprotein signal peptides from both Gram-positive and Gram-negative bacteria, aligned to the PROSITE profile PS51257/PROKAR_LIPOPROTEIN. Lipid attachment occurs at the completely conserved cysteine in position 35. Note that individual sequences may be shorter or longer than 35 amino acids; in the logo, they have been stretched or shortened to fit the model. Picture from PROSITE [28] made with WebLogo [146]

For SPs translocated by the Tat pathway, a few dedicated prediction methods are also available. In addition to the twin-arginine motif in the N-terminal region that gave them their name, they also differ from Sec SPs by being on average longer and less hydrophobic [64]. The available servers are TatFind [65] (*see* Table 1), which is based on a regular expression combined with a set of simple rules concerning hydrophobicity and charge, TatP [66] (*see* Table 1), which is based on a regular expression combined with two ANNs, and the newer HMM-based PRED-TAT [67] (*see* Table 1). In addition, three motifs are available in the family and domain databases: the PROSITE profile TAT (*see* Table 1), the Pfam profile TAT_signal (*see* Table 1), and the TIGRFAMs profile TAT_signal_seq (*see* Table 1). A logo of sequences aligned to the PROSITE profile is shown in Fig. 1. Note that none of these methods makes any distinction between Gram-positive and Gram-negative bacteria—if the Tat SPs indeed differ between the two bacterial groups, there should be room for improvement in the prediction.

6 Prediction of Secretion Without Signal Peptides

In Gram-negative bacteria, secretion without an N-terminal, cleaved SP happens to proteins belonging to secretion systems of types I, III, IV, and VI [68, 69]. For Gram-positive bacteria, the phenomenon appears to be less important, but there are some examples of proteins exported via, for example, the Wss, holin, and SecA2 pathways [69, 70]. This has sometimes been referred to as *nonclassical secretion* [70–72], but at least in Gram-negative bacteria, this term should properly be reserved for secretion that happens independently of any of the numbered secretion systems, for example, by membrane vesicles [73].

SecretomeP 2.0 [70] (*see* Table 1) is a general secretion predictor from 2005 designed to handle secretion without SPs. At the time, it was not easy to locate experimentally confirmed examples of this, so the SecretomeP authors took a different approach, based on the idea that secreted proteins must be expected to share certain features independent of the pathway used to secrete them. Thus, the positive training data set simply consisted of classically secreted proteins with the SP removed. A large number of structural and functional features calculated from the amino acid sequence were then tested for their predictive power for SCL, and the most promising features were used to train an ANN. For Gram-negative bacteria, only four features were selected: amino acid composition, arginine content, instability index, and predicted disorder. These features led to a sensitivity of 88% (on the truncated examples) and a specificity of 96%. Unfortunately, the ability of SecretomeP to predict type I, III, IV, and VI secretion has never been tested. It is possible that effectors of types III, IV, and VI secretion systems, which get injected directly into the host cytoplasm, do not share the extracellular characteristics that SecretomeP is trained to recognize.

The competing method SecretP 2.0 [71] (*see* Table 1), which is also based on feature selection, just using SVMs instead of ANNs, used another, more problematic approach to the problem of data set generation: the positive training set for so-called nonclassical secretion consisted of proteins that were annotated for secretion but had no annotated SP in UniProt [11]. The problematic aspect of this is that a lack of annotated SP may simply reflect an incomplete annotation rather than a real absence of SP, meaning that SecretP in fact could be based on classical SP-containing proteins. This suspicion is confirmed by the fact that SignalP was found to predict seven out of nine supposedly nonclassically secreted proteins in Gram-negative bacteria [71].

Despite its name, SecretP 2.1 [72] (*see* Table 1) is not an updated version of SecretP 2.0; it attempts to answer a completely different question: Given that a protein from a Gram-negative bacterium is secreted, it yields a prediction about which secretion system is responsible for it. SecretP 2.1 distinguishes between proteins of the type I, II, III, IV, V, or VII secretion system—type VI and type VIII data were also collected during data set construction, but too few were found for prediction. Before the actual prediction method was developed, the authors constructed sequence similarity networks for each of the six groups and found that proteins of the type I, V, and VII secretion systems formed a few large clusters with a high average number of links, while the others formed many small clusters or singletons. This can be taken to mean that homology-based prediction would be well suited for predicting type I, V, and VII secretion. However, the authors chose to make a global-property-based method, based on amino acid composition and autocorrelation of physicochemical parameters as inputs to a system of SVMs, to make the multicategory classification. The reported overall accuracy is approximately 90%. Note that the server is not suited for genomic-scale analysis; it only accepts one sequence at a time.

When it comes to secretion-system-specific predictors, very little attention has been directed toward type I and VI secretion. One method for machine-learning-based prediction of type I secretion has been published [74], but it is not available as a web server, and furthermore the prediction is limited to those type I secreted proteins that contain RTX repeats. I am not aware of any attempt to specifically predict type VI secretion; maybe the number of known examples is too small to develop a prediction method.

Type IV secretion has been somewhat better investigated, with several bioinformatics analyses published specifically concerning the phenomenon in two species, *Legionella pneumophila* and

Coxiella burnetii [75–77]. Two SVM-based machine learning methods with a broader scope are available as web servers, T4EffPred [78] (*see* Table 1) and T4SEpre [79] (*see* Table 1). T4EffPred is a global-property-based predictor that uses entire sequences as input, while T4SEpre uses the C-terminal 100 amino acids (shorter window lengths were also tested but did not perform as well). Within this window, T4SEpre uses a PWM-like method for encoding sequences, together with predicted secondary structures and solvent accessibility. Authors of both methods reported the performance on type IVa and type IVb effectors separately, but the T4SEpre authors additionally reported that the method trained on IVa and tested on IVb or vice versa did work to some degree, showing that the signals for the two pathways are not completely different.

The prediction of type III effectors, on the other hand, has received a lot of attention in recent years, and the rest of this section will be dedicated to type III secretion. Several experimental results point to the signal for secretion being within the N-terminal part of the protein [80], but it is not clear whether the signal is read at the mRNA level or the protein level. The mRNA hypothesis is predominantly based on the observation that some proteins retained their ability to be secreted after two balanced frameshift mutations completely changed the amino acid sequence of the N-terminal part of the protein [81].

Two methods published in the same journal issue in 2009 introduced the use of machine learning for type III effector prediction: the SVM-based SIEVE [82] (*see* Table 1) and the Naïve Bayes-based EffectiveT3 [83] (*see* Table 1). The EffectiveT3 authors tested several machine learning methods, including SVM, but found that Naïve Bayes gave the best performance. In both of these methods, a feature selection procedure was carried out, but the feature sets were quite different—SIEVE used evolutionary conservation, phylogenetic profiles, G + C content of the encoding gene, and the sequence of the N-terminal part of the protein, while EffectiveT3 used amino acid composition and occurrence of short degenerate motifs such as "polar–hydrophobic–polar." Later the same year, an unnamed method was published by Löwer and Schneider [84] (*see* Table 1) using a moving window input, which was processed by an ANN or an SVM. The website gives the user the opportunity to choose between ANN and SVM, but the authors estimated that the ANN had the best generalization capability.

Four newer methods have implemented web servers. These are BPBAac [85] (*see* Table 1), which uses a PWM-like approach, T3_MM [86] (*see* Table 1), which is based on first-order Markov models, and the hybrid methods BEAN [87, 88] and pEffect [89] (*see* Table 1), both of which use homology to known effectors to enhance prediction.

Performances are difficult to compare, especially since the authors disagree on which negative set to use. SIEVE used all proteins in each represented organism that were not in the positive set, yielding a ratio of positive to negative examples of approximately 1:120, while the authors of EffectiveT3, BPBAac, T3_MM, and BEAN constructed balanced training sets where the ratio was 1:2 by sampling annotated proteins without annotation about type III secretion. Löwer and Schneider used an even more balanced set, where the ratio was approximately 1:1. While the SIEVE approach will tend to underestimate performance because there could be undiscovered positive examples in the negative data, the balanced-data-set approach measures performance in an unrealistic setting. Furthermore, as the SIEVE authors correctly remark, "any method of filtering negative examples to provide a more confidently nonsecreted set might introduce significant biases in the data set that could render classification trivial" [80]. The pEffect authors took a radically different approach and included eukaryotic as well as bacterial data in their negative set, which makes sense if you imagine the method used in a metagenomics setting. The resulting ratio of positive to negative examples was 1:30 after homology reduction.

One caveat about measuring performance is illustrated by the BPBAac and T3_MM methods, which originate from the same group. Although they both reported reasonably high performances (sensitivity 91% and 90%, specificity 97% and 91%), they disagreed wildly about predictions in genomes of *Salmonella* strains that had not been part of the training set: Only around 25% of the predictions by the more conservative method, BPBAac, were shared by T3_MM [86].

The pEffect authors reported a sensitivity of 95% at a precision of 87%, which is better than EffectiveT3, BPBAac, T3_MM, and BEAN 2.0 on a data set that was not used to train any of the methods.

An interesting aspect of these methods—perhaps more interesting than their predictive performances—is what they can tell us about the actual signal for type III secretion. The first lesson is that the signal is apparently universal across different systematic groups within Gram-negative bacteria. Several of the authors of these methods tested this by a phylogenetically informed cross-validation, i.e., testing the performance on one systematic group of bacteria with a method trained on the others [82, 83, 85, 86].

The fact that it is possible to predict type III secretion to some degree from amino acids speaks against the mRNA hypothesis, although nobody actually tested the hypothesis by training a nucleotide-sequence-based predictor and comparing its performance to an amino-acid-sequence-based predictor. As mentioned earlier, the first version of SIEVE used G + C content as an input feature, but the authors later found that this feature could be removed without a major impact on performance, and the current version of SIEVE uses only amino acid sequences as input [80]. An alternative explanation for the observations that led to the mRNA hypothesis could be that the signal is so unspecific that a spurious reading frame has a relatively high chance of producing a type III secretion signal. This was verified by the authors of EffectiveT3 and BPBAac, who tested their predictors on artificially created frameshifted sequences and found that 10–14% of frame-shifted positive examples were predicted positive [83, 85].

Since the signal is so weakly defined, it is reasonable to ask whether sequence order of amino acids plays any role or whether it is just the composition of the N-terminal region. SIEVE and BPBAac both use the actual sequence as input, while EffectiveT3, T3_MM, and BEAN use the composition of amino acids or amino acid pairs. The better performance of BPBAac over T3_MM using the same data set could be taken as a sign that the positions of amino acids do have a role to play.

Another question is whether the signal is actually N-terminal, as many of the methods assume. The EffectiveT3 authors tested this by training on the 15 C-terminal residues instead of N-terminal and found no performance above the random expectation [83]. The EffectiveT3 and SIEVE authors and Löwer and Schneider all investigated how many N-terminal residues were required for prediction and found no improvement by moving beyond 30 (except in the case of plant pathogens predicted by EffectiveT3, where the limit seemed to be 50) [82–84]. In contrast, the BPBAac authors used 100 N-terminal positions and found that to be superior to 50 [85]. BEAN version 1 used 51 N-terminal positions, but in BEAN 2.0 two additional windows were added comprising positions 52–121 from the N-terminus and 1–50 from the C-terminus [87, 88]. The authors state that their results show that C-terminal signals sometimes play a role, but since the two windows were added simultaneously, it might as well be the 52–121 window that made the difference (which would be in accordance with the BPBAac result). The pEffect method, by contrast, uses the entire sequence as input, and the authors write: "Our work reveals that signals for recognition and transport of effectors are distributed over the entire protein sequence instead of being confined to the N-terminus" [89]. However, the evidence for this conclusion is not very strong, since it is no surprise that homology between known effectors is distributed across the entire sequence. To draw conclusions on the placement of the signal, it is necessary to consider the nonhomology part of pEffect, the so-called de novo prediction, and that declined when fragments were used instead of the whole sequence. To obtain a real answer to the question, it would be necessary to train an N-terminal version of the de novo part of pEffect to see whether it performed equally well.

7 Prediction of Transmembrane Topology

In Gram-negative bacteria, transmembrane proteins come in two flavors: α-helix proteins, almost exclusively in the cytoplasmic membrane, and β-barrel proteins, exclusively in the outer membrane. The prediction of TMHs has a long history in bioinformatics. Initially, the basis for the prediction was simply a plot of the hydrophobicity, averaged in a sliding window over the sequence [1, 90]. A slightly more advanced approach was represented by TOP-PRED in 1992 [91], which combined hydrophobicity analysis with a count of the number of positively charged residues in each loop to choose the topological model that best conformed to the "positive-inside rule" [92].

The prediction of transmembrane β-strands is more complicated, not only because of the smaller number of known examples, but also because of their shorter length and lower hydrophobicity. Typically, only those amino acid side chains facing the lipid phase are hydrophobic, while those facing the pore of the β-barrel tend to be polar. The first attempt at a transmembrane β-barrel (TMBB) topology prediction method, published in 1985 [93], focused not on hydrophobicity patterns but instead on predicting the turns separating the strands. However, the following year a method for structural prediction that accounted for amphipathic β-strands was published [94].

Later, machine learning methods were used to predict membrane protein topology, i.e., which parts of a sequence are inside, transmembrane, and outside. In particular, the HMM technology has been popular in this area because it makes it possible to model the so-called grammar of a problem: if a TMH or a strand follows an inside loop, it must be followed by an outside loop, and vice versa. This is typically modeled by a cyclic HMM, having submodels for helices/strands, inside loops, and outside loops. In the rest of this section, TMH prediction will be described first, followed by TMBB prediction.

The best known HMM for TMH prediction is TMHMM [95] (*see* Table 1), but also HMMTOP [96] (*see* Table 1) has enjoyed widespread usage. A comparative analysis in 2001 found TMHMM to be the best performing TMH predictor [97]. Newer surveys covering more recently published predictors unfortunately do not provide quantitative performance comparisons [98–100].

Since hydrophobicity is a feature of both SPs and TMHs, the two are easily confused by prediction methods. TMHMM often falsely predicts an SP as a TMH, and versions 1–3 of SignalP would often predict a TMH close to the N-terminus as an SP. Newer topology prediction methods, such as the HMM-based Phobius [101], Philius, which is based on Dynamic Bayesian Networks [102], the ANN-based MEMSAT3 [103], the SVM-based MEMSAT-SVM

[104], and the ANN + HMM-based SPOCTOPUS [105] (*see* Table 1), deal with this problem by modeling both these signals. However, the paper on version 4 of SignalP [51] reports a better discrimination between SPs and TMHs than all these methods, and it is worth noting that the performance difference is larger for bacterial sequences than for eukaryotic sequences. This probably reflects the fact that the signal peptide models in Phobius, Philius, MEMSAT, and SPOCTOPUS are not divided into organism types, which causes the results to be biased toward the organism group with the most data (eukaryotes).

Another confounding factor is the fact that multispanning membrane proteins sometimes have so-called reentrant loops—segments of the sequence that dip into the membrane but do not span it, leaving the membrane on the same side at which they entered. Reentrant loops are not very frequent; only 137 examples from bacteria are currently reported in UniProt, one of them with experimental evidence. OCTOPUS [106] (*see* Table 1) and SPOCTOPUS make an attempt at predicting reentrant loops.

The use of profiles of homologous sequences generated by BLAST or PSI-BLAST (*see* Subheadings 2 and 3) in the training and prediction of TMH recognition methods has been shown to enhance predictive performance by approximately ten percentage units [107]. Methods that use profiles include PRODIV-TMHMM [107], PolyPhobius [108], MEMSAT3, MEMSAT-SVM, OCTOPUS, and SPOCTOPUS.

An interesting alternative method is SCAMPI [109] (*see* Table 1) that does not use machine learning or statistics on a training set to calculate its parameters; instead, the parameters are based on a series of experiments where all 20 possible amino acids have been inserted at various positions into a model TMH [110]. These experiments have been used to calculate an apparent free energy contribution, ΔG_{app}, that is used as an analog to a hydrophobicity scale. The overall ΔG_{app} for each sequence window is calculated and used as input to an HMM-like model with only two free parameters to be estimated from the training data. The SCAMPI authors reported a performance comparable to the best machine learning methods.

Consensus methods for TMH prediction have been shown to perform better than any of the constituent methods. An early effort in this direction was BPROMPT from 2003 [111] (*see* Table 1). The newer server TOPCONS [112, 113] (*see* Table 1) offers a consensus prediction of both TMHs and SPs based on OCTOPUS, SPOCTOPUS, PolyPhobius, Philius, and SCAMPI. TOPCONS reports 83% correctly predicted topologies on a benchmark set. The downside of TOPCONS is the running time, increased by the fact that four of the five predictors are based on profiles that first must be constructed from a database search. An alternative consensus server, based only on methods that do not require profiles, is TOPCONS-single [114] (*see* Table 1), which does approximately six percentage units worse than TOPCONS, but 70 times faster.

In TMBB topology prediction, the first machine learning method was an ANN published in 1998 [115], but the early 2000s saw a surge in the publication of HMMs, including HMM-B2TMR [116], PRED-TMBB [117, 118] (*see* Table 1), and ProfTMB [119, 120] (*see* Table 1). A comparative study in 2005 [121] found that HMM-based predictors performed better than ANN- and SVM-based predictors but achieved the best performance by constructing a consensus of HMM-B2TMR, PRED-TMBB, and ProfTMB, together with the ANN-based methods B2TMPRED [122] (*see* Table 1) and TBBpred [123] (*see* Table 1). The resulting method, ConBBPRED (*see* Table 1), is available as a server; however, it is very impractical to use since it does not launch the constituent methods but requires the user to obtain the various predictions and input those to the server manually.

BOCTOPUS [124, 125] (*see* Table 1) is a newer hybrid method, where four window-based SVMs calculate scores for each residue in the inner loop, outer loop, pore-facing transmembrane, and lipid-facing transmembrane, respectively. These scores are fed to a filter that delineates the barrel domain (if any) and then to an HMM that calculates the final topology.

Some other methods have focused on the task of discriminating between outer membrane TMBB proteins and other proteins rather than predicting the correct topology, and here, the cyclic HMM is not necessarily the best approach. BOMP [126] (*see* Table 1) from 2004 is one example that does not use machine learning but simple statistical measures, including a C-terminal pattern (regular expression) that is found in many outer membrane proteins. HHomp [127] (*see* Table 1) uses a collection of profile HMMs, built from outer membrane proteins of known structure, to search for matches to the query sequence. It is thus based on the idea that most TMBB proteins are evolutionarily related, and for predicted TMBB proteins, it additionally provides a classification into a number of functional subgroups. BetAware [128, 129] (*see* Table 1) uses an ANN to scan entire sequences and predict whether they are TMBB proteins. The second version of BetAware includes a probabilistic model, a so-called Grammatical-Restrained Hidden Conditional Random Field, for predicting the topology, but it is only invoked when the ANN output indicates that the query is a TMBB protein.

Another direction is represented by the methods transFold [130, 131] (*see* Table 1) and TMBpro [132] (*see* Table 1), which aim to deliver more information than just the TMBB topology (1D structure). They both predict which residues will pair with each other in the β-strand (2D structure), and TMBpro additionally predicts a full set of coordinates of the protein (3D structure). However, they cannot be used to screen a data set for outer membrane proteins since they both assume that the submitted protein (only one at a time is accepted) is a TMBB protein.

Methods that are best at one task are not necessarily the same ones that are best at another task. In the paper on BOCTOPUS version 2 [125], the authors compare their performance to a number of other methods and find that BOCTOPUS 2 is better at predicting the correct topology than a number of other recent methods (PRED-TMBB, ProfTMB, and BetAware; transFold and TMBpro are not included in the benchmark), but HHomp and BetAware yield better discrimination. Note that TMBB prediction is still more difficult than TMH prediction; BOCTOPUS 2 reports 69% correct TMBB topologies, in contrast with the 83% correct TMH topologies reported by TOPCONS.

8 Multicategory Predictors

The first software to attempt a classification of proteins into multiple SCLs was PSORT [133]. It was basically a signal-based method, incorporating the previously mentioned early methods for prediction of SPs [45, 46] and TMHs [90], but it also used amino acid composition, especially for recognizing outer membrane proteins.

For bacteria, PSORT has been superseded by PSORTb [3, 42, 134] (*see* Table 1), which is now in version 3. Version 1 was for Gram-negative bacteria only, but version 2 included Gram-positive bacteria. Version 3 additionally offers predictions for Archaea and the so-called problematic bacteria, which stain either Gram-positive, although they have an outer membrane (such as genus *Deinococcus*), or Gram-negative, although they have no outer membrane (phylum Tenericutes).

PSORTb is a hybrid method, incorporating signal-based, global-property-based, and homology-based predictions. The signal-based component comprises recognition of SPs and TMHs and a database of motifs (regular expressions) derived from PROSITE, which are found to be exclusive to specific SCLs. The global properties component is SVM-based; in version 1 its input consisted of amino acid composition only, but in versions 2 and 3, a collection of overrepresented subsequences is used. The homology-based component is a simple BLAST with direct annotation transfer. Finally, a Bayesian network is used to integrate the outputs from the components and arrive at a final prediction.

The final prediction, however, may be "unknown." PSORTb values precision over recall, so it prefers to deliver no prediction rather than a prediction with weak evidence. It may also arrive at two SCLs, signifying that the protein is predicted to function in both compartments or belong to the interface between the compartments.

The SCLs predicted by PSORTb 3 in some cases extend beyond the standard five categories for Gram-negative bacteria; there are new subcategory SCLs, such as fimbrial, flagellar, and host-associated. The reported precision of PSORTb 3 on Gram-negative bacteria (the five main categories only) is 97%, with a recall of 94%. This was tested by fivefold cross-validation with a data set that was homology reduced, but only down to 80% identity. Note that this high recall only applies to bacterial species that are well represented in the data set or closely related to such species; a benchmark paper reported that in *Campylobacter jejuni*, PSORTb 3 returned "unknown" for as many as 47% of the protein sequences [135].

Another predictor that owes its high performance to homology is the SCL predictor built into the prediction workbench Proteome Analyst [15, 33] (*see* Table 1). It uses a combination of direct and indirect annotation transfer by retrieving up to three hits from the Swiss-Prot part of UniProt by BLAST and then parsing the "subcellular location" field, the keywords, and the cross-referenced InterPro entries. The retrieved words are then processed by a Naïve Bayes classifier. Other machine learning methods (ANN and SVM) were also tried, and although they could enhance performance by a few percent, the authors decided to stick with Naïve Bayes in order to be able to provide explanations for the individual predictions. The PSORTb 3 paper [134] reports that PSORTb 3.0 and Proteome Analyst 3.0 have comparable precisions but make complementary predictions, so that a combined analysis with both methods has the highest coverage overall.

Kuo-Chen Chou's group has published a long series of predictors for protein SCL (*see*, e.g., [18]). They prefer to publish one website per organism group instead of providing one website with an option for selecting organism group, and furthermore they tend to change the name for each new version instead of adding a version number. For bacteria, the relevant predictors are named Gneg-PLoc [19], Gpos-PLoc [20], Gneg-mPLoc [21], Gpos-mPLoc [22], iLoc-Gneg [23], and iLoc-Gpos [24] (*see* Table 1). The PLoc/mPLoc/iLoc servers are hybrid methods, mostly relying on indirect homology annotation through the Gene Ontology (GO) terms of database hits. GO [136] is an ordered system (a directed acyclic graph) of controlled terms that describe the biological process, molecular function, and cellular component of proteins. In the prediction servers, GO terms are extracted from all database hits with a pairwise identity above a certain cutoff, and then a *k*-nearest neighbor classifier is applied to the high-dimensional vectors of occurrences of GO terms. If no hits are found, or if the found hits have no GO annotation, a profile-based global property approach is used. The nature of this approach varies between PLoc, mPLoc, and iLoc. However, the corresponding papers contain no information about how often the global property approach is needed, and the performance of this approach has never been reported separately. The overall accuracy, measured by jackknife (leave-one-out cross-validation) is reported to be 87% for Gneg-PLoc (with eight categories; the standard five plus fimbrium, flagellum, and nucleoid), 86% for Gneg-mPLoc, and 91% for iLoc-Gneg. Note that the PLoc and mPLoc servers only accept one sequence per submission, while iLoc accepts up to 50.

The new feature of the mPLoc and iLoc servers relative to the PLoc servers is the ability to predict multiple SCLs, i.e. predict whether a protein can exist in more than one cellular compartment. This can be of importance in eukaryotes, where many proteins, for example, may shuttle between the cytoplasm and the nucleus, but it is of somewhat minor importance in bacteria. The data set for Gneg-mPLoc and iLoc-Gneg contained only 64 such examples out of 1,392 homology-reduced proteins.

PSLpred [137] (*see* Table 1) is a hybrid method specific to Gram-negative bacteria. It uses amino acid composition, dipeptide composition, physicochemical parameters, and PSI-BLAST against a database of proteins with known SCL and integrates the features by SVM. Interestingly, the authors found that the PSI-BLAST module on its own performed worse than the other modules. The final reported overall accuracy was 91% on the PSORTb 2 data.

Another hybrid approach is LocTree3 [138] (*see* Table 1), which has two components: a PSI-BLAST [10] homology search with direct transfer of SCL annotation and a global property approach corresponding to LocTree2 [139]. If no homology hits are found with an E-value better than a specified cutoff, the LocTree2 method is applied. It consists of a decision tree of SVMs trained with a so-called profile kernel, basically using the occurrence of short substrings in profiles made by PSI-BLAST as input. When delivering a prediction, LocTree3 reports whether the evidence is based on a homology or on LocTree2. For bacteria LocTree2 has a reported performance (overall accuracy) of 86%, and LocTree3 has 90%. Interestingly, in the LocTree3 paper, the measured accuracy for PSORTb 3.0 was only 57%. This huge difference from PSORTb's own reported performance is hard to explain, but it may reflect different views on the exactly correct way to parse UniProt's SCL annotations.

It is claimed that LocTree2/3 can predict SCLs for all domains of life, but it seems less well suited for Gram-positive bacteria since it offers no opportunity to choose between Gram-positives and Gram-negatives. Thus, it may predict categories like periplasm and outer membrane for Gram-positive bacteria, while totally failing to predict cell wall.

The methods described so far in this section have all been wholly or partly homology-based. However, there is also the global-property-based CELLO [13] (*see* Table 1), an SVM-based predictor for eukaryotes, Gram-negative bacteria, and

Gram-positive bacteria. The SVMs are organized in a two-level system, where the first level contains a number of SVMs trained on various sequence encodings, and the second layer is a so-called jury SVM that decides on the prediction based on the outputs of the first-layer SVMs. The sequences are encoded by total amino acid composition, dipeptide composition, and amino acid composition (in some cases with a reduced alphabet) in a number of partitions of each sequence. The performance for Gram-negative bacteria is reported to be 95% overall accuracy without homology reduction and 83% with homology reduction (down to 30% identity).

Another predictor that is not homology-based is SOSUI-GramN [140] (*see* Table 1), which, as the name suggests, is specific to Gram-negative bacteria. It is not based on machine learning but on various physicochemical properties measured over an entire sequence and in the N- and C-terminal regions. Tests for these measured parameters are then arranged in a complex decision tree that takes into account that, for example, secretion can happen through various pathways.

The authors of the consensus method MetaLocGramN [135] (*see* Table 1) benchmarked four methods, PSORTb 3, CELLO, SOSUI-GramN, and PSLpred, and found PSORTb 3 to be the best in overall performance. However, PSORTb 3 had a very low sensitivity for the extracellular category, where PSLpred was found to be better. The consensus method MetaLocGramN is based not only on these four predictors but also on a number of signal-based predictors focusing on SPs, TMHs, TMBBs, and type III secretion signals. To integrate all the predictors, an ANN was tried, but it did not work consistently better than the constituent methods. Instead, the finished method is based on statistical feature selection and logistic regression, and the result is better than in PSORTb 3, especially for the extracellular category, measured on a new data set. Unfortunately, the MetaLocGramN web server depends on all the constituent methods being up and alive. At the time of writing, SOSUI-GramN was temporarily unavailable, which made MetaLocGramN hang indefinitely.

9 Discussion

As is apparent from this chapter, the many possible ways of approaching the SCL prediction problem has resulted in a large number of available prediction servers. Comparing their performances can be complicated, and all their authors tend to claim superior performance for their particular method. Add to this that usability is sometimes limited (some web servers allow only one or a few sequences in each submission), that response times vary considerably, and that there are almost as many different output formats as there are servers, and you get a rather frustrating situation. Even the definitions of SCLs may vary from server to server; for example, a peripheral membrane protein may be defined as belonging to the membrane or to the compartment it protrudes into.

This situation is clearly not ideal for the user, who might prefer a "one-stop shop" to go to for all sequence-based prediction needs, an equivalent of UniProt or InterPro. But this kind of confusion is probably inevitable in a field that is evolving so fast. Scientific competition is basically beneficial, and competing groups should certainly not be discouraged from publishing their predictors independently. That being said, prediction servers ought to follow certain standards concerning usability, definitions, and formats.

Personally, I must admit to having added to the complexity through my involvement in the servers SignalP, LipoP, and TatP (*see* Subheading 5). In hindsight, we should have published LipoP and TatP not as separate servers but as functionalities within the SignalP server. Hopefully, the next version of SignalP will be able to predict all these types of SP in one user interface.

The multicategory prediction methods report quite impressive performances, and, as described in Subheading 1, the error rates for prediction may now be lower than the error rates for high-throughput experiments. However, it is important to keep in mind that these performances are achieved by analyzing the annotations of homologs found by sequence similarity searches. I see three problems with this. First, predictions for novel organisms and metagenomics samples with few known homologs will necessarily be harder than for the organisms the training and test sets were built from, so coverage and precision for such organisms will be considerably lower than the reported performances. Second, the annotations used for prediction are themselves error prone and not necessarily derived from experiments. Especially when relying on keywords and GO terms, there is a real danger of circular reasoning, where annotations based on predictions are used as a basis for new predictions, which then may enter the databases as annotations. Third, homology-based predictions do not reflect real biological knowledge about the protein sorting process in the way a successful signal-based predictor does.

As machine learning algorithms continue to evolve, new classes of algorithms should also be expected to be applied to predictions of SCL. In fields such as image processing and speech recognition, novel types of ANNs—deep and recurrent neural networks—have been used extensively in recent years [141, 142], and they are beginning to be employed in bioinformatics as well, for example, to predict protein secondary structures [143] or alternative splicing [144]. The previously mentioned TMBpro (*see* Subheading 7) is an example of a method based on recurrent ANNs. The advantage of recurrent ANNs is that they are naturally designed to handle sequential data, so sequences are not chopped up into apparently unrelated windows, and they can potentially learn long-range correlations. A first attempt at using a recurrent ANN in multicategory SCL prediction was published recently [145]—so far, only for eukaryotic data, and without a web server, but the results seem promising. Coupled with a so-called convolutional layer—basically a series of PWMs of varying width through which the input sequences were presented—the network was able to learn from the data where in each sequence to focus its attention. Performance was much better than that of other methods working only on sequence, and on the same level as advanced homology-based methods. This technology represents a new kind of compromise between signal-based and global-property-based methods since it is apparently able to find sorting signals in sequences, even though it is given the sequences and their SCL categories only during training. It will be very interesting to see where this and other novel technologies will take SCL prediction in the coming years.

References

1. Kyte J, Doolittle RF (1982) A simple method for displaying the hydropathic character of a protein. J Mol Biol 157:105–132
2. von Heijne G (1983) Patterns of amino acids near signal-sequence cleavage sites. Eur J Biochem 133:17–21
3. Gardy JL, Laird MR, Chen F et al (2005) PSORTb v.2.0: expanded prediction of bacterial protein subcellular localization and insights gained from comparative proteome analysis. Bioinformatics 21:617–623
4. Rey S, Gardy J, Brinkman F (2005) Assessing the precision of high-throughput computational and laboratory approaches for the genome-wide identification of protein subcellular localization in bacteria. BMC Genomics 6:162
5. Nielsen H (2016) Predicting subcellular localization of proteins by bioinformatic algorithms. In: Bagnoli F, Rappuoli R (eds) Protein export in gram-positive bacteria. Current topics in microbiology and immunology. Springer, Berlin, Heidelberg
6. Nakashima H, Nishikawa K (1994) Discrimination of intracellular and extracellular proteins using amino acid composition and residue-pair frequencies. J Mol Biol 238:54–61
7. Andrade MA, O'Donoghue SI, Rost B (1998) Adaptation of protein surfaces to subcellular location. J Mol Biol 276:517–525
8. Reinhardt A, Hubbard T (1998) Using neural networks for prediction of the subcellular location of proteins. Nucleic Acids Res 26:2230–2236
9. Hua S, Sun Z (2001) Support vector machine approach for protein subcellular localization prediction. Bioinformatics 17:721–728
10. Altschul SF, Madden TL, Schaffer AA et al (1997) Gapped BLAST and PSI-BLAST: a new generation of protein database search programs. Nucleic Acids Res 25:3389
11. The UniProt Consortium (2015) UniProt: a hub for protein information. Nucleic Acids Res 43:D204–D212
12. Nair R, Rost B (2002a) Sequence conserved for subcellular localization. Protein Sci 11:2836–2847
13. Yu C-S, Chen Y-C, Lu C-H, Hwang J-K (2006) Prediction of protein subcellular localization. Proteins 64:643–651
14. Nair R, Rost B (2002b) Inferring sub-cellular localization through automated lexical analysis. Bioinformatics 18(Suppl 1):S78–S86
15. Lu Z, Szafron D, Greiner R et al (2004) Predicting subcellular localization of proteins using machine-learned classifiers. Bioinformatics 20:547–556
16. Shatkay H, Höglund A, Brady S et al (2007) SherLoc: high-accuracy prediction of protein subcellular localization by integrating text and protein sequence data. Bioinformatics 23:1410–1417
17. Briesemeister S, Blum T, Brady S et al (2009) SherLoc2: a high-accuracy hybrid method for

predicting subcellular localization of proteins. J Proteome Res 8:5363–5366

18. Chou K-C, Shen H-B (2010) Cell-PLoc 2.0: an improved package of web-servers for predicting subcellular localization of proteins in various organisms. Nat Sci 2:1090–1103
19. Chou K-C, Shen H-B (2006) Large-scale predictions of gram-negative bacterial protein subcellular locations. J Proteome Res 5:3420–3428
20. Shen H-B, Chou K-C (2007) Gpos-PLoc: an ensemble classifier for predicting subcellular localization of gram-positive bacterial proteins. Protein Eng Des Sel 20:39–46
21. Shen H-B, Chou K-C (2010) Gneg-mPLoc: a top-down strategy to enhance the quality of predicting subcellular localization of gram-negative bacterial proteins. J Theor Biol 264:326–333
22. Shen H-B, Chou K-C (2009) Gpos-mPLoc: a top-down approach to improve the quality of predicting subcellular localization of gram-positive bacterial proteins. Protein Pept Lett 16:1478–1484
23. Xiao X, Wu Z-C, Chou K-C (2011) A multi-label classifier for predicting the subcellular localization of gram-negative bacterial proteins with both single and multiple sites. PLoS One 6:e20592
24. Wu Z-C, Xiao X, Chou K-C (2012) iLoc-Gpos: a multi-layer classifier for predicting the subcellular localization of singleplex and multiplex gram-positive bacterial proteins. Protein Pept Lett 19:4–14
25. Stormo GD, Schneider TD, Gold L, Ehrenfeucht A (1982) Use of the "perceptron" algorithm to distinguish translational initiation sites in *E. coli*. Nucleic Acids Res 10:2997–3011
26. Schneider TD, Stephens RM (1990) Sequence logos: a new way to display consensus sequences. Nucleic Acids Res 18:6097–6100
27. Krogh A, Brown M, Mian IS et al (1994) Hidden Markov models in computational biology: applications to protein modeling. J Mol Biol 235:1501–1531
28. Sigrist CJA, de Castro E, Cerutti L et al (2013) New and continuing developments at PROSITE. Nucleic Acids Res 41:D344–D347
29. Finn RD, Bateman A, Clements J et al (2014) Pfam: the protein families database. Nucleic Acids Res 42:D222–D230
30. Haft DH, Selengut JD, Richter RA et al (2013) TIGRFAMs and genome properties in 2013. Nucleic Acids Res 41:D387–D395
31. Mitchell A, Chang H-Y, Daugherty L et al (2015) The InterPro protein families database: the classification resource after 15 years. Nucleic Acids Res 43:D213–D221
32. Rish I (2001) An empirical study of the naive Bayes classifier. In: IJCAI 2001 workshop Empir Methods Artif Intell. IBM, New York, pp 41–46
33. Szafron D, Lu P, Greiner R et al (2004) Proteome analyst: custom predictions with explanations in a web-based tool for high-throughput proteome annotations. Nucleic Acids Res 32:W365–W371
34. Briesemeister S, Rahnenführer J, Kohlbacher O (2010) Going from where to why—interpretable prediction of protein subcellular localization. Bioinformatics 26:1232–1238
35. Hertz JA, Krogh AS, Palmer RG (1991) Introduction to the theory of neural computation. Westview Press, Redwood City, CA
36. Noble WS (2006) What is a support vector machine? Nat Biotechnol 24:1565–1567
37. Hobohm U, Scharf M, Schneider R, Sander C (1992) Selection of representative protein data sets. Protein Sci 1:409–417
38. Höglund A, Dönnes P, Blum T et al (2006) MultiLoc: prediction of protein subcellular localization using N-terminal targeting sequences, sequence motifs and amino acid composition. Bioinformatics 22:1158–1165
39. Sander C, Schneider R (1991) Database of homology-derived protein structures and the structural meaning of sequence alignment. Proteins 9:56–68
40. Nielsen H, Engelbrecht J, von Heijne G, Brunak S (1996) Defining a similarity threshold for a functional protein sequence pattern: the signal peptide cleavage site. Proteins 24:165–177
41. Nielsen H, Wernersson R (2006) An overabundance of phase 0 introns immediately after the start codon in eukaryotic genes. BMC Genomics 7:256
42. Gardy JL, Spencer C, Wang K et al (2003) PSORT-B: improving protein subcellular localization prediction for gram-negative bacteria. Nucleic Acids Res 31:3613–3617
43. Baldi P, Brunak S, Chauvin Y et al (2000) Assessing the accuracy of prediction algorithms for classification: an overview. Bioinformatics 16:412–424
44. Gorodkin J (2004) Comparing two K-category assignments by a K-category correlation coefficient. Comput Biol Chem 28:367–374

45. von Heijne G (1986) A new method for predicting signal sequence cleavage sites. Nucleic Acids Res 14:4683–4690
46. McGeoch DJ (1985) On the predictive recognition of signal peptide sequences. Virus Res 3:271–286
47. von Heijne G, Abrahmsén L (1989) Species-specific variation in signal peptide design: implications for protein secretion in foreign hosts. FEBS Lett 244:439–446
48. Nielsen H, Brunak S, Engelbrecht J, von Heijne G (1997) Identification of prokaryotic and eukaryotic signal peptides and prediction of their cleavage sites. Protein Eng 10:1–6
49. Nielsen H, Krogh A (1998) Prediction of signal peptides and signal anchors by a hidden Markov model. Proc Int Conf Intell Syst Mol Biol 6:122–130
50. Bendtsen JD, Nielsen H, von Heijne G, Brunak S (2004) Improved prediction of signal peptides: SignalP 3.0. J Mol Biol 340:783–795
51. Petersen TN, Brunak S, von Heijne G, Nielsen H (2011) SignalP 4.0: discriminating signal peptides from transmembrane regions. Nat Methods 8:785–786
52. Menne KML, Hermjakob H, Apweiler R (2000) A comparison of signal sequence prediction methods using a test set of signal peptides. Bioinformatics 16:741–742
53. Klee E, Ellis L (2005) Evaluating eukaryotic secreted protein prediction. BMC Bioinformatics 6:1–7
54. Choo K, Tan T, Ranganathan S (2009) A comprehensive assessment of N-terminal signal peptides prediction methods. BMC Bioinformatics 10:S2
55. Zhang X, Li Y, Li Y (2009) Evaluating signal peptide prediction methods for gram-positive bacteria. Biologia (Bratisl) 64:655–659
56. Hiller K, Grote A, Scheer M et al (2004) PrediSi: prediction of signal peptides and their cleavage positions. Nucleic Acids Res 32:W375–W379
57. Gomi M, Sonoyama M, Mitaku S (2004) High performance system for signal peptide prediction: SOSUIsignal. Chem-Bio Inform J 4:142–147
58. Frank K, Sippl MJ (2008) High-performance signal peptide prediction based on sequence alignment techniques. Bioinformatics 24: 2172–2176
59. Broome-Smith JK, Gnaneshan S, Hunt LA et al (1994) Cleavable signal peptides are rarely found in bacterial cytoplasmic membrane proteins. Mol Membr Biol 11:3–8
60. Juncker AS, Willenbrock H, von Heijne G et al (2003) Prediction of lipoprotein signal peptides in gram-negative bacteria. Protein Sci 12:1652–1662
61. Rahman O, Cummings SP, Harrington DJ, Sutcliffe IC (2008) Methods for the bioinformatic identification of bacterial lipoproteins encoded in the genomes of gram-positive bacteria. World J Microbiol Biotechnol 24:2377–2382
62. Fariselli P, Finocchiaro G, Casadio R (2003) SPEPlip: the detection of signal peptide and lipoprotein cleavage sites. Bioinformatics 19:2498–2499
63. Bagos PG, Tsirigos KD, Liakopoulos TD, Hamodrakas SJ (2008) Prediction of lipoprotein signal peptides in gram-positive bacteria with a hidden Markov model. J Proteome Res 7:5082–5093
64. Cristóbal S, de Gier J-W, Nielsen H, von Heijne G (1999) Competition between Sec- and TAT-dependent protein translocation in *Escherichia coli*. EMBO J 18:2982–2990
65. Rose RW, Brüser T, Kissinger JC, Pohlschröder M (2002) Adaptation of protein secretion to extremely high-salt conditions by extensive use of the twin-arginine translocation pathway. Mol Microbiol 45:943–950
66. Bendtsen JD, Nielsen H, Widdick D et al (2005a) Prediction of twin-arginine signal peptides. BMC Bioinformatics 6:167
67. Bagos PG, Nikolaou EP, Liakopoulos TD, Tsirigos KD (2010) Combined prediction of Tat and Sec signal peptides with hidden Markov models. Bioinformatics 26: 2811–2817
68. Binnewies TT, Bendtsen JD, Hallin PF et al (2005) Genome update: protein secretion systems in 225 bacterial genomes. Microbiology 151:1013–1016
69. Desvaux M, Hébraud M, Talon R, Henderson IR (2009) Secretion and subcellular localizations of bacterial proteins: a semantic awareness issue. Trends Microbiol 17:139–145
70. Bendtsen JD, Kiemer L, Fausbøll A, Brunak S (2005b) Non-classical protein secretion in bacteria. BMC Microbiol 5:58
71. Yu L, Guo Y, Li Y et al (2010a) SecretP: identifying bacterial secreted proteins by fusing new features into Chou's pseudo-amino acid composition. J Theor Biol 267:1–6
72. Yu L, Luo J, Guo Y et al (2013) In silico identification of gram-negative bacterial secreted proteins from primary sequence. Comput Biol Med 43:1177–1181
73. Lloubes R, Bernadac A, Houot L, Pommier S (2013) Non classical secretion systems. Res Microbiol 164:655–663
74. Luo J, Li W, Liu Z et al (2015) A sequence-based two-level method for the prediction of

type I secreted RTX proteins. Analyst 140: 3048–3056

75. Burstein D, Zusman T, Degtyar E et al (2009) Genome-scale identification of *Legionella pneumophila* effectors using a machine learning approach. PLoS Pathog 5:e1000508
76. Chen C, Banga S, Mertens K et al (2010) Large-scale identification and translocation of type IV secretion substrates by *Coxiella burnetii*. Proc Natl Acad Sci U S A 107:21755–21760
77. Lifshitz Z, Burstein D, Peeri M et al (2013) Computational modeling and experimental validation of the *Legionella* and *Coxiella* virulence-related type-IVB secretion signal. Proc Natl Acad Sci U S A 110:E707–E715
78. Zou L, Nan C, Hu F (2013) Accurate prediction of bacterial type IV secreted effectors using amino acid composition and PSSM profiles. Bioinformatics 29:3135–3142
79. Wang Y, Wei X, Bao H, Liu S-L (2014) Prediction of bacterial type IV secreted effectors by C-terminal features. BMC Genomics 15:50
80. McDermott JE, Corrigan A, Peterson E et al (2011) Computational prediction of type III and IV secreted effectors in gram-negative bacteria. Infect Immun 79:23–32
81. Anderson DM, Schneewind O (1997) A mRNA signal for the type III secretion of Yop proteins by *Yersinia enterocolitica*. Science 278:1140–1143
82. Samudrala R, Heffron F, McDermott JE (2009) Accurate prediction of secreted substrates and identification of a conserved putative secretion signal for type III secretion systems. PLoS Pathog 5:e1000375
83. Arnold R, Brandmaier S, Kleine F et al (2009) Sequence-based prediction of type III secreted proteins. PLoS Pathog 5:e1000376
84. Löwer M, Schneider G (2009) Prediction of type III secretion signals in genomes of gram-negative bacteria. PLoS One 4:e5917
85. Wang Y, Zhang Q, Sun M, Guo D (2011) High-accuracy prediction of bacterial type III secreted effectors based on position-specific amino acid composition profiles. Bioinformatics 27:777–784
86. Wang Y, Sun M, Bao H, White AP (2013) T3_MM: a Markov model effectively classifies bacterial type III secretion signals. PLoS One 8:e58173
87. Dong X, Zhang Y-J, Zhang Z (2013) Using weakly conserved motifs hidden in secretion signals to identify type-III effectors from bacterial pathogen genomes. PLoS One 8:e56632
88. Dong X, Lu X, Zhang Z (2015) BEAN 2.0: an integrated web resource for the identification and functional analysis of type III secreted effectors. Database 2015:bav064
89. Goldberg T, Rost B, Bromberg Y (2016) Computational prediction shines light on type III secretion origins. Sci Rep 6:34516
90. Klein P, Kanehisa M, DeLisi C (1985) The detection and classification of membrane-spanning proteins. Biochim Biophys Acta 815:468–476
91. von Heijne G (1992) Membrane protein structure prediction: hydrophobicity analysis and the positive-inside rule. J Mol Biol 225: 487–494
92. von Heijne G, Gavel Y (1988) Topogenic signals in integral membrane proteins. Eur J Biochem 174:671–678
93. Paul C, Rosenbusch JP (1985) Folding patterns of porin and bacteriorhodopsin. EMBO J 4:1593–1597
94. Vogel H, Jähnig F (1986) Models for the structure of outer-membrane proteins of *Escherichia coli* derived from raman spectroscopy and prediction methods. J Mol Biol 190:191–199
95. Krogh A, Larsson B, von Heijne G, Sonnhammer EL (2001) Predicting transmembrane protein topology with a hidden Markov model: application to complete genomes. J Mol Biol 305:567–580
96. Tusnády GE, Simon I (2001) The HMMTOP transmembrane topology prediction server. Bioinformatics 17:849–850
97. Möller S, Croning MDR, Apweiler R (2001) Evaluation of methods for the prediction of membrane spanning regions. Bioinformatics 17:646–653
98. Elofsson A, von Heijne G (2007) Membrane protein structure: prediction versus reality. Annu Rev Biochem 76:125–140
99. Punta M, Forrest LR, Bigelow H et al (2007) Membrane protein prediction methods. Methods 41:460–474
100. Tusnády GE, Simon I (2010) Topology prediction of helical transmembrane proteins: how far have we reached? Curr Protein Pept Sci 11:550–561
101. Käll L, Krogh A, Sonnhammer EL (2004) A combined transmembrane topology and signal peptide prediction method. J Mol Biol 338: 1027–1036
102. Reynolds SM, Käll L, Riffle ME et al (2008) Transmembrane topology and signal peptide prediction using dynamic Bayesian networks. PLoS Comput Biol 4:e1000213

103. Jones DT (2007) Improving the accuracy of transmembrane protein topology prediction using evolutionary information. Bioinformatics 23:538–544
104. Nugent T, Jones DT (2009) Transmembrane protein topology prediction using support vector machines. BMC Bioinformatics 10:159
105. Viklund H, Bernsel A, Skwark M, Elofsson A (2008) SPOCTOPUS: a combined predictor of signal peptides and membrane protein topology. Bioinformatics 24:2928–2929
106. Viklund H, Elofsson A (2008) OCTOPUS: improving topology prediction by two-track ANN-based preference scores and an extended topological grammar. Bioinformatics 24:1662–1668
107. Viklund H, Elofsson A (2004) Best α-helical transmembrane protein topology predictions are achieved using hidden Markov models and evolutionary information. Protein Sci 13:1908–1917
108. Käll L, Krogh A, Sonnhammer EL (2005) An HMM posterior decoder for sequence feature prediction that includes homology information. Bioinformatics 21:i251–i257
109. Bernsel A, Viklund H, Falk J et al (2008) Prediction of membrane-protein topology from first principles. Proc Natl Acad Sci 105: 7177–7181
110. Hessa T, Meindl-Beinker NM, Bernsel A et al (2007) Molecular code for transmembrane-helix recognition by the Sec61 translocon. Nature 450:1026–1030
111. Taylor PD, Attwood TK, Flower DR (2003) BPROMPT: a consensus server for membrane protein prediction. Nucleic Acids Res 31:3698–3700
112. Bernsel A, Viklund H, Hennerdal A, Elofsson A (2009) TOPCONS: consensus prediction of membrane protein topology. Nucleic Acids Res 37:W465–W468
113. Tsirigos KD, Peters C, Shu N et al (2015) The TOPCONS web server for consensus prediction of membrane protein topology and signal peptides. Nucleic Acids Res 43:W401–W407
114. Hennerdal A, Elofsson A (2011) Rapid membrane protein topology prediction. Bioinformatics 27:1322–1323
115. Diederichs K, Freigang J, Umhau S et al (1998) Prediction by a neural network of outer membrane β-strand protein topology. Protein Sci 7:2413–2420
116. Martelli PL, Fariselli P, Krogh A, Casadio R (2002) A sequence-profile-based HMM for predicting and discriminating β barrel membrane proteins. Bioinformatics 18:S46–S53
117. Bagos P, Liakopoulos T, Spyropoulos I, Hamodrakas S (2004a) A hidden Markov model method, capable of predicting and discriminating beta-barrel outer membrane proteins. BMC Bioinformatics 5:29
118. Bagos PG, Liakopoulos TD, Spyropoulos IC, Hamodrakas SJ (2004b) PRED-TMBB: a web server for predicting the topology of β-barrel outer membrane proteins. Nucleic Acids Res 32:W400–W404
119. Bigelow HR, Petrey DS, Liu J et al (2004) Predicting transmembrane beta-barrels in proteomes. Nucleic Acids Res 32:2566–2577
120. Bigelow H, Rost B (2006) PROFtmb: a web server for predicting bacterial transmembrane beta barrel proteins. Nucleic Acids Res 34:W186–W188
121. Bagos P, Liakopoulos T, Hamodrakas S (2005) Evaluation of methods for predicting the topology of beta-barrel outer membrane proteins and a consensus prediction method. BMC Bioinformatics 6:7
122. Jacoboni I, Martelli PL, Fariselli P et al (2001) Prediction of the transmembrane regions of β-barrel membrane proteins with a neural network-based predictor. Protein Sci 10:779–787
123. Natt NK, Kaur H, Raghava GPS (2004) Prediction of transmembrane regions of β-barrel proteins using ANN- and SVM-based methods. Proteins 56:11–18
124. Hayat S, Elofsson A (2012) BOCTOPUS: improved topology prediction of transmembrane β barrel proteins. Bioinformatics 28:516–522
125. Hayat S, Peters C, Shu N et al (2016) Inclusion of dyad-repeat pattern improves topology prediction of transmembrane β-barrel proteins. Bioinformatics 32:1571–1573
126. Berven FS, Flikka K, Jensen HB, Eidhammer I (2004) BOMP: a program to predict integral β-barrel outer membrane proteins encoded within genomes of gram-negative bacteria. Nucleic Acids Res 32:W394–W399
127. Remmert M, Linke D, Lupas AN, Söding J (2009) HHomp—prediction and classification of outer membrane proteins. Nucleic Acids Res 37:W446–W451
128. Savojardo C, Fariselli P, Casadio R (2011) Improving the detection of transmembrane β-barrel chains with N-to-1 extreme learning machines. Bioinformatics 27:3123–3128
129. Savojardo C, Fariselli P, Casadio R (2013) BETAWARE: a machine-learning tool to detect and predict transmembrane beta-barrel proteins in prokaryotes. Bioinformatics 29:504–505

130. Waldispühl J, Berger B, Clote P, Steyaert J-M (2006a) transFold: a web server for predicting the structure and residue contacts of transmembrane beta-barrels. Nucleic Acids Res 34:W189–W193
131. Waldispühl J, Berger B, Clote P, Steyaert J-M (2006b) Predicting transmembrane β-barrels and interstrand residue interactions from sequence. Proteins 65:61–74
132. Randall A, Cheng J, Sweredoski M, Baldi P (2008) TMBpro: secondary structure, β-contact and tertiary structure prediction of transmembrane β-barrel proteins. Bioinformatics 24:513–520
133. Nakai K, Kanehisa M (1991) Expert system for predicting protein localization sites in gram-negative bacteria. Proteins 11:95–110
134. Yu NY, Wagner JR, Laird MR et al (2010b) PSORTb 3.0: improved protein subcellular localization prediction with refined localization subcategories and predictive capabilities for all prokaryotes. Bioinformatics 26:1608–1615
135. Magnus M, Pawlowski M, Bujnicki JM (2012) MetaLocGramN: a meta-predictor of protein subcellular localization for gram-negative bacteria. Biochim Biophys Acta 1824:1425–1433
136. Ashburner M, Ball CA, Blake JA et al (2000) Gene ontology: tool for the unification of biology. Nat Genet 25:25–29
137. Bhasin M, Garg A, Raghava GPS (2005) PSLpred: prediction of subcellular localization of bacterial proteins. Bioinformatics 21:2522–2524
138. Goldberg T, Hecht M, Hamp T et al (2014) LocTree3 prediction of localization. Nucleic Acids Res 42:W350–W355
139. Goldberg T, Hamp T, Rost B (2012) LocTree2 predicts localization for all domains of life. Bioinformatics 28:i458–i465
140. Imai K, Asakawa N, Tsuji T et al (2008) SOSUI-GramN: high performance prediction for sub-cellular localization of proteins in gram-negative bacteria. Bioinformation 2:417–421
141. Krizhevsky A, Sutskever I, Hinton GE (2012) ImageNet classification with deep convolutional neural networks. In: Pereira F, Burges CJC, Bottou L, Weinberger KQ (eds) Advances in neural information processing systems, vol. 25, Curran Associates, Inc., Red Hook, NY, pp 1097–1105
142. Dahl GE, Yu D, Deng L, Acero A (2012) Context-dependent pre-trained deep neural networks for large-vocabulary speech recognition. IEEE Trans Audio Speech Lang Process 20:30–42
143. Magnan CN, Baldi P (2014) SSpro/ACCpro 5: almost perfect prediction of protein secondary structure and relative solvent accessibility using profiles, machine learning and structural similarity. Bioinformatics 30:2592–2597
144. Xiong HY, Alipanahi B, Lee LJ et al (2015) The human splicing code reveals new insights into the genetic determinants of disease. Science 347:1254806
145. Sønderby SK, Sønderby CK, Nielsen H, Winther O (2015) Convolutional LSTM networks for subcellular localization of proteins. In: Dediu A-H, Hernández-Quiroz F, Martín-Vide C, Rosenblueth DA (eds) Algorithms for computational biology, Lecture notes in computer science, vol 9199. Springer International Publishing, New York, pp 68–80
146. Crooks GE, Hon G, Chandonia J-M, Brenner SE (2004) WebLogo: a sequence logo generator. Genome Res 14:1188–1190

Chapter 3

Cell Fractionation

Melissa Petiti, Laetitia Houot, and Denis Duché

Abstract

Protein function is generally dependent on its subcellular localisation. In Gram-negative bacteria such as *Escherichia coli,* a protein can be targeted to five different compartments: the cytoplasm, the inner membrane, the periplasm, the outer membrane and the extracellular medium. Different approaches can be used to determine the protein localisation within a cell such as in silico identification of protein signal sequences and motifs, electron microscopy and immunogold labelling, optical fluorescence microscopy, and biochemical technics. In this chapter, we describe a simple and efficient method to isolate the different compartments of *Escherichia coli* by a fractionation method and to determine the presence of the protein of interest. For inner membrane proteins we propose a method to discriminate between integral and peripheral membrane proteins.

Key words Spheroplast, Peptidoglycan, Osmotic shock, Freeze and thaw, Protein solubilisation, Membrane, Subcellular localisation

1 Introduction

Many Gram-negative bacteria secrete extracellular proteins such as hydrolytic enzymes or toxins. Secretion can occur through specific macrocomplex systems composed of a more or less large number of proteins located in the cell envelope. Identifying the localisation of these proteins is therefore an important task to address the assembly and the molecular mechanism of these secretion systems.

Four subcellular compartments compose Gram-negative bacteria, five if we consider the extracellular medium in which effectors are delivered. These different compartments are the cytoplasm, the inner membrane (IM), the periplasm in which the peptidoglycan layer extends, and the outer membrane (OM) [1, 2]. Isolation an characterization of effectors from the extracellular medium will be described in Chapter 31 of this book. In this chapter, we will first describe a simple and efficient method of recovering proteins from the periplasm and to generate spheroplasts from *Escherichia coli* cells.

Laure Journet and Eric Cascales (eds.), *Bacterial Protein Secretion Systems: Methods and Protocols*, Methods in Molecular Biology, vol. 1615, DOI 10.1007/978-1-4939-7033-9_3, © Springer Science+Business Media LLC 2017

Then we will present a method to recover the cytoplasmic and the membrane fractions from the spheroplasts by several cycles of freezing and thawing. Finally, we describe how treatments with specific buffers can give insight into protein–membrane associations.

2 Materials

2.1 Cell Fractionation

1. TES buffer: 200 mM Tris–HCl, pH 8.0, 0.5 mM EDTA (ethylenediaminetetraacetic acid), 0.5 M sucrose.
2. Lysosyme 10 mg/mL (freshly prepared solution).
3. DNase 1 10 mg/mL.
4. $MgCl_2$ 1 M (stock solution).
5. 100× phenylmethylsulfonyl (PMSF) 0.1 M in absolute ethanol. Store at −20 °C.
6. Beckman Coulter (Brea, CA) Optima TLX ultracentrifuge with TLA 55K rotor or equivalent.

2.2 Protein Solubilisation

1. Urea 2 M.
2. NaCl 0.5 M.
3. Triton X-100 1% (v/v).
4. Sodium carbonate 100 mM pH 11.5, ice cold.
5. Trichloroacetic acid (TCA) 10% (v/v). The stock solution [TCA 100% (w/v)] is stored at 4 °C in a brown bottle.
6. Acetone 90% (v/v) in ultrapure water stored at −20 °C in a brown bottle.

3 Methods

3.1 Cell Fractionation/Spheroplast Formation

In this section, we detail step by step spheroplast preparation from *Escherichia coli* cells (*see* **Note 1**) using a method based on lysozyme/EDTA treatment [3] and a mild osmotic shock [4–6] (*see* **Note 2**).

1. Grow a 3 mL starter culture overnight in a lysogeny broth medium at 37 °C with required antibiotics.
2. Inoculate a 20 mL culture at OD_{600} = 0.05 and incubate at 37 °C until the optical density of the culture is around 0.8. If necessary, induce protein production under the required conditions (*see* **Note 3**).
3. Take 1 mL of culture and centrifuge for 5 min at 5000 × *g* to pellet the cells. Discard the supernatant. Resuspend the pellet in an appropriate volume of sodium dodecyl sulphate polyacrylamide gel electrophoresis (SDS-PAGE) loading buffer. This fraction will be referred as the total cell fraction (T).

The next steps will be performed at 4 °C, and all buffers must be cooled in advance on ice before use.

4. Centrifuge the remaining culture for 5 min at 5000 × *g* at 4 °C (*see* **Note 4**). Discard the supernatant.
5. Gently resuspend the cell pellet in 200 μL of TES buffer (described in materials section, *see* **Note 5**). Do not vortex and do not pipette, only resuspend the cell pellet by inverting the tube.
6. Add 8 μL of a freshly prepared solution of lysozyme (10 mg/mL in TES buffer) and mix gently by shaking the tube.
7. Add 720 μL TES buffer diluted twice in water (v/v) and incubate for 30 min on ice. Gently mix the suspension to perform the osmotic shock by gently inverting and rolling the tube (*see* **Note 6**).
8. Centrifuge at 5000 × *g* for 5 min at 4 °C. Keep the pellet as the spheroplast fraction, IM + cytoplasm (+OM), and the supernatant as the periplasmic fraction, P.
9. Resuspend the spheroplast fraction in 1 mL of TES buffer diluted twice in water (v/v) containing 2 mM PMSF, 2 mM $MgCl_2$ and 10 μg/mL DNase 1 (*see* **Notes 7** and **8**).
10. Lyse the spheroplasts by performing four cycles of freezing and thawing, from −273 °C (liquid azote) to 37 °C (*see* **Note 9**).
11. Remove unbroken cells and cell debris by centrifugation at 2000 × *g* for 5 min. Keep the supernatant as cytoplasmic and membrane fraction.
12. Centrifuge the supernatant at 120,000 × *g*, 4 °C, for 45 min. Use a Beckman Coulter Optima TLX ultracentrifuge and a TLA55 fixed-angle rotor for small-volume ultracentrifugation or something similar. The pellet is kept as the membrane fraction and the supernatant is conserved as the cytoplasmic fraction.
13. The membrane fraction is suspended in 1 mL TES buffer diluted twice in water (v/v) or in the desired buffer (*see* **Note 10**). Separation of IM and OM is described in Chapter 6 of this book.
14. At this step, fractions (OD_{600} = 0.2–0.4) could be tested for the presence of the protein of interest by SDS-PAGE and western blot analysis with required antibodies. As a control, the same fractions can be tested for the presence of specific IM, OM, cytoplasm, or periplasm markers.

3.2 Protein Solubilisation (See Note 11)

1. Prepare 1 mL membrane fractions as previously described.
2. Aliquot the membrane fractions into five samples, 200 μL each.

3. Centrifuge at 120,000 × *g* for 45 min at 4 °C to pellet the membranes as described previously (*see* Subheading 3.1, **step 12**).
4. Resuspend each pellet in 200 μL 0.5 M NaCl, 2 M urea, 100 mM sodium carbonate, pH 11.5 ice cold, or 1% (v/v) of Triton X-100 to compare the five extraction conditions.
5. Incubate at least 1 h at 4 °C with agitation.
6. Centrifuge the suspensions at 120,000 × *g* for 45 min at 4 °C. Carefully collect the different supernatants and transfer to new tubes.
7. Resuspend each pellet in SDS-PAGE loading buffer and keep as the membrane-associated protein fractions.
8. Add 10% TCA (final concentration) to supernatant samples, and incubate for at least 1 h at 4 °C to allow protein precipitation (*see* **Notes 12** and **13**).
9. Centrifuge at 18,000 × *g* for 30 min at 4 °C.
10. Wash the pellets with 200 μL of acetone 90% (pre-chilled solution).
11. Centrifuge at 18,000 × *g* for 10 min at 4 °C.
12. Carefully discard the supernatant and air-dry the pellet containing the extracted membrane proteins at room temperature for 5–10 min (*see* **Note 14**).
13. Resuspend pellets in SDS-PAGE loading buffer and keep as extracted membrane protein fractions.
14. Perform a western blot analysis to identify the extraction condition that is suitable for the protein of interest.

4 Notes

1. Spheroplasts are cells resulting from a loss of the bacterial cell wall. The OM has been altered but the cytoplasm remains delimited by the IM [7].
2. *Escherichia coli* cells are first incubated in a concentrated sucrose solution containing EDTA. Sucrose makes the medium hypertonic, while EDTA chelates divalent cations and destabilises the OM. Then lysozyme is added to cleave the periplasmic peptidoglycan layer. However, peptidoglycan hydrolysis is not total, and a mild osmotic shock is required to maximize the procedure. Then the periplasmic content of the cell is separated from the spheroplasts by centrifugation.
3. The volume of the cell culture can be adapted according to the downstream application.
4. Pre-cool centrifuge before use.

5. The TES buffer is responsible for OM destabilisation. 0.5 M sucrose makes the medium hypertonic, 0.5 mM EDTA, 200 mM Tris–HCl, and pH 8 affect the membrane structure by removing the lipopolysaccharide coat from the cells [8].
6. This mild osmotic shock provokes a sudden influx of water in the periplasmic space and increases the distance between polysaccharide chains of the peptidoglycan. This facilitates the lysozyme binding and the degradation of the peptidoglycan [8].
7. PMSF is a serine protease inhibitor. However, it has been shown that the addition of trypsin inhibitor after proteolysis is not required to prevent further digestion because trypsin digestion is very specific.
8. During spheroplast lysis, DNA is released in the medium, adheres to membranes and makes the preparation difficult to handle. To circumvent this issue, DNase 1 is added to lysates. As DNase I activity requires magnesium, Mg^{2+} is added in excess in order to overtake chelation by the EDTA present in the TES buffer.
9. Three to five cycles of freezing and thawing are an efficient and simple method to disrupt spheroplasts [9]. However, spheroplasts can also be disrupted by sonication. In this case, sonicate spheroplast suspension twice for 30 s. Keep the suspension cold during sonication. A Branson Microtip Sonifier 450 (BRANSON Ultrasonics Corp., Danbury, CT) can be used with a microtip probe.
10. Membrane fraction resuspension can be difficult. Passing the sample through the needle of a syringe several times can optimise this step.
11. When studying a poorly characterised protein, it is important to compare different extraction conditions to optimise protein solubilisation. The use of an appropriate solubilisation buffer can provide information about protein localisation in cells and even be used to differentiate between integral and peripheral membrane proteins. Thus, a high salt buffer allows the extraction of peripheral proteins associated with membranes by electrostatic interactions. It is common to use 2 M urea to extract peripheral proteins that associate with membranes by hydrophobic bonds [10]. Triton-X100 is the most commonly used detergent for the solubilisation of integral IM proteins [11]. It is worth noting that, during the preparation of membrane fractions, membranes tend to re-anneal by an unknown mechanism, resulting in the formation of closed membrane vesicles that might trap some proteins of interest. In this case, alkaline carbonate buffer can be added to convert membrane vesicles to open membrane sheets and release trapped proteins into the supernatant [12]. Membranes can subsequently be cleared from the sample by centrifugation.

12. This step can be done overnight in a cold room under rotary agitation.
13. Be careful; pellets are not always visible after centrifugation. Carefully place the tubes in the centrifuge to determine the location of the future pellet.
14. Never let the pellet air-dry completely because this will dramatically impede resuspension.

Acknowledgments

This work was supported by the Centre National de la Recherche Scientifique and the Agence National de la Recherche (ANR-14-CE09-0023).

References

1. Kaback HR (1972) Transport across isolated bacterial cytoplasmic membranes. Biochim Biophys Acta 265:367–416
2. Kellenberger E, Ryther A (1958) Cell wall and cytoplasmic membrane of *Escherichia coli*. J Biophys Biochem Cytol 25:323–326
3. Neu HC, Heppel LA (1964) The release of Ribonuclease into the medium when *Escherichia coli* cells are converted to spheroplasts. J Biol Chem 239:3893–3900
4. French C, Keshavarz-Moore E, Ward JM (1996) Development of a simple method for the recovery of recombinant proteins from the *Escherichia coli* periplasm. Enzym Microb Technol 19:332–338
5. Skerra A, Plückthun A (1991) Secretion and *in vivo* folding of the F_{ab} fragment of the antibody McPC603 in *Escherichia coli*: influence of disulphides and cis-prolines. Protein Eng 4:971–979
6. Nossal NG, Heppel LA (1966) The release of enzymes by osmotic shock from *Escherichia coli* in exponential phase. J Biol Chem 241:3055–3062
7. Kaback HR (1971) Bacterial membranes. Methods Enzymol 22:99–120
8. Witholt B, Heerikhuizen HV, De Leij L (1976) How does lysozyme penetrate through the bacterial outer membrane. Biochim Biophys Acta 443:534–544
9. Mowbray J, Moses V (1976) The tentative identification in *Escherichia coli* of a multienzyme complex with glycolytic activity. Eur J Biochem 66:25–36
10. Schook W, Puszkin P, Bloom W, Ores C, Kochwa S (1979) Mechanochemical properties of brain clathrin: interactions with actin and alpha-actinin and polymerization into basketlike structures or filaments. Proc Natl Acad Sci U S A 76:116–120
11. Schnaitman CA (1971) Solubilization of the cytoplasmic membrane of *Escherichia coli* by triton X-100. J Bacteriol 108:545–552
12. Fujiki Y, Fowler S, Shio H, Hubbard AL, Lazarow PB (1982) Polypeptide and phospholipid composition of the membrane of rat liver peroxisomes: comparison with endoplasmic reticulum and mitochondrial membranes. J Cell Biol 93:103–110

Chapter 4

Defining Lipoprotein Localisation by Fluorescence Microscopy

Maria Guillermina Casabona, Mylène Robert-Genthon, Didier Grunwald†, and Ina Attrée

Abstract

In recent years it has become evident that lipoproteins play crucial roles in the assembly of bacterial envelope-embedded nanomachineries and in the processes of protein export/secretion. In this chapter we describe a method to determine their precise localisation, for example inner versus outer membrane, in Gram-negative bacteria using human opportunistic pathogen *Pseudomonas aeruginosa* as a model. A fusion protein between a given putative lipoprotein and the red fluorescent protein mCherry must be created and expressed in a strain expressing cytoplasmic green fluorescent protein (GFP). Then the peripheral localisation of the fusion protein in the cell can be examined by treating cells with lysozyme to create spheroplasts and monitoring fluorescence under a confocal microscope. Mutants in the signal peptide can be engineered to study the association with the membrane and efficiency of transport. This protocol can be adapted to monitor lipoprotein localisation in other Gram-negative bacteria.

Key words Lipoprotein, Localisation, Cell envelope, Spheroplasts, Bacterial secretion, Fluorescence microscopy

1 Introduction

Lipoproteins are involved in a variety of processes, such as biogenesis of the cell envelope and signalling [1, 2], and they can play important roles in virulence [3]. They are usually hydrophilic proteins present in both Gram-positive and Gram–negative bacteria that are anchored to membranes thanks to a lipid moiety attached to their invariable cysteine residue in their N-terminal region [4]. In Gram-negative bacteria, lipoproteins are present in both inner and outer membranes (IM and OM, respectively). The OM-localised lipoproteins are transported through the periplasm by the well-characterised lipoprotein outer membrane localization (Lol) system (reviewed in [5] and [6]). Lipoproteins can be predicted by

†To the memory of Didier Grunwald, who passed away recently.

Laure Journet and Eric Cascales (eds.), *Bacterial Protein Secretion Systems: Methods and Protocols*, Methods in Molecular Biology, vol. 1615, DOI 10.1007/978-1-4939-7033-9_4, © Springer Science+Business Media LLC 2017

bioinformatics tools, such as DOLOP [7, 8] (*see* also Chapter 1 of this book) and through their characteristic N-terminus signal sequence composed of hydrophobic and uncharged residues called a *lipobox*, which has a consensus (V/L)XXC motif. The lipobox is both the lipidation site and the maturation site recognised by the lipoprotein signal peptidase II, which cleaves the signal peptide just upstream of the conserved cysteine [7, 8].

The *Pseudomonas aeruginosa* envelope is a dynamic multi-layer structure that harbours many multi-protein assemblies, such as extracellular appendages and secretion systems, crucial for bacterial survival in different environments, bacterial pathogenesis and bacterial adaptation. In *P. aeruginosa* there are 175 predicted lipoproteins, most of them of unknown function and mostly predicted to localise to the OM [9]. A recent study using shotgun proteomics also identified many lipoproteins as being anchored to the IM [10].

The literature contains many examples of lipoproteins that are essential in the assembly and function of bacterial secretion systems (SSs). For example, in the Type IV SS (T4SS), VirB7 is a 5.5 kDa lipoprotein that is anchored to the OM and exposed to periplasmic space [11, 12]. VirB7 is essential for the stability of other Vir proteins during the assembly of functional T-complex transport machinery in *Agrobacterium tumefaciens* through dimerisation with VirB9 [12, 13].

Another example is a family of lipoproteins, the so-called pilotins, found in both T2SS and T3SS machineries [14]. The pilotin ExsB is an OM lipoprotein of the T3SS of *P. aeruginosa* that plays an important role in secretion as well as in virulence in vivo by stabilising the β-barrel porin-secretin in the OM [15, 16]. ExsB counterparts in *Salmonella* and *Yersinia* strains also display a decrease in T3 protein secretion and T3SS-associated virulence.

In the T2SS, it has been shown that pilotin lipoproteins have chaperone-like properties since they can protect proteins from proteolytic degradation. Their correct localisation and insertion are needed for survival under stress conditions [17] and for correct functioning of the SS [18].

More recently, two lipoproteins were described as being crucial in the T6SS-1 of *P. aeruginosa*: TssJ1 [19] and TagQ [20]. TssJ1 is conserved throughout the T6SS and is localised to the OM, where it interacts with the TssM-L complex embedded within the IM and stabilises the membrane-bound complex essential for secretion [10, 21–23]. TagQ, on the other hand, is unique to *P. aeruginosa* T6SS-1 and is also vital for competing activity of T6SS-1-expressing strains. TagQ, together with three other envelope-localised proteins, TagS, TagT and TagR, plays a role in post-translational regulation of T6SS-1 activity. It has been proposed that the Tag module senses the signal that triggers the assembly and activation of the T6SS-1. Moreover, it has been shown that correct localisation and anchoring to the OM of TagQ are indispensable for the localisation of the periplasmic protein TagR, which directly promotes signal transduction through a phosphorylation pathway [20, 24] and modulates the activity of the T6S machinery. Therefore, central to understanding the role of lipoproteins in a bacterial SS is the determination of their localisation in the cell envelope and interplay with other components of the given secretion apparatus.

In the past, mostly biochemical approaches such as cell fractionation have been used to determine the localisation of these proteins in cell membranes [18, 20]. In recent years, fluorescence microscopy using superfolder green fluorescent protein (sfGFP) and mCherry has been widely developed for monitoring the behaviour of bacterial proteins in vivo [25–28]. Here we describe a method for defining *P. aeruginosa* lipoprotein localisation in vivo by constructing a fusion protein between the protein of interest and mCherry and monitoring its localisation by means of confocal microscopy. In Gram-negative bacteria, the localisation of the lipoprotein can then be studied by creating spheroplasts by means of lysozyme treatment. As shown in Fig. 1, three different possibilities, such as IM, periplasm and OM, will be distinguishable under a confocal microscope analysis.

2 Materials

All solutions must be prepared using ultrapure water (prepared by purifying deionised water to attain a sensitivity of 18 MΩ cm at 25 °C). Prepare and store all reagents at room temperature unless indicated otherwise. Waste disposal regulations should be followed carefully when disposing of waste materials.

2.1 Preparation of Bacteria

1. *P. aeruginosa* strain producing cytoplasmic GFP (*see* **Note 1**).
2. A plasmid producing a fusion of mCherry to the target protein of interest (*see* **Note 2**).
3. 1% (w/v) agarose in phosphate saline buffer (PBS), sterile.
4. Luria–Bertani (LB) broth (Becton, Dickinson and Co., Franklin Lakes, NJ). Autoclave. Supplement with appropriate antibiotics.
5. Inducer for gene expression: 20% (w/v) arabinose stock solution. Sterilise by filtration. Store at 4 °C.
6. Incubator shaker.
7. Slides and coverslips.
8. Benchtop centrifuge.

2.2 Preparation of Spheroplasts

1. 1% (w/v) agarose in PBS, sterile.
2. LB broth, sterile, supplemented with appropriate antibiotics.

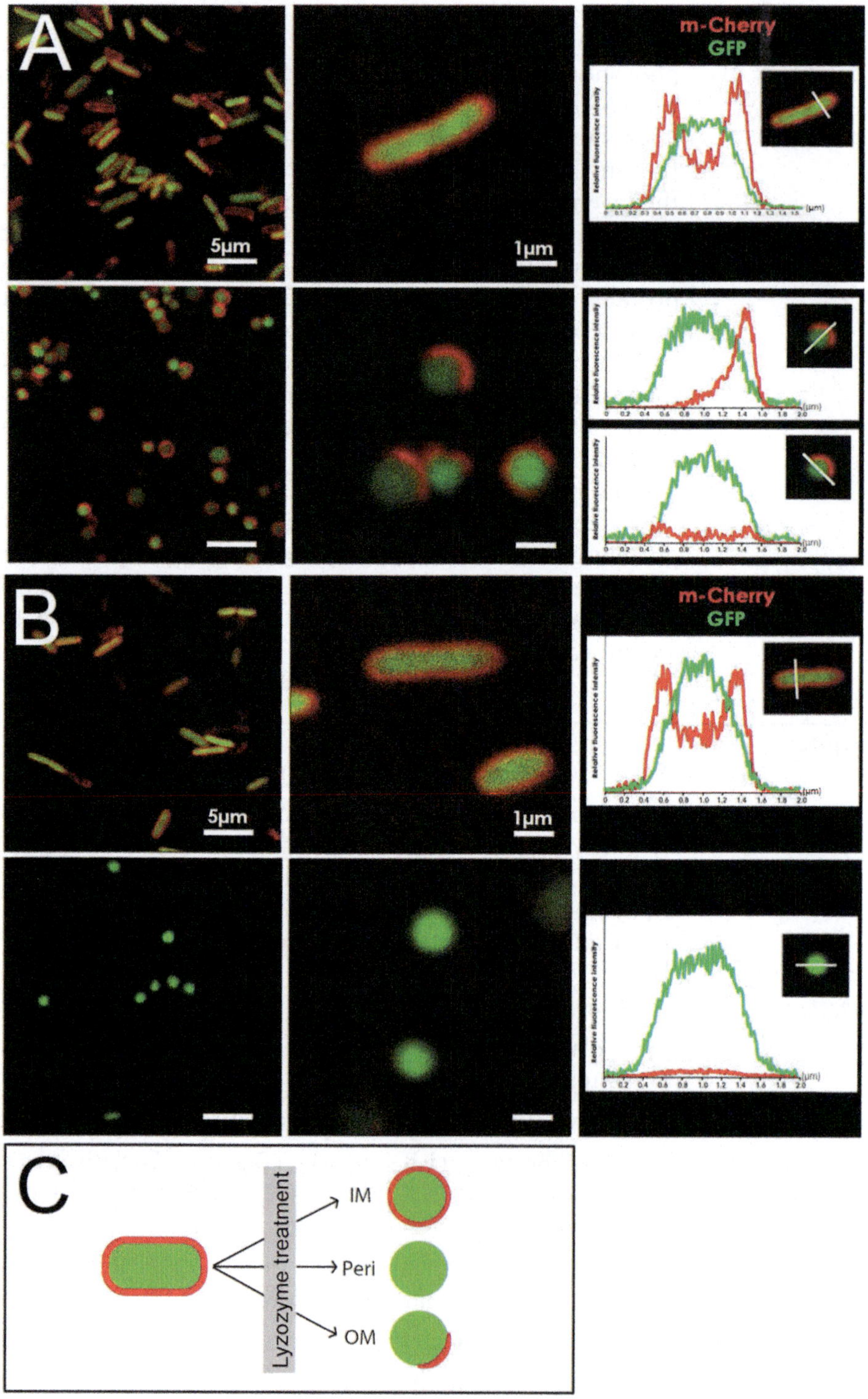

Fig. 1 Outer membrane localisation of T6SS signalling lipoprotein TagQ assessed by confocal microscopy. (**a**) *P. aeruginosa* PAO1 expressing cytoplasmic GFP and TagQ-mCherry were imaged directly from exponentially grown cultures (*upper panel*) or after treatment with lysozyme (*lower panel*). Note the *red* labelling in the periphery of bacteria and in partially detached OM in spheroplast preparations. Inserts represent a plot

3. TM buffer: 10 mM Tris-acetate, pH 8.2, 200 mM $MgSO_4$. Sterilise by filtration. Store at 4°C.
4. TSM buffer: 50 mM Tris-acetate, pH 8.2, 8% (w/v) sucrose, 10 mM $MgSO_4$. Sterilise by filtration. Store at 4°C.
5. Lysozyme: 20 mg/mL and 2 mg/mL stock solutions. Sterilise by filtration. Store at −20°C. Avoid freezing and thawing of sample.
6. Incubator shaker.
7. Slides and coverslips.
8. Benchtop centrifuge.

2.3 Imaging

1. Confocal microscope: TCS-SP2 (Leica Microsystems, Mannheim, Germany) using a DMRE (Leica) upright microscope. For confocal acquisitions, fluorescence is collected through a spectral detection mode using mechanical filtering. The different channels are acquired sequentially to avoid fluorescence leakage.
2. Software for image processing like FiJi free software [29].

3 Methods

The following protocols have been developed for the analysis of lipoproteins of *P. aeruginosa*, such as TagQ. The preparation of spheroplasts is adapted from a study on sorting signals of lipoproteins in the *P. aeruginosa* strain PAO1 [28], in combination with our previous expertise in fluorescent confocal imaging [30]. These studies are often further validated by biochemical approaches, such as separation of bacterial membranes by centrifugation on sucrose gradients followed by immunodetection [10, 18, 20] (*see* also Chapter 6).

3.1 Preparation of Bacteria for Confocal Microscopy

1. Start an overnight culture of freshly transformed GFP-expressing *P. aeruginosa* producing the mCherry-tagged protein of interest in 3 mL LB broth supplemented with appropriate antibiotics for plasmid maintenance at 37 °C with shaking at 300 rpm/min.

Fig. 1 (continued) profile of the fluorescence intensity on selected objects obtained by FiJi free software. (**b**) *P. aeruginosa* PAO1 expressing cytoplasmic GFP and the TagQ-mCherry protein lacking cysteine residue within lipobox motif. The growth conditions, spheroplast preparation, imaging and image treatment were done as for the wild-type TagQ, represented in **a**. Note the periphery localisation of $TagQ_{\Delta C}$-mCherry in bacteria and absence of labelling in spheroplasts, indicating the loss of the protein during lysozyme treatment. (**c**) Schema showing putative localisation of bacterial lipoproteins assessed by analysing mCherry protein fusions by fluorescent microscopy before and after lysozyme treatment. Three options are represented: IM, OM and periplasmic localisation

2. Use the overnight culture to inoculate 3 mL LB broth supplemented with antibiotics to an OD_{600} of 0.15.
3. Culture up to mid-log phase of growth at 37 °C with shaking and induce for 2 h by adding arabinose to a final concentration to be determined for each protein (0.01–0.25%) (*see* **Note 3**).
4. Harvest 1 mL of culture by a centrifugation step at 6000 × *g* for 5 min at room temperature.
5. Wash cells with 100 μL fresh LB broth without antibiotics.
6. Depose 5 μL bacteria on the centre of a clean glass slide.
7. Mix by pipetting gently with 5 μL 1% warm agarose (do not exceed 60°C; *see* **Note 4**) and immediately cover the sample using a coverslip and apply gentle and uniform pressure to have a monolayer of bacteria. The sample is now ready to be analysed under the microscope.

3.2 Preparation of Spheroplasts

1. Follow **steps 1–3** in Subheading 3.1.
2. Harvest 1 mL culture by a centrifugation step at 6000 × *g* for 5 min at room temperature.
3. Resuspend the bacterial pellet in 30 μL ice-cold TM buffer.
4. Add lysozyme to a final concentration of 500 μg/mL (*see* **Note 3**).
5. Incubate 30 min at room temperature without shaking.
6. Centrifuge at 1000 × *g* for 5 min at room temperature.
7. Discard supernatant and gently resuspend spheroplasts in 50 μL ice-cold TSM buffer and let stand on ice. Be careful not to shake the preparation.
8. To analyse the spheroplasts under the microscope, gently depose 5 μL on a clean slide, add 2 μL warm 1% agarose, and cover with a coverslip. At this step, avoid pressing hard on the coverslip, which may lead to an explosion of spheroplasts. Observe images shortly after the mounting of the slide (*see* **Notes 4** and **5**).

3.3 Imaging and Image Analysis

1. Start the microscope and let it warm up.
2. Observe the specimens with an oil immersion objective (×63, NA: 1.4), and analyse them by confocal laser scanning microscopy (CLSM) using a TCS-SP2 operating system (Leica) working with an upright microscope.
3. Adjust the pinhole to Airy 1 (around 550 nm optical sections).
4. Excite GFP and mCherry fluorescence and collect sequentially (400 Hz line by line) using 488 nm for GFP and 543 nm for mCherry excitation. Collect fluorescence emissions between

500 and 535 nm for GFP and from 575 to 650 nm for mCherry (*see* **Note 6**). Images (512 × 512 pixels) are obtained with various electronic zooms (×8, ×32). The optical sections are acquired at a focal position corresponding to a median position in the analysed objects.

5. To determine the relative position for green and red fluorescence, display a plot profile of the fluorescence intensity along the line in the images. Those histograms can be obtained using FiJi free software (*see* Fig. 1).

4 Notes

1. To generate *P. aeruginosa* expressing cytoplasmic GFP, constitutive promotor driving the *gfp* expression pX2-*gfp* [31] is transferred to pmini-CTX1 using *Eco*RI-*Hind*III digestion [20]. pmini-CTX1 is an integrative plasmid that carries an *oriT* for conjugation-mediated plasmid transfer [32].
2. To generate fusions to mCherry, the gene encoding mCherry is amplified and cloned into *Xba*I and *Sac*I sites of pJN105 [33], giving pJN-mCherry. Then the DNA fragment containing the ribosome binding site and the gene of interest without the stop codon is amplified by polymerase chain reaction and cloned upstream of mCherry using the *Eco*RI and *Xba*I sites [20]. Site-directed mutations within the lipobox sequence are created using the QuikChangeII Site-Directed Mutagenesis kit. pJN105-derived plasmids are introduced in *P. aeruginosa* by transformation [34].
3. As stated in the previous section, the methods presented here were developed and optimised for *P. aeruginosa*. Note that different concentrations of inducer and lysozyme might be necessary for other bacteria, and this will have to be adjusted accordingly. Notably, if the concentration of inducer is too high, there might be aggregation of the fusion protein, and this can be spotted when imaging: the protein will form clusters mostly in the pole of the bacterial cell. In some cases, the growth rate of bacteria overexpressing the lipoprotein-mCherry fusion protein could be greatly reduced due to the toxic effect of the fusion protein. Concentration of lysozyme below optimal conditions (to be determined experimentally for each bacterium) might result only in partial destruction of peptidoglycan and consequently in a heterogeneous population of objects, leading to incomplete OM detachment (concentrations tested from 75 to 500 μg/mL; data not shown). Magnesium and sucrose also affect spheroplast stability and must be present during imaging.

4. The agarose for fixing bacteria needs to be melted using an incubator at 60–65 °C with shaking. Avoid boiling the solution as this will result in a more concentrated agarose solution. Do not reuse this solution more than twice, and make sure to keep it sterile. A 1% agarose solution cannot be too warm since it causes fragility of the OM, which will result in leaking of the green cytoplasmic fluorescence (red ghost bacteria).
5. In the same way, spheroplasts are fragile, and to avoid explosion with warm agarose, microscope slide preparation must be done immediately before observations. To avoid drying of the spheroplast sample: first add pre-warmed 1% agarose and then samples. A small volume of agarose (2 μL) is added with 5 μL of spheroplast sample to preserve their delicate shape. If the bacteria or spheroplasts are left on the slide for too long, they might dry out and get damaged, which will affect fluorescence detection and image quality. Cover with the coverslip immediately to avoid solidification of agarose. Apply gentle and uniform pressure to obtain a monolayer of bacteria. Use a tissue to avoid marks on the coverslip.
6. It is recommended, while adjusting the focus and looking for an interesting field with the oculars, to use GFP fluorescence or the transmitted light. Indeed, mCherry is especially photosensitive, and the intensity of the mercury arc lamp (HBO-50W) at this wavelength excitation, which is largely more phototoxic than the laser excitation used for the confocal acquisition, can induce an important photobleaching.

Acknowledgements

We thank Dr. K. M. Sall for initiating studies on TagQ and TssJ1 fusion proteins and Dr. S. Elsen for help in plasmid generation. MGC was supported by a PhD grant from the French Cystic Fibrosis Association Vaincre la Mucovisidose. The microscopy facility is supported by the Biosciences and Biotechnology Institute of Grenoble (BIG), CEA-Grenoble and the grant to Laboratoire of Excellence, LabEx GRAL (ANR-10-LABX-49-01).

References

1. Farris C, Sanowar S, Bader MW, Pfuetzner R, Miller SI (2010) Antimicrobial peptides activate the Rcs regulon through the outer membrane lipoprotein RcsF. J Bacteriol 192(19):4894–4903
2. Leverrier P, Declercq JP, Denoncin K, Vertommen D, Hiniker A, Cho SH et al (2011) Crystal structure of the outer membrane protein RcsF, a new substrate for the periplasmic protein-disulfide isomerase DsbC. J Biol Chem 286(19):16734–16742
3. Aliprantis AO, Yang RB, Mark MR, Suggett S, Devaux B, Radolf JD et al (1999) Cell activation and apoptosis by bacterial lipoproteins through toll-like receptor-2. Science 285(5428):736–739

4. Pugsley AP (1993) The complete general secretory pathway in gram-negative bacteria. Microbiol Rev 57(1):50–108
5. Zuckert WR (2014) Secretion of bacterial lipoproteins: through the cytoplasmic membrane, the periplasm and beyond. Biochim Biophys Acta 1843(8):1509–1516
6. Konovalova A, Silhavy TJ (2015) Outer membrane lipoprotein biogenesis: Lol is not the end. Philos Trans R Soc Lond Ser B Biol Sci 370(1679)
7. Babu MM, Priya ML, Selvan AT, Madera M, Gough J, Aravind L et al (2006) A database of bacterial lipoproteins (DOLOP) with functional assignments to predicted lipoproteins. J Bacteriol 188(8):2761–2773
8. Madan Babu M, Sankaran K (2002) DOLOP--database of bacterial lipoproteins. Bioinformatics 18(4):641–643
9. Remans K, Vercammen K, Bodilis J, Cornelis P (2010) Genome-wide analysis and literature-based survey of lipoproteins in Pseudomonas aeruginosa. Microbiology 156(Pt9):2597–2607
10. Casabona MG, Vandenbrouck Y, Attree I, Coute Y (2013) Proteomic characterization of Pseudomonas aeruginosa PAO1 inner membrane. Proteomics 13(16):2419–2423
11. Fernandez D, Dang TA, Spudich GM, Zhou XR, Berger BR, Christie PJ (1996) The Agrobacterium tumefaciens virB7 gene product, a proposed component of the T-complex transport apparatus, is a membrane-associated lipoprotein exposed at the periplasmic surface. J Bacteriol 178(11):3156–3167
12. Fernandez D, Spudich GM, Zhou XR, Christie PJ (1996) The Agrobacterium tumefaciens VirB7 lipoprotein is required for stabilization of VirB proteins during assembly of the T-complex transport apparatus. J Bacteriol 178(11):3168–3176
13. Christie PJ, Cascales E (2005) Structural and dynamic properties of bacterial type IV secretion systems (review). Mol Membr Biol 22(1–2):51–61
14. Collin S, Guilvout I, Nickerson NN, Pugsley AP (2011) Sorting of an integral outer membrane protein via the lipoprotein-specific Lol pathway and a dedicated lipoprotein pilotin. Mol Microbiol 80(3):655–665
15. Izore T, Perdu C, Job V, Attree I, Faudry E, Dessen A (2011) Structural characterization and membrane localization of ExsB from the type III secretion system (T3SS) of Pseudomonas aeruginosa. J Mol Biol 413(1):236–246
16. Perdu C, Huber P, Bouillot S, Blocker A, Elsen S, Attree I et al (2015) ExsB is required for correct assembly of the Pseudomonas aeruginosa type III secretion apparatus in the bacterial membrane and full virulence in vivo. Infect Immun 83(5):1789–1798
17. Guilvout I, Chami M, Engel A, Pugsley AP, Bayan N (2006) Bacterial outer membrane secretin PulD assembles and inserts into the inner membrane in the absence of its pilotin. EMBO J 25(22):5241–5249
18. Viarre V, Cascales E, Ball G, Michel GP, Filloux A, Voulhoux R (2009) HxcQ liposecretin is self-piloted to the outer membrane by its N-terminal lipid anchor. J Biol Chem 284(49):33815–33823
19. Aschtgen MS, Bernard CS, De Bentzmann S, Lloubes R, Cascales E (2008) SciN is an outer membrane lipoprotein required for type VI secretion in enteroaggregative *Escherichia coli*. J Bacteriol 190(22):7523–7531
20. Casabona MG, Silverman JM, Sall KM, Boyer F, Coute Y, Poirel J et al (2013) An ABC transporter and an outer membrane lipoprotein participate in posttranslational activation of type VI secretion in Pseudomonas Aeruginosa. Environ Microbiol 15(2):471–486
21. Durand E, Nguyen VS, Zoued A, Logger L, Pehau-Arnaudet G, Aschtgen MS et al (2015) Biogenesis and structure of a type VI secretion membrane core complex. Nature 523(7562):555–560
22. Felisberto-Rodrigues C, Durand E, Aschtgen MS, Blangy S, Ortiz-Lombardia M, Douzi B et al (2011) Towards a structural comprehension of bacterial type VI secretion systems: characterization of the TssJ-TssM complex of an *Escherichia coli* pathovar. PLoS Pathog 7(11):e1002386
23. Rao VA, Shepherd SM, English G, Coulthurst SJ, Hunter WN (2011) The structure of Serratia marcescens Lip, a membrane-bound component of the type VI secretion system. Acta Crystallogr Sect D 67(Pt 12):1065–1072
24. Basler M, Ho BT, Mekalanos JJ (2013) Tit-for-tat: type VI secretion system counterattack during bacterial cell-cell interactions. Cell 152(4):884–894
25. Alcock F, Baker MA, Greene NP, Palmer T, Wallace MI, Berks BC (2013) Live cell imaging shows reversible assembly of the TatA component of the twin-arginine protein transport system. Proc Natl Acad Sci U S A 110(38):E3650–E3659
26. Guillon L, El Mecherki M, Altenburger S, Graumann PL, Schalk IJ (2012) High cellular organization of pyoverdine biosynthesis in Pseudomonas aeruginosa: clustering of PvdA at the old cell pole. Environ Microbiol 14(8):1982–1994

27. Imperi F, Visca P (2013) Subcellular localization of the pyoverdine biogenesis machinery of Pseudomonas aeruginosa: a membrane-associated "siderosome". FEBS Lett 587(21):3387–3391
28. Lewenza S, Mhlanga MM, Pugsley AP (2008) Novel inner membrane retention signals in Pseudomonas aeruginosa lipoproteins. J Bacteriol 190(18):6119–6125
29. Schindelin J, Arganda-Carreras I, Frise E et al (2012) Fiji: an open-source platform for biological-image analysis. Nat Methods 9(7):676–682
30. De Bentzmann S, Giraud C, Bernard CS, Calderon V, Ewald F, Plesiat P et al (2012) Unique biofilm signature, drug susceptibility and decreased virulence in Drosophila through the Pseudomonas aeruginosa two-component system PprAB. PLoS Pathog 8(11):e1003052
31. Thibault J, Faudry E, Ebel C, Attree I, Elsen S (2009) Anti-activator ExsD forms a 1:1 complex with ExsA to inhibit transcription of type III secretion operons. J Biol Chem 284(23):15762–15770
32. Hoang TT, Kutchma AJ, Becher A, Schweizer HP (2000) Integration-proficient plasmids for Pseudomonas aeruginosa: site-specific integration and use for engineering of reporter and expression strains. Plasmid 43(1):59–72
33. Newman JR, Fuqua C (1999) Broad-host-range expression vectors that carry the L-arabinose-inducible *Escherichia coli* araBAD promoter and the araC regulator. Gene 227(2):197–203
34. Chuanchuen R, Narasaki CT, Schweizer HP (2002) Benchtop and microcentrifuge preparation of Pseudomonas aeruginosa competent cells. Biotechniques 33(4):760. 2–3

Chapter 5

Identification of Lipoproteins Using Globomycin and Radioactive Palmitate

Nienke Buddelmeijer

Abstract

Bacterial lipoproteins are characterized by fatty acids that are covalently attached to their amino terminus via posttranslational modification in the cytoplasmic membrane. Three enzymatic steps are involved in the synthesis of mature triacylated lipoprotein: prolipoprotein converts into diacylglyceryl-prolipoprotein that in turn converts into apolipoprotein, which is finally converted into mature triacylated lipoprotein. Here we describe the detection of one of these intermediate forms of lipoprotein, diacylglyceryl-prolipoprotein, using ^{3}H-palmitate labeling and inhibition by globomycin and detection by fluorography.

Key words Tris-tricine gel electrophoresis, Globomycin, ^{3}H-palmitate labeling, Fluorography

1 Introduction

Lipoproteins are abundant proteins of the bacterial cell envelope [1]. They are modified at their amino terminus by fatty acids that are derived from membrane phospholipids [2]. The mature part of lipoproteins is highly variable, resulting in a wide variety of biological functions. In proteobacteria and actinobacteria, the modification pathway is composed of two acyltransferases and one peptidase [3]. Lipoproteins are synthesized in the cytoplasm as prolipoproteins that are inserted into the membrane via The general secretion (Sec) or twin-arginine translocation (Tat) machineries. The enzyme phosphatidylglycerol:prolipoprotein diacylglyceryl transferase (Lgt) adds a diacylglyceryl moiety from phosphatidylglycerol onto the invariant cysteine in the so-called lipobox located in the amino terminal region of the protein, resulting in the formation of diacylglyceryl-prolipoprotein. In the second step, lipoprotein signal peptidase (Lsp) cleaves the signal peptide from diacylglyceryl-prolipoprotein, resulting in a free α-amino group on diacylglyceryl-cysteine. The third and last step in lipoprotein modification is

Laure Journet and Eric Cascales (eds.), *Bacterial Protein Secretion Systems: Methods and Protocols*, Methods in Molecular Biology, vol. 1615, DOI 10.1007/978-1-4939-7033-9_5, © Springer Science+Business Media LLC 2017

N-acylation catalyzed by apolipoprotein *N*-acyltransferase (Lnt). The acyl donor is phosphatidylethanolamine, of which the *sn*-1 acyl group is transferred onto the α-amino group of apolipoprotein, resulting in mature triacylated protein. The intermediate forms of small lipoproteins or peptides can be analyzed using gel electrophoresis techniques in combination with specific inhibitors or mutant strains. Globomycin and its derivatives specifically inhibit Lsp leading to an accumulation of diacylglyceryl-prolipoprotein in the cytoplasmic membrane [4]. This intermediate carries a diacylglyceryl moiety that in case of *Escherichia coli* is composed of C16:0 and C18:cis-11 fatty acids and still has the signal peptide attached [5, 6]. The migration of this form of lipoprotein on a sodium dodecyl sulfate (SDS)-polyacrylamide gel is slower than prolipoprotein, apolipoprotein, and mature lipoprotein.

2 Materials

2.1 ^{3}H-Palmitate Labeling of E. coli Cultures

1. Luria Broth Miller (LB) medium: 5 g yeast extract, 10 g peptone, 10 g NaCl per liter in ultrapure water. Sterilize by autoclaving for 20 min at 120 °C.
2. *E. coli* strains (*see* **Note 1**).
3. 5 mCi/mL of 9,10-^{3}H(N)-palmitate in ethanol. Specific activity: 30–60 Ci(1,11–2,22 TBq)/mmol.
4. Incubator.

2.2 Inhibition of Lsp by Globomycin

1. Luria Broth Miller (LB) medium: 5 g yeast extract, 10 g peptone, 10 g NaCl per liter in ultrapure water. Sterilize by autoclaving for 20 min at 120 °C.
2. *E. coli* strains (*see* **Note 1**).
3. Globomycin: use final concentration of 160 μg/mL for *E. coli* K12 strains and 5 μg/mL for B strains (*see* **Note 2**).
4. Incubator.

2.3 Immune Precipitation of Lipoproteins

1. Antibodies specific to lipoprotein of interest (*see* **Note 3**).
2. 100% tricloroacetic acid (TCA).
3. Acetone.
4. Solubilization buffer: 25 mM Tris–HCl, pH 8.0, 1% SDS, 1 mM ethylenediaminetetraacetic acid (EDTA).
5. Immune precipitation buffer: 50 mM Tris–HCl, pH 8.0, 150 mM NaCl, 1 mM EDTA, 2% Triton X-100.
6. Protein G-Sepharose.
7. Wash buffer I: 50 mM Tris–HCl, pH 8.0, 1 M NaCl, 1% Triton-X-100.

8. Wash buffer II: 10 mM Tris–HCl pH 8.0.
9. Tabletop microcentrifuge.

2.4 Tris-Tricine Gel Electrophoresis

1. Mini-gel caster system and SDS-polyacrylamide gel electrophoresis (PAGE) migration apparatus.
2. Cathode buffer (top, 10×): 1 M Tris, 1 M Tricine, 1% SDS pH 8.25. Do not adjust pH.
3. Anode buffer (bottom, 10×): 1 M Tris–HCl, pH 8.9.
4. Gel buffer (3×): 3 M Tris, 1 M HCl, 0.3% SDS.
5. Acrylamide (49.5%T, 3%C or 6%C). Store at 4 °C.
6. Ammonium persulfate (APS): 10% solution in water. Store at −20 °C.
7. *N, N, N, N'*-tetramethyl-ethylenediamine (TEMED). Store at 4 °C.
8. SDS-PAGE running buffer: 25 mM Tris, 250 mM glycine, 0.1% SDS. Do not adjust pH.
9. SDS loading buffer (3×): 150 mM Tris–HCl, pH 6.8, 6% SDS, 0.3% bromophenol blue, 30% glycerol.
10. Water bath at 100 °C.
11. i-Butanol.
12. Vacuum gel-drying system.
13. Amplify solution.
14. X-ray film.

3 Methods

3.1 ^{3}H-Palmitate Labeling of E. coli Cultures

1. Grow cultures of *E. coli* in LB Miller medium at 37 °C.
2. Add 100 μCi/mL ^{3}H-palmitate to the cell culture at early exponential phase (OD_{600} of 0.2), and let cultures grow for 2 h (*see* **Note 4**).

3.2 Inhibition of Lsp by Globomycin

Add globomycin to the cell culture after 1 h of growth in the presence of ^{3}H-palmitate and let cultures grow for an additional 1 h (*see* **Note 5**).

3.3 Immune Precipitation of Lipoproteins

This protocol is based on ref. [7].

1. Add a final concentration of 10% TCA to 1 mL of cell culture to precipitate all proteins.
2. Centrifuge-precipitated proteins for 1 min in a tabletop centrifuge.
3. Wash the pellets twice with 1 mL ice-cold (−20 °C) acetone.

4. Briefly dry protein pellets.
5. Resuspend pellets in 50 μL of solubilization buffer and boil samples for 2 min. Let cool down.
6. Add 450 μL immune precipitation buffer (*see* **Note 6**).
7. Centrifuge samples for 10 min in a tabletop centrifuge.
8. Take 200 μL from the top and add 300 μL immune precipitation buffer.
9. Add antibodies and incubate on ice overnight.
10. Add 100 μL Protein G-Sepharose slush and incubate on ice for 20 min.
11. Centrifuge 1 min at 4 °C in tabletop centrifuge and wash pellet twice with Wash Buffer I (*see* **Note 7**).
12. Wash once with Wash Buffer II.
13. Resuspend slush in 100 μL SDS loading buffer and boil for 2 min to release proteins from Protein G-Sepharose.
14. Centrifuge samples for 5 min in tabletop centrifuge and use supernatant for gel electrophoresis (*see* **Note 8**).

3.4 Tris-Tricine Gel Electrophoresis

Tris-Tricine gels are especially useful for separating small proteins and peptides (less than 30 kDa) [8].

1. Prepare a mini-gel format separating gel (16%) by mixing 5 mL gel buffer, 6 mL acrylamide solution, and 4 mL water (total volume 15 mL). Add 5 μL TEMED and 50 μL APS and cast gel in an 8.6 × 6.8 × 0.75 cm gel holder. Allow space for stacking gel and overlay with water or i-butanol.
2. Prepare stacking gel (4%) by mixing 3.3 mL gel buffer, 1 mL acrylamide solution, and 5.7 mL water (total volume 10 mL). Add 7.5 μL TEMED and 75 μL APS and cast gel and insert comb immediately.
3. Load samples on gel along with protein standard. Electrophorese at 30 V until the samples have entered the stacking gel and continue at 200 V till the dye front reaches the bottom of the gel (*see* **Note 9**).
4. After migration of samples, pry gel plates open and briefly wash gel in water.
5. Transfer to amplify solution and let soak for 10 min under agitation.
6. Dry gel in gel dryer under vacuum for 60 min at 80 °C (*see* **Note 10**).
7. Transfer the gel into a cassette and put an X-ray film over it. Expose the film for 10 days at −80 °C. Let the cassette warm up to room temperature before developing the X-ray film (Fig. 1).

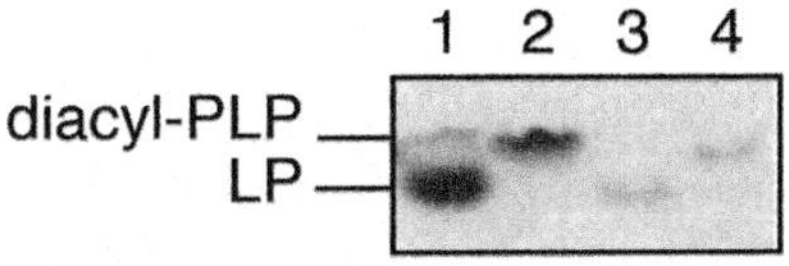

Fig. 1 Accumulation of diacyl-prolipoprotein (diacyl-PLP) in envelope of *E. coli* B cells labeled with ^{3}H palmitate upon treatment with globomycin (5 μg/mL). Without globomycin (lanes 1 and 3), with globomycin (lanes 2 and 4), anti-Lpp immunoprecipitates in lane 3 (sample from lane 1) and lane 4 (sample from lane 2). Figure adapted from Fig. 5 of [10] corresponding to lanes 7 through 10, with permission

4 Notes

1. Various wild-type strains of *E. coli* can be used.
2. Globomycin is commercially produced and derivatives have been described [9]. The concentration of these derivatives needs to be determined empirically.
3. Braun's lipoprotein (Lpp) of *E. coli* (78 amino acids) has been the reference lipoprotein to study modification and cellular localization. Antibodies against other bacterial lipoproteins have been used. It is recommended to use proteins of small size or peptide fragments (10 kDa) to facilitate identification of intermediate forms of lipoprotein modification.
4. For *E. coli* K12 strains such as MC4100 or MG1655, a final OD_{600} of 0.6–0.8 is obtained.
5. Extensive exposure to globomycin leads to cell lysis owing to the inhibition of essential enzyme Lsp and, as a result, the accumulation of Lpp in the cytoplasmic membrane while still being cross-linked to the peptidoglycan.
6. Triton X-100 solubilizes lipoproteins from membranes.
7. Pellet is antigen–antibody–sepharose slush.
8. The amount of sample to charge needs to be determined empirically and depends on the antigen and antibody used.
9. Electrophoresis of 16% Tris-Tricine gels takes longer than regular SDS-PAGE gels. Calculate 2–3 h for a mini-gel format.
10. Remove vacuum from gel before switching off pump to avoid cracking of gel.

References

1. Kovacs-Simon A, Titball RW, Michell SL (2010) Lipoproteins of bacterial pathogens. Infect Immun 79:548–561
2. Lai J-S, Philbrick WM, Wu HC (1980) Acyl moieties in phospholipids are the precursors for the fatty acids in murein lipoprotein in *Escherichia coli*. J Biol Chem 255: 5384–5387
3. Buddelmeijer N (2015) The molecular mechanism of bacterial lipoprotein modification--how, when and why? FEMS Microbiol Rev 39:246–261
4. Inukai M, Takeuchi K, Shimizu K, Arai M (1978) Mechanism of action of globomycin. J Antibiot 31:1203–1205
5. Cronan JE Jr, Rock CO (1996) Biosynthesis of membrane lipids. In: Neidhardt FC (ed) *Escherichia coli* and *Salmonella*: molecular and cellular biology. ASM, Washington, DC
6. Hantke K, Braun V (1973) Covalent binding of lipid to protein. Diglyceride and amide-linked fatty acid at the N-terminal end of the murein-lipoprotein of the *Escherichia coli* outer membrane. Eur J Biochem 34:284–296
7. Kumamoto CA, Gannon PM (1988) Effects of *Escherichia coli secB* mutations on pre-maltose binding protein conformation and export kinetics. J Biol Chem 263:11554–11558
8. Schagger H (2006) Tricine-SDS-PAGE. Nat Protoc 1:16–22
9. Kiho T et al (2004) Structure-activity relationships of globomycin analogues as antibiotics. Bioorg Med Chem 12:337–361
10. Hussain M, Ichihara S, Mizushima S (1980) Accumulation of glyceride-containing precursor of the outer membrane lipoprotein in the cytoplasmic membrane of *Escherichia coli* treated with globomycin. J Biol Chem 255:3707–3712

Chapter 6

Defining Membrane Protein Localization by Isopycnic Density Gradients

Rhys A. Dunstan, Iain D. Hay, and Trevor Lithgow

Abstract

In many bacteria, membrane proteins account for around one-third of the proteome and can represent much more than half of the mass of a membrane. Classic techniques in cell biology can be applied to characterise bacterial membranes and their membrane protein constituents. Here we describe a protocol for the purification of outer and inner membranes from *Escherichia coli*. The procedure can be applied with minor modifications to other bacterial species, including those carrying capsular polysaccharide attached to the outer membrane.

Key words Sucrose density gradient, Membrane biogenesis, Beta-barrel proteins, Cytoplasmic membrane, Lipoproteins

1 Introduction

Gram-negative bacteria are characterised by two membranes. The cytoplasmic (inner) membrane is a phospholipid bilayer into which alpha-helical transmembrane proteins are integrated and onto which peripheral membrane proteins are attached by lipid-mediated or protein–protein interactions [1, 2].

The lipid phase of the outer membrane is a bilayer with an outer leaflet predominantly composed of lipopolysaccharides and an inner leaflet of phospholipids [3, 4]. This, together with a massively high ratio of protein:lipid [5, 6], gives the outer membrane a far greater buoyant density than the inner membrane. In 1975 Yamato et al. reported a reproducible procedure to disrupt *Escherichia coli* cells in a French press, recover a relatively pure membrane fraction, and segregate it into inner and outer membrane fractions by sucrose isopycnic gradient ultracentrifugation [7]. This methodology was validated using marker enzyme assays: measuring partial reactions of oxidative phosphorylation for the inner membrane and finding that phospholipase A activity segregated into a higher-density fraction. These fractions were

Laure Journet and Eric Cascales (eds.), *Bacterial Protein Secretion Systems: Methods and Protocols*, Methods in Molecular Biology, vol. 1615, DOI 10.1007/978-1-4939-7033-9_6, © Springer Science+Business Media LLC 2017

therefore deemed to represent the inner and outer membranes, and electron microscopy was used to assess their relative morphology and the homogeneity of the membrane preparations [7].

We now have a mature knowledge of the protein and lipid composition of the inner and outer membranes for a range of bacterial species. By fractionating sucrose gradients and analysing the fractions by SDS-PAGE and immunoblotting with antibodies, it is possible to follow the co-purification of specific proteins of interest with either membrane. Proteomics has been applied to assess and validate the purity of membrane fractions, so that the method has been applied to membrane purification from bacterial species including *Campylobacter* [8], *Caulobacter* [9], *Citrobacter* [10], *Cornebacterium* [11], *Neisseria* [12], *Proteus* [13], *Pseudomonas* [14] and *Salmonella* [15]. The method has been applied to demonstrate effects on the outer membrane protein composition when the beta-barrel assembly machinery is diminished [16], when LPS biosynthesis is diminished [17], or to dissect the assembly pathways for beta-barrel and alpha-helical membrane proteins into the outer membrane [18].

Here we detail an optimised protocol for the purification of outer and inner membrane fractions from *E. coli*.

2 Materials

2.1 Membrane Purification

1. Lysogeny broth (LB) medium: 10 g tryptone, 5 g yeast extract and 5 g NaCl, make up to 1 L with distilled water. Autoclave and store at room temperature.
2. 1 M Tris stock solution: Add 121.1 g Tris to 800 mL distilled water. Adjust pH to 7.5 with HCl, then dilute to 1 L with distilled water. Autoclave and store at room temperature.
3. 500 mM ethylenediaminetetraacetic acid (EDTA) stock solution: Add 186.1 g EDTA (disodium ethylenediamine tetraacetate, $2H_2O$) to 800 mL distilled water. Adjust pH to 8.0 with NaOH, then dilute to 1 L with distilled water. Autoclave and store at room temperature (*see* **Note 1**).
4. Ultra-pure sucrose.
5. Tris buffer: 10 mM Tris–HCl, pH 7.5 (dilute 10 mL 1 M Tris stock solution to 1 L with distilled water).
6. 100 mg/mL lysozyme: Add 1 g lysozyme to 10 mL distilled water, prepare 500 μL aliquots, and store at −20 °C.
7. 100 mM phenylmethylsulfonyl fluoride (PMSF): Dissolve 174 mg PMSF in 10 mL isopropanol. Prepare 500 μL aliquots and store at −20 °C (*see* **Note 2**).
8. EDTA buffer: 1.65 mM EDTA, pH 7.5 (dilute 660 μL EDTA stock solution to 200 mL of distilled water). Store at room temperature.

9. Tris-Sucrose (TS) buffer: 10 mM Tris–HCl, pH 7.5, 0.75 M sucrose (dissolve 51.3 g sucrose in 200 mL Tris buffer). Store at 4 °C.
10. TES buffer: 3.3 mM Tris–HCl, pH 7.5, 1.1 mM EDTA, 0.25 M sucrose (add 1 volume TS buffer to 2 volumes 1.65 mM EDTA buffer). Approximately 40 mL for each sample will be required. Store at 4 °C.
11. 5 mM EDTA, pH 7.5 (dilute 5 mL EDTA stock solution to 500 mL with distilled water).
12. 25% sucrose solution: 25% (w/w) sucrose, 5 mM EDTA, pH 7.5 (add 7.5 g sucrose to 22.5 mL of EDTA). Pass through a 0.45 μM filter and store at 4 °C.
13. Use Emulsiflex (AVESTIN, Ottawa, ON, Canada) or other cell disruptor or similar.
14. Centrifuge with Sorvall SS34 (Thermo Fisher Scientific, Inc., Waltham, MA) tubes and rotor (or something similar with ability to spin up to approximately 50 mL at up to 15,000 × *g*).
15. Ultracentrifuge with Beckman 70.1 Ti tubes (Beckman Coulter Inc., Brea, CA) (open-top thick-wall polycarbonate tubes) and rotor (or something similar with ability to spin approximately 100,000 × *g*).
16. Wheaton teflon tissue grinder/dounce (Millville, NJ).

2.2 Sucrose Density Fractionation

1. 5 mM EDTA, pH 7.5 (see earlier steps). Store at room temperature.
2. Ultra-pure sucrose.
3. Sucrose EDTA fractions: 35–60% (w/w) sucrose, 5 mM EDTA, pH 7.5 solutions. For example, to make 50% (w/w), add 15 g sucrose to 15 mL 5 mM EDTA. Pass through a 0.45 μM filter and store at 4 °C.
4. Displacing sucrose solution: 70% (w/w) sucrose, 5 mM EDTA, pH 7.5 (add 140 g sucrose to 60 mL EDTA) (*see* **Note 3**).
5. Beckman Coulter SW 40 Ti tubes (disposable plastic tubes) and rotor.
6. Density Gradient Fractionation System (Teledyne Isco, Lincoln, NE, USA)

2.3 Isolation of Membranes after Density Fractionation

1. TES buffer (see earlier steps).
2. 25% (w/w) sucrose 5 mM EDTA, pH 7.5 solution (see earlier).
3. Beckman 70.1 Ti tubes (open-top thick-wall polycarbonate tubes) and rotor.

3 Methods

3.1 Membrane Purification

1. Grow 5 mL O/N starter culture in LB at 37 °C from a single colony.
2. Dilute culture 1:100 in 400 mL LB with antibiotics as required and grow until OD_{600} = ~1.0 (*see* **Note 4**).
3. Pellet cells by centrifugation for 5 min at 5000 × *g* 4 °C (*see* **Note 5**).
4. Resuspend cells in 10 mM Tris–HCl, pH 7.5 (~200 mL).
5. Repeat centrifugation and resuspend pellet in 10 mL TS.
6. Add 50 μg/mL lysozyme (5 μL stock) and 2 mM PMSF (200 μL stock) to break down peptidoglycan layer and inhibit host serine proteases respectively.
7. Slowly add 2 volumes (20 mL) 1.65 mM EDTA, pH 7.5, to destabilise outer membrane for lysis.
8. Incubate 10 min on ice.
9. Lyse cells using an AVESTIN Emulsiflex; 2–3 passes at ~15,000 psi will be required to fully lyse cells.
10. Centrifuge cell lysate at 15,000 × *g*, 20 min at 4 °C, to remove the cell debris.
11. Collect the supernatant and pellet total membranes by ultracentrifugation at 38,000 revolutions per minute (rpm) (132,000 × *g*), 45 min, 4 °C (70.1 Ti rotor—use ~8 mL in each tube).
12. Resuspend membrane pellets in 1 mL TES using a dounce.
13. Pool membranes and make up to ~8 mL with TES and centrifuge 38,000 rpm, 45 min, 4 °C.
14. Resuspend membrane pellet in a small volume (~400 μL) of 25% sucrose in 5 mM EDTA, pH 7.5, using a dounce, and either snap freeze in liquid nitrogen for storage at −80 °C or continue to sucrose density fractionation.

3.2 Sucrose Density Fractionation

1. Immediately before use carefully prepare a six-step sucrose gradient from 60–35% w/w (1.9 mL of 60%, 55%, 50%, 45%, 40%, 35%) in SW40 tubes. A sharp interphase between layers should be clearly visible (*see* **Note 6**).
2. Layer 400 μL of total membranes on top of a 60–35% w/w gradient.
3. Spin in an ultracentrifuge using SW40 rotor for 17 h. at 34,000 rpm (205,000 × *g*), 4 °C (*see* **Note 7**).
4. Isolate 1 mL fractions (use ISCO fractionator with 70% sucrose, 5 mM EDTA, pH 7.5 as displacing fluid, or carefully pipette 1 mL fractions off top of gradient). Store each fraction at −80 °C until use (*see* **Note 8**).

5. To visualise on SDS-PAGE with coomassie brilliant blue staining, load 30 μL of each fraction, or ~15 μL, for analysis by western immunoblotting with antibodies against known inner and outer membrane protein (e.g., $F_1\beta$ and BamA respectively).

3.3 Isolation of Membranes after Density Fractionation

1. To isolate membranes from specific fractions, add TES buffer (to a final volume of ~8 mL) to each fraction of interest or to pooled fractions and pellet by ultracentrifugation, 1.5 h, 4 °C, 38,000 rpm (70 Ti .1 rotor).
2. Resuspend each fraction in ~100 μL 25% (w/w) sucrose, 5 mM EDTA, pH 7.5, with a dounce, and store membranes at −80 °C (*see* **Note 9**).

4 Notes

1. EDTA will not be soluble until pH reaches 8.0. Use vigorous stirring and heat (if needed).
2. PMSF may crystallise in solution, vortex thoroughly before use.
3. Vigorous stirring and heat may be required to fully dissolve sucrose.
4. Protocol can be adjusted for optimal conditions for protein expression or conditional shutdowns, for example, if needed.
5. Everything (buffers, tubes, etc.) should be at 4 °C from now on.
6. Carefully pipette the sucrose gradient solutions to the edge of the tube as close to the meniscus layer as possible. This will prevent disruption between sucrose fractions' interface.
7. Owing to the different nature of the outer membrane between bacterial strains (e.g., capsulated, non-capsulated), the duration and speed of centrifugation may need to be adjusted. For example, when performing sucrose gradients on total membranes from the capsulated *Klebsiella pneumonia* we routinely centrifuge for 40 h at 33,300 rpm (196,000 × *g*).
8. When collecting by drops using the fractionator, approximately 20 drops is equivalent to 1 mL. Alternatively, fractions can be isolated by carefully piercing the bottom of the tube and allowing the fraction to drip out the bottom; or, if visible, individual layers can be directly isolated from the tube by piercing the side of the tube with a syringe and sucking the corresponding layers out. Depending on the amount of protein added to the gradient, the denser outer membrane should predominate in the bottom section of the gradient and may be present as a white band; the lighter inner membrane will be present in the top section of the gradient and will generally be more diffuse and may have a reddish appearance.

9. The volume of TES used will vary with the amount of membranes isolated or desired membrane concentration for later experiments.

References

1. Dalbey RE, Wang P, Kuhn A (2011) Assembly of bacterial inner membrane proteins. Annu Rev Biochem 80:161–187
2. Okuda S, Tokuda H (2011) Lipoprotein sorting in bacteria. Annu Rev Microbiol 65:239–259
3. Kamio Y, Nikaido H (1976) Outer membrane of salmonella typhimurium: accessibility of phospholipid head groups to phospholipase c and cyanogen bromide activated dextran in the external medium. Biochemistry 15(12):2561–2570
4. Smit J, Kamio Y, Nikaido H (1975) Outer membrane of salmonella typhimurium: chemical analysis and freeze-fracture studies with lipopolysaccharide mutants. J Bacteriol 124(2):942–958
5. Osborn MJ, Gander JE, Parisi E, Carson J (1972) Mechanism of assembly of the outer membrane of salmonella typhimurium. Isolation and characterization of cytoplasmic and outer membrane. J Biol Chem 247(12):3962–3972
6. Schnaitman CA (1970) Protein composition of the cell wall and cytoplasmic membrane of *Escherichia coli*. J Bacteriol 104(2):890–901
7. Yamato I, Anraku Y, Hirosawa K (1975) Cytoplasmic membrane vesicles of *Escherichia coli*. A simple method for preparing the cytoplasmic and outer membranes. J Biochem 77(4):705–718
8. Hobb RI, Fields JA, Burns CM, Thompson SA (2009) Evaluation of procedures for outer membrane isolation from *Campylobacter jejuni*. Microbiology 155(Pt 3):979–988
9. Anwari K, Webb CT, Poggio S, Perry AJ, Belousoff M, Celik N, Ramm G, Lovering A, Sockett RE, Smit J, Jacobs-Wagner C, Lithgow T (2012) The evolution of new lipoprotein subunits of the bacterial outer membrane BAM complex. Mol Microbiol 84(5):832–844
10. Selkrig J, Mosbahi K, Webb CT, Belousoff MJ, Perry AJ, Wells TJ, Morris F, Leyton DL, Totsika M, Phan MD, Celik N, Kelly M, Oates C, Hartland EL, Robins-Browne RM, Ramarathinam SH, Purcell AW, Schembri MA, Strugnell RA, Henderson IR, Walker D, Lithgow T (2012) Discovery of an archetypal protein transport system in bacterial outer membranes. Nat Struct Mol Biol 19(5):506–510
11. Marchand CH, Salmeron C, Bou Raad R, Meniche X, Chami M, Masi M, Blanot D, Daffe M, Tropis M, Huc E, Le Marechal P, Decottignies P, Bayan N (2012) Biochemical disclosure of the mycolate outer membrane of *Corynebacterium glutamicum*. J Bacteriol 194(3):587–597
12. Masson L, Holbein BE (1983) Physiology of sialic acid capsular polysaccharide synthesis in serogroup B *Neisseria meningitidis*. J Bacteriol 154(2):728–736
13. Siegmund-Schultze N, Kroll HP, Martin HH, Nixdorff K (1991) Composition of the outer membrane of *Proteus mirabilis* in relation to serum sensitivity in progressive stages of cell form defectiveness. J Gen Microbiol 137(12):2753–2759
14. Jagannadham MV, Abou-Eladab EF, Kulkarni HM (2011) Identification of outer membrane proteins from an Antarctic bacterium *Pseudomonas syringae* Lz4W. Mol Cell Proteomics 10(6):M110 004549
15. Bishop RE, Gibbons HS, Guina T, Trent MS, Miller SI, Raetz CR (2000) Transfer of palmitate from phospholipids to lipid a in outer membranes of gram-negative bacteria. EMBO J 19(19):5071–5080
16. Charlson ES, Werner JN, Misra R (2006) Differential effects of yfgL mutation on *Escherichia coli* outer membrane proteins and lipopolysaccharide. J Bacteriol 188(20):7186–7194
17. Steeghs L, de Cock H, Evers E, Zomer B, Tommassen J, van der Ley P (2001) Outer membrane composition of a lipopolysaccharide-deficient *Neisseria meningitidis* mutant. EMBO J 20(24):6937–6945
18. Dunstan RA, Hay ID, Wilksch JJ, Schittenhelm RB, Purcell AW, Clark J, Costin A, Ramm G, Strugnell RA, Lithgow T (2015) Assembly of the secretion pores GspD, Wza and CsgG into bacterial outer membranes does not require the Omp85 proteins BamA or TamA. Mol Microbiol 97(4):616–629

Chapter 7

Cell Surface Exposure

Anna Konovalova

Abstract

Surface-exposed proteins of Gram-negative bacteria are represented by integral outer membrane beta-barrel proteins and lipoproteins. No computational methods exist for predicting surface-exposed lipoproteins, and therefore lipoprotein topology must be experimentally tested. This chapter describes three distinct but complementary methods for the detection of surface-exposed proteins: cell surface protein labeling, accessibility to extracellular protease and antibodies.

Key words Biotinylation, PEGylation, Surface proteolysis, Whole-cell dot blot, Protein topology

1 Introduction

Cells of Gram-negative bacteria are surrounded by an additional membrane, known as the outer membrane (OM) [1]. The outer surface of the OM is composed of lipopolysaccharide (LPS) and decorated with proteins. Until recently, it was thought that only integral beta-barrel proteins (referred to as outer membrane proteins or OMPs) were surface exposed as they often display long extracellular loops. In contrast, OM lipoproteins, peripheral proteins tethered to the OM by N-terminal lipids, were thought to be found exclusively in the inner leaflet of the OM facing the aqueous periplasm [2]. However, in recent years a number of surface-exposed lipoproteins that face the cell exterior instead of the periplasm have been identified (*see* for reviews [3–5]).

In the case of OMPs, topology and extracellular loops can be easily predicted computationally based on the presence of hydrophobic ß strands [6–8]. In contrast, lipoproteins are a very diverse group of proteins that do not share sequence or structure similarities. Many surface-exposed lipoproteins have no obvious transmembrane domains and are assembled on the cell surface by novel mechanisms [3–5]. Therefore, lipoprotein surface exposure or

Laure Journet and Eric Cascales (eds.), *Bacterial Protein Secretion Systems: Methods and Protocols*, Methods in Molecular Biology, vol. 1615, DOI 10.1007/978-1-4939-7033-9_7, © Springer Science+Business Media LLC 2017

topology in the OM cannot be predicted and must be experimentally tested.

Several methods have been developed for the detection of surface-exposed proteins. All of them are based either on the availability of a functional group for protein modification/labeling or protein accessibility to extracellular proteases or antibodies.

Protein modification utilizes reagents that are able to react efficiently with certain functional groups and form a covalent bond [9]. Reagents with N-hydroxysuccinimide (NHS)-esters are often the first choice. They target primary amines (available in side chains of lysine residues or the N-terminus of the protein if not modified). Lysines are relatively abundant within protein sequences and often surface exposed and, thus, accessible in protein structures. The other commonly used reagents are based on maleimides, which react with the cysteine sulfhydryl groups. Cysteines are not commonly found in protein sequences, a fact that can be taken advantage of using genetically introduced cysteine codons to study detailed protein topology (*see* Chapter 8 for more information).

Protein labeling reagents vary greatly in their properties, such as size, hydrophobicity, and detection methods (*see* Table 1 for examples). Reagents for selective surface labeling should be cell impermeable. When working with Gram-negative bacteria, the unique permeability properties of the OM must be taken into consideration [10]. The OM is an asymmetric membrane with phospholipids in the inner leaflet and LPS in the outer leaflet. LPS is negatively charged and bridged with divalent cations, such as Mg^{2+} and Ca^{2+} (*see* **Note 1**). These lateral interactions seal the OM and make it impermeable to hydrophobic compounds. On the other hand, the OM also contains protein channels that allow diffusion of nutrients and small molecules. Therefore, small hydrophilic reagents can enter the periplasm and label proteins on both sides of the membrane. Because of these properties of the OM, the criteria for the selection of reagents for cell surface labeling are often the opposite of those given in the product instruction manuals that were most often developed for the labeling of eukaryotic cells.

Table 1
Protein labeling reagents for selective surface labeling

Functional group to be labeled	Name	Polar	Molecular weight, Da
Primary amine	NHS-LC-LC-Biotin	–	567.70
	NHS-PEG(n)-Biotin	+	Available in 1–10 kDa range
Sulfhydryl	Mal-PEG(n)-Biotin	+	Available in 1–10 kDa range

In our lab we have used hydrophobic NHS-LC-LC-biotin to selectively label surface-exposed lipoproteins in *Escherichia coli* [11, 12]. In addition, hydrophilic reagents that are significantly larger than the diffusion limit (600 Da for *E. coli* [13]) will also preferentially label cell surfaces. The OM in other bacteria may have different properties, so reagents should be validated every time for their cell surface selectivity. For example, not all Gram-negative bacteria will have such a highly asymmetric OM and hence are not as resistant to hydrophobic compounds as *E. coli*. The sensitivity to detergents serves as a good indication of the presence of phospholipids in the outer leaflet. If this is the case, using hydrophobic reagents is not recommended. This rule also applies to *E. coli* mutants with defects in OM biogenesis and maintenance. In addition, when the permeability properties of the OM are unknown, it is better to use high-molecular-weight reagents to avoid the generation of false positive results.

Protein labeling reagents allow the detection of modified proteins by the addition of a biotin group, a long-chain polyethylene glycol (PEG) linker, or a combination of these. Protein biotinylation allows for immunoblot detection using anti-biotin antibodies or streptavidin conjugates. However, because all the cell surface proteins are labeled, the specific protein of interest must often be purified (or at least enriched) prior to detection. Alternatively, biotinylated proteins can be affinity purified using a streptavidin resin and then probed with protein-specific antibodies. Working with high-molecular-weight PEG linkers provides an advantage for the direct detection of the labeled protein in a cell lysate based on the size shift during immunoblot analysis using protein-specific antibodies.

Protein accessibility to extracellular proteases, also known as surface proteolysis, is another common method of studying protein surface exposure [14–16]. It is based on the addition of proteases with broad specificity, such as trypsin or proteinase K, to intact cells. Because proteases cannot enter cells, only surface-accessible proteins or domains of proteins will be cleaved. Antibodies that recognize the protein of interest are then used to detect cleavage using immunoblotting.

Proteins can be inherently protease resistant because of their tight folding or lack of protease cleavage sites or because they are protected by interactions with other proteins. Therefore, negative results of protease shaving experiments are hard to interpret. One way to address this problem is to test whether the protein of interest is protease sensitive in the cell lysate done under nondenaturing conditions, for example, using a mild detergent lysis solution (*see* **Note 2**).

It is crucial to use proper controls to ensure that proteases do not target periplasmic proteins for the following reasons. First, proteins are important components of the OM and contribute significantly to its stability. Complete proteolysis of surface domains can destabilize the OM, an event made apparent by the degradation of periplasmic proteins. For this reason, titration experiments are advisable to find the optimal protease concentration. Second, both trypsin and proteinase K retain their activity in sodium dodecyl sulfate (SDS) [17, 18] and therefore they can digest proteins in cell lysates during preparation of the samples for immunoblotting. This also leads to the generation of false positives. To avoid this, protease inhibitors should be added and excess protease removed prior to cell lysis in SDS loading buffer.

A number of assays that utilize antibodies added extracellularly to whole cells are used to study protein surface exposure [19–21]. These include dot blots, whole-cell enzyme-linked immunosorbent assay (ELISA), immunofluorescence, and flow-cytometry. Just like proteases, antibodies cannot enter intact cells and, hence, will bind to only surface-accessible epitopes.

Antibodies to be used in these assays should satisfy two requirements. First, they should be able to recognize the native protein. Many antibodies that bind to denatured protein during immunoblot procedures cannot bind to native proteins because binding epitopes are hidden in the protein structure. This is especially important when antibodies were raised against denatured proteins or a peptide. Second, antibodies should be polyclonal. For example, if a transmembrane protein contains a surface-exposed and periplasmic domain, then the use of monoclonal antibodies that recognize the periplasmic domain will lead to false negatives. However, experiments with polyclonal and monoclonal (or epitope-specific) antibodies can provide valuable insights into protein topology [11, 22, 23]. A protocol for a dot blot assay is described in what follows. This assay is easy and inexpensive and requires no dedicated equipment.

Each of the aforementioned methods has its own advantages and disadvantages, and ideally, a combination of methods should be used. One of the common limitations is that the ability to detect proteins on cell surfaces depends not only on protein localization but also on sequence and structure because these features determine the presence and accessibility of groups for labeling, protease sites, or epitopes for antibody binding. In addition, surface proteins can be physically occluded from detection by interactions with other proteins, hidden within long sugar chains of LPS, an S layer, or extracellular matrix. On the other hand, these methods can also generate false positives. For example, surface labeling and proteolysis can destabilize the OM, giving the reagent and proteases access to the periplasm. Many immunodetection techniques require cell fixation, which can also lead to a number of artifacts [24]. Therefore, it is very important to incorporate careful controls for OM integrity when designing these experiments.

As a general recommendation, proteins with known periplasmic topology should be used as negative controls. These include either soluble periplasmic proteins or lipoproteins with experimentally verified periplasmic localization. Ideally, such proteins should not be a part of a bigger protein complex and should be readily detectable with specific antibodies. If such protein-specific antibodies are not available, it is possible to use heterologously expressed proteins as a controls, for example periplasmically localized fluorescent proteins (such as mCherry or superfolding GPF [25]), maltose binding protein, glutathione S-transferase, or other proteins for which antibodies are commercially available. However, it is important to remember that protein overexpression can also lead to aberrant results. Therefore, when using such heterologous proteins, one should verify that they do not have a negative impact on cell growth and OM permeability. In addition, a variant of a protein of interest that would localize it to different compartments (e.g., by swapping signal sequences) can also serve as a negative control.

2 Materials

2.1 Cell Surface Labeling Based on Modification of Primary Amines

1. NHS reagents (*see* Table 1 for product information). NHS reagents are moisture sensitive. Store desiccated at 4 °C, and equilibrate to room temperature (RT) before opening. Prepare 25 mM stock solution according to product instructions immediately before use.
2. Labeling buffer containing no primary amines, such as phosphate buffered saline (PBS): 10 mM Na_2HPO_4, 1.8 mM KH_2PO_4, 2.7 mM KCl, 137 mM NaCl, pH 8.0.
3. Quenching solution: 1 M glycine or 1 M Tris–HCl, pH 8.0.

2.2 Cell Surface Labeling Based on Modification of Sulfhydryls

1. Maleimide reagents (*see* Table 1 for product information). Maleimide reagents are moisture sensitive. Store desiccated at 4 °C, and equilibrate to RT before opening. Prepare 25 mM stock solution according to product instructions immediately before use.
2. Labeling buffer containing no sulfhydryls, such as Tris-buffered saline (TBS): 50 mM Tris–HCl, 150 mM NaCl, pH 7.0 or PBS: 10 mM Na_2HPO_4, 1.8 mM KH_2PO_4, 2.7 mM KCl, 137 mM NaCl, pH 7.0.
3. Tris(2-carboxyethyl)phosphine (TCEP) solution: 500 mM (optional).

2.3 Cell Surface Proteolysis

1. Proteinase K solution: 20 mg/mL.
2. Reaction buffer, TBS: 50 mM Tris–HCl, 150 mM NaCl, 5 mM $CaCl_2$, pH 8.0.

3. Phenylmethylsulfonyl fluoride (PMSF): 500 mM in ethanol.
4. SDS loading buffer 1×: 50 mM Tris–HCl, pH 6.8, 2% SDS, 10% glycerol, 0.002% bromophenol blue.

2.4 Whole-Cell Dot Blot Assay

1. Nitrocellulose membrane.
2. PBS: 10 mM Na_2HPO_4, 1.8 mM KH_2PO_4, 2.7 mM KCl, 137 mM NaCl, pH 7.0.
3. EDTA: 0.5 M.
4. Blocking buffer: PBS with 2% nonfat dried milk.
5. Antibodies for detection of protein of interest as well as a negative control.
6. Appropriate horseradish peroxidase (HRP)-conjugated secondary antibodies.
7. Chemiluminescent substrate for HRP detection.

3 Methods

3.1 Cell Surface Labeling Based on Modification of Primary Amines

This protocol is valid for any NHS-based reagent.

1. Collect exponentially growing cells by centrifugation (*see* **Note 1**).
2. Wash cells three times with ice-cold PBS to remove amine-containing culture medium.
3. Resuspend cells to 10^{10} cells/mL.
4. Add NHS reagent to a final concentration of 2.5 mM.
5. Incubate at RT for 30 min.
6. Add one-tenth of the volume of quenching solution.
7. Collect the cells by centrifugation.
8. Wash cells twice with PBS supplemented with 100 mM glycine or directly in 100 mM Tris–HCl to quench and remove excess reagent.
9. Analyze by immunoblotting or follow with protein purification if needed (*see* **Note 2**).

3.2 Cell Surface Labeling Based on Modification of Sulfhydryls

This protocol is valid for any maleimide-based reagent.

1. Collect exponentially growing cells by centrifugation (*see* **Note 1**).
2. Wash cells three times with ice-cold TBS or PBS.
3. Resuspend cells to 10^{10} cells/mL.
4. (Optional) If protein contains oxidized (disulfide bonded) cysteines, treat cells with 5 mM TCEP in TBS or PBS, pH 7.0, for 30 min at RT. Wash cells twice with TBS or PBS to remove excess TCEP (*see* **Note 3**).

5. Add reagent to a final concentration of 2.5 mM.
6. Incubate at RT for 30 min.
7. Collect cells by centrifugation.
8. Wash cells twice with PBS or TBS to remove excess reagent.
9. Analyze by immunoblotting or follow with protein purification if needed (*see* **Note 2**).

3.3 Cell Surface Proteolysis

1. Prepare 2× dilutions of proteinase K in range of 20–1.25 mg/mL in the reaction buffer.
2. Collect exponentially growing cells by centrifugation (*see* **Note 1**).
3. Resuspend cells to 10^{10} cells/mL in reaction buffer. Use 90 μL cell suspension for each reaction.
4. Add 10 μL corresponding proteinase K solution or 10 μL reaction buffer (untreated control). Incubate at RT for 30 min.
5. Preheat SDS loading buffer in a 96 °C thermoblock or boiling water bath.
6. Add 1 μL PMSF stock solution to inactivate proteinase K.
7. Collect cells by centrifugation and wash twice with reaction buffer supplemented with 5 mM PMSF to remove excess proteinase K.
8. Resuspend cells in 100 μL preheated SDS loading buffer. Boil immediately for at least 10 min.
9. Analyze by immunoblotting (*see* **Note 2**).

3.4 Whole-Cell Dot Blot Assay

1. Collect exponentially growing cells by centrifugation (*see* **Note 1**).
2. Resuspend cells to 10^9 cells/mL in PBS. Split into two tubes.
3. Add EDTA to final concentration of 10 mM to one of the tubes and sonicate on ice four times for 30 s to prepare cell lysate (*see* **Note 4**).
4. Spot 2 μL of cell suspension or cell lysate on a nitrocellulose membrane and air-dry (approx. 5 min).
5. Place membrane in blocking solution. Incubate with gentle shaking for 30 min at RT.
6. Add appropriate amount of primary antibodies (*see* **Note 5**). Incubate with gentle shaking for 1 h at RT.
7. Wash membrane five times for 3 min with PBS.
8. Add a blocking buffer containing secondary antibodies. Incubate with gentle shaking for 1 h at RT.
9. Wash membrane five times for 3 min with PBS.
10. Use chemiluminescent substrate and develop according to standard immunoblot procedure.

4 Notes

1. Using a culture medium supplemented with cations helps to reinforce the OM and prevent permeability. If using media with low cation concentration (e.g., lysogeny broth (LB)), add 10 mM $MgSO_4$ and 5 mM $CaCl_2$. In addition, cations can be added to the reaction buffers of any procedures described without interference.
2. If negative results are obtained, analyzing the protein accessibility for labeling or protease cleavage might be necessary. To prepare gentle cell lysate, add BugBuster® 10× protein extraction reagent (EMD Millipore) to the cell suspension. Unlike the original BugBuster, this reagent does not add salts or buffer components and can be used with labeling/protease assays without interference.
3. β-mercaptoethanol and dithiothreitol (DTT) contain sulfhydryl groups and are incompatible with maleimide labeling. TCEP does not contain sulfhydryl groups and therefore can be used to reduce disulfide bonds prior to labeling.
4. Using detergents for cell lysis (such as BugBuster reagent) will interfere with protein binding to nitrocellulose membrane. Prepare lysate by sonication. Adding EDTA helps to disperse LPS and form membrane vesicles with mixed orientation. Sometimes it is necessary to readjust the concentration of the primary antibodies for a dot blot assay. As a general recommendation, start by using a concentration three times higher than that used for an immunoblot procedure.

References

1. Silhavy TJ, Kahne D, Walker S (2010) The bacterial cell envelope. Cold Spring Harb Perspect Biol 2(5):a000414
2. Okuda S, Tokuda H (2011) Lipoprotein sorting in bacteria. Annu Rev Microbiol 65:239–259
3. Zuckert WR (2014) Secretion of bacterial lipoproteins: through the cytoplasmic membrane, the periplasm and beyond. Biochim Biophys Acta 1843(8):1509–1516
4. Konovalova A, Silhavy TJ (2015) Outer membrane lipoprotein biogenesis: lol is not the end. Philos Trans R Soc Lond Ser B Biol Sci 370(1679)
5. Wilson MM, Bernstein HD (2015) Surface-exposed lipoproteins: an emerging secretion phenomenon in gram-negative bacteria. Trends Microbiol
6. Freeman TC Jr, Wimley WC (2010) A highly accurate statistical approach for the prediction of transmembrane beta-barrels. Bioinformatics 26(16):1965–1974
7. Singh NK, Goodman A, Walter P, Helms V, Hayat S (2011) TMBHMM: a frequency profile based HMM for predicting the topology of transmembrane beta barrel proteins and the exposure status of transmembrane residues. Biochim Biophys Acta 1814(5):664–670
8. Hayat S, Elofsson A (2012) BOCTOPUS: improved topology prediction of transmembrane beta barrel proteins. Bioinformatics 28(4):516–522
9. Hermanson GT (2013) Bioconjugate techniques, 3rd edn. Academic press, London, pp 1–1146

10. Nikaido H (2003) Molecular basis of bacterial outer membrane permeability revisited. Microbiol Mol Biol Rev 67(4):593–656
11. Konovalova A, Perlman DH, Cowles CE, Silhavy TJ (2014) Transmembrane domain of surface-exposed outer membrane lipoprotein RcsF is threaded through the lumen of beta-barrel proteins. Proc Natl Acad Sci U S A 111(41):E4350–E4358
12. Cowles CE, Li Y, Semmelhack MF, Cristea IM, Silhavy TJ (2011) The free and bound forms of Lpp occupy distinct subcellular locations in *Escherichia coli*. Mol Microbiol 79(5):1168–1181
13. Rosenbusch JP (1990) Structural and functional properties of porin channels in *E. coli* outer membranes. Experientia 46(2): 167–173
14. Wilson MM, Anderson DE, Bernstein HD (2015) Analysis of the outer membrane proteome and secretome of *Bacteroides fragilis* reveals a multiplicity of secretion mechanisms. PLoS One 10(2):e0117732
15. Pugsley AP, Kornacker MG, Ryter A (1990) Analysis of the subcellular location of pullulanase produced by *Escherichia coli* carrying the pulA gene from *Klebsiella pneumoniae* strain UNF5023. Mol Microbiol 4(1):59–72
16. Pinne M, Haake DA (2009) A comprehensive approach to identification of surface-exposed, outer membrane-spanning proteins of *Leptospira interrogans*. PLoS One 4(6):e6071
17. Porter WH, Preston JL (1975) Retention of trypsin and chymotrypsin proteolytic activity in sodium dodecyl sulfate solutions. Anal Biochem 66(1):69–77
18. Hilz H, Wiegers U, Adamietz P (1975) Stimulation of proteinase K action by denaturing agents: application to the isolation of nucleic acids and the degradation of 'masked' proteins. Eur J Biochem 56(1):103–108
19. Pinne M, Haake D (2011) Immunofluorescence assay of leptospiral surface-exposed proteins. J Vis Exp 53
20. Blom K, Lundin BS, Bolin I, Svennerholm A (2001) Flow cytometric analysis of the localization of helicobacter pylori antigens during different growth phases. FEMS Immunol Med Microbiol 30(3):173–179
21. Matsunaga J, Werneid K, Zuerner RL, Frank A, Haake DA (2006) LipL46 is a novel surface-exposed lipoprotein expressed during leptospiral dissemination in the mammalian host. Microbiology 152(Pt 12):3777–3786
22. Moeck GS, Bazzaz BS, Gras MF, Ravi TS, Ratcliffe MJ, Coulton JW (1994) Genetic insertion and exposure of a reporter epitope in the ferrichrome-iron receptor of *Escherichia coli* K-12. J Bacteriol 176(14):4250–4259
23. Newton SM, Klebba PE, Michel V, Hofnung M, Charbit A (1996) Topology of the membrane protein LamB by epitope tagging and a comparison with the X-ray model. J Bacteriol 178(12):3447–3456
24. Schnell U, Dijk F, Sjollema KA, Giepmans BN (2012) Immunolabeling artifacts and the need for live-cell imaging. Nat Methods 9(2): 152–158
25. Dinh T, Bernhardt TG (2011) Using superfolder green fluorescent protein for periplasmic protein localization studies. J Bacteriol 193(18):4984–4987

Chapter 8

Probing Inner Membrane Protein Topology by Proteolysis

Maxence S. Vincent and Eric Cascales

Abstract

Inner membrane proteins are inserted into the membrane via α-helices. These helices do not only constitute membrane anchors but may mediate specific interactions with membrane protein partners or participate in energetic processes. The number, location, and orientation of these helices is referred to as *topology*. Bitopic membrane proteins that consist of a single membrane-embedded domain connecting two soluble domains are distinguished from polytopic ones that consist of multiple membrane-spanning helices connected by extramembrane domains. Defining inner membrane protein topology could be achieved by different methods. Here we describe a protease accessibility assay that makes it possible to define topology based on digestion profiles.

Key words Membrane protein, Inner membrane, Insertion, Topology, Transmembrane segment, Bitopic, Polytopic, Proteolysis, Protease, Proteinase K, Carboxypeptidase Y

1 Introduction

Bacterial secretion systems are multiprotein machines that catalyze the traffic of protein substrates across the cell envelope [1]. Most secretion systems described so far assemble large channels composed of inner and outer membrane proteins [1]. In secretion systems, inner membrane proteins are crucial for the assembly of platforms for pilus polymerization, substrate recruitment and selection, or energetic purposes [1]. Defining inner membrane protein topology, a term referring to the number, position, and orientation of transmembrane helices (TMHs), is therefore an important step to characterize these proteins. Depending on the number and position of these TMHs, inner membrane proteins are categorized into bitopic and polytopic proteins (Fig. 1). Bitopic membrane proteins consist of a single membrane-embedded domain connecting two soluble domains located in two different compartments. The TMH of bitopic proteins could be located at the N- or C-terminus. By contrast, polytopic membrane proteins consist of multiple TMHs connected by extramembrane domains called *loops*. Bitopic proteins

Laure Journet and Eric Cascales (eds.), *Bacterial Protein Secretion Systems: Methods and Protocols*, Methods in Molecular Biology, vol. 1615, DOI 10.1007/978-1-4939-7033-9_8, © Springer Science+Business Media LLC 2017

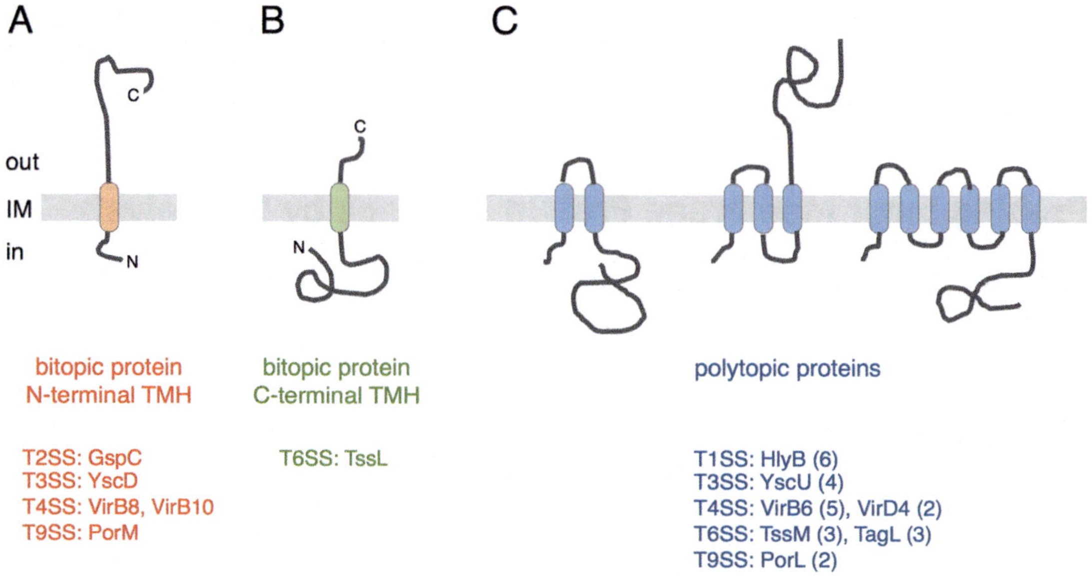

Fig. 1 Nomenclature of inner membrane proteins with selected topologies. Shown are the topologies of a bitopic protein with an N-terminal TMH (**a**), a bitopic protein with a C-terminal TMH (**b**) and polytopic proteins with different numbers of TMH (**c**). Representative examples of inner membrane (IM) proteins with these topologies associated with bacterial secretion systems are listed. For polytopic proteins, the number of transmembrane segments are indicated in brackets

with N-terminal TMHs are relatively common, and this category includes GcpC, YscD, VirB10 and PorM, subunits associated with the Type II (T2SS), Type III (T3SS), Type IV (T4SS) and Type IX (T9SS) secretion systems [2–5]. Bitopic proteins with C-terminus TMHs, also called C-tail proteins, are rare. In secretion systems, only the Type VI secretion system (T6SS) TssL protein has been demonstrated to adopt this topology [6, 7]. Polytopic membrane proteins are also commonly associated with secretion systems, and this category includes the HlyB, YscU, VirB6, TssM and PorL proteins that are respectively associated with the T1SS, T3SS, T4SS, T6SS and T9SS [5, 8–12].

Inner membrane protein TMH position and orientation can be predicted using computational methods based on hydrophobicity patterns and the "positive-inside" rule (*see* Chapter 2). Several approaches have also been developed to experimentally define protein topology [13, 14], including the *pho-lac* dual reporter system (*see* Chapter 10) and the substituted cysteine accessibility method (*see* Chapter 9). In this chapter, we will describe a third approach based on the accessibility of extramembrane, soluble domains to exogenous proteases. In addition to assessing inner membrane topology, protease accessibility assays are also of interest for testing the in vitro translocation of proteins [7, 15] and for testing whether proteins are subjected to conformational changes in vivo (*see* Chapter 22) [16–18].

2 Material

2.1 Cell Growth and Spheroplast Preparation

1. Lysogeny broth (LB) or recommended medium to grow the strain of interest.
2. TNS buffer: 20 mM Tris–HCl, pH 8.0, 100 mM NaCl, 30% sucrose: Dissolve 0.243 g Tris(hydroxymethyl) aminomethane, 0.684 g NaCl, and 30 g sucrose in sterile distilled water (final volume of 100 mL). Adjust pH to 8.0 with 1 M HCl.
3. TN buffer: 20 mM Tris–HCl, pH 8.0, 100 mM NaCl: Dissolve 0.243 g Tris(hydroxymethyl) aminomethane and 0.684 g NaCl in 100 mL sterile distilled water. Adjust pH to 8.0 with 1 M HCl.
4. 0.5 M ethylenediaminetetraacetic acid (EDTA), pH 8.0: Dissolve 1.86 g EDTA (disodium salt) in 10 mL sterile distilled water. Adjust pH to 8.0 with 10 M NaOH.
5. Lysozyme stock solution (100×), 10 mg/mL lysozyme: Dissolve 10 mg goose egg lysozyme in 1 mL sterile distilled water. Store at −20 °C.
6. Incubator.
7. Spectrophotometer to measure bacterial density.
8. Labtop centrifuge.

2.2 Protease Accessibility Assay

1. Triton X-100 stock solution, 10% Triton X-100: Mix 1 mL 100% Triton X-100 with 9 mL sterile distilled water (*see* **Note 1**). Store at room temperature.
2. Carboxypeptidase Y stock solution (100×), 10 mg/mL carboxypeptidase Y: Dissolve 10 mg purified carboxypeptidase Y in 1 mL sterile distilled water. Store at −20 °C.
3. Proteinase K stock solution (100×), 10 mg/mL proteinase K: Dissolve 10 mg purified proteinase K in 1 mL sterile distilled water. Store at −20 °C.
4. Cocktail of protease inhibitors (Complete, F. Hoffmann-La Roche AG, Basel, Switzerland, or equivalent).
5. Phenylmethylsulfonyl fluoride (PMSF) stock solution (100×), 100 mM PMSF: Dissolve 17.4 mg PMSF in 1 mL absolute ethanol (*see* **Note 2**). Store at −20 °C.
6. 50% trichloroacetic acid (TCA) solution: Dissolve 50 g TCA in 30 mL distilled water. Complete to 100 mL with distilled water (*see* **Note 3**).
7. Acetone.
8. Vortex.

2.3 Sample Analysis by SDS-PAGE and Immunodetection

1. Sodium dodecyl sulfate (SDS)-polyacrylamide gel electrophoresis (PAGE) loading buffer: 60 mM Tris–HCl, pH 6.8, 2% SDS, 10% glycerol, 5% β-mercaptoethanol, 0.01% bromophenol blue.
2. Water bath at 96 °C.
3. Mini-gel caster system and SDS-PAGE apparatus.
4. Protein blotting apparatus.
5. Antibodies for protein immunodetection.

3 Method

3.1 Cell Growth and Spheroplast Preparation (See Note 4)

1. Grow 30 mL culture in appropriate medium to allow cell growth and production of protein of interest (*see* **Note 5**).
2. Collect cells by centrifugation at 5000 × *g* for 10 min at 4 °C.
3. Discard supernatant and gently resuspend cell pellet in optical density at 600 nm (OD_{600}) of 12 in ice-cold TNS buffer. Incubate on ice for 10 min.
4. Add EDTA at 1 mM final concentration (*see* **Note 6**). Incubate on ice for 5 min.
5. Add lysozyme at final concentration of 100 μg/mL and incubate on ice for 15–40 min (*see* **Note 7**).
6. Dilute sample twice with ice-cold TN buffer, mix by gently inverting the tube, and keep on ice for 10 min.
7. Collect spheroplasts by centrifugation at 10,000 × *g* for 5 min at 4 °C.
8. Gently resuspend spheroplasts to an OD_{600} of 6 in ice-cold TN buffer.

3.2 Protease Accessibility (See Notes 8 and 9)

1. Divide cell suspension into five samples, numbered 1–5. Sample 1 will remain untreated.
2. Add 1% (final concentration) of Triton X-100 in samples 3 and 5 to lyse spheroplasts (*see* **Note 9**). Mix by vortexing and incubation 10 min on ice.
3. Add carboxypeptidase Y (100 μg/mL final concentration from 10 mg/mL stock solution) in tubes 2 and 3. Incubate for 30 min on ice.
4. Add proteinase K (100 μg/mL final concentration from 10 mg/mL stock solution) in tubes 4 and 5. Incubate for 30 min on ice.
5. Quench the proteolysis reaction by adding PMSF and inhibitor cocktail in tubes 1–5. Incubate for 5 min on ice.

6. Add 0.5 volume 50% TCA in tubes 1–5. Mix by vortexing and incubate for 20 min on ice.
7. Collect precipitated material by centrifugation at 20,000 × *g* for 20 min at 4 °C.
8. Discard supernatant and add 500 μL acetone. Vortex.
9. Collect precipitated proteins by centrifugation at 20,000 × *g* for 20 min at 4 °C.
10. Discard supernatant and keep tubes open until pellet is dry (*see* **Note 10**).

3.3 Sample Analysis by SDS-PAGE and Immunodetection

1. Resuspend pellet in SDS-PAGE loading buffer by throughout vortexing.
2. Boil samples in water bath for 5–10 min (*see* **Note 11**).
3. Proceed to SDS-PAGE and immunoblotting using your favorite protocol.

A schematic example of expected results for topology mapping using proteolysis is shown in Fig. 2.

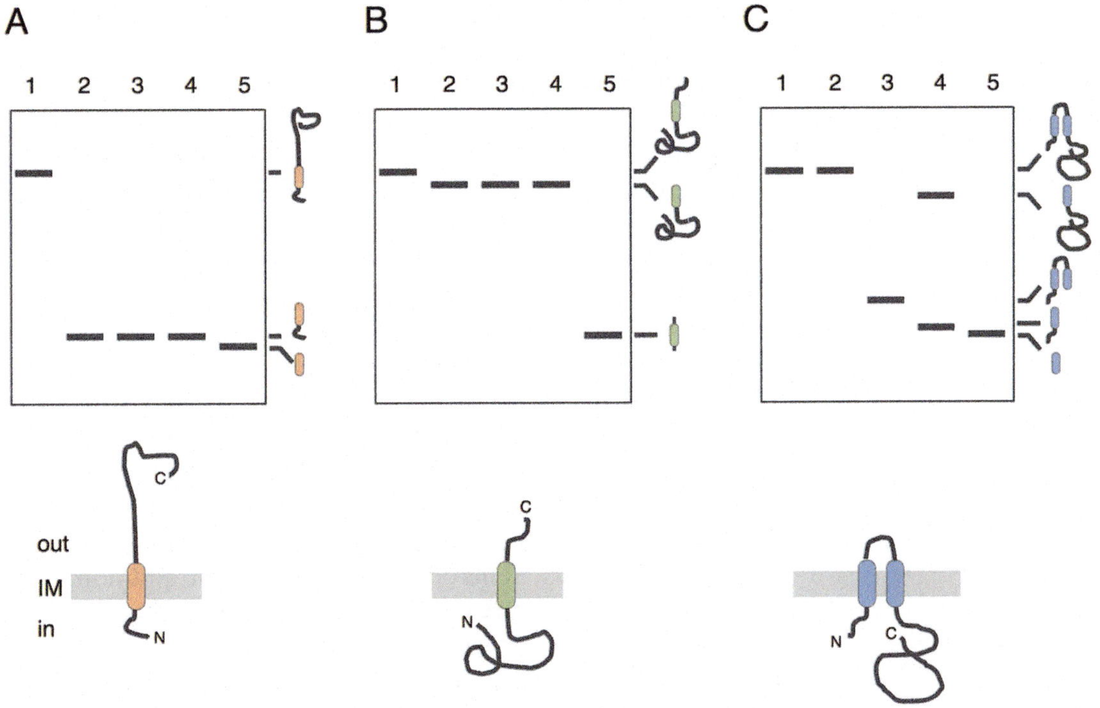

Fig. 2 Schematic representation of expected results. The expected immunoblot results for inner membrane proteins with the topology shown below are schematically represented. Samples 1–5 are shown (*1*, untreated sample; *2*, carboxypeptidase Y; *3*, carboxypeptidase Y on Triton X-100-lysed spheroplasts; *4*, proteinase K; *5*, proteinase K on Triton X-100-lysed spheroplasts). The representation of the protein degradation products corresponding to the immunodetection bands are shown on the right of each "blot"

4 Notes

1. Triton X-100 is a detergent used to lyse cells and solubilize a subset of membrane proteins. It is a viscous solution and therefore should be pipetted slowly and with care.
2. PMSF is a serine protease inhibitor with a short half-life. Owing to its instability in solution, it is recommended to prepare fresh solution on the spot.
3. TCA is highly irritating. Thus, it should be handled with care (gloves, laboratory suit, and glasses).
4. For Gram-negative bacteria, spheroplasts should be prepared to provide access of protease to periplasmic side of inner membrane. For Gram-positive bacteria, grow, harvest, and resuspend cells as specified in **steps 1–2** of Subheading 3.1 and then proceed to **step 8** of Subheading 3.1.
5. Use the appropriate medium to grow the cells. If the expression of the gene coding the protein of interest needs to be induced, add the inducer at the appropriate concentration.
6. This concentration of EDTA is commonly used for disturbing the lipopolysaccharide layer of the outer membrane in *E. coli* cells. Other bacterial strains may need higher concentrations of EDTA.
7. Lysozyme concentration and incubation time should be adapted to the bacterial strain used in the assay. Efficient spheroplast preparation of most Gram-negative bacteria requires incubation on ice for 15–40 min.
8. Protease accessibility should be tested with two proteases: one processive exopeptidase hydrolyzing from the C-terminus of the protein (e.g., carboxypeptidase Y) and one endopeptidase with low or broad specificity (e.g., trypsin, papain, proteinase K). When using the calcium-dependent proteinase K, add 0.1 mM $CaCl_2$ to the TN buffer. For simplification, this protocol describes an assay with carboxypeptidase Y and proteinase K.
9. Appropriate controls include protease accessibility assays with lysed spheroplasts. Spheroplasts are lysed by the addition of 1% Triton X-100. The presence of Triton X-100 in the assay buffer does not interfere with most proteases.
10. If available, the pellet could be dried using a SpeedVac (Thermo Fisher Scientific, Inc., Waltham, MA) vacuum concentrator (or equivalent).
11. A number of highly hydrophobic polytopic inner membrane proteins precipitate in SDS-PAGE loading buffer when boiled. For the first assay, keep the concentrating gel during the immunoblot to verify that the protein is not retained in the well.

Acknowledgements

Work in EC laboratory is supported by the Centre National de la Recherche Scientifique, the Aix-Marseille Université, and grants from the Agence Nationale de la Recherche (ANR-14-CE14-0006-02 and ANR-15-CE11-0019-01). MSV is a recipient of a doctoral fellowship from the French Ministère de l'Enseignement Supérieur et de la Recherche.

References

1. Costa TR, Felisberto-Rodrigues C, Meir A, Prevost MS, Redzej A, Trokter M, Waksman G (2015) Secretion systems in gram-negative bacteria: structural and mechanistic insights. Nat Rev Microbiol 13:343–359
2. Bleves S, Lazdunski A, Filloux A (1996) Membrane topology of three Xcp proteins involved in exoprotein transport by *Pseudomonas aeruginosa*. J Bacteriol 178:4297–4300
3. Ross JA, Plano GV (2011) A C-terminal region of *Yersinia pestis* YscD binds the outer membrane secretin YscC. J Bacteriol 193:2276–2289
4. Das A, Xie YH (1998) Construction of transposon Tn3phoA: its application in defining the membrane topology of the *Agrobacterium tumefaciens* DNA transfer proteins. Mol Microbiol 27:405–414
5. Vincent MS, Canestrari MJ, Leone P, Stathopulos J, Ize B, Zoued A, Cambillau C, Kellenberger C, Roussel A, Cascales E. (2017) Characterization of the *Porphyromonas gingivalis* Type IX Secretion trans-envelope PorKLMNP core complex. J Biol Chem. 292:3252–3261.
6. Aschtgen MS, Zoued A, Lloubès R, Journet L, Cascales E (2012) The C-tail anchored TssL subunit, an essential protein of the enteroaggregative *Escherichia coli* Sci-1 type VI secretion system, is inserted by YidC. Microbiology 1:71–82
7. Pross E, Soussoula L, Seitl I, Lupo D, Kuhn A (2016) Membrane targeting and insertion of the C-tail protein SciP. J Mol Biol 428:4218–4227.
8. Gentschev I, Goebel W (1992) Topological and functional studies on HlyB of *Escherichia coli*. Mol Gen Genet 232:40–48
9. Allaoui A, Woestyn S, Sluiters C, Cornelis GR (1994) YscU, a Yersinia enterocolitica inner membrane protein involved in Yop secretion. J Bacteriol 176:4534–4542
10. Jakubowski SJ, Krishnamoorthy V, Cascales E, Christie PJ (2004) Agrobacterium tumefaciens VirB6 domains direct the ordered export of a DNA substrate through a type IV secretion system. J Mol Biol 341:961–977
11. Ma LS, Lin JS, Lai EM (2009) An IcmF family protein, ImpLM, is an integral inner membrane protein interacting with ImpKL, and its walker a motif is required for type VI secretion system-mediated Hcp secretion in agrobacterium tumefaciens. J Bacteriol 191:4316–4329
12. Logger L, Aschtgen MS, Guérin M, Cascales E, Durand E (2016) Molecular dissection of the interface between the type VI secretion TssM cytoplasmic domain and the TssG baseplate component. J Mol Biol 428:4424–4437
13. Traxler B, Boyd D, Beckwith J (1993) The topological analysis of integral cytoplasmic membrane proteins. J Membr Biol 132:1–11
14. van Geest M, Lolkema JS (2000) Membrane topology and insertion of membrane proteins: search for topogenic signals. Microbiol Mol Biol Rev 64:13–33
15. Cunningham K, Lill R, Crooke E, Rice M, Moore K, Wickner W, Oliver D (1989) SecA protein, a peripheral protein of the *Escherichia coli* plasma membrane, is essential for the functional binding and translocation of proOmpA. EMBO J 8:955–959
16. Larsen RA, Thomas MG, Postle K (1999) Protonmotive force, ExbB and ligand-bound FepA drive conformational changes in TonB. Mol Microbiol 31:1809–1824
17. Germon P, Ray MC, Vianney A, Lazzaroni JC (2001) Energy-dependent conformational change in the TolA protein of *Escherichia coli* involves its N-terminal domain, TolQ, and TolR. J Bacteriol 183:4110–4114
18. Cascales E, Christie PJ (2004) Agrobacterium VirB10, an ATP energy sensor required for type IV secretion. Proc Natl Acad Sci U S A 101:17228–17233

Chapter 9

Mapping of Membrane Protein Topology by Substituted Cysteine Accessibility Method (SCAM™)

Mikhail Bogdanov

Abstract

A described simple and advanced protocol for the substituted-cysteine accessibility method as applied to transmembrane (TM) orientation (SCAM™) permits a topology analysis of proteins in their native state and can be universally adapted to any membrane system to either systematically map an uniform topology or identify and quantify the degree of mixed topology. In this approach, noncritical individual amino acids that are thought to reside in the putative extracellular or intracellular loops of a membrane protein are replaced one at a time by cysteine residue, and the orientation with respect to the membrane is evaluated using a pair of membrane-impermeable nondetectable and detectable thiol-reactive labeling reagents.

Key words Membrane protein, Topology, Cysteine, Maleimides, SCAM™

1 Introduction

1.1 Membrane Protein Topology and Topogenesis

It has been estimated that the vast majority of membrane proteins adopt α-helical bundle (Fig. 1a) structure, which contains transmembrane domains (TMDs) that span the membrane in a zigzag but not in a "one-by-one" fashion (Fig. 1b). A fundamental aspect and primary structural element of the structure of integral membrane proteins is membrane protein topology. Membrane protein topology refers to the two-dimensional structural information of a membrane protein and describes the way a polypeptide chain is arranged in the membrane, i.e., the number of TMDs and their orientation in the membrane that indicates the sidedness of extramembrane domains (EMDs) [1–4]. Although the topology of membrane proteins provides low-resolution structural information, it can be a starting point for different biochemical experiments or modeling of three-dimensional structures. Membrane protein topology and assembly are governed by structural principles and topological rules and directed by topogenic signals and sequences in the nascent polypeptide chain that are recognized and decoded

Laure Journet and Eric Cascales (eds.), *Bacterial Protein Secretion Systems: Methods and Protocols*, Methods in Molecular Biology, vol. 1615, DOI 10.1007/978-1-4939-7033-9_9, © Springer Science+Business Media LLC 2017

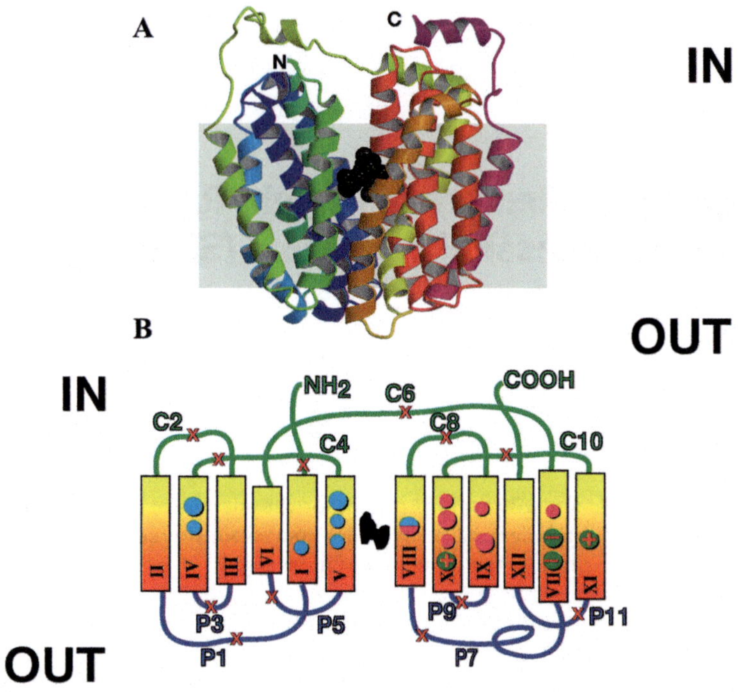

Fig. 1 Three-dimensional view of membrane protein topology of *E. coli* lactose permease (LacY) probed by SCAM™. (**a**) A highly resolved α-helical bundle of lactose permease reproduced from [44] with permission from American Association for the Advancement of Science. (**b**) X marks diagnostic single cysteines introduced strategically one at a time to probe sidedness of this protein by SCAM™

not only by the protein insertion and translocation machineries (translocon) but also by the given lipid profile [4]. Thus each membrane protein may contain different combinations of topogenic signals (positively and negatively charged residues) and sequences (the number of charged flanking residues, the hydrophobicity and length of the TMDs, EMDs with potential phosphorylation and glycosylation sites) that cooperate sequentially or differently with the translocon components, the protein itself, and a given lipid profile to finalize its membrane topology [4].

1.2 Method of Choice

Topology studies offer guidance on membrane protein structure and function. Given the enormous number of sequences available in genome-sequencing projects, it is not realistic to assume that the structures of all encoded proteins will be generated by crystallographic approaches, especially for membrane proteins. Moreover, purification, crystallization, and structure determination of hydrophobic membrane proteins remain a challenge. Also the exact boundaries between TMD ends and EMDs remain largely unknown. For most highly resolved membrane proteins, the hydrophobic thicknesses of TMDs does not seem to match the lipid bilayer thickness expected or experimentally determined from the acyl chain length of the surrounding lipids. Fortunately, substituted-cysteine accessibility method as applied to transmembrane (TM) orientation (SCAM™) can be used to define the boundaries between membrane-embedded regions and the loop regions exposed to the aqueous phase of membrane proteins in their native environment, thereby supplementing high-resolution structural information [5]. Although X-ray crystallography produces highly detailed structural information on membrane proteins, crystal structures may be distorted owing to purification and crystallization constraints. Information on interactions with other proteins and the lipid environment are also lost during purification since these crucial interactions are substituted by protein–detergent ones. Heterologous expression in a host strain with a lipid composition different from that of the native host can also result in a loss of proper lipid–protein interactions, which can affect topological organization and function. Crystal structures themselves are also static, and a structural basis for dynamic TM events is not approachable either by crystallography or even nuclear magnetic resonance (NMR) [6, 7]. Therefore, biochemical topological analyses at low resolution can be invaluable for characterizing membrane proteins, for which high-resolution structures are not yet available, and for building a reliable mechanistic model even in the absence of highly resolved protein structures [5, 8]. For all these reasons and limitations, noncrystallographic approaches have been developed and employed to determine the lower-resolution topological arrangement of TMDs in full-length membrane proteins [5, 9–15] and understand its relationship to the mechanism of the membrane protein insertion process or biological function [2–4, 15]. To verify predicted membrane protein topology models, the existence of all putative TMDs and EMDs must be verified and the hydrophilic loops must be localized to one or the other side of the membrane. These strategies are quite varied but utilize the impermeability of the membrane bilayer to hydrophilic molecules, the difference in properties between the compartments separated by the membrane, and the incorporation into proteins of a wide variety of reporter groups whose orientation is presumed to reflect the topology of the protein. Experimental topology mapping techniques include but are not limited to cysteine scanning, glycosylation mapping, insertion of proteolytic sites, foreign antigenic epitopes, and glycosylation motifs, or they can be as complex as fusions of C-terminally truncated proteins to enzymatic and fluorescent topology reporters [5, 13, 16]. Together, these tools document EMD residue or inserted tag accessibility and, therefore, the topology of a membrane protein. Reporter domains should ideally lack intrinsic topogenic information, be readily identified, and passively and efficiently follow topogenic information presented by the nascent target protein fragment. Nevertheless, translational gene fusion approaches assume that the folding of the C-terminally truncated proteins does not depend on C-terminal sequence and therefore could not always faithfully assign the predicted TM topology for many polytopic membrane proteins [17]. In this context, SCAM™ has emerged as the method of choice because of its relative simplicity, reliability, and feasibility [5]. In this approach SCAM [18] was adapted to map and assign the TM topology of polytopic membrane proteins (denoted by SCAM™) [19]. SCAM™ is still relatively labor intensive, but it is the most informative and least invasive topology mapping method and the most useful technique thus far developed for topology studies. The method demonstrates that reporter groups can be as simple as a single amino acid substitution. Aside from its simplicity, the advantage of this approach is that the topology is documented in the context of full-length membrane protein molecules and chemical modification can be carried out using whole cells, thereby avoiding problems related to the conversion of cells into membrane vesicles with a uniform orientation. This contrasts with genetic methods that infer the topology from the disposition of reporter molecules fused to fragments of the target protein and therefore completely ignore long-range interhelical and interloop interactions. SCAM™ was also further developed to map a uniform, dual, mixed, or unusual membrane protein topology in intact cells, isolated membrane vesicles, or liposomes using a two-step labeling protocol [10, 19–24].

1.3 Justifying SCAM™ Legacy and Advantages

The reasons for the broad application of SCAM™ are both conceptual and practical. The strategic use of SCAM™ for mapping membrane topology has made it possible to circumvent the many limitations of alternative methods used to map a topology of integral membrane proteins:

1. Since only single amino acid changes are made, cysteine chemistry has the highest resolution in that water accessibility of individual cysteine residues can be determined.
2. Cysteine has no or little preference for a particular secondary structure. Owing to only minmal changes in the primary sequence, structural perturbation of introduced cysteine mutations is essentially absent or much milder than in other methods commonly used to determine topology.
3. Detection of engineered cysteine modifications is simple, and analysis is done by chemical modification using a broad range of commercially available reagents that differ in charge, size, mass, and hydrophilicity.
4. These reagents can be used to detect protein sulfhydryl groups with sensitivities in the femtomole range.
5. Chemical modification can be done using intact cells.

6. This method is capable of distinguishing the accessibility of residues separated by only three or four residues and is useful in the precise fine-structure mapping of TMD ends in polytopic membrane proteins.
7. This method is generally applicable to different membrane systems.

1.4 Application of SCAM™

In most cases SCAM™ provides topological information after the orientation of proteins within membranes is established [5, 19]. The application of this approach allowed either detailed mapping or significant refinement of the topology of a variety of integral membrane proteins, including a more accurate mapping of the ends of TMDs of protein topology that was been established by other methods [12–14]. However SCAM™ is not only an alternative approach to low-resolution determination of membrane protein structure; it also constitutes an attractive independent approach to dynamic studies of membrane proteins. The dynamic aspects of protein structure as a function of the physiological state of a cell is best probed in whole cells or membranes. SCAM™ has been used to monitor dynamic conformational and topological changes accompanied by substrate binding and release during enzyme turnover and function [25, 26]. The labeling of single-cysteine replacements of a major component (SecG) of the SecYEG translocon with 4-acetamido-4′-maleimidylstilbene-2,2′-disulfonic acid (AMS) either at rest or during adenosine triphosphate (ATP)-dependent preprotein translocation clearly demonstrated that a cytoplasmic region of SecG undergoes topology inversion [27].

Although the labeling patterns derived from SCAM™ assays usually reflect a steady-state topology of a membrane protein, semiquantitative analysis of the surface accessibility of individual cysteines introduced into extramembrane loops can be carried out at various stages of protein assembly. In this case, SCAM™ can be used to provide topological information during membrane insertion, folding, and assembly of proteins. Cysteine accessibility during bacteriorhodopsin translation was monitored by pulse-chase radiolabeling and modification by AMS to determine the order and timing of insertion of TM segments into the membrane of *Halobacterium salinarum* [28]. In this in vivo assay, the rate of insertion of TMDs into the *H. salinarum* cytoplasmic membrane was monitored by rapid modification of unique cysteines in extracellular EMDs of the protein with AMS, resulting in a shift in mobility of the protein in sodium dodecyl sulfate (SDS)-polyacrylamide gel electrophoresis (PAGE). SCAM™ has also been utilized to establish a packing geometry of a pilin VirB2 subunit and its ATP-dependent in and out of membrane dynamics and reorganization within the T-pilus and T4SS secretion channel [29].

The applicability of SCAM™ has been succesfully extended to the study of TM toplogy of membrane proteins assembled in different lipid environments [19, 20, 30]. By combining SCAM™ with mutants of *E. coli* in which membrane phospholipid composition can be systematically controlled, the role of phospholipids as determinants of membrane protein topological organization was established [3, 4, 10]. This approach was essential to test TM protein conformation that is sensitive to lipid composition [22] and directly monitor any conformational changes that are associated with changes in phospholipid composition either in vivo [10, 19, 21, 31] or in vitro [23, 32]. The use of cysteine-specific probes in combination with "lipid" mutants makes this approach a powerful means of deriving a molecular understanding of a highly dynamic topogenesis process from relatively static experimental data such as endpoint topologies of membrane proteins [2–4]. The ability to change lipid composition after the assembly of a membrane protein (by either resupplying or diluting a desired lipid) demonstrated the potential for polytopic membrane proteins to change their topological organization after insertion and assembly in the membrane [10, 19, 21, 31]. Thus, SCAM™ became a unique technique to establish a detailed mechanistic understanding of how lipid–protein interactions [10] and interactions with protein itself [22] contribute to overall TM topogenesis [4].

Thus, SCAM™ is a powerful, well-tested, and popular technique for examining static and dynamic membrane protein topologies.

1.5 Overview and General Rationale of Topology Mapping Using SCAM™

1.5.1 SCAM™

This approach is based on the introduction of cysteine residues one at a time in the putative EMDs of otherwise cysteine-less membrane proteins of interest followed by chemical modification with a membrane-impermeable thiol-specific probes either before or after compromising cell membrane integrity to determine cysteine membrane sidedness. Accessibility in whole cells establishes extracellular location, while accessibility only after cell disruption establishes intracellular location (*see* **Notes 1** and **2**). The accessibility of EMDs flanking a TMD then establishes the orientation of the TMD with respect to the plane of the membrane bilayer.

1.5.2 Developing of a Working Topology Model and Selection of Diagnostic Cysteines for SCAM™

SCAM™ is based on the generation of a library of independent single-cysteine mutants in which unique reactive residues are strategically "implanted" at desired positions to probe their sidedness. The process of choosing suitable residues for replacement by a cysteine is often empirically determined, and the rationale for deciding which residues to substitute is aided by different machine-learning topology predictors. The physicochemical constraints imposed by the lipid environment and the known hydrophobicity of individual amino acids provide a method using hydropathy plots to predict the topology of a membrane protein.

Secondary structure predicted by computer-aided hydropathy analysis (which so far is 60–70% reliable) is an initial starting point for the likelihood that a particular residue is in an EMD. The available algorithms utilize also all available structure and topology data, information from aligned homologous sequences, topological rules, and bioinformatical evidence for topology prediction in a probabilistic framework provided by the hidden Markov model. The most recent state-of-the-art topology predictor, the Scale-Based Method for Prediction of Integral Membrane Proteins (SCAMPI) performed simultaneously with the best statistics-based Hydrophobicity Plot-Based Topology Predictor (TopPredΔG) shows the highest accuracy (85%) [33]. However, still in many cases different algorithms produce different predictions, and very often such predictive methods generate a misleading topology. A misassigned or misoriented TMDs and EMDs are still occurred because the primary sequence and overall hydrophobicity of TMDs are not the only determinants of membrane integration. Although simple hydrophobicity is the predominant factor determining the insertion efficiency, analysis of whole genomic data by the TM tendency scale revealed an overlap of TMs and soluble sequences in the so-called semihydrophobic range [34]. This raises the possibility that a significant number of proteins will have sequences that lie close to an equilibrium between membrane insertion and exclusion that can result in switching between TM and non-TM states depending on environmental and physiological conditions or assuming a dual or mixed topologies [3, 21, 22]. Ideally predictors should exploit all molecular interactions of a nascent membrane protein (water–protein, translocon–protein, lipid bilayer–protein). Despite the fact that some topogenic signals in polytopic membrane proteins have been identified, the cellular mechanisms (e.g., translocon, lipid bilayer, membrane potential) for response to these signals are not fully understood. The difference in response may reflect differences in the mechanisms determining TM orientation [4]. Obviously, besides interactions of the nascent polypeptide chain with itself, water, membrane interface, translocon, and neighboring lipids, the electrophoretic and electrostatic properties such as the asymmetry of the lipid bilayer itself are important.

Although SCAMPI accommodates a biological ΔG scale that reflects the thermodynamic cost of translocon-to-bilayer and water-to-membrane partitionings of TMDs and EMDs during membrane integration, long-range interactions between TMDs, unanticipated intersubunit and intraprotein (salt bridges between charged residues within the hydrophobic core) interactions, post-translational phosphorylation and glycosylation, retrograde translocation and insertion events, specific lipid–protein interactions, and the unique membrane lipid composition of different organisms and intracellular organells are some of the variables that are not always addressed by predictive methods. Therefore, hydropathy analysis of the sequence of a polytopic membrane may only reveal potential TMDs and EMDs and their relative orientation and sidedness as a starting point for designing biochemical experiments to establish topological organization. The choice of sites, i.e., where to introduce a cysteine, can be further guided by various other available algorithms and refined by a comparison of predictions with experimentally determined structures. Several databases of TM topologies are available and can be used to evaluate the reliability of predicted topologies. The recently developed Consensus Constrained TOPology prediction (CCTOP; http:// cctop.enzim.ttk.mta.hu) server provides TM topology prediction and utilizes ten different topology prediction methods and incorporates topology information from different experimental sources. The TM topologies in these databases were determined experimentally by means of X-ray crystallography, NMR, gene fusions, SCAM™, glycosylation scanning, and other biochemical methods [35].

1.5.3 Mutation Strategy, Host and Vector Selections, Construction of Plasmids Expressing Single-Cysteine Derivatives

The pre-requisite for the method is the generation of the cysteine-free template of a protein of interest. The expression of a functional "cysteine-less" template has permitted the use of scanning cysteine mutagenesis and thiol modification techniques for mapping membrane topology. All amino acid substitutions should be verified by DNA sequencing and functional analysis and expression level by Western blotting of each derivative should be carried out if possible. Ideally target gene expression should be under control of an inducible promoter such as OP_{tac} or repressor (TetR) regulating the promoter (*PLtetO-1*) to minimize a drastic overexpression or continuous expression of potentially disruptive gene products (*see* **Note 3**).

A prerequisite for each cysteine replacement is retention of function that provides assurance of retention of near native structure. The native cysteine residues are usually changed into alanine or serine residues which are small, commonly found in membrane proteins and appear to be tolerated at most positions thus rendering an active protein. A residue that does not tolerate substitution by Cys makes a crucial contribution to maintaining the structure of the site, and/or to the folding and function of target protein. Replacement of charged residues is generally not advised because these have a high probability of being topogenic signals or may be involved in long-range charge-charge intramolecular or intermolecular (lipid-protein) interactions. If the protein contains stretches of residues of intermediate hydrophobicity that cannot unambiguously be identified as membrane spanning TMD, substitutions should be made approximately every ten residues. The cysteine-less protein serves also as a negative labeling control to assure that residues such as lysine and histidine are not labeled by the reagents (*see* **Note 1**). Ideally the protein under study should be devoid of all native cysteine residues because these residues may also react with thiol-modifying reagents or they may form disulfide bonds with the engineered cysteines and prevent their interaction with modifying thiol-specific reagents. However the templates containing endogenous natural cysteines can be still utilized in SCAM™ if they do not react with the thiol-specific reagents due to membrane residency [31]. Some of which may be unreactive toward sulfhydryl reagents because they are in disulfide-bonded pairs within EMDs. However these "silent" thiols can be "awaken" by reducing agents and even utilized successfully as diagnostic residues in topology mapping [19].

To obtain a minimal topological map, a single-cysteine replacement in each of the putative EMDs should be expressed from a plasmid and analyzed in an appropriate host. The host strain for plasmid expression should be deleted from the target protein gene if it contains native cysteines and is expressed at levels high enough to be detected in the assay. Since the target protein is expressed from a multicopy plasmid, it is often possible to analyze a protein in its normal host without deletion of the native protein. Since SCAM™ is based on the controlled membrane permeability of sulfhydryl reagents, the results of SCAM™ analysis are valid only if the modifying thiol-specific reagent is membrane impermeable and cells are intact. Significant differences in the membrane permeability of different host strains toward different maleimides [5, 31] should be considered and tested during host selection procedure (*see* **Note 4**).

1.5.4 Cell Growth and Regulated Expression of Single-Cysteine Derivatives

Cells are first grown overnight at 37 °C in Luria–Bertani (LB) medium supplemented with appropriate antibiotic and then subcultured in the morning to an OD_{600} of around 0.05 in the LB medium supplemented with appropriate antibiotic to maintain plasmid-encoding single-cysteine replacements in cysteineless protein. Carrying plasmids encoding target protein under appropriate promoters control are induced usually by growth in the presence of inducer for at least several generations (OD_{600} reading ~0.5–0.6) to reach a logarithmic growth phase of cells.

1.5.5 General Protocol for SCAM™

Following the expression of monocysteine mutants, cells are subjected to SCAM™ analysis by in vivo labeling with different detectable and nondetectable thiol-reactive reagents [5, 9]. Maleimide-based thiol reagents, which are available in a wide variety of forms, are particularly suited for SCAM™ [5]. Maleimide reacts with the ionized form of a thiol group (thiolate anion) (Fig. 2a), and this reaction requires a water molecule as a proton acceptor. In most cases the unreactive cysteine residues are located within the membrane hydrophobic core or in a sterically hindered environment [5]. The reaction rate of different thiols is controlled primarily by their surface exposure and proximal environment. Most experiments utilize biotinylated, radioactive, fluorescent [5],

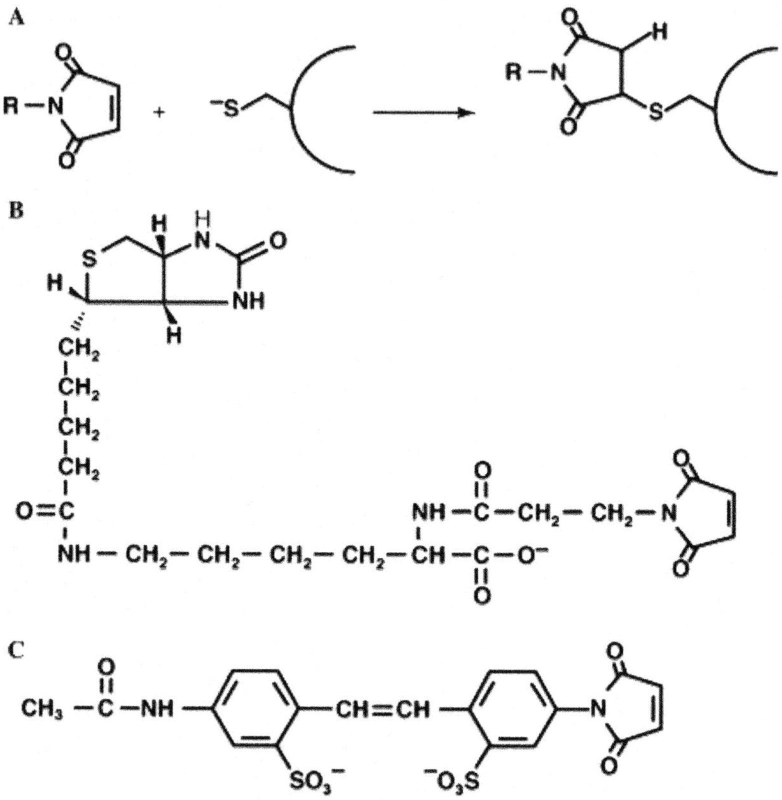

Fig. 2 A chemical structure of thiol-modifying reagents and their reaction with a thiol. (**a**) Reaction of a maleimide with the thiolate of a protein cysteine to form a covalent adduct via nucleophilic addition to the double bond of the maleimide ring. Maleimides are virtually unreactive until they encounter an available ionized thiol group. For most water-exposed cysteine residues in proteins, the pKa of the thiol of cysteine lies in the range of 8–9 and the formation of cysteinyl thiolate anions is optimum in aqueous rather in a nonpolar environment, where the pKa of the thiol of cysteine is around 14. Therefore, the reaction rate of different sulfhydryls is controlled primarily by their water exposure, making the residues that reside in regions of TMDs unfavorable for the generation of thiolate anions. Thus, the labeling characteristics of intramembrane (unreactive) and extramembrane (reactive) cysteines should be consistent with their localization in either a nonpolar or polar environment, respectively. (**b**) Structure of biotin-containing labeling reagent MPB. (**c**) Structure of blocking nondetectable reagent AMS. Figure is reproduced from [5] with permission from Elsevier

or mass-tagged derivatives of maleimide as alkylating reagents [5, 11]. For the mapping of membrane topologies, these maleimides may be used in combination with highly impermeant, nontagged maleimide derivatives, which are used as blocking agents. Nondetectable reagents, such as AMS, do not cross the cytoplasmic bacterial membrane but penetrate the outer membrane and therefore only modify cysteines exposed outside of the cytoplasmic membrane [5] (*see* **Note 4**).

Impermeable thiol reagents that can be easily detected after modification of target proteins are essential for the successful application of SCAM™. Biotin-linked maleimides, such as 3-(N-maleimido-propionyl) biocytin (MPB) (Fig. 2b) is particularly useful owing to its low membrane permeability properties. Following the SDS-PAGE of target proteins isolated by immunoprecipitation or affinity tags, biotinylated proteins are easily detected using avidin-horseradish peroxidase (HRP) and chemiluminescence. MPB essentially constitutes a universal, multipurpose, thiol-specific probe capable of detecting protein SH groups within the femtomole range [36]. SCAM™ using thiol-specific membrane–impermeable MPB (Fig. 2) has been extensively employed to probe the topological organization of many membrane proteins [5]. A maleimide-based thiol reagent is added and the modification reaction terminated by adding a 50- to 100-fold excess of β-mercaptoethanol (β-ME), dithiothreitol, or cysteine to inactivate the unreacted maleimide [5].

The general design of labeling experiments to distinguish between cysteines located in an extracellular or intracellular EMD is outlined in Fig. 3. Single-cysteine replacements are expressed in the appropriate host, and the cells are harvested and suspended in modification buffer. In this assay, sonication of cells is used to disrupt cell membranes, making both extracellular (periplasmic) and cytoplasmic cysteines accessible to MPB, whereas cysteines located within a TM domain are still protected from labeling [10, 19–21]. In this approach, monocysteines are expressed, and reactivity with MPB in intact cells (extracellular exposure) or only after cell disruption by sonication (cytoplasmic exposure) was used to establish TM orientation (Fig. 3). Derivatization of cysteines in whole cells will indicate extracellular exposure, while derivatization only during sonication will indicate a cytoplasmic exposure since desintegration allows access of thiol-specific reagent to both sides of a membrane. Indeed cysteine residing within extracellular EMDs was labeled whether or not cells were disrupted by sonication, while cysteine residing within cytoplasmic EMDs was protected from labeling and is labeled only after cell disruption (Fig. 3, two top panels), which exposes previously inaccessible cysteine residues. It is important to note that the extent of biotinylation should be the same before and after sonication for an extracellular cysteine. If sonication results in an increase in biotinylation, this may indicate a mixed topology (please see following section). A major problem is the variability between samples because of mechanical loss during the work-up or because of differences in the expression level of individual cysteine replacements. Although single-cysteine replacements could affect protein expression, conclusions are based on a comparison of the extent of labeling in whole cells and

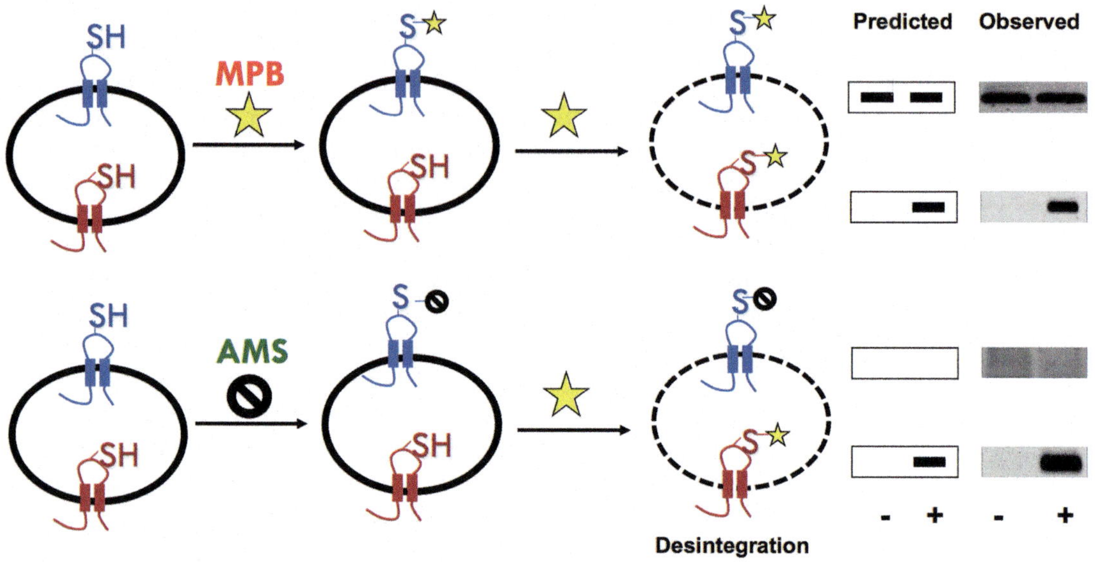

Fig. 3 General strategy for SCAM™ using impermeable MPB and impermeant transparent AMS to probe the sidedness of EMDs. A target membrane protein (only one TMD hairpin is shown) containing a single cysteine exposed either to the extracellular (*blue*, periplasmic) or intracellular (*red*, cytoplasmic) side of the membrane is expressed in host cells. Half of the cells are reacted with a detectable thiol reagent MPB to specifically label the externally exposed cysteine (first row from top) and the other half are reacted with a nondetectable thiol reagent AMS (third and fourth rows) to protect external cysteines in subsequent labeling steps. Both halves of the cells are either kept intact (–) or disintegrated (+) by sonication to expose and label previously inaccesible cytoplasmic cysteine (second and fourth rows). Labeling by MPB that can be blocked completely by pretreatment with AMS is independent indicative of a periplasm-facing residue (third row). Labeling by MPB that cannot be blocked by such AMS treatment is independent indicative of a residue that is facing the cytoplasmn (fourth row). A target protein was immunoprecipitated and resolved by SDS-PAGE, and biotinylated protein was detected using avidin-HRP and chemiluminescence (*right panel*)

disrupted cells for the same protein. This approach simplifies the interpretation of data obtained with a series of protein derivatives that may express at different levels since conclusions about topology are based on the relative reactivity of cysteines in the same sample before and after cell disruption. Since sample pairs are analyzed on the same western blot, no signal intensity normalization is required. The level of expression of any given derivative will affect the absolute intensity of labeling but not the ratio of the labeling between the sample pairs.

Although central to the method is the use of detectable thiol-specific reagents to differentiate intracellular from extracellular EMDs, confirmation of labeling of external water-exposed cysteines by MPB can be achieved by first blocking putative external cysteines in intact cells with a thiol-specific reagent that is transparent in the detection phase of the procedure. Such a preblocking step also allows selective labeling of luminal (exposed to cytoplasm) cysteines after cell disruption and, therefore, a detection of previously inaccessible cytoplasmic cysteine residues. A set of impermeable blocking reagents that effectively react with thiols exposed to solvent but are transparent in the detection phase of the procedure is available for SCAM™ [5] (*see* **Note 4**). One such reagent is AMS (Fig. 2c), which is membrane-impermeable due to its size, two charged sulfonate groups, and high solubility in water. As a blocking reagent AMS is widely used because of its demonstrated inability to cross the cytoplasmic membrane of bacteria or the plasma membrane of mammalian cells. Intact cells are treated either with or without AMS followed by labeling with MPB of intact cells and cells during disruption (Fig. 3 two bottom panels). Excess AMS is removed prior to any subsequent treatments by several pelletings and washing with buffer.

Thiol reagents react with cysteine residues present in all other proteins in the membrane. Immunoprecipitation of the membrane protein of interest or a rapid purification step is necessary to eliminate other labeled proteins. A biotinylated protein can be recovered from cell lysates directly with streptavidin-agarose beads and then detected by western blotting using a target-specific antibody [5]. Antigen–antibody complexes can be isolated using Pansorbin (*Staphylococcus aureus* cells), protein A agarose, or protein A/G Sepharose beads [5, 22]. If antibodies specific to the protein under study are not available, then epitope tags, such as Myc, or affinity tags, such as 6× His [5], can be incorporated at the C-terminus of the target protein (*see* **Note 5**) for either immunoprecipitation or isolation by Ni^{2+} chelated affinity resin packed into microcolumns or attached to agarose beads [5, 37]. Of course, protein function or topology should not be compromised by the presence of the tag (*see* **Note 5**). Following modification with AMS and MPB and isolation, the target protein is resolved by SDS-PAGE, blotted to a nitrocellulose membrane, and detected by western blotting. Biotinylation of exposed cysteine residues of whole cells (periplasmic exposure) confirmed in AMS-pretreated cells or only during sonication (both periplasmic and cytoplasmic exposure) is detected using a Fluor-S MaxTM MultiImager (Bio-Rad) or compatible imaging system and signals are quantified using available software.

1.5.6 Application of SCAM™ to Identification of Mixed and Dual Topologies

Topological protein heterogeneity generated cotranslationally for native proteins has been observed [38, 39]. Manipulation of protein domains has resulted often in the coexistence of multiple topological arrangements for the same protein within the same membrane with different ratios of properly oriented and inverted topological isoforms [40]. Membrane proteins can also display dual topologies dependent on membrane lipid composition, thereby providing molecular insight into how some proteins might exhibit multiple topological organizations within the same membrane or some alternative organization in different membranes [21]. How does one discriminate extracellular and intracellular EMDs of membrane proteins that adopt mixed or dual TM topologies within the same membrane?

A preblocking step allows selective labeling of luminal (exposed to cytoplasm) cysteines after cell permeabilization or disruption and therefore the detection of mixed topologies coexisting within the same cell membrane [21, 39]. The dual topology or degree of mixed topology can be assessed with a two-step protocol as shown on Fig. 3. Intact cells are treated either with or without AMS followed by labeling with MPB during disruption by sonication. AMS treatment will either prevent any biotinylation with whole cells expressing protein with periplasmic-facing monocysteine residue or only reduce the amount of biotinylation observed in disrupted cells. Labeling by MPB that can be blocked completely by pretreatment with AMS is indicative of a periplasmic-facing residue (Fig. 3). Labeling by MPB that cannot be blocked by AMS pretreatment is indicative of a residue that is facing the cytoplasm and uniform topology if biotinylated intensities of either periplasmic or cytoplasmic monocysteines in intact and preblocked cells are equal to the total intensity of biotinylation in unblocked cells labeled during desintegration. If a protein adopts a dual topology, only 50% of diagnostic cysteine at periplasmic EMD would be protected from labeling by MPB (Fig. 4, panel A). AMS pretreatment can also reduce the amount of biotinylation observed in disrupted

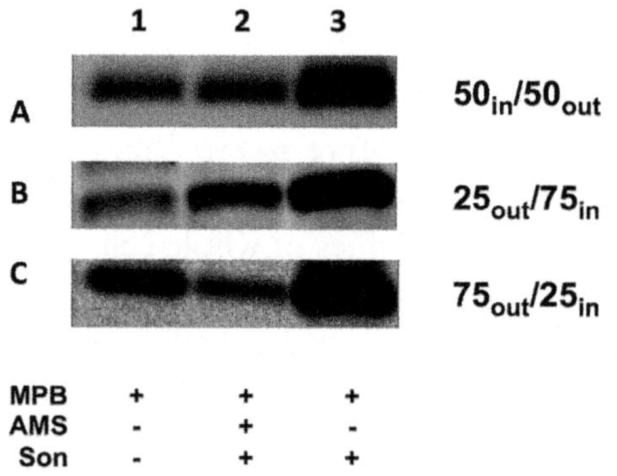

Fig. 4 Detection of dual topologies and mixed topologies by SCAM™. Topology of EMD containing a single-cysteine replacement exposed to the extracellular (periplasmic) side of the membrane was analyzed using a two-step labeling protocol. Intact cells were either labeled with MPB (1) or pretreated (2) or not with AMS (3) followed by treatment of sonicated cells with MPB (2 and 3). The varoius labeling of periplasmic monocysteine during sonication after pretreatment of intact cells with AMS indicates a dual (*panel A*) or mixed (*panels B* and *C*) topology. The difference in the extent of biotinylation before and after preblocking with AMS can be utilized to measure the percentages of oppositely oriented (In and Out) populations of membrane protein, as shown on *right*

cells. Reduced biotinylation of periplasmic cysteine by MPB in whole cells not pretreated with AMS (Fig. 4, first row of panels A–C), along with the biotinylation of monocysteine during sonication after pretreatment of intact cells with AMS (Fig. 4, second row of panels B and C), indicates that some population of the protein molecules is inserted in an inverted orientation. Thus, in the case of mixed topology, biotinylation will occur with whole cells not pretreated with AMS and to a greater or lesser extent after cell disruption of cells pretreated with AMS, indicating a different ratio of conversely oriented topological isoforms. The sum of biotinylated intensities in intact (first row of panels A–C) and preblocked cells (second row of panels A–C) should be equal to the total intensity of biotinylation in unblocked cells labeled during desintegration (third row of panels A–C). The protective effect of AMS is almost complete [19, 22] and has been succesfully used to quantify the degree of mixed topology [21, 22].

To confirm the versatility of this method, single-cysteine variants of polytopic membranes were directly labeled with MPB and their accessibility to AMS was scored in intact [19], permeabilized [19, 31, 39], disintegrated cells [10, 20–22, 30], oriented membrane vesicles [19], or proteoliposomes [23, 24].

2 Materials

2.1 Construction of Plasmids Expressing Single-Cysteine Derivatives

Quickchange site-directed mutagenesis kits (Stratagene) or equivalent.

2.2 Growth of *E. coli* Strains

1. Luria–Bertani (LB) medium.
2. Stock of appropriate antibiotic at desired concentration.
3. Stock of approproate inducer at desired concentration (either 500 mM isopropyl-ß-D-thiogalactoside (IPTG) or 20% arabinose or anhydrotetracycline (aTc) (1 mg/mL)).

2.3 SCAM™

1. Buffer A: 100 mM HEPES-KOH buffer, 250 mM sucrose, 25 mM $MgCl_2$, 0.1 mM KCl, adjusted to pH 7.5 or the same buffer adjusted to pH 9.0.
2. 10 mM MPB freshly dissolved in dimethyl sulfoxide (DMSO). Final concentration of DMSO used to dissolve MPB should never exceed 0.5% (*see* **Note 6**).
3. 100 mM aqueous stock solution of AMS.
4. 2 M β-ME.
5. Ultrasonic sonifier.
6. TLA-100 ultracentrifuge (Beckman Coulter, Indianapolis, IN) equipped with TLA-55 rotor or equivalent.

7. Microfuge polyallomer tubes (natural tint, capacity 1.5 mL).
8. Pierce™ spin columns with screw cap (Thermo Fisher Scientific, Waltham, MA) or equivalent.
9. Tabletop centrifuge.

2.4 Membrane Protein Solubilization

1. Solubilization buffer: 50 mM Tris–HCl, pH 8.1, 2% SDS, 1 mM ethylenediaminetetraacetic acid (EDTA).
2. Vortex equipped with microtube foam rack for multiple polyallomer tubes.

2.5 Immunoprecipitation (IP)

1. IP1 buffer: 50 mM Tris–HCl, pH 8.1, 0.15 M NaCl, 1 mM EDTA, 2% Lubrol-PX, 0.4% SDS (*see* **Note 7**).
2. IP2 buffer: 50 mM Tris–HCl, pH 8.1, 1 M NaCl, 1 mM EDTA, 2% Lubrol-PX, 2%, 0.4% SDS.
3. Protein A/G-agarose affinity resin.
4. Purification buffer: 20 mM Tris–HCl, pH 7.4, 300 mM NaCl, 25 mM imidazole, 10% (v/v) glycerol, supplemented with 2% Lubrol-PX or other appropriate nonionic detergent (*see* **Notes 5** and 7).

2.6 SDS-PAGE and Western Blotting

1. 2× SDS gel loading (sample) buffer: 10 mM Tris–HCl, pH 6.8, 5.6% (w/v) SDS, 200 mM dithiothreitol, 10% (w/v) glycerol, 0.01% bromophenol blue.
2. Precast gels for SDS-PAGE, 12.5% polyacrylamide.
3. 0.45 μm nitrocellulose transfer membranes.
4. Blocking buffer: 5% bovine serum albumin (BSA) in Tris-buffered saline (TBS: 10 mM Tris–HCl, pH 7.4, 0.9% NaCl).
5. Avidin linked to horseradish peroxidase (avidin-HRP) reconstituted to concentration of 2 mg/mL according to manufacturer.
6. Chemiluminescent substrates for detection of HRP.
7. Semidry blotting system.
8. Imaging system such as Fluor-S Max™ MultiImager (Bio-Rad Laboratories, Hercules, CA) equipped with CCD camera and Nikon 50 mm 1:1.4 AD (F 1.4) lens at ultrasensitive chemiluminescence setting to cool camera to −33 °C. Alternatively, autoradiography films can be used.

3 Methods

3.1 Labeling with Maleimide Derivatives

1. Harvest 100 mL mid-log phase cells expressing a single-cysteine derivative of protein of interest by centrifugation and suspend cell pellets in 3 mL buffer A adjusted to pH 7.5. Divide sample into three equal aliquots (0.75 mL) in Microfuge polyallomer tubes.

2. One cell aliquot is incubated for 30 min by rotating in dark at room temperature at 25 °C with AMS at a final concentration of 5 mM (37.5 μL 100 mM aqueous stock solution) to block periplasmic water-accessible cysteine residues from outside of cells. Remove excess unreacted AMS by two cycles of centrifugation and resuspension in 0.75 mL buffer A.
3. Treat one set of samples with MPB at final concentration of 100 μM (7.5 μL 10 mM stock solution) (*see* **Note 4**) for 5 min at room temperature to label cysteines exposed to extracellular (periplasmic) side of inner membrane. To increase reactivity of diagnostic cysteine residues (particularly those that might be in sterically hindered EMDs), the reaction with MPB can be carried out at pH 9 (*see* **Note 1**). Quench reaction by addition of β-ME to 20 mM (7.5 μL 2 M stock solution). After labeling, cells are sonicated for 1 min using amplitude of 15%.
4. To simultaneously label cysteines exposed to both sides of cell membrane, subject third sample to sonication for 1 min in presence of MPB at final concentration of 100 μM on ice. Incubate for 4 min at room temperature and quench reaction by addition of β-ME to 20 mM (7.5 μL 2 M stock solution).
5. To label previously inaccessible unblocked cytoplasmic cysteines, remaining sample pretreated with AMS from **step 2** is biotinylated by adding MPB at final concentration of 100 μM (7.5 μL 10 mM stock solution) during sonication for 1 min followed by incubation for another 4 min at room temperature before quenching reaction by addition of β-ME to 20 mM (7.5 μL 2 M stock solution).
6. All sonicated samples are centrifuged at 4 °C at 65,000 × *g* for 10 min followed by resuspension of isolated membranes in 100 μL Buffer A containing 20 mM β-ME by vigorous vortexing for 2 h at room temperature.

3.2 Sample Solubilization

Solubilize isolated membranes with the appropriate detergent or detergent mixture, such as SDS alone, Triton X-100 alone, SDS and Triton-X-100, CHAPS, octylglucoside, deoxycholate, cholate and Tween 20, β-D-dodecylmaltoside, nonidet P-40, or sodium deoxycholate [5] (*see* **Note 7**). Resuspended pellets derived from sonicated cells are solubilized by addition of an equal volume (100 μL) of solubilization buffer followed by vigorous vortexing for 15 min at room temperature, incubation at 37 °C for 15 min, and an additional 15 min vortexing at room temperature. If solubilization buffer with SDS is used, dilute the sample with 0.3 mL cold IP1 buffer containing nonionic detergent to neutralize the denaturing properties of SDS, vortex resulting 0.5 mL sample for 1 min, centrifuge at 4 °C at 20,800 × *g* for 10 min, and clear by centrifugation in a prechilled (4 °C) tabletop centrifuge at 20,800 × *g* for 10 min.

3.3 Isolation of Derivatized Target Proteins.

1. Transfer supernatants to cups of spin columns.
2. Add appropriate target-protein-specific polyclonal or monoclonal antibodies (*see* **Note 5**).
3. Incubate overnight at 4 °C with rocking.
4. Add 30 μL of a suspension of protein A/G-agarose affinity resin.
5. Incubate at 4 °C on a rocking platform for 90 min.
6. Wash agarose resin with 0.5 mL IP1 by vortexing spin column for 1 min and subsequent centrifugation in prechilled (4 °C) tabletop centrifuge at 20,800 × *g* for 1 min. Discard flow-through.
7. Repeat washing step with 0.5 mL IP2.
8. Repeat washing step with 0.5 mL 10 mM Tris–HCl, pH 8.1.
9. Add SDS sample buffer.
10. Vortex vigorously for 15 min at room temperature.
11. Incubate at 37 °C for 15 min.
12. Vortex vigorously for 15 min at room temperature.
13. Centrifuge spin column on a tabletop centrifuge at 20,800 × *g* for 1 min to collect flow-through with solubilized proteins into microfuge polyallomer tube.

3.4 SDS-PAGE, Western Blot Analysis, and Staining with Avidin-HRP

1. Subject immunoprecipitated samples to SDS-PAGE.
2. Transfer proteins to nitrocellulose membranes by electroblotting as described in detail in previously published SCAM ™ protocol [9].
3. Incubate nitrocellulose membrane overnight with blocking buffer.
4. Wash membrane with TBS buffer containing 0.3% BSA for 10 min.
5. Add Avidin–HRP at a final dilution of 1:5000–10,000 from a 2 mg/mL stock solution, in TBS buffer containing 0.3% BSA.
6. Incubate for at least 1 h.
7. Wash membrane twice with TBS buffer containing 0.3% BSA for 15 min each.
8. Wash membrane twice with TBS/Nonidet P40 buffer.
9. Wash membrane once with TBS buffer.
10. Incubate membrane for 3 min with chemiluminescent substrates.
11. Visualize biotinylated proteins using imaging system.

3.5 Data Analysis and Interpretation

The criteria used for determining the location of an introduced cysteine are as follows. Labeling of a cysteine residue with a membrane-impermeable sulfhydryl reagent before disruption of whole cells is indicative of an extracellular (periplasmic) cysteine residue provided accessibility to a cytoplasmically localized control protein (*see* **Note 4**) and a cysteineless derivative of the target protein are negative (*see* **Note 1**). An absence of labeling in whole cells but labeling during cell disruption indicates a cytoplasmic location for a cysteine-containing EMD. The only valid comparison in intensity is between whole-cell and sonicated sets (images treated identically) run on the same gel.

No labeling with a sulfhydryl reagent before or during cell disruption implies localization to a hydrophobic membrane environment or unfavorable local orientation/positioning of introduced thiol groups, which may prevent access by the reagent or result in an increase of the pKa of the thiol group as discussed subsequently (*see* **Note 1**). The percentage change in alkylation between samples biotinylated with MPB only and those labeled in an AMS-protected manner during disintegration should be used to detect and quantitate an amount of oppositely oriented membrane proteins adopting dual or mixed topologies (Figs. 3 and 4).

4 Notes

1. Caution must be used in assigning an intramembrane location to a cysteine residue because it is unreactive to hydrophilic thiol reagent in both intact and disrupted cells. No definitive conclusion can be drawn for the location of a domain based on a lack of reaction of the cysteine. Lack of or low levels of labeling may result from any of the following reasons: (1) steric hindrance due to local secondary structure, (2) internalization into the compact fold of the protein, (3) lack of ionization of the thiol group owing to a hydrophobic environment, (4) local environment with the same charge as the thiol reagent, or (5) increased pKa of the thiol due to the high negative charge density of neighboring residues or anionic lipids [5, 9]. Periplasmic EMDs tend to be shorter (sometimes only three amino acids in length) than cytoplasmic EMDs. Therefore, there may be little or no protrusion of these loops into the extracellular space, thereby preventing reaction of the cysteine residues in these locations with relatively bulky reagents [30]. Alkylating reagents appear to react better with cysteines toward the middle of extended hydrophilic loops than near the TMD interfacial domain. Cysteine scanning across a domain or determining the accessibility of neighbors is an effective means for identifying useful replacement sites and differentiating between local effects and unreactive TMDs. Scanning can be coupled with alkaline treatment. Since the formation of cysteinyl thiolate anions is favored by increasing the solution pH (optimum pH 8.0–9.0), increasing the pH

during labeling will favor the reaction [5, 9, 10, 30] (Fig. 2a). However, maleimides are known to react also with primary amines at pH values above 7.5. An effective control to rule out nonthiol residue modification is to use a cysteineless template of the target protein. Unfavorable orientation of a thiol group owing to a local secondary structure may restrict or prevent access by large thiol reagents. Increasing the reaction buffer pH would not only favor alkylation of an extramembrane cysteine but also disrupt the local restrictive secondary structure while truly membrane-embedded cysteines would not be expected to react. EMDs that are sterically hindered or exhibit an elevated pKa can be derivatized by increasing the pH up to 10.5 without compromising membrane integrity [10, 30]. However, appropriate controls, such as making known cytoplasmically exposed domains inaccessible and removing any cysteineless target protein label, should be used.

2. Conclusions based on the full reactivity of diagnostic residues should also be drawn with caution since cysteine residues facing a hydrophilic pore or near a substrate-binding site may be within a TM segment but chemically reactive because of water channels or pockets.

3. The araB system suffers from cell-to-cell heterogeneity when expression levels vary greatly within the cell population, thereby making physiological interpretations problematic because of the induction of promoter yielded mixed populations of uninduced and fully induced cells. Contrary P*LtetO-1* is tightly repressible by the Tet repressor in the absence of inducer and can be induced with the nontoxic inducer aTc, which has an increased affinity for repressor TetR and thus can provide tight and homogeneous cell-to-cell expression. Overexpression of membrane proteins under the control of araB or T7 promoter is not recommended or should be carried out with cation. If the araB system is still utilized a short induction time (1 h) at a low concetration (0.05%) of arabinose is recommended. The pET expression systems is widely used because of its ability to produce large quantities of a desired protein when activated. However, the use of a pET vector in SCAM™ is not recommended for the expression of target membrane proteins because the overloading of cells with a highly overexpressed target protein can "jam" a translocon and trigger the accumulation of membrane protein into newly made intracytoplasmic membranes (Lu, Zheng, and Bogdanov, unpublished).

4. SCAM™ is based on the controlled membrane permeability of sulfhydryl reagents. Membranes, either in their native state or due to experimental manipulation, can be slightly permeable to labeling reagents. Therefore, optimal labeling conditions must be established for each reagent and host. The results of

SCAM™ analysis are valid only if the modifying reagent is thiol-specific, the membrane is impermeable, cells are intact, and cell disruption does not expose sterically hindered or water inaccessible cysteine residues [5]. Various reagents, including MPB, will cross membranes in a concentration-, time-, and temperature-dependent manner, and permeability varies with the genetic background of the host cells [5, 30, 31]. Therefore, conditions must be empirically determined to minimize the derivatization of intracellular cysteines. The membrane permeability of a thiol-specific labeling reagent can be tested, and labeling conditions (concentration, time, and temperature) can be established by quantification of the degree of labeling of an abundant cytoplasmic protein that is rich in surface-exposed cysteine residues. *E. coli* ß-galactosidase and other cytosolic bacterial markers, such as glutathione or elongation factor Tu, have been used to access membrane permeability [5]. In such cases, a labeling experiment with both intact and permeabilized cells is carried out, except that soluble proteins rather than the membrane fraction are retained by immunoprecipitation and analysis. Whole and disrupted cells are treated with various concentrations of reagent, from 10 μM to 1 mM, at temperatures ranging from 0 to 25 °C, and for various lengths of time, from 5 min to 1 h. In most cases, a low concentration of MPB (100 μM) and relatively short incubation period (5 min) at room temperature favor biotinylation of extramembrane thiol groups. Significant differences in the permeability of different host strains emphasize the need to screen host strains for reagent permeability prior to initiating experiments [5, 31]. Thiol reagents are available that contain a biotin group, a fluorescent group, or a radiolabel allowing the detection of labeled proteins by avidin-HRP and indirect chemilumiscence detection, fluorescence, or autoradiography [5, 37]. A methylpolyethylene glycol-maleimide 5000 (Mal-PEG) adds 5 kDa to the target protein and therefore can be used successfully as a mobility shift reagent [41, 42]. In this type of assay, intact cells are pretreated first with impermeant sulfhydryl reagents AMS or sodium (2-sulfonatoethyl) methanethiosulfonate (MTSES) that can modify only extracellular thiol groups. The thiol groups of cytoplasmic or membrane-embedded cysteine residues are not accessible to AMS or MTSES, and proteins can be alkylated with Mal-PEG after full denaturation by SDS/Urea/EDTA, which renders all unprotected cysteine residues accessible to modification. The application of fluorescent maleimides (UV-excitable Oregon Green 488 maleimide carboxylic acid (OGM) in SCAM™ eliminates the western blotting procedure [37]. Infrared fluorescent dye IRDye800-maleimide (LI-COR) delivers enhanced sensitivity, dissolves freely in water, and modifies free thiols efficiently at physiological pH and therefore can be utilized in SCAM™ for visualization in target proteins by a LI-COR system [43]. Detectable thiol-reactive labeling reagents with high membrane permeabilities can also be used to label a test protein. Reagents that can cross membranes, such as N-ethylmaleimide (used as ^{14}C–labeled form), modify all extramembrane cysteines irrespective of their sidedness. These detectable maleimides added to samples after preincubation with AMS or MTSES cannot modify prelabeled surface-exposed cysteine residues, resulting in a different labeling pattern. However, detectable maleimides with limited permeabilities, biotin-linked (MPB) or UV-excitable (OGM) maleimides, are the most commonly used thiol-specific reagents owing to their low membrane permeability and simple detection [5, 9, 37].

5. The introduction of 6× or 10× His tags should be used with caution because such tags may affect the topology of small integral membrane proteins with weak topological determinants because their intrinsic flexibility makes them prone to different rearrangements [11, 41]. SCAM™ should map the topology of such proteins in their untagged form [41]. Negatively charged residues become potent translocation signals when they are present in high numbers, flank a marginally hydrophobic TMD, or lie within a window of six residues from the end of a highly hydrophobic TMD [3, 4]. Therefore, Myc or FLAG tags may affect the topology of small-membrane proteins owing to the presence of four and five negatively charged residues in their sequences (EQKLISEEDL and DYKDDDDK, respectively). The topology of these proteins can be affected by the insertion of tags or by their position (Gordon and Bogdanov, unpublished).
6. Maleimide-based reagents are often sensitive to hydrolysis, and so reagents that have not been stored or handled properly may no longer be reactive. Therefore, reagents should be prepared immediately prior to use and should be protected from light where possible.
7. Prechilled (4 °C) 50 mM Tris–HCl (pH 8.1) should be used to prepare IP buffer with an SDS:nonionic detergent ratio of 1:5, which is instrumental for immunoprecipitation of highly hydrophobic integral membrane proteins. Lubrol-PX can be substituted by Thesit^R (Honeywell Fluka™ USA) or lauryldimethylamine-oxide (LDAO) at exactly the same concentration. Triton X-100 is not recommended for immunoprecipiation or affinity purification of very hydrophobic multispanning membrane proteins due to their severe aggregation after solubilization by this detergent alone or even in mixture with SDS (Bogdanov, unpublished observation). LDAO can be advantageous in the purification of small integral membrane proteins that span membranes two or four times (Bogdanov, unpublished observation).

References

1. von Heijne G (2006) Membrane-protein topology. Nat Rev Mol Cell Biol 7:909–918
2. Bogdanov M, Xie J, Dowhan W (2009) Lipid-protein interactions drive membrane protein topogenesis in accordance with the positive inside rule. J Biol Chem 284:9637–9641
3. Dowhan W, Bogdanov M (2009) Lipid-dependent membrane protein topogenesis. Annu Rev Biochem 78:515–540
4. Bogdanov M, Dowhan W, Vitrac H (2014) Lipids and topological rules governing membrane protein assembly. Biochim Biophys Acta 1843:1475–1488
5. Bogdanov M, Zhang W, Xie J, Dowhan W (2005) Transmembrane protein topology mapping by the substituted cysteine accessibility method (SCAM™): application to lipid-specific membrane protein topogenesis. Methods 36:148–171
6. Fleishman SJ, Unger VM, Ben-Tal N (2006) Transmembrane protein structures without X-rays. Trends Biochem Sci 31:106–113
7. Lacapere JJ, Pebay-Peyroula E, Neumann JM, Etchebest C (2007) Determining membrane protein structures: still a challenge! Trends Biochem Sci 32:259–270
8. Bochud A, Ramachandra N, Conzelmann A (2013) Adaptation of low-resolution methods for the study of yeast microsomal polytopic membrane proteins: a methodological review. Biochem Soc Trans 41:35–42
9. Bogdanov M, Heacock PN, Dowhan W (2010) Study of polytopic membrane protein topological organization as a function of membrane lipid composition. Methods Mol Biol 619:79–101
10. Bogdanov M, Xie J, Heacock P, Dowhan W (2008) To flip or not to flip: lipid-protein charge interactions are a determinant of final membrane protein topology. J Cell Biol 182:925–935
11. Nasie I, Steiner-Mordoch S, Gold A, Schuldiner S (2010) Topologically random insertion of EmrE supports a pathway for evolution of inverted repeats in ion-coupled transporters. J Biol Chem 285:15234–15244
12. Zhu Q, Casey JR (2007) Topology of transmembrane proteins by scanning cysteine accessibility mutagenesis methodology. Methods 41:439–450
13. Islam ST, Lam JS (2013) Topological mapping methods for alpha-helical bacterial membrane proteins--an update and a guide. Microbiology 2:350–364
14. Lee H, Kim H (2014) Membrane topology of transmembrane proteins: determinants and experimental tools. Biochem Biophys Res Commun 453:268–276
15. Liapakis G (2014) Obtaining structural and functional information for GPCRs using the substituted-cysteine accessibility method (SCAM). Curr Pharm Biotechnol 15:980–986
16. van Geest M, Lolkema JS (2000) Membrane topology and insertion of membrane proteins: search for topogenic signals. Microbiol Mol Biol Rev 64:13–33
17. van Geest M, Lolkema JS (1999) Transmembrane segment (TMS) VIII of the Na(+)/citrate transporter CitS requires downstream TMS IX for insertion in the *Escherichia coli* membrane. J Biol Chem 274:29705–29711
18. Karlin A, Akabas MH (1998) Substituted-cysteine accessibility method. Methods Enzymol 293:123–145
19. Bogdanov M, Heacock PN, Dowhan W (2002) A polytopic membrane protein displays a reversible topology dependent on membrane lipid composition. EMBO J 21:2107–2116
20. Bogdanov M, Heacock P, Guan Z, Dowhan W (2010) Plasticity of lipid-protein interactions in the function and topogenesis of the membrane protein lactose permease from *Escherichia coli*. Proc Natl Acad Sci U S A 107:15057–15062
21. Bogdanov M, Dowhan W (2012) Lipid-dependent generation of a dual topology for a membrane protein. J Biol Chem 287:37939–37948
22. Vitrac H, Bogdanov M, Heacock P, Dowhan W (2011) Lipids and topological rules of membrane protein assembly: balance between long- and short-range lipid-protein interactions. J Biol Chem 286:15182–15194
23. Vitrac H, Bogdanov M, Dowhan W (2013) In vitro reconstitution of lipid-dependent dual topology and postassembly topological switching of a membrane protein. Proc Natl Acad Sci U S A 110:9338–9343
24. Vitrac H, Bogdanov M, Dowhan W (2013) Proper fatty acid composition rather than an ionizable lipid amine is required for full transport function of lactose permease from *Escherichia coli*. J Biol Chem 288:5873–5885
25. Tang XB, Casey JR (1999) Trapping of inhibitor-induced conformational changes in the erythrocyte membrane anion exchanger AE1. Biochemistry 38:14565–14572
26. Hu YK, Kaplan JH (2000) Site-directed chemical labeling of extracellular loops in a membrane protein. The topology of the Na,K-ATPase alpha-subunit. J Biol Chem 275:19185–19191

27. Nagamori S, Nishiyama K, Tokuda H (2002) Membrane topology inversion of SecG detected by labeling with a membrane-impermeable sulfhydryl reagent that causes a close association of SecG with SecA. J Biochem 132:629–634
28. Dale H, Angevine CM, Krebs MP (2000) Ordered membrane insertion of an archaeal opsin in vivo. Proc Natl Acad Sci U S A 97:7847–7852
29. Kerr JE, Christie PJ (2010) Evidence for VirB4-mediated dislocation of membrane-integrated VirB2 pilin during biogenesis of the agrobacterium VirB/VirD4 type IV secretion system. J Bacteriol 192:4923–4934
30. Xie J, Bogdanov M, Heacock P, Dowhan W (2006) Phosphatidylethanolamine and monoglucosyldiacylglycerol are interchangeable in supporting topogenesis and function of the polytopic membrane protein lactose permease. J Biol Chem 281:19172–19178
31. Zhang W, Bogdanov M, Pi J, Pittard AJ, Dowhan W (2003) Reversible topological organization within a polytopic membrane protein is governed by a change in membrane phospholipid composition. J Biol Chem 278:50128–50135
32. Wang X, Bogdanov M, Dowhan W (2002) Topology of polytopic membrane protein subdomains is dictated by membrane phospholipid composition. EMBO J 21:5673–5681
33. Bernsel A, Viklund H, Falk J, Lindahl E, von Heijne G, Elofsson A (2008) Prediction of membrane-protein topology from first principles. Proc Natl Acad Sci U S A 105: 7177–7181
34. Zhao G, London E (2006) An amino acid "transmembrane tendency" scale that approaches the theoretical limit to accuracy for prediction of transmembrane helices: relationship to biological hydrophobicity. Protein Sci 15:1987–2001
35. Dobson L, Remenyi I, Tusnady GE (2015) CCTOP: a consensus constrained TOPology prediction web server. Nucleic Acids Res 43:W408–W412
36. Bayer EA, Zalis MG, Wilchek M (1985) 3-(N-Maleimido-propionyl)biocytin: a versatile thiol-specific biotinylating reagent. Anal Biochem 149:529–536
37. Berezuk AM, Goodyear M, Khursigara CM (2014) Site-directed fluorescence labeling reveals a revised N-terminal membrane topology and functional periplasmic residues in the *Escherichia coli* cell division protein FtsK. J Biol Chem 289:23287–23301
38. Moss K, Helm A, Lu Y, Bragin A, Skach WR (1998) Coupled translocation events generate topological heterogeneity at the endoplasmic reticulum membrane. Mol Biol Cell 9:2681–2697
39. Woodall NB, Yin Y, Bowie JU (2015) Dual-topology insertion of a dual-topology membrane protein. Nat Commun 6:8099
40. Gafvelin G, von Heijne G (1994) Topological "frustration" in multispanning *E. coli* inner membrane proteins. Cell 77:401–412
41. Nasie I, Steiner-Mordoch S, Schuldiner S (2013) Topology determination of untagged membrane proteins. Methods Mol Biol 1033:121–130
42. Gelis-Jeanvoine S, Lory S, Oberto J, Buddelmeijer N (2015) Residues located on membrane-embedded flexible loops are essential for the second step of the apolipoprotein N-acyltransferase reaction. Mol Microbiol 95:692–705
43. Liu Y, Basu A, Li X, Fliegel L (2015) Topological analysis of the Na+/H+ exchanger. Biochim Biophys Acta 1848:2385–2393
44. Abramson J, Smirnova I, Kasho V, Verner G, Kaback HR, Iwata S (2003) Structure and mechanism of the lactose permease of *Escherichia coli*. Science 301:610–615

Chapter 10

Defining Membrane Protein Topology Using *pho-lac* Reporter Fusions

Gouzel Karimova and Daniel Ladant

Abstract

Experimental determination of membrane protein topology can be achieved using various techniques. Here we present the *pho-lac* dual reporter system, a simple, convenient, and reliable tool to analyze the topology of membrane proteins in vivo. The system is based on the use of two topological markers with complementary properties, the *Escherichia coli* β-galactosidase LacZ, which is active in the cytoplasm, and the *E. coli* alkaline phosphatase PhoA, which is active in the bacterial periplasm. Specifically, in this *pho-lac* gene system, the reporter molecule is a chimera composed of the mature PhoA that is in frame with the β-galactosidase α-peptide, LacZα. Hence, when targeted to the periplasm, the PhoA-LacZα dual reporter displays high alkaline phosphatase activity but no β-galactosidase activity. Conversely, when located in the cytoplasm, PhoA-LacZα has no phosphatase activity but exhibits high β-galactosidase activity in *E. coli* cells expressing the ω fragment of LacZ, LacZω (via the α-complementation phenomenon). The dual nature of the PhoA-LacZα reporter allows a simple way to normalize both enzymatic activities to obtain readily interpretable information about the subcellular location of the fusion site between the membrane protein under study and the reporter. In addition, the PhoA-LacZα reporter permits utilization of dual-indicator agar plates to easily discriminate between colonies bearing cytoplasmic fusions, periplasmic fusions, or out-of-frame fusions. In total, the *phoA-lacZα* fusion reporter approach is a straightforward and rather inexpensive method of characterizing the topology of membrane proteins in vivo.

Key words Membrane proteins, Membrane topology, Dual reporter system, Phosphatase, β-galactosidase

1 Introduction

Membrane proteins are key players in the vast majority of cellular processes [1–3]. Understanding how these proteins perform their functions often starts by determining their topology, i.e., the number of transmembrane segments (TMSs), their location, and their orientation relative to the membrane [1]. In this chapter, only proteins employing α-helices to span the membrane are discussed.

Many accurate computational methods are now available for predicting membrane protein topology, the best ones achieving a prediction accuracy exceeding 80% [4–7] (*see* also Chapter 1).

Laure Journet and Eric Cascales (eds.), *Bacterial Protein Secretion Systems: Methods and Protocols*, Methods in Molecular Biology, vol. 1615, DOI 10.1007/978-1-4939-7033-9_10, © Springer Science+Business Media LLC 2017

Yet predicted topologies need to be experimentally validated. Here we present the *pho-lac* dual reporter system, a simple, convenient, and reliable tool to analyze the topology of membrane proteins in vivo.

Manoil and Beckwith were the first to describe a reporter gene fusion approach to studying the topological organization of integral membrane proteins [8]. In a practical way, the membrane protein under study is genetically fused in frame to the N-terminus of a reporter molecule that displays a characteristic phenotype depending on its subcellular (i.e., cytosolic or extracellular) location. Here we will focus on two *Escherichia coli* proteins, alkaline phosphatase (PhoA) and β-galactosidase (LacZ), that are widely used for reporter fusion technology as a pair of topological markers with complementary properties. PhoA is a dimeric, Zn^{2+}-dependent enzyme, which is active only when located in the periplasm (reviewed in [9, 10]). In contrast, LacZ is a large homotetrameric enzyme that is active in the bacterial cytoplasm [10–12].

Initially, the PhoA and LacZ reporters were used separately to determine the topological organization of many proteins expressed in *E. coli* cells (reviewed in [10]), but several important drawbacks were reported [13–18]. A main problem is that in such separate fusion approaches, the topological location of a given residue of the target protein is determined by comparing two enzymatic activities, the alkaline phosphatase activity of the PhoA fusion and the β-galactosidase activity of the LacZ fusion. To enable a straightforward comparison, the reported enzymatic activities should be normalized to the rate of fusion protein synthesis [13, 15, 16]. This requires time-consuming experiments, for example, pulse labeling, immunodetection, and quantification of incorporated radioactivity [13, 15].

To overcome this limitation, Alexeyev and Winkler designed a dual reporter system in which the PhoA and LacZ reporters are combined [19]. In their *pho-lac* gene system, the reporter molecule is a chimeric protein composed of the α-peptide of LacZ (aa 4–60), which is fused in frame to the C-terminus of the mature PhoA (aa 22–472). This PhoA-LacZα chimeric protein displays high phosphatase activity when targeted to the periplasm and has no β-galactosidase activity. In contrast, when the PhoA-LacZα reporter is directed to the cytoplasm, it has no phosphatase activity but displays high β-galactosidase activity because the α-fragment of LacZ is able to interact with a truncated, inactive variant of LacZ (the so-called ω fragment, such as *LacZ*ΔM15) to restore its enzymatic activity (due to the phenomenon of α-complementation) [20, 21]. Importantly, the authors of the system suggested a simple way to normalize PhoA and LacZ enzymatic activities without determining protein synthesis rates [19]. Their assumption was that the expression level of a particular membrane protein/PhoA-LacZα fusion would affect the activity levels of both phosphatase and β-galactosidase enzymes but should not change their relative ratio. Therefore, when the PhoA-LacZα reporter is fused to various residues of a target protein to generate a set of fusion molecules, the phosphatase and β-galactosidase enzymatic activities of each hybrid protein can be normalized (usually to the highest value observed within the set of fusions) to obtain a so-called normalized activity ratio (NAR). Thus, this ratio provides readily interpretable information about the subcellular location of the particular fusion point (i.e., the residue of the membrane protein after which the PhoA-LacZα is fused) [19].

In this system, both enzymatic activities of the PhoA-LacZα reporter can be measured quantitatively using standard colorimetric substrates, ortho-nitrophenyl-β-D-galactoside (ONPG) for β-galactosidase and p-nitrophenyl phosphate (pNPP) for phosphatase [22, 23]. Remarkably, the PhoA-LacZα reporter activities can be directly visualized on indicator plates containing compatible PhoA- and LacZ-specific chromogenic substrates. On such agar plates, the cells expressing high levels of phosphatase activity are able to convert the X-Pho (5-bromo-4-chloro-3-indolyl-phosphate) substrate into a blue-colored, precipitated compound, while the cells expressing high levels of β-galactosidase activity can be revealed with the Red-Gal (6-chloro-3-indolyl-β-D-galactoside) substrate that is enzymatically converted into an insoluble red chromophore.

In the laboratory, to apply the dual PhoA-LacZα reporter system, we constructed a low copy number vector, pKTop (4279 bp, ori p15A), that expresses the PhoA-LacZα reporter under the transcriptional control of a *lac* promoter and carries a kanamycin resistance selectable marker [24]. Briefly, the *phoA-lacZα* cassette designed as in [19] was constructed by a polymerase chain reaction (PCR) overlapping technique and inserted into the vector pKNT25 (by replacing the T25 ORF). A multicloning site located upstream of *phoA-lacZα* makes it possible to create in-frame fusions at the N-terminus of the PhoA-LacZα dual reporter (Fig. 1) (for more details, *see* ref. [24]).

The *phoA-lacZα* reporter system has been successfully applied to the analysis of the topology of polytopic membrane proteins involved in a wide range of cellular functions [19, 25–30]. In our hands, the *phoA-lacZα* dual system has been used to experimentally validate the topological organization of various *E. coli* cell division proteins [24, 28], the *Staphylococcus aureus* multicomponent GraXSR-VraFG signal transduction system [29], and the Pil proteins involved in the biogenesis of type IV pilus from *Neisseria meningitidis* [30].

To illustrate the methodology of utilization of the *phoA-lacZα* dual reporter system, we describe here the use of the pKTop vector, which expresses the *phoA-lacZα* dual reporter, to characterize the topology of the *E. coli* YmgF protein, a 72-residue-long membrane polypeptide that associates to the cell division machinery [24].

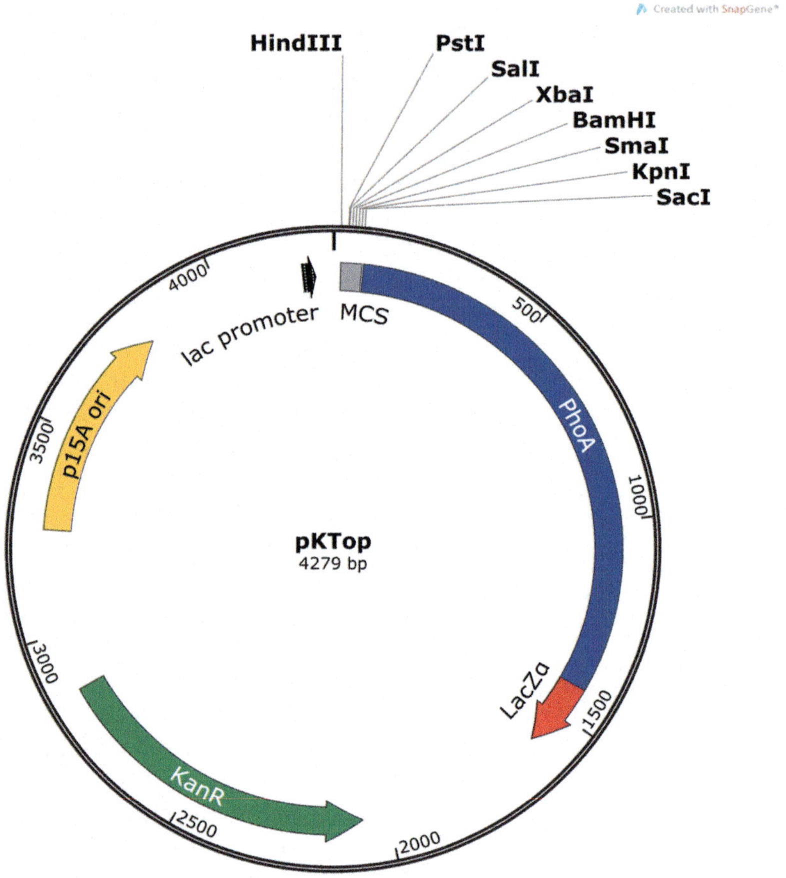

Fig. 1 Schematic representation of pKTop vector used for studying membrane protein topology with PhoA-LacZα dual reporter method. The figure shows the position of the *phoA-lacZα* reporter gene, a *lac* promoter, kanamycin resistance gene, origin of replication (p15A), and multicloning site (MCS) that makes it possible to create fusions at the N-terminus of PhoA-LacZα dual reporter. The plasmid can also be used to create an Exo III-generated library of 3′-truncated variants of target protein gene fused to *phoA-lacZα*, as described in Subheading 3.3. The map is created with SnapGene software (from GSL Biotech)

2 Materials

2.1 Bacterial Growth Media, Strain and Plasmid Construction

1. Luria–Bertani (LB) broth.
2. LB agar plates (*see* **Note 1**).
3. Kanamycin: 50 mg/mL in water. Store at −20 °C.
4. Glucose: 20% w/v in water. Store at room temperature (RT).
5. Isopropyl β-D-1-thiogalactopyranoside (IPTG): 100 mM in water. Store at −20 °C.

6. 6-chloro-3-indolyl-β-D-galactoside (Red-Gal, synonyms: Rose-Gal, Salmon-Gal): 25%w/v in DMSO (dimethyl sulfoxide) or DMF (dimethylformamide). Store in the dark at −20 °C (*see* **Note 2**).
7. 5-bromo-4-chloro-3-indolyl-phosphate (X-Pho): 100 μg/mL in water. Store in dark at −20 °C.
8. *E. coli* K12 strains capable of α-complementation of LacZ: XL1-blue, TG1, DH5α, DH10B, etc. (*see* **Notes 3** and **4**).
9. A plasmid carrying the *phoA-lacZα* dual reporter cassette, such as pKTop [24].
10. Common molecular biology enzymes: restriction enzymes, DNA modification enzymes, ligases, PCR polymerases, etc.
11. Molecular biology kits for purification of plasmid DNA, PCR, and DNA fragments, etc.
12. Equipment for growth of bacterial culture: thermostat, incubation shaker.
13. PCR thermal cycler.

2.2 β-Galactosidase Assay

1. M63 medium: 100 mM KH_2PO_4, 15 mM $(NH4)_2SO_4$, 1.7 mM Fe_2SO_4, 1 mM $MgSO_4$, pH 7.0 (*see* **Note 5**).
2. Chloroform.
3. 20% sodium dodecyl sulfate (SDS).
4. β-galactosidase assay buffer (PM2): 70 mM Na_2HPO_4, 30 mM $NaHPO_4$, 1 mM $MgSO_4$, 0.2 mM $MnSO_4$, pH 7.0 (*see* **Note 6**).
5. ONPG: 0.4% w/v in PM2 buffer without 2-mercaptoethanol. Store at −20 °C (*see* **Note 7**).
6. 1 M Na_2CO_3.
7. Microplate reader (or microphotometer).
8. Microtiter plates.

2.3 Phosphatase Assay

1. Wash buffer for phosphatase assay (WB): 10 mM Tris–HCl, pH 8.0, 10 mM $MgSO_4$.
2. Iodoacetamide: 500 mM in water. Prepare a fresh solution and store in dark until ready to use at 2–8 °C (*see* **Note 8**).
3. Phosphatase assay buffer (PM1): 1 M Tris–HCl, pH 8.0, 0.1 mM $ZnCl_2$, 1 mM iodoacetamide.
4. p-nitrophenyl phosphate (pNPP) (*see* **Note 9**).
5. 1 M Tris–HCl, pH 8.0.
6. 2N NaOH.
7. Microfuge.
8. Microplate reader (or microphotometer).
9. Microtiter plates.

3 Methods

Readers are expected to be familiar with experimental tools that are generally used in classical molecular biology, for example, PCR amplification, DNA digestion, ligation, transformation, and plasmid DNA purification [23, 31].

3.1 Selection of pho-lac Fusion Sites Within Target Membrane Protein

1. Predict in silico the membrane topology of the protein of interest. For consensus prediction of membrane protein topological organization, it is important to explore methods based on various prediction algorithms [32] (*see* **Note 10** and Chapter 2).

   For example, as shown in Fig. 2a, all four predictors present YmgF as a protein with two TMSs separated by a short periplasmic loop and with both the N- and C-terminal extremities in the cytosol.

2. Choose the fusion points for the PhoA-LacZα reporter. At least one reporter fusion is required for each extramembranous domain of the protein under study. The PhoA-LacZα reporter should be preferentially fused to the C-terminal ends of the extramembranous loops between the potential TMSs [10, 33, 34].

   For experimental validation of the predicted YmgF topology, we selected seven distinct codons (D8, M21, K32, E39, N48, L57, and Q72) as sites for the PhoA-LacZα reporter insertion based on the predicted topology models (*see* Fig. 2a).

3. Construct the recombinant plasmids expressing the target membrane protein/*phoA-lacZα* fusions following the three commonly used methods described in Subheadings 3.2, 3.3, and 3.4 (*see* **Note 11**).

3.2 C-Terminal Fusion Approach

1. Amplify by PCR the DNA fragments carrying the predetermined 3′-truncated target protein gene with specific PCR primers harboring appropriate restriction sites for subcloning in pKTop plasmid. These PCR primers should be designed in such a way that the amplified DNA fragment is fused in frame to the downstream *phoA-lacZα* reporter gene (i.e., to generate the expected C-terminal translation fusion).
2. Digest the PCR-amplified DNA fragments with appropriate restriction enzymes.
3. Ligate the digested PCR-amplified DNA fragments into the *phoA-lacZα* reporter vector linearized with the same restriction enzymes.
4. Transform the ligation mixture into *E. coli* competent cells (*see* **Notes 3, 4,** and **12**).

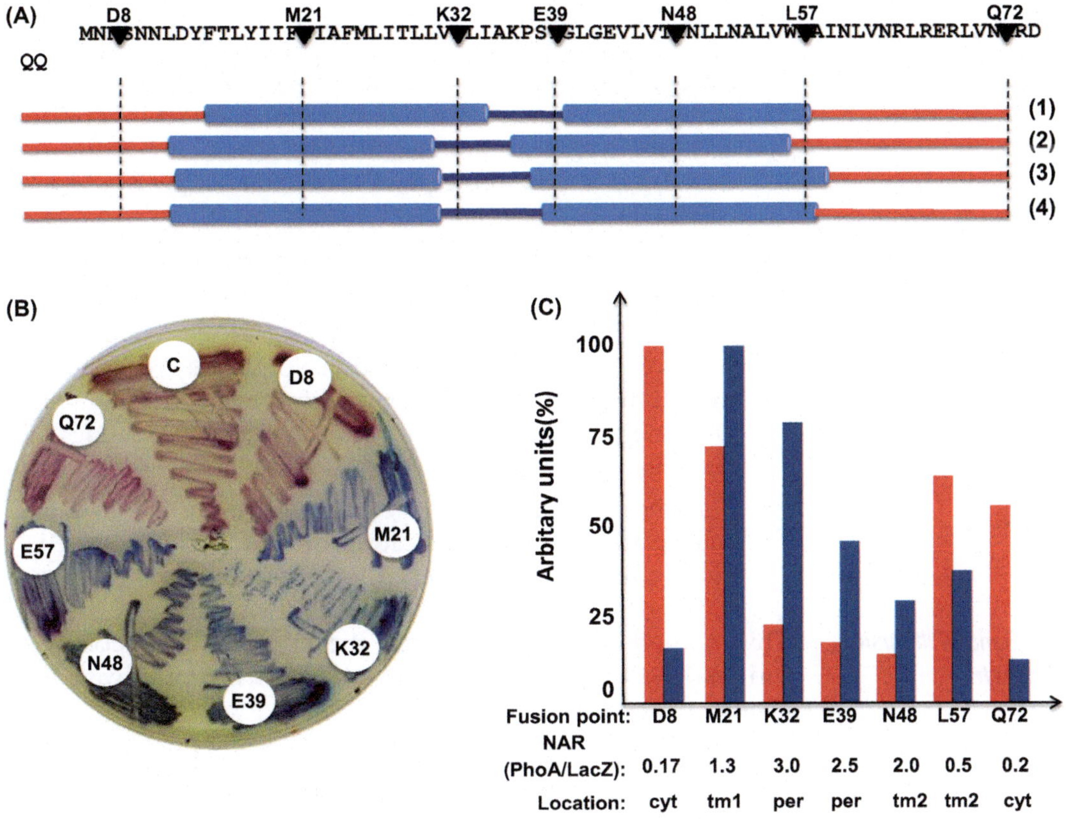

Fig. 2 YmgF topology analysis. (**a**) In silico predicted topological models of YmgF. Predictions were made using four different methods: (1) PSIPRED, (2) Topcons, (3) Phobius, and (4) TopPred. The *red lines* indicate the cytosolic domains, *light blue* TMSs (TMS1 and TMS2), and the *deep blue line* the periplasmic part. The small *black arrowheads* at the *top* of the sequences indicate the positions of the different sites selected to construct fusions with the Pho-LacZα reporter. (**b**) Experimental determination of YmgF membrane topology. *E. coli* DH5α cells expressing different YmgF/Pho-LacZα fusions (position of *insertion* indicated in label) were plated on an indicator medium containing the two chromogenic substrates Red-Gal (for β-galactosidase activity) and X-Pho (for phosphatase activity). *Blue* coloration of the colonies (high phosphatase activity) indicates a membrane or periplasmic location of the fusion point. Red coloration of the colonies (high β-galactosidase activity) indicates cytosolic location of fusion point. Control cells (i.e., *E. coli* DH5α/pKTop) are indicated by the C label. (**c**) Quantitative enzymatic assays of various YmgF/PhoA-LacZα fusions. The bar chart (*blue bar*, phosphatase activity; *red bar*, β-galactosidase activity) represents the relative PhoA and LacZ enzymatic activities, measured on liquid cultures of DH5α expressing the indicated YmgF/Pho-LacZα fusions (position of *insertion* indicated on the abscissa). The normalized activities (NAR) are plotted below as is the deduced subcellular localization of the residues at the position of the Pho-LacZα insertion (cyt: cytoplasmic; per: periplasmic; tm: TMS)

3.3 Nested Deletion Approach

This method makes it possible to obtain a library of randomly generated 3′-truncations of the target protein gene to be fused to the *phoA-lacZα* reporter cassette [19, 25–27, 35].

1. Clone the gene encoding the target membrane protein upstream of the *phoA-lacZα* reporter gene (e.g., between the *Pst*I and *Xba*I sites of pKTop).

2. Digest the resulting plasmid with a restriction enzyme that cuts between the target gene and *phoA-lacZα* cassette to generate a 5′-overhang end (or blunt end) at the 3′-end of the target gene sequence (e.g., *Xba*I, *BamH*I, or *Sma*I of pKTop), and then with a second enzyme that generates a 3′-overhang end at the 5′-end of the *phoA-lacZα* gene (e.g., *Kpn*I or *Sac*I of pKTop).
3. Add Exo III nuclease to progressively digest the target gene from its 3′-end [36]. Remove aliquots at regular intervals to yield a pool of random 3′-target gene truncations.
4. After removing single-stranded regions by a treatment with the Mung Bean nuclease, treat with the Klenow fragment of DNA polymerase I (in the presence of deoxynucleotides), and recircularize the plasmids with T4 DNA ligase.
5. Transform mixture into competent cells (*see* **Notes 3, 4,** and **12**) and plate on dual-indicator medium as described in Subheading 3.5 (*see* **Note 13**).

3.4 Sandwich Fusion Approach

This method allows one to insert the PhoA-LacZα reporter into various loops (cytosolic or periplasmic) of the otherwise intact protein [19, 25–27].

1. Design PCR primers to allow for an insertion of the *phoA-lacZα* cassette in frame with both the upper and downstream portions of the target gene.
2. Clone the PCR-amplified *phoA-lacZα* cassette at selected restriction sites within the target gene.
3. The selected restriction sites in the target gene may already preexist within the native sequence or, alternatively, they must be introduced by site-directed mutagenesis. Caution should be exercised during construction of recombinant plasmids that express a hybrid protein between the target membrane protein (or its fragment) and the PhoA-LacZα reporter (*see* **Notes 12** and **14**).

3.5 Analyzing Clones on Dual Substrate Plates

Screening for topological locations can be done directly on agar plates that contain the specific chromogenic substrates for both PhoA- and LacZ-enzymatic activities (*see* **Notes 15** and **16).**

1. Select the *E. coli* cells (e.g., DH5α or other suitable strains; *see* **Notes 3, 4,** and **12**) transformed **with the constructed plasmids** on LB agar plates containing the appropriate antibiotic (e.g., kanamycin for pKTop) and glucose (0.1–0.2%) and incubate for 20–24 h at 30 °C to reduce the expression of the recombinant *phoA-lacZα* fusions (*see* **Note 12**).
2. Streak a single clone of each *phoA-lacZα* fusion under study on a fresh dual-indicator plate, containing kanamycin

(50 μg/mL), Red-Gal (80 μg/mL), X-Pho (100 μg/mL), IPTG (1 mM).

3. Incubate plate for 20–24 h at 30°–37 °C (for an example, *see* Fig. 2b).

3.6 Growth of Bacterial Culture for Enzymatic Assays

Many different protocols for PhoA and LacZ activity assays have been described in the literature [22, 23, 31, 37]. We use a simplified version of these protocols, in which the *E. coli* cells are permeabilized with chloroform and SDS. The protocols can be easily adapted for use in a 96-well microtiter plate format.

1. Pick a single colony from a fresh LB/kanamycin/glucose plate that has been inoculated with DH5α (pKTop-x) cells. Transfer the colony into 5 mL LB broth with kanamycin (50 μg/mL) and glucose (0.1%). Grow overnight at 37 °C with shaking (150–200 rpm).
2. The next day, dilute overnight culture 1:100 in LB broth fresh medium with kanamycin and let it grow at 37 °C with shaking (150–200 rpm) for 2.5–3 h (to reach mid-exponential phase of growth).
3. Add 1 mM IPTG to each culture, to induce expression of hybrid X/PhoA-LacZα protein, and incubate for an additional hour with aeration at 37 °C (*see* **Note 17**).

3.7 Assay of β-Galactosidase Activity

1. Centrifuge 1.2 mL of the bacterial culture in an Eppendorf tube (e.g., for 5 min at 7000 rpm (4500 g) at RT in a tabletop microfuge) and resuspend the pellet in 1.2 mL M63 medium. Transfer 200 μL of bacterial suspension to microtiter plate well and measure optical density (OD_{595}) of cells at 595–600 nm (*see* **Note 18**).
2. To permeabilize cells, add 100 μL chloroform and 100 μL 0.05% SDS to 1 mL of the washed cells, vortex for 10 s, and incubate for 5 min at 37 °C. Then place the tubes on ice for 5 min.
3. After the chloroform has settled, transfer 50 μL of the upper phase of the bacterial suspension to a microtiter plate well.
4. To start the reaction, add 100 μL of the reaction mixture, which contains PM2 buffer and ONPG (0.15%), to the bacterial suspension, and incubate at RT until a yellow color develops. To stop the reaction, add 50 μL 1 M Na_2CO_3. Record incubation time. Record OD_{600} and OD_{405} for each sample.
5. Calculate enzymatic activity in relative units (A) according to the following formula:

   $A = 1000 \times (OD_{405}\text{sample} - OD_{405}\text{control well})/(OD_{595}\text{sample} - OD_{595}\text{control well})/t\ (\text{min})$ of incubation.

3.8 Assay of Phosphatase Activity

1. Centrifuge 1.2 mL of the bacterial culture in Eppendorf tube (e.g., for 5 min at 7000 rpm at RT in a tabletop microfuge).
2. Wash cells in cold WB and resuspend pellet in 1.2 mL cold PM1 buffer. Transfer 200 μL of the bacterial suspension to a microtiter plate well and measure optical density (OD_{595}) of cells at 595–600 nm.
3. To permeabilize the cells, add 100 μL chloroform and 100 μL 0.05% SDS to 1 mL of the washed cells, vortex for 10 s, and incubate for 5 min at 37 °C. Then place tubes on ice for 5 min. After the chloroform has settled, transfer 100 μL of the upper phase of the bacterial suspension to a microtiter plate well.
4. To start the reaction, add 50 μL of the pNPP solution (0.15% in 1 M Tris–HCl, pH 8.0) to the bacterial suspension and incubate at RT until yellow color develops. Add 50 μL 2 N NaOH to stop the reaction. Record incubation time. Record OD_{405} for each sample.
5. Calculate enzymatic activity in relative units (A) according to the following formula: $A = 1000 \times (OD_{405}\text{sample} - OD_{405}\text{control well})/(OD_{595}\text{ sample} - OD_{595}\text{control well})/t$ (min) of incubation.
6. After obtaining both enzymatic activities for each tested x/*phoA-lacZα* fusion (from **steps 5** in Subheadings 3.7 and 3.8), the NAR is calculated as follows:

NAR = (PhoA activity/Highest PhoA activity)/(LacZ activity/Highest LacZ activity), where Highest PhoA or Highest LacZ are the corresponding maximum activities measured within the set of analyzed fusions (*see* Fig. 2c for an example).

As shown in Fig. 2c, the dual PhoA-LacZα experimental approach confirmed that YmgF is a polypeptide possessing two TMSs separated by a short periplasmic loop and with both termini exposed to the cytosol. The membrane association of YmgF was further corroborated by subcellular fractionation of cells expressing YmgF-GFP fusion, which was found to be fully associated with the bacterial membrane fraction [24]. Altogether, these results showed that YmgF is an integral membrane protein.

4 Notes

1. LB agar plates are prepared by adding, just before pouring and when necessary, kanamycin (50 μg/mL), Red-Gal (80 μg/mL), X-Pho (100 μg/mL), 1 mM IPTG, or glucose (0.1–0.2%).
2. Magenta-Gal (synonym: Red-β-D-Gal, 5-Bromo-6-chloro-3-indolyl-β-D-galactopyranoside) can be used instead of Red-Gal at the same final concentration (80 μg/mL). Stock

solution 25% w/v in DMF or DMSO. Store at −20 °C and protect from light. Importantly, Magenta-Gal is generally more expensive than Red-Gal.

3. In *E. coli*, the expression of the endogenous *phoA* gene is inhibited by moderate concentrations of inorganic phosphate in growth media. In LB broth, a high-phosphate medium (ca 4–6 mM) [38], endogenous phosphatase activity of various widely used *E. coli* strains (DH5, JM101, JM109, etc.) is shown to be quite low (about 3–5 U) [39]. Therefore, the *phoA-lacZα* dual reporter system can be used in any *E. coli* strain that carries the wild-type *phoA* gene and is capable of β-galactosidase α-complementation (i.e., it should express the ω fragment of LacZ such as the common variant produced by *E. coli* LacZΔM15).
4. One study revealed that *E. coli* strain DH5α is indeed *phoA*-deficient (Δ*phoA*) [39]. This makes the strain especially useful for studies in which the *phoA-lacZα* dual reporter technology is applied.
5. In the laboratory, we prepare 2× stock solution.
6. 2-mercaptoethanol (100 mM) can be added to PM2 buffer to approximately double the level of the β-galactosidase enzymatic activity. But 2-mercaptoethanol is considered toxic, causing irritation to the respiratory tract, nasal passageways, skin, etc., **so it can be omitted**.
7. The ONPG stock solution can be frozen and thawed several times.
8. Iodoacetamide is included in the buffers to block activation of cytoplasmic PhoA by oxidation during cell lysis [16, 22].
9. In the laboratory, we use SIGMAFAST pNPP tablets. To prepare a stock solution (0.5%), add one tablet to 1 mL 1 M Tris–HCl, pH 8.0. Store at −20 °C.
10. Diverse accurate predictors for potential α-helical transmembrane proteins are freely available online. We routinely use PSIPRED [40], Topcons [41], Phobius [42], and TopPred 1.10 [43].
11. If the target protein is a relatively large polypeptide (e.g., polytopic membrane proteins), the nested deletion approach and the sandwich fusion approach may be more suitable.
12. Many factors can affect the success of a recombinant plasmid construction, including, for example, the potential toxicity of the expressed gene product, the plasmid copy number, and the genotype of the bacterial host strains. In the laboratory, to minimize problems during the construction of recombinant plasmids, in which the expression of a cloned gene is driven by a relatively strong *lac* promoter (such as in pKTop), the

corresponding ligation mixtures are transformed into *LacI*$_q$ *E. coli* competent cells (i.e., overexpressing the LacI repressor), routinely, into XL1 blue. The transformed cells are grown at 30 °C, for 24–32 h, on LB agar plates containing appropriate antibiotic(s) and supplemented with glucose (0.1–0.2%) that, because of the catabolic repression phenomenon, will reduce the basal transcription of the *lac* promoter and, therefore, diminish the expression of the membrane protein/PhoA-LacZα fusions. Any standard transformation protocol may be used [23, 31, 37]. Routinely, we use a simple $CaCl_2$ procedure to prepare *E. coli* competent cells [31, 37]. This method, which yields a competency level > 10^6 cfu/mg, is adequate for most needs.

13. Statistically, only one-third of recombinant plasmids are expected to encode in-frame fusions between target protein truncations and the PhoA-LacZα reporter. The cells bearing plasmids with out-of-frame fusions can be easily detected on dual-indicator agar plates since they should stay colorless.
14. It is important to note that the PhoA-LacZα reporter could influence the folding of the target protein when inserted internally. Often, C-terminal PhoA-LacZα fusions have higher enzymatic activities than the corresponding sandwich fusions, indicating lower expression levels or more steric problems in the latter [19].
15. Bacteria expressing high levels of phosphatase activity, i.e., when a PhoA-LacZα reporter is localized in periplasm, will turn blue as a result of conversion of the X-Pho substrate into a blue-colored precipitated product. Conversely, cells that express high levels of β-galactosidase activity, i.e., when the PhoA-LacZα reporter is localized in the cytoplasm, will become red owing to the conversion of the Red-Gal substrate into an insoluble red compound. Fusions of the PhoA-LacZα reporter within TMSs usually result in a purple pigmentation of colonies (i.e., a combination of both red and blue colorations as a mixture of cytoplasmic and periplasmic fusions). In addition, for ExoIII-generated *phoA-lacZα* fusion library constructions, utilization of such dual-indicator agar plates allows for easy detection (and elimination) of colorless colonies that harbor noninformative, out-of-frame fusions.
16. It should be noted that, whereas true in-frame fusions to cytoplasmic domains develop red coloration in 12–16 h with *E. coli* TG1, out-of-frame fusions are also able to develop red coloration after a longer incubation period (> in 30–48 h). The precise timing is likely to be different for different *E. coli* strains and would depend on the growth rate [19].
17. It is possible to grow subcultures from overnight inoculations to the mid-exponential phase in the presence of IPTG, i.e., increasing time of induction (3.5–4 h instead of 1 h).

18. Adjust the aliquot volume for OD measurements depending on the equipment used. The volume of cells used in a reaction may depend on the level of expected enzymatic activity.

Acknowledgment

This work was supported by Institut Pasteur and Centre National de la Recherche Scientifique (CNRS UMR 3528, Biologie Structurale et Agents Infectieux).

References

1. von Heijne G (2006) Membrane-protein topology. Nat Rev Mol Cell Biol 7:909–918
2. Islam ST, Lam JS (2013) Topological mapping methods for α-helical bacterial membrane proteins - an update and a guide. Microbiologyopen 2:350–364
3. Dobson L, Remenyi I, Tusnady GE (2015a) The human transmembrane proteome. Biol Direct 10:1–18
4. Chen CP, Rost B (2002) State-of-the-art in membrane protein prediction. Appl Bioinforma 1:21–35
5. Tusnady GE, Simon I (2010) Topology prediction of helical transmembrane proteins: how far have we reached? Curr Protein Pept Sci 11:550–561
6. Dobson L, Remenyi I, Tusnady GE (2015b) CCTOP: a consensus constrained TOPology prediction web server. Nucleic Acids Res 43:W408–W412
7. Peters C, Konstantinos D, Shu N et al (2016) Improved topology prediction using the terminal hydrophobic helices rule. Bioinformatics 32:1158–1162
8. Manoil C, Beckwith J (1986) A genetic approach to analyzing membrane protein topology. Science 233:1403–1408
9. Manoil C, Mekalanos JJ, Beckwith J (1990) Alkaline-phosphatase fusions - sensors of subcellular location. J Bacteriol 172:515–518
10. van Geest M, Lolkema JS (2000) Membrane topology and insertion of membrane proteins: search for topogenic signals. Microbiol Mol Biol Rev 64:13–33
11. Lee C, Inouye H, Brickman ER et al (1989) Genetic studies on the inability of beta-galactosidase to be translocated across the *Escherichia coli* cytoplasmic membrane. J Bacteriol 171:4609–4616
12. Silhavy TJ, Shuman HA, Beckwith J et al (1977) Use of gene fusions to study outer membrane protein localization in *Escherichia coli*. Proc Natl Acad Sci U S A 74:5411–5415
13. Bibi E, Beja O (1994) Membrane topology of multidrug resistance protein expressed in *Escherichia coli*. N-terminal domain. J Biol Chem 269:19910–19915
14. Boyd D, Manoil C, Beckwith J (1987) Determinants of membrane protein topology. Proc Natl Acad Sci U S A 84:8525–8529
15. Boyd D, Manoil C, Froshauer S et al (1990) Use of gene fusions to study membrane-protein topology. In: Gierash LM, King J (eds) Protein folding: deciphering the second half of the genetic code. AAAS Books, Washington
16. Manoil C (1990a) Analysis of protein localization by use of gene fusions with complementary properties. J Bacteriol 172:1035–1042
17. San Millan JL, Boyd D, Dalbey R et al (1989) Use of phoA fusions to study the topology of the *Escherichia coli* inner membrane protein leader peptidase. J Bacteriol 171:5536–5541
18. Silhavy TJ, Beckwith JR (1985) Uses of lac fusions for the study of biological problems. Microbiol Rev 49:398–418
19. Alexeyev MF, Winkler HH (1999) Membrane topology of the Rickettsia prowazekii ATP/ADP translocase revealed by novel dual pho-lac reporters. J Mol Biol 285:1503–1513
20. Langley KE, Villarejo MR, Fowler AV et al (1975) Molecular basis of beta-galactosidase alpha-complementation. Proc Natl Acad Sci U S A 72:1254–1257
21. Ullmann A, Jacob F, Monod J (1967) Characterization by in vitro complementation of a peptide corresponding to an operator-proximal segment of the beta-galactosidase structural gene of *Escherichia coli*. J Mol Biol 24:339–343

22. Manoil C (1990b) Analysis of membrane protein topology using alkaline phosphatase and beta-galactosidase gene fusions. Methods Cell Biol 34:35–47
23. Miller JH (1992) A short course in bacterial genetics: a laboratory manual and handbook for Escherichia coli and related bacteria. Cold Spring Harbor Laboratory Press, Cold Spring Harbor, New York
24. Karimova G, Robichon C, Ladant D (2009) Characterization of YmgF, a 72-residue inner membrane protein that associates with the *Escherichia coli* cell division machinery. J Bacteriol 191:33–46
25. Islam ST, Taylor VL, Qi M et al (2010) Membrane topology mapping of the O-antigen flippase (Wzx), polymerase (Wzy), and ligase (WaaL) from Pseudomonas aeruginosa PAO1 reveals novel domain architectures. MBio 1:e00189–e00110
26. Korres H, Verma NK (2004) Topological analysis of glucosyltransferase GtrV of Shigella flexneri by a dual reporter system and identification of a unique reentrant loop. J Biol Chem 279:22469–22476
27. Nair AH, Korres H, Verma NK (2011) Topological characterisation and identification of critical domains within glucosyltransferase IV (GtrIV) of Shigella flexneri. BMC Biochem 12:1–14
28. Karimova G, Davi M, Ladant D (2012) The beta-lactam resistance protein Blr, a small membrane polypeptide, is a component of the *Escherichia coli* cell division machinery. J Bacteriol 194:5576–5588
29. Falord M, Karimova G, Hiron A et al (2012) GraXSR proteins interact with the VraFG ABC transporter to form a five-component system required for cationic antimicrobial peptide sensing and resistance in *Staphylococcus aureus*. Antimicrob Agents Chemother 56:1047–1058
30. Georgiadou M, Castagnini M, Karimova G et al (2012) Large-scale study of the interactions between proteins involved in type IV pilus biology in Neisseria meningitidis: characterization of a subcomplex involved in pilus assembly. Mol Microbiol 84:857–873
31. Green MR, Sambrook J (2012) Molecular cloning : a laboratory manual, 4th edn. Cold Spring Harbor Laboratory Press, Cold Spring Harbor, New York
32. Nilsson J, Persson B, von Heijne G (2000) Consensus predictions of membrane protein topology. FEBS Lett 486:267–269
33. Boyd D, Traxler B, Beckwith J (1993) Analysis of the topology of a membrane protein by using a minimum number of alkaline phosphatase fusions. J Bacteriol 175:553–556
34. Cassel M, Seppala S, von Heijne G (2008) Confronting fusion protein-based membrane protein topology mapping with reality: the *Escherichia coli* ClcA H+/Cl- exchange transporter. J Mol Biol 381:860–866
35. Sugiyama JE, Mahmoodian S, Jacobson GR (1991) Membrane topology analysis of *Escherichia coli* mannitol permease by using a nested-deletion method to create mtlA-phoA fusions. Proc Natl Acad Sci U S A 88: 9603–9607
36. Henikoff S (1987) Unidirectional digestion with exonuclease III in DNA sequence analysis. Methods Enzymol 155:156–165
37. Sambrook J, Russell DW (2006) The condensed protocols from molecular cloning : a laboratory manual. Cold Spring Harbor Laboratory Press, Cold Spring Harbor, New York
38. Schurig-Briccio LA, Farias RN, Rintoul MR et al (2009) Phosphate-enhanced stationary-phase fitness of *Escherichia coli* is related to inorganic polyphosphate level. J Bacteriol 191:4478–4481
39. Rodriguez-Quinones F, Benedi VJ (2003) *Escherichia coli* strain DH5α is a suitable host for the study of phoA insertions. Focus 15:110–112
40. Jones DT (2007) Improving the accuracy of transmembrane protein topology prediction using evolutionary information. Bioinformatics 23:538–544
41. Tsirigos KD, Peters C, Shu L et al (2015) The TOPCONS web server for combined membrane protein topology and signal peptide prediction. Nucleic Acids Res 43:W401–W407
42. Käll L, Krogh A, Sonnhammer ELL (2007) Advantages of combined transmembrane topology and signal peptide prediction - the Phobius web server. Nucleic Acids Res 35:W429–W432
43. Claros MG, von Heijne G (1994) TopPred II: an improved software for membrane protein structure predictions. Comput Appl Biosci 10:685–686

Chapter 11

In Vivo and In Vitro Protein–Peptidoglycan Interactions

Gang Li and S. Peter Howard

Abstract

Bacteria have developed a number of trans-envelope systems to transport molecules or assemble organelles across bacterial envelopes. However, bacterial envelopes contain a rigid netlike peptidoglycan structure that protects cells from osmotic lysis. Trans-envelope systems thus must interact with the peptidoglycan barrier to generate gaps or anchor structures to the peptidoglycan scaffold. Here we describe methods to use in vivo cross-linking and in vitro co-sedimentation to study protein–peptidoglycan interactions in Gram-negative bacteria. In particular, we address important considerations to ensure the specificity of the interactions in question.

Key words Trans-envelope systems, Peptidoglycan, Cross-linking, Co-sedimentation, Muramic acid assay

1 Introduction

Bacterial cells have a unique peptidoglycan layer in the cell envelope [1]. The rigid netlike peptidoglycan structure determines cell shape and protects bacteria from osmotic lysis. However, it also acts as a barrier for transport of proteins or assembly of large envelope-spanning complexes [2]. Local hydrolysis or remodeling of peptidoglycan is therefore necessary to generate gaps to assemble trans-envelope structures. In addition, macromolecular complexes may use peptidoglycan as a structural extension to anchor securely to a cell envelope [3]. To date, peptidoglycan-interacting components have been identified in a wide range of trans-envelope systems including type II secretion (T2SS), type IV pilus (T4P), flagella, type III secretion (T3SS), type IV secretion (T4SS), and type VI secretion (T6SS).

Here we describe an in vivo cross-linking approach to studying interactions between a protein component of the T2SS (ExeA) and peptidoglycan in *Aeromonas hydrophila* [4]. Bacterial cells are incubated with the cleavable cross-linking reagent 3,3′-dithiobis[sulfosuccinimidylpropionate] (DTSSP), which has two amine-reactive groups at the ends of an eight-atom spacer

Laure Journet and Eric Cascales (eds.), *Bacterial Protein Secretion Systems: Methods and Protocols*, Methods in Molecular Biology, vol. 1615, DOI 10.1007/978-1-4939-7033-9_11, © Springer Science+Business Media LLC 2017

arm. DTSSP can covalently link two primary amines (from ExeA and peptidoglycan in this case) if they are in close proximity. After cross-linking, peptidoglycan sacculi are isolated by a modified small-volume version of the sodium dodecyl sulfate (SDS)-boiling method to remove noncovalently associated proteins [5]. Purified peptidoglycan samples are then treated with β-mercaptoethanol to release cross-linked proteins to be analyzed by SDS-PAGE (polyacrylamide gel electrophoresis) and immunoblot.

We also describe a co-sedimentation (pulldown) assay that uses purified peptidoglycan sacculi to investigate the binding of proteins of interest to peptidoglycan [6]. This is a direct and straightforward method to detect protein–petidoglycan interactions in vitro. We provide a protocol to prepare highly pure peptidoglycan sacculi using SDS-boiling and enzymic treatments [5]. A colorimetric method is used to quantitate peptidoglycan by measuring lactic acids released from the muramic acid residues of peptidoglycan upon acidic and alkaline hydrolysis [7]. We pay particular attention to the hydrophobic nature of purified peptidoglycan sacculi (presumably originating from lipid-linked disaccharide precursors that remain at the ends of glycan strands) [8]. Special experimental procedures are followed to ease peptidoglycan handling and overcome nonspecific interactions in co-sedimentation experiments.

2 Materials

2.1 In Vivo Cross-Linking

1. Phosphate buffered saline (PBS): 150 mM NaCl, 40 mM sodium phosphate, pH 7.5.
2. Sodium citrate buffer: 5 mM sodium citrate, pH 5.0.
3. DTSSP solution: 10 mM DTSSP in sodium citrate buffer. Dissolve DTSSP in sodium citrate buffer. Prepare fresh prior to each experiment.
4. Stop solution (20×): 1 M Tris–HCl, pH 8.0.
5. SDS solution (2×): 8% SDS in water.
6. Boiling-water bath.
7. Ultracentrifuge capable of 130,000 × *g*.

2.2 Purification of Peptidoglycan

1. SDS solution: 8% SDS in water.
2. Tris–HCl buffer (100×): 1 M Tris–HCl, pH 7.0.
3. α-amylase stock solution (100×): 10 mg/mL α-amylase in 10 mM Tris–HCl, pH 7.0. Store at −20 °C.
4. Pronase stock solution (100×): 20 mg/mL pronase in 10 mM Tris–HCl, pH 7.0. Store at −20 °C. Incubate stock solution at 60 °C for 2 h to inactivate possible muramidase contamination before use.

5. Magnetic stirrer with incorporated hotplate.
6. Ultracentrifuge capable of 130,000 × *g*.

2.3 Muramic Acid Assay of Peptidoglycan

1. H_2SO_4 hydrolysis solution: 5 M H_2SO_4 in water.
2. NaOH neutralization solution: 10 M NaOH in water.
3. Concentrated H_2SO_4 (18.8 M).
4. $CuSO_4$ solution: 4% (w/v) $CuSO_4$•$5H_2O$ in water.
5. 4-phenylphenol solution: 1.5% (w/v) 4-phenylphenol in ethanol. Store at −20 °C.
6. Muramic acid standard solutions: 0–1 mM muramic acid in water. Store at −20 °C.
7. Spectrophotometer capable of 570 nm wavelength.
8. Sulfuric-acid-resistant cuvettes (glass or quartz).
9. Boiling-water bath.
10. Fume hood and personal protective equipment.

2.4 Co-sedimentation Assay

1. Tween 20 stock solution (10×): 0.5% Tween 20 in water. Store at 4 °C.
2. Binding buffer (10×): 400 mM sodium phosphate, pH 6.5.
3. Bovine albumin stock solution (100×): 1 mg/mL bovine albumin in water. Store at −20 °C.
4. Refrigerated microcentrifuge.

3 Methods

3.1 In Vivo Cross-Linking

1. Grow bacterial strains under conditions in which the proteins of interest are produced and functional. For each cross-linking experiment, 5–10 mL culture is generally sufficient. *A. hydrophila* strains cultured in buffered Luria-Bertani (LB) medium are used in this protocol [4]. In addition to the wild-type *A. hydrophila* strain, cells expressing ExeA variants that contain substitution mutations in the putative peptidoglycan binding domain are included as controls (*see* **Note 1**).
2. Pellet cells by centrifugation at 6000 × *g* for 5 min. Wash twice with PBS and resuspend cells in PBS. Adjust cell suspension to an OD_{600} of 2.0. Transfer 1 mL cells to microcentrifuge tubes for cross-linking. To avoid cold shock that may affect cell envelope structure and physiology, perform the preceding steps at room temperature.
3. Add fresh 10 mM DTSSP solution to a final concentration of 0.5 mM (or a range of 0.1–1 mM in initial experiments). Mix immediately by inverting the tube several times. Incubate the

mixtures at room temperature for 5 min or for a range of 2–10 min in initial experiments (*see* **Note 2**).

4. Add 1 M Tris–HCl, pH 8.0 solution to a final concentration of 50 mM to stop the cross-linking reaction and quench excessive cross-linker. Incubate the mixtures at room temperature for 15 min.
5. Take aliquots of each cross-linked sample (whole-cell samples) for later SDS-PAGE analysis. Store at −20 °C.
6. Add the rest of the cross-linked samples dropwise to an equal volume of 8% SDS solution in glass tubes preincubated in a boiling-water bath. Incubate the samples for 15 min with intervals of vigorous vortexing. Cool the samples to room temperature (may leave on bench overnight).
7. Pellet peptidoglycan by ultracentrifugation at 130,000 × *g* at room temperature for 1 h (*see* **Note 3**).
8. Resuspend pellets in 0.5 mL of water by vigorous vortexing. Repeat SDS-boiling and centrifugation steps two additional times.
9. Resuspend final pellets in 0.1 mL water (peptidoglycan samples).
10. Mix aliquots of peptidoglycan samples with 2× SDS-PAGE sample buffer containing 0 or 10% β-mercaptoethanol. Heat samples to 95 °C for 5 min.
11. Analyze whole-cell samples and peptidoglycan samples by routine SDS-PAGE and immunoblot to detect proteins of interest (*see* **Note 4**).

3.2 Purification of Peptidoglycan from Gram-Negative Bacteria

1. Grow *A. hydrophila* or *E.coli* cells to late exponential phase in LB or other appropriate medium. Typically 1 L culture yields 2–5 mg purified peptidoglycan.
2. Pellet bacterial cells in 250–500 mL centrifuge bottles at 6000 × *g* at 4 °C for 10 min. Resuspend cells in 20 mL ice-cold water per liter bacterial culture.
3. Slowly pour resuspended cells into an equal volume of 8% SDS solution stirred in a beaker in a boiling-water bath. Incubate mixture with continuous stirring for 1 h. Cool sample to room temperature (may leave stirring overnight at room temperature).
4. Pellet the peptidoglycan by ultracentrifugation at 130,000 × *g* at room temperature for 1 h (*see* **Note 3**).
5. Resuspend peptidoglycan in 20 mL water (room temperature) with vigorous vortexing. Pour sample into boiling 8% SDS solution and incubate for 15 min.

6. Wash the peptidoglycan four times with room temperature water, centrifuging 1 h at 130,000 × *g* for each wash to remove residual SDS (*see* **Note 5**).
7. Resuspend peptidoglycan in 10 mL water. Add 10 mM Tris–HCl, pH 7.0, and 0.1 mg/mL α-amylase. Incubate at 37 °C for 2 h to hydrolyze glycogen trapped in peptidoglycan sacculi.
8. Add 0.2 mg/mL preincubated pronase and incubate at 60 °C for 90 min to hydrolyze proteins associated with peptidoglycan.
9. Add mixture to an equal volume of boiling 8% SDS solution and incubate for 15 min in a boiling-water bath.
10. Wash peptidoglycan four times with water at room temperature as described earlier.
11. Resuspend purified peptidoglycan in 2 mL water. Store at 4 °C. Do not freeze (*see* **Note 6**).

3.3 Muramic Acid Assay of Peptidoglycan

Warning: This colorimetric method uses strong acid and alkaline solutions. Wear suitable personal protective equipment and follow laboratory safety guidelines. Alternative methods using high-performance liquid chromatography or quantitative aminosugar analysis are described elsewhere [5, 9].

1. Add 80 μL peptidoglycan in water (*see* **Note 7**) to an equal volume of 5 M H_2SO_4 solution and incubate at 90 °C for 2 h to hydrolyze the peptidoglycan. Include muramic acid solutions (0–1 mM) in parallel to generate a standard curve.
2. Add 360 μL water and 140 μL 10 M NaOH solution. Incubate at 37 °C for 30 min to release lactic acids from muramic acid residues of peptidoglycan.
3. Transfer 300 μL of the hydrolyzed samples to clean glass tubes in duplicate. Add 2 mL concentrated H_2SO_4 (18.8 M). Cap tubes and vortex to mix (*see* **Note 8**).
4. Incubate glass tubes in a boiling-water bath for 5 min. Cool tubes in a room-temperature water bath.
5. Add 20 μL 4% $CuSO_4$ solution and 40 μL 1.5% 4-phenylphenol solution. Cap and vortex.
6. Incubate samples at 30 °C for 30 min to develop a blue color that is stable for at least 1 h.
7. Measure absorbance at 570 nm in glass or quartz cuvettes (*see* **Note 9**). Average the readings of duplicate samples and determine the muramic acid concentration using the standard curve. The peptidoglycan preparation generally contains 0.5–2 mM muramic acid.

3.4 Co-sedimentation of Peptidoglycan and Proteins of Interest

1. For co-sedimentation of peptidoglycan and purified ExeA protein, the following binding conditions are used: 100 μM muramic acid units of peptidoglycan sacculi, 0.05% Tween 20, 40 mM sodium phosphate buffer, pH 6.5, 10 μg/mL bovine albumin, and variable amounts of ExeA in a reaction volume of 150 μL. The co-sedimentation procedure is optimized to overcome hydrophobic aggregation of peptidoglycan sacculi and nonspecific binding of proteins. Add each component in the following order.
2. Add peptidoglycan and water to 1.5 mL microcentrifuge tubes.
3. Add Tween 20 stock solution (10×) and vortex.
4. Add sodium phosphate binding buffer (10×) and mix.
5. Add bovine albumin stock solution (100×) and mix.
6. Add purified proteins of interest and mix (*see* **Note 10**).
7. Incubate the mixtures at 4 °C for 1 h.
8. Pellet the peptidoglycan sacculi and associated proteins in a microcentrifuge at top speed (21,000 × *g*) at 4 °C for 1 h.
9. Add 1× SDS-PAGE sample buffer to the peptidoglycan pellets. Incubate the samples at 95 °C for 5 min. Vortex vigorously to resuspend peptidoglycan and release associated proteins.
10. Analyze the mixture samples (before centrifugation), supernatant samples, and pellet samples by SDS-PAGE and immunoblot.

4 Notes

1. A general consideration in cross-linking experiments is that the cross-linking reaction may result in artefactual linkage of non-complexed components. It is therefore critical to include suitable controls to exclude such a possibility. The most desirable controls are mutant proteins that contain deletion or substitution mutations in the peptidoglycan binding motif.
2. The cross-linker concentration and reaction time will need to be optimized for different experiments.
3. Store and centrifuge peptidoglycan–SDS mixtures at room temperature to avoid precipitation of SDS at low temperatures.
4. Cross-linked proteins migrate as higher-molecular-weight complexes in SDS-PAGE. Proteins cross-linked to peptidoglycan sacculi are not able to enter gels unless the cross-linkers are cleaved by β-mercaptoethanol treatment prior to electrophoresis. *See* [4] for analysis of protein–peptidoglycan cross-linking in detail.
5. Resuspend the peptidoglycan pellets first in a small volume of water by vortexing. Add water to full volume before ultracentrifugation.

6. Purified peptidoglycan sacculi form large aggregates when frozen and thawed. It is difficult to break up the aggregates by vortexing without using sonication, which in turn fragments the peptidoglycan and reduces yields following centrifugation. The sacculi also aggregate in the presence of buffers or salts. It is thus preferable to store peptidoglycan sacculi in pure water at 4 °C.
7. It is important that the peptidoglycan samples do not contain chloride salts (e.g., NaCl), which can result in the release of hydrogen chloride gas during heating with concentrated H_2SO_4 [10].
8. The tubes do not need to be sealed in an airtight manner. However, exercise caution when handling concentrated H_2SO_4.
9. Check if the cuvettes are sulfuric-acid compatible. The cuvettes need to be dry. If the cuvettes are wet, precipitation may occur and interfere with accurate absorbance readings. If sufficient numbers of cuvettes are not available, rinse the cuvettes with 1 mL concentrated H_2SO_4 (18.8 M) with a pipette before adding the next sample.
10. We do not suggest adding crude bacterial lysate because the lysate may contain peptidoglycan hydrolases.

References

1. Höltje JV (1998) Growth of the stress-bearing and shape-maintaining murein sacculus of *Escherichia coli*. Microbiol Mol Biol Rev 62:181–203
2. Dijkstra AJ, Keck W (1996) Peptidoglycan as a barrier to transenvelope transport. J Bacteriol 178:5555–5562
3. Scheurwater EM, Burrows LL (2011) Maintaining network security: how macromolecular structures cross the peptidoglycan layer. FEMS Microbiol Lett 318:1–9
4. Howard SP, Gebhart C, Langen GR, Li G, Strozen TG (2006) Interactions between peptidoglycan and the ExeAB complex during assembly of the type II secretin of *Aeromonas hydrophila*. Mol Microbiol 59:1062–1072
5. Glauner B (1988) Separation and quantification of muropeptides with high-performance liquid chromatography. Anal Biochem 172:451–464
6. Li G, Howard SP (2010) ExeA binds to peptidoglycan and forms a multimer for assembly of the type II secretion apparatus of *Aeromonas hydrophila*. Mol Microbiol 76:772–781
7. Hoijer MA, Melief MJ, van Helden-Meeuwsen CG, Eulderink F, Hazenberg MP (1995) Detection of muramic acid in a carbohydrate fraction of human spleen. Infect Immun 63:1652–1657
8. Typas A, Banzhaf M, Gross CA, Vollmer W (2011) From the regulation of peptidoglycan synthesis to bacterial growth and morphology. Nat Rev Microbiol 10:123–136
9. Clarke AJ (1993) Compositional analysis of peptidoglycan by high-performance anion-exchange chromatography. Anal Biochem 212:344–350
10. Sulfuric Acid (2015) The Columbia Encyclopedia, 6th edn. http:// www.encyclopedia.com. Accessed 26 Jan 2016

Chapter 12

Measure of Peptidoglycan Hydrolase Activity

Yoann G. Santin and Eric Cascales

Abstract

Most gene clusters encoding multiprotein complexes of the bacterial cell envelope, such as conjugation and secretion systems, Type IV pili, and flagella, bear a gene encoding an enzyme with peptidoglycan hydrolase activity. These enzymes are usually glycoside hydrolases that cleave the glycan chains of the peptidoglycan. Their activities are spatially controlled to avoid cell lysis and to create localized rearrangement of the cell wall. This is assured by interaction with the structural subunits of the apparatus. Here we describe protocols to test the peptidoglycan hydrolase activity of these proteins in vitro and in solution.

Key words Cell wall, Localized degradation, Peptidoglycan, Lytic transglycosylase, Remazol blue

1 Introduction

The peptidoglycan is a mesh-like structure that provides the shape and protection against external pressure to bacterial cells. It is composed of glycan chains resulting from the polymerization of *N*-acetylmuramic acid (MurNAc)-*N*-acetylglucosamine (GlcNAc) disaccharides. These chains are linked by peptide stems that differ from one species to another. With pores of approximately 2 nm, the cell wall constitutes a physical barrier for the passage of macromolecules and for the assembly of cell-envelope-spanning complexes [1–3]. Most trans-envelope multiprotein machineries therefore have evolved dedicated enzymes that locally degrade the cell wall to provide sufficient space for their assembly and insertion without compromising the bacterial shape and survival [3, 4]. These enzymes usually cleave the β-1,4 bond between the *N*-acetylmuramic acid and the *N*-acetylglucosamine of the glycan chains and form nonreducing 1,6-anhydromuropeptides characteristic of lytic transglycosylases (LTGs) [4–7]. Genes encoding these enzymes are found associated in Type III secretion, Type IV secretion, or flagellum gene clusters [3, 4, 6]. The best-studied specialized LTGs are FlgJ and SltF, which are associated with flagellar assembly [8–10], and EtgA, VirB1, and TagX/MltE,

Laure Journet and Eric Cascales (eds.), *Bacterial Protein Secretion Systems: Methods and Protocols*, Methods in Molecular Biology, vol. 1615, DOI 10.1007/978-1-4939-7033-9_12,

which are necessary for the biogenesis of Type III (T3SS), Type IV (T4SS), and Type VI (T6SS) secretion systems, respectively [6, 8–18]. However, these enzymes could be deleterious for bacterial cells, and therefore their activity needs to be restricted to the site of assembly in order to avoid breaches in the cell wall. Studies have provided evidence that LTGs are recruited via specific interactions to subunits of the machine [18–21] and, in a few cases, that these interactions stimulate LTG activity [18, 21, 22].

Methods have been developed and used to test whether putative LTGs have peptidoglycan hydrolase activities. An indirect approach is to clone the gene encoding the putative LTG to a signal sequence in order to address the protein to the periplasm of *E. coli* and follow the cell growth after induction as overproduction of the LTG causes cell lysis [23, 24]. More direct protocols have been developed using purified LTGs, including zymogram [25, 26]. However, this technique, which consists of subjecting purified LTGs to sodium dodecyl sulfate-polyacrylamide gel electrophoresis (SDS-PAGE) in a gel supplemented with purified peptidoglycan, has limits, such as the refolding of the protein after migration. Additional approaches, performed in solution, do not require the denaturation and refolding steps. These turbidometric assays, detailed in what follows, are methods for following the activity of purified LTGs on peptidoglycan or peptidoglycan labeled with remazol brilliant blue (RBB) dye [27, 28]. The peptidoglycan assay relies on the decrease of absorbance of the peptidoglycan solution [27], whereas the RBB assay relies on the release of the dye captured in the peptidoglycan net [28] in the presence of the LTG. In addition, more precise approaches, such as the analysis of peptidoglycan degradation products released after incubation of the peptidoglycan with the purified protein by reverse-phase high-performance liquid chromatography coupled to mass spectrometry [29, 30], make it possible to define the site of cleavage of the enzyme.

2 Material

2.1 Peptidoglycan Purification

1. 8% SDS solution: Resuspend 8 g SDS resuspended in 100 mL sterile distilled water.
2. 20 mM Tris–HCl, pH 8.0, 100 mM NaCl: Dissolve 2.43 g Tris(hydroxymethyl) aminomethane and 5.84 g NaCl in 1 L sterile distilled water. Adjust pH to 8.0 with 1 M HCl.
3. 20 mM Tris–HCl, pH 7.2, 50 mM NaCl: Dissolve 2.43 g Tris(hydroxymethyl) aminomethane and 2.92 g NaCl in 1 L sterile distilled water. Adjust pH to 7.2 with 1 M HCl.
4. 0.5 M NaCl: Dissolve 29.22 g NaCl in 1 L sterile distilled water.

5. α-amylase stock solution (100×): 20 mg/mL α-amylase in 20 mM Tris–HCl, pH 7.2. Store at −20 °C.
6. Pronase stock solution (100×): 20 mg/mL pronase in 20 mM Tris–HCl, pH 7.2. Incubate pronase stock solution for 1 h at 56 °C. Store at −20 °C.
7. French press, Emulsiflex apparatus or any apparatus to disrupt bacterial cells.
8. Vortex.
9. Water bath at 96 °C.
10. Incubate at 37 °C.
11. Ultracentrifuge (Beckman Coulter (Brea, CA) with TLA100.3 and TLA100.4 rotors, or equivalent).

2.2 Turbidometric Analyses of Peptidoglycan Degradation

1. 60 mM MES, pH 6.0, 180 mM NaCl buffer: Dissolve 11.71 g 2-(*N*-morpholino)ethanesulfonic acid and 10.52 g NaCl in 1 L sterile distilled water.
2. Purified protein to be tested.
3. Lysozyme stock solution: 10 mg/mL egg-white lysozyme in sterile distilled water.
4. Incubate at 37 °C.
5. Spectrophotometer.

2.3 Peptidoglycan Labeling with Remazol Brilliant Blue

1. 400 mM NaOH: Dissolve 16 g NaOH in 1 L sterile distilled water.
2. RBB stock solution (10×): Dissolve 1.566 g RBB R (Sigma-Aldrich, St. Louis, MO) in 10 mL sterile distilled water.
3. 1 M HCl: Dilute 10 mL HCl 37% solution (10 M) with 90 mL distilled water.
4. Phosphate buffered saline (PBS) buffer: Dissolve 1.44 g Na_2HPO_4, 0.24 g KH_2PO_4, 0.2 g KCl, and 8 g NaCl in 1 L sterile distilled water. Adjust pH to 7.4 with 1 M HCl.
5. Incubate at 37 °C.
6. Vortex.
7. Ultracentrifuge (Beckman with TLA100.3 rotor, or equivalent).

2.4 RBB-Labeled Peptidoglycan Degradation Assay

1. PBS buffer: Dissolve 1.44 g Na_2HPO_4, 0.24 g KH_2PO_4, 0.2 g KCl, and 8 g NaCl in 1 L sterile distilled water. Adjust pH to 7.4 with 1 M HCl.
2. Purified protein to be tested.
3. Ethanol 96° or absolute.
4. Lysozyme stock solution: 10 mg/mL egg-white lysozyme in sterile distilled water.

5. Incubate at 37 °C.
6. Ultracentrifuge (Beckman with TLA100.3 rotor, or equivalent).
7. Spectrophotometer.

3 Methods

3.1 Peptidoglycan Purification

The peptidoglycan purification protocol is adapted from [31, 32].

1. Grow cells in 400 mL of appropriate medium until culture reaches an A_{600} of about 1–1.2.
2. Harvest cells by centrifugation at 10,000 × *g* for 20 min at 4 °C. Resuspend cells in 20 mL 20 mM Tris–HCl, pH 8.0, 100 mM NaCl. Break cells by three passages at French press or using an Emulsiflex apparatus.
3. Pellet cell envelopes by centrifugation at 400,000 × *g* (90,000 rpm in a Beckman TLA-100.4 rotor) for 45 min at 4 °C. Resuspend cells in 10 mL 0.5 M NaCl.
4. Add 10 mL 8% SDS and incubate for 1 h at 96 °C.
5. Leave solution at room temperature overnight.
6. Pellet peptidoglycan by ultracentrifugation at 400,000 × *g* at 25 °C for 45 min (*see* **Note 1**).
7. Resuspend peptidoglycan fraction in 10 mL 0.5 M NaCl and add 10 mL 8% SDS. Incubate for 30 min at 96 °C.
8. Pellet peptidoglycan by ultracentrifugation at 400,000 × *g* at 25 °C for 30 min and resuspend peptidoglycan in 10 mL water.
9. Repeat **step 8** twice.
10. Resuspend peptidoglycan in 10 mL 20 mM Tris–HCl, pH 7.2, 50 mM NaCl supplemented with 200 μg/mL α-amylase and 200 μg/mL pronase. Incubate overnight at 37 °C.
11. Add 10 mL 8% SDS and incubate for 1 h at 96 °C.
12. Pellet peptidoglycan by ultracentrifugation at 400,000 × *g* at 25 °C for 30 min and resuspend peptidoglycan in 10 mL water.
13. Repeat **step 12** twice.
14. Resuspend peptidoglycan pellet in 1 mL water. Store at 4 °C.

3.2 Turbidometric Analyses of Peptidoglycan Degradation

1. Dilute 125 μL of the purified peptidoglycan suspension obtained at **step 14** in Subheading 3.1 with 875 μL 60 mM MES, pH 6.0, 180 mM NaCl, and incubate at 37 °C for 30 min. Use three tubes for each reaction to measure peptidoglycan hydrolysis in triplicate.
2. Measure the A_{600} for each tube (*see* **Note 2**).

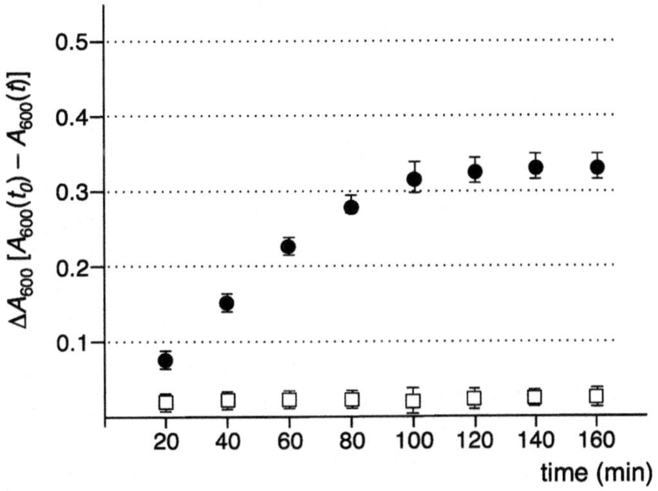

Fig. 1 LTG activity measured by peptidoglycan hydrolysis assay. A representative example of peptidoglycan degradation is shown. Purified peptidoglycan was incubated with buffer (*open square*) or purified LTG (*closed circles*), and the absorbance at 600 nm (A_{600}) was measured every 20 min. The difference of absorbance at time *zero* (t_0) minus the absorbance at time t (ΔA_{600}) was plotted against time (in minutes)

3. Add 2–5 nmol of protein to be tested to each tube and incubate at 37 °C (*see* **Note 3**).
4. Measure the A_{600} every 10 min and plot the difference of absorbance (initial absorbance subtracted from the absorbance at time t) against time (*see* **Note 4**).

A typical example of the turbidometric peptidoglycan assay is shown in Fig. 1.

3.3 Peptidoglycan Labeling with Remazol Brilliant Blue

The peptidoglycan labeling protocol is adapted from [28].

1. Mix 250 μL of the purified peptidoglycan fraction obtained at **step 14** in Subheading 3.1 with 250 μL 400 mM NaOH and incubate for 30 min at 37 °C.
2. Add RBB dye to mixture at final concentration of 25 mM. Vortex and incubate mixture overnight at 37 °C.
3. Add 500 μL 1 M HCl and mix by vortexing.
4. Pellet peptidoglycan by ultracentrifugation at 400,000 × *g* at 25 °C for 30 min and resuspend peptidoglycan in 2 mL water.
5. Repeat **step 3** twice.
6. Resuspend peptidoglycan pellet in 250 μL PBS buffer. Store at 4 °C.

3.4 RBB-Labeled Peptidoglycan Degradation Assay

1. Dilute 10 μL RBB-labeled peptidoglycan obtained at **step 5** in Subheading 3.3 with 90 μL PBS buffer and incubate at 37 °C for 30 min. Use nine tubes for each reaction to measure peptidoglycan hydrolysis in triplicate, at three different times.

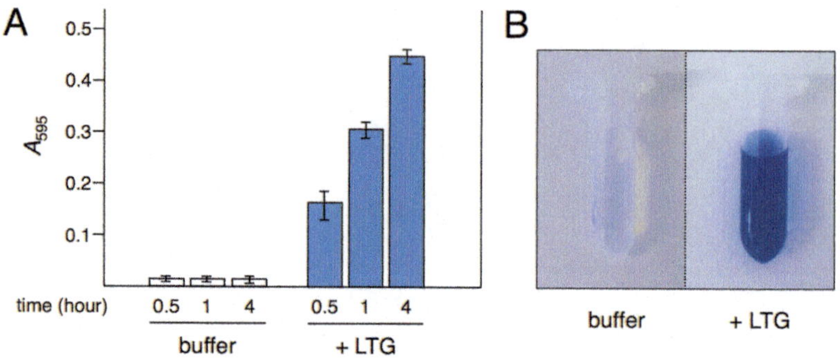

Fig. 2 LTG activity measured by RBB release assay. A representative example of peptidoglycan degradation is shown. (**a**) RBB-labeled peptidoglycan was incubated with buffer (*open bars*) or purified LTG (+LTG, *blue bars*) and the absorbance at 595 nm (A_{595}) of the supernatant was measured after 0.5, 1, and 4 h of incubation. (**b**) Photographs of supernatant fractions of RBB-labeled peptidoglycan incubated with buffer (*left tube*) or purified LTG (+LTG) (*right tube*) after 4 h of incubation

2. Add 0.2–0.5 nmol of protein to be tested to mixture and incubate at 37 °C (*see* **Note 3**). This step corresponds to time zero.
3. Add 100 μL ethanol in three tubes 30 min after time zero to quench reaction.
4. Pellet peptidoglycan by ultracentrifugation at 400,000 × *g* at 25 °C for 30 min.
5. Measure A_{595} of the supernatant.
6. At 1 and 4 h after time zero, repeat **steps 3–5**.

A typical example of the dye release assay is shown in Fig. 2.

4 Notes

1. Do not incubate at 4 °C to avoid SDS precipitation.
2. Typically, an A_{600} ~ 0.4–0.7 is measured from peptidoglycan purified from *E. coli*.
3. Control assays include incubation of the peptidoglycan suspension with (i) buffer and (ii) purified lysozyme. Ideally, additional controls include incubation of the peptidoglycan with (a) the protein to be tested but bearing amino-acid substitutions in the catalytic site (if known or predicted) and (b) the wild-type protein in presence of 100 μM of bulgecin A, an inhibitor of lytic transglycosylases [33].
4. The initial rate of the hydrolysis reaction (in AU/min/mol) can be calculated from the slope of the initial linear curve.

Acknowledgements

Work in EC laboratory is supported by the Centre National de la Recherche Scientifique, the Aix-Marseille Université, and grants from the Agence Nationale de la Recherche (ANR-14-CE14-0006-02 and ANR-15-CE11-0019-01).

References

1. Demchick P, Koch AL (1996) The permeability of the wall fabric of *Escherichia coli* and *Bacillus subtilis.* J Bacteriol 178:768–773
2. Scheurwater E, Reid CW, Clarke AJ (2008) Lytic transglycosylases: bacterial space-making autolysins. Int J Biochem Cell Biol 40:586–591
3. Scheurwater EM, Burrows LL (2011) Maintaining network security: how macromolecular structures cross the peptidoglycan layer. FEMS Microbiol Lett 318:1–9
4. Koraimann G (2003) Lytic transglycosylases in macromolecular transport systems of gram-negative bacteria. Cell Mol Life Sci 60:2371–2388
5. Höltje JV (1996) Lytic transglycosylases. EXS 75:425–429
6. Zahrl D, Wagner M, Bischof K, Bayer M, Zavecz B, Beranek A, Ruckenstuhl C, Zarfel GE, Koraimann G (2005) Peptidoglycan degradation by specialized lytic transglycosylases associated with type III and type IV secretion systems. Microbiology 151:3455–3467
7. van Heijenoort J (2011) Peptidoglycan hydrolases of *Escherichia coli.* Microbiol Mol Biol Rev 75:636–663
8. de la Mora J, Ballado T, González-Pedrajo B, Camarena L, Dreyfus G (2007) The flagellar muramidase from the photosynthetic bacterium *Rhodobacter sphaeroides.* J Bacteriol 189:7998–8004
9. de la Mora J, Osorio-Valeriano M, González-Pedrajo B, Ballado T, Camarena L, Dreyfus G (2012) The C terminus of the flagellar muramidase SltF modulates the interaction with FlgJ in *Rhodobacter sphaeroides.* J Bacteriol 194:4513–4520
10. Nambu T, Minamino T, Macnab RM, Kutsukake K (1999) Peptidoglycan-hydrolyzing activity of the FlgJ protein, essential for flagellar rod formation in *Salmonella typhimurium.* J Bacteriol 181:1555–1561
11. Mushegian AR, Fullner KJ, Koonin EV, Nester EW (1996) A family of lysozyme-like virulence factors in bacterial pathogens of plants and animals. Proc Natl Acad Sci U S A 93:7321–7326
12. Kohler PL, Hamilton HL, Cloud-Hansen K, Dillard JP (2007) AtlA functions as a peptidoglycan lytic transglycosylase in the *Neisseria gonorrhoeae* type IV secretion system. J Bacteriol 189:5421–5428
13. Zhong Q, Shao S, Mu R, Wang H, Huang S, Han J, Huang H, Tian S (2011) Characterization of peptidoglycan hydrolase in Cag pathogenicity island of *Helicobacter pylori.* Mol Biol Rep 38:503–509
14. García-Gómez E, Espinosa N, de la Mora J, Dreyfus G, González-Pedrajo B (2011) The muramidase EtgA from enteropathogenic *Escherichia coli* is required for efficient type III secretion. Microbiology 157:1145–1160
15. Arends K, Celik EK, Probst I, Goessweiner-Mohr N, Fercher C, Grumet L, Soellue C, Abajy MY, Sakinc T, Broszat M, Schiwon K, Koraimann G, Keller W, Grohmann E (2013) TraG encoded by the pIP501 type IV secretion system is a two-domain peptidoglycan-degrading enzyme essential for conjugative transfer. J Bacteriol 195:4436–4444
16. Laverde Gomez JA, Bhatty M, Christie PJ (2014) PrgK, a multidomain peptidoglycan hydrolase, is essential for conjugative transfer of the pheromone-responsive plasmid pCF10. J Bacteriol 196:527–539
17. Weber BS, Hennon SW, Wright MS, Scott NE, de Berardinis V, Foster LJ, Ayala JA, Adams MD, Feldman MF (2016) Genetic dissection of the Type VI secretion system in *Acinetobacter* and identification of a novel peptidoglycan hydrolase, TagX, required for its biogenesis. MBio
18. Santin YG, Cascales E (2016) Domestication of a housekeeping transglycosylase for assembly of a type VI secretion system. EMBO Rep 18(1):138–149
19. Höppner C, Carle A, Sivanesan D, Hoeppner S, Baron C (2005) The putative lytic transglycosylase VirB1 from *Brucella suis* interacts with the type IV secretion system core components VirB8, VirB9 and VirB11. Microbiology 151:3469–3482
20. Creasey EA, Delahay RM, Daniell SJ, Frankel G (2003) Yeast two-hybrid system survey

of interactions between LEE-encoded proteins of enteropathogenic *Escherichia coli*. Microbiology 149:2093–2106

21. Burkinshaw BJ, Deng W, Lameignère E, Wasney GA, Zhu H, Worrall LJ, Finlay BB, Strynadka NC (2015) Structural analysis of a specialized type III secretion system peptidoglycan-cleaving enzyme. J Biol Chem 290:10406–10417
22. Herlihey FA, Osorio-Valeriano M, Dreyfus G, Clarke AJ (2016) Modulation of the lytic activity of the dedicated autolysin for flagellum formation SltF by flagellar rod proteins FlgB and FlgF. J Bacteriol 198:1847–1856
23. Engel H, Kazemier B, Keck W (1991) Murein-metabolizing enzymes from *Escherichia coli*: sequence analysis and controlled overexpression of the slt gene, which encodes the soluble lytic transglycosylase. J Bacteriol 173:6773–6782
24. Lommatzsch J, Templin MF, Kraft AR, Vollmer W, Höltje JV (1997) Outer membrane localization of murein hydrolases: MltA, a third lipoprotein lytic transglycosylase in *Escherichia coli*. J Bacteriol 179:5465–5470
25. Leclerc D, Asselin A (1989) Detection of bacterial cell wall hydrolases after denaturing polyacrylamide gel electrophoresis. Can J Microbiol 35:749–753
26. Bernadsky G, Beveridge TJ, Clarke AJ (1994) Analysis of the sodium dodecyl sulfate-stable peptidoglycan autolysins of select gram-negative pathogens by using renaturing polyacrylamide gel electrophoresis. J Bacteriol 176:5225–5232
27. Fibriansah G, Gliubich FI, Thunnissen AM (2012) On the mechanism of peptidoglycan binding and cleavage by the endo-specific lytic transglycosylase MltE from *Escherichia coli*. Biochemistry 51:9164–9177
28. Uehara T, Parzych KR, Dinh T, Bernhardt TG (2010) Daughter cell separation is controlled by cytokinetic ring-activated cell wall hydrolysis. EMBO J 29:1412–1422
29. Scheurwater EM, Clarke AJ (2008) The C-terminal domain of *Escherichia coli* YfhD functions as a lytic transglycosylase. J Biol Chem 283:8363–8373
30. Clarke AJ (1993) Compositional analysis of peptidoglycan by high-performance anion-exchange chromatography. Anal Biochem 212:344–350
31. Leduc M, Joseleau-Petit D, Rothfield LI (1989) Interactions of membrane lipoproteins with the murein sacculus of *Escherichia coli* as shown by chemical crosslinking studies of intact cells. FEMS Microbiol Lett 51:11–14
32. Cascales E, Lloubès R (2004) Deletion analyses of the peptidoglycan-associated lipoprotein Pal reveals three independent binding sequences including a TolA box. Mol Microbiol 51:873–885
33. Imada A, Kintaka K, Nakao M, Shinagawa S (1982) Bulgecin, a bacterial metabolite which in concert with beta-lactam antibiotics causes bulge formation. J Antibiot (Tokyo) 35:1400–1403

Chapter 13

Protein–Protein Interaction: Bacterial Two-Hybrid

Gouzel Karimova, Emilie Gauliard, Marilyne Davi, Scot P. Ouellette, and Daniel Ladant

Abstract

The bacterial two-hybrid (BACTH, for "Bacterial Adenylate Cyclase-Based Two-Hybrid") system is a simple and fast genetic approach to detecting and characterizing protein–protein interactions in vivo. This system is based on the interaction-mediated reconstitution of a cyclic adenosine monophosphate (cAMP) signaling cascade in *Escherichia coli*. As BACTH uses a diffusible cAMP messenger molecule, the physical association between the two interacting chimeric proteins can be spatially separated from the transcription activation readout, and therefore it is possible to analyze protein–protein interactions that occur either in the cytosol or at the inner membrane level as well as those that involve DNA-binding proteins. Moreover, proteins of bacterial origin can be studied in an environment similar (or identical) to their native one. The BACTH system may thus permit a simultaneous functional analysis of proteins of interest—provided the hybrid proteins retain their activity and their association state. This chapter describes the principle of the BACTH genetic system and the general procedures to study protein–protein interactions in vivo in *E. coli*.

Key words Two-hybrid system, Protein interaction assay, Membrane protein, cAMP signaling, Chimeric proteins

1 Introduction

Two-hybrid systems are genetic assays that permit the detection and characterization of protein–protein interactions in vivo. This approach was pioneered by Fields and Song, who described the original yeast two-hybrid system [1]. All of the two-hybrid techniques discussed in what follows are based on the co-expression, in the same cell, of two hybrid proteins that, upon interaction, produce a phenotypic or selective trait [2]. In the bacterial two-hybrid (BACTH, for "bacterial adenylate cyclase-based two-hybrid") system, the readout of the interactions relies on the complementation between two fragments from the adenylate cyclase of *Bordetella pertussis* to reconstitute a cyclic adenosine monophosphate (cAMP) signaling cascade in *Escherichia coli* [3]. Because it exploits a cAMP signaling cascade, the BACTH system can be easily applied to

Laure Journet and Eric Cascales (eds.), *Bacterial Protein Secretion Systems: Methods and Protocols*, Methods in Molecular Biology, vol. 1615, DOI 10.1007/978-1-4939-7033-9_13, © Springer Science+Business Media LLC 2017

study interactions between membrane proteins [4], and it has indeed been widely used to characterize the assembly of bacterial secretion systems. These specialized nanomachines, which bacteria use to secrete a wide variety of compounds (e.g., small molecules, sugars, proteins, DNA), consist of up to tens of proteins that assemble in the bacterial membranes in multimolecular complexes. The BACTH system has been instrumental in characterizing the molecular interactions between these different components of various secretion systems [5–9]. This chapter describes the principle of this genetic system and outlines the main procedures to study protein–protein interactions in vivo in *E. coli*.

1.1 Principle of Bacterial Adenylate Cyclase-Based Two-Hybrid System

The BACTH bacterial two-hybrid system is a simple and fast approach to detecting and characterizing protein–protein interactions in vivo. It offers all the advantages of working with *E. coli* and is readily accessible to many researchers having basic knowledge in standard microbiological and molecular biology techniques (e.g., plasmid preparation, bacterial transformation, polymerase chain reaction (PCR)).

The BACTH system is based on the interaction-mediated reconstitution of adenylate cyclase enzyme activity in an *E. coli cya* mutant, defective in its endogenous adenylate cyclase [3, 10]. It exploits the fact that the catalytic domain of adenylate cyclase (CyaA) from *B. pertussis* [11] consists of two complementary fragments, T25 and T18, that are not active when physically separated (Fig. 1a). When these two fragments are fused to interacting polypeptides, X and Y, heterodimerization of the hybrid proteins results in functional complementation between the T25 and T18 fragments and, therefore, in cAMP synthesis (Fig. 1b). Cyclic AMP produced by the reconstituted chimeric enzyme binds to the catabolite activator protein (CAP). The cAMP/CAP complex is a pleiotropic regulator of gene transcription in *E. coli* [12]. It turns on the expression of several resident genes, including genes of the *lac* and *mal* operons involved in lactose and maltose catabolism (Fig. 1c). Consequently, bacteria become able to utilize lactose or maltose as the unique carbon source and can be easily distinguished on indicator or selective media [3, 10].

1.2 General Procedure

Detection of in vivo interactions between two proteins of interest with the BACTH system requires the co-expression of these proteins as fusions with the T25 and T18 fragments in bacteria that are lacking its endogenous adenylate cyclase activity (*E. coli cya*). This is achieved by using two compatible vectors, one expressing the T25 fusion (pKT25 or pKNT25), the other expressing the T18 fusion (pUT18 or pUT18C) [10, 13]. The bacteria are cotransformed with the two recombinant plasmids and plated on either indicator or selective media to reveal the resulting Cya^{+} phenotype (Fig. 2). The efficiency of complementation between the two

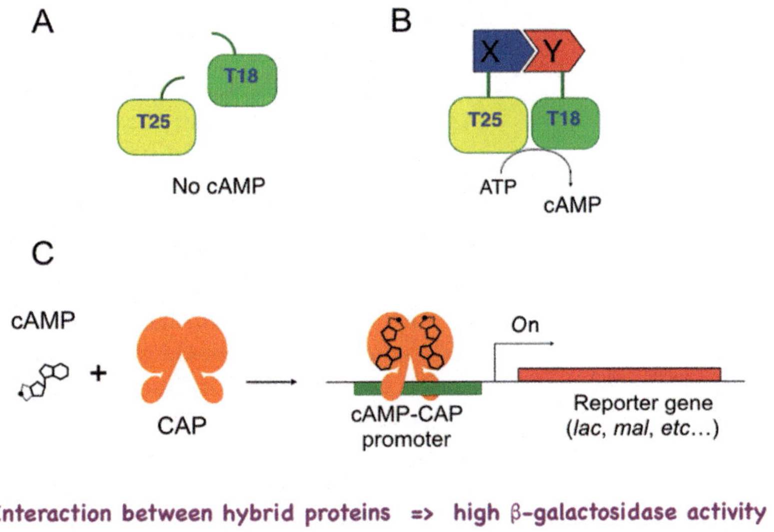

Fig. 1 Principle of BACTH system. (**a**) When the two fragments of *B. pertussis* adenylate cyclase, T25 and T18, are coexpressed as separate polypeptides, they cannot assemble and there is no enzyme activity. (**b**) When the T25 and T18 fragments are coexpressed as fusions with polypeptides X and Y that can interact, the association of the T25-X and T18-Y hybrid proteins reconstitutes the adenylate cyclase activity. (**c**) Cyclic AMP synthesized by the reconstituted enzyme binds to the catabolite activator protein (CAP), and the cAMP/CAP complex can associate with specific promoter DNA and activates transcription of catabolite operons (such as *lac* operon or *mal* regulon)

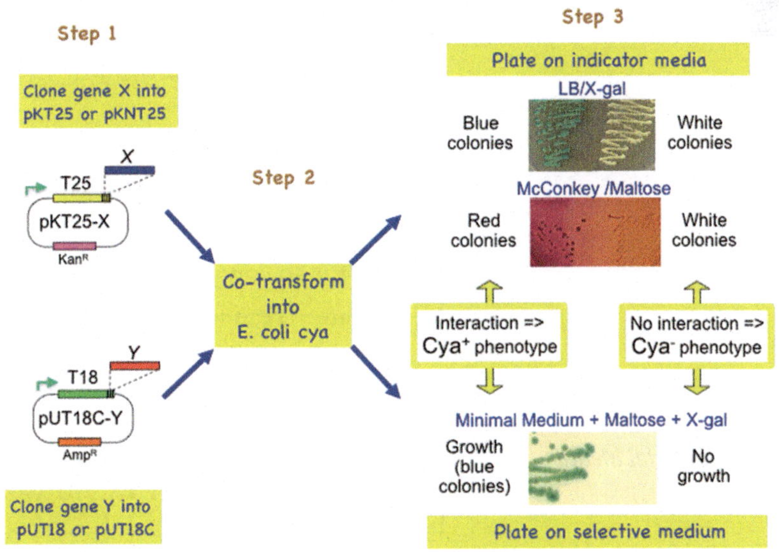

Fig. 2 Analysis of protein–protein interactions with the bacterial two-hybrid system. See text for detailed explanations

hybrid proteins can be further quantified by measuring cAMP levels (a direct measure of the reconstituted adenylate cyclase enzymatic activity) or by assaying the β-galactosidase enzymatic activities in bacterial extracts [3, 10], an easy and robust assay that is directly correlated with the cAMP produced in the cells since the expression of β-galactosidase is positively regulated by cAMP/CAP. The hybrid proteins expressed in *E. coli* can also be characterized using diverse biochemical approaches, for example, immunodetection, immunoprecipitation, and copurification.

The BACTH system has been used by many different laboratories to detect and characterize interactions between a wide variety of bacterial, eukaryotic, or viral proteins [3, 13–16]. An attractive aspect of this genetic assay is that, because it uses a cAMP signaling cascade, the interaction between the hybrid proteins does not need to take place near the transcription machinery, as is the case with the yeast two-hybrid system or many other bacterial two-hybrid systems [1, 2]. For this reason, the BACTH system is particularly appropriate for studying interactions between membrane proteins because these interactions cannot be easily tested with transcription-based two-hybrid systems [4, 14, 17].

2 Materials

2.1 Equipment

1. Equipment for DNA cloning and bacterial transformation.
2. Incubator for plates and shaking liquid cultures.
3. 2.2 mL 96-well storage plate or deep-well storage plate, sterile.
4. 1.2 mL polypropylene 96-well storage block or glass tubes, sterile.
5. Microporous tape sheet, e.g., AirPore (Qiagen Co., Helden, Germany).
6. Multichannel pipettor.
7. Shaker (for shaking deep-well 96-well blocks).
8. Microplate reader, e.g., by Tecan Co. (Männedorf, Switzerland), or equivalent plate reader.
9. Equipment and reagents for western blotting (optional).

2.2 Bacterial Media

1. Luria–Bertani (LB) broth: 10 g NaCl, 10 g tryptone, and 10 g yeast extract, adjust pH to 7.0 with NaOH, add deionized H_2O to a final volume of 1 L, and autoclave.
2. LB plates: add 15 g agar per liter of LB broth and autoclave. Allow the medium to cool to less than 45 °C, then add the antibiotics and pour the plates.

3. LB/X-Gal plates: To prepare LB/X-Gal plates, the LB/agar medium (above) is autoclaved, allowed to cool to less than 45 °C, and supplemented, just before pouring plates, with 40 μg/mL of the X-Gal (5-bromo-4-chloro-3-indolyl-β-D-galactopyranoside) chromogenic substrate and appropriate antibiotics. Isopropyl-β-D-thiogalactopyranoside (IPTG) (final concentration of 0.5 mM) is usually also added to the medium to induce full expression of hybrid proteins as well as that of the β-galactosidase reporter enzyme.
4. MacConkey/maltose medium: 40 g MacConkey agar are dissolved in 1 L distilled water and autoclaved (*see* **Note 1**). A stock solution of glucose-free maltose (20% in water) is sterilized by filtration. Maltose (1% final concentration) as well as antibiotics (ampicillin at 100 μg/mL and kanamycin at 50 μg/mL) are added to the autoclaved MacConkey medium just before pouring plates. IPTG (final concentration of 0.5 mM) is usually added to the medium to induce full expression of hybrid proteins.
5. 5× M63/maltose minimal medium: 10 g $(NH_4)_2SO_4$, 68 g KH_2PO_4, 2.5 mg $FeSO_4{\cdot}7H_2O$, add deionized H_2O to final volume of 1 L, adjust pH to 7.0 with KOH and autoclave. When necessary, vitamin B1 is added to a final concentration of 1 μg/mL and casamino acids at 50 μg/mL.
6. M63/maltose plates: Autoclave 15 g agar in 800 mL H_2O. Then add 200 mL sterile 5× M63 medium, 0.2–0.4% maltose, and the appropriate antibiotics at half the usual concentrations (i.e., 50 μg/mL ampicillin, 25 μg/mL kanamycin) just before pouring plates.

2.3 Solutions for β-Galactosidase Assays

1. β-galactosidase assay medium (PM2): 70 mM Na_2HPO_4, 30 mM NaH_2PO_4, 1 mM $MgSO_4$, 0.2 mM $MnSO_4$, pH 7.0. Add 100 mM β-mercaptoethanol just before use (*see* **Note 2**).
2. Substrate solution: ONPG, o-nitrophenol-β-galactoside, solution of 4 mg/mL in PM2 medium without β-mercaptoethanol (store at −20 °C).
3. Stop solution: 1 M Na_2CO_3.
4. Chloroform.
5. Sodium dodecyl sulfate (SDS) 0.1%: Dissolve 0.1 g SDS into 100 mL H_2O.

2.4 BACTH Reporter Strains, Plasmids, and Antibodies

1. *E. coli* reporter strain carrying a deletion of the *cya* gene (*see* **Note 3**).
2. Set of compatible vectors allowing genetic fusions of proteins of interest at either the N- or the C-terminus of the T25 fragment (pKT25 and pKNT25) or of the T18 fragment (pUT18 and pUT18C) (*see* **Note 4**).

3. Anti-CyaA monoclonal antibody (3D1, sc-13582; Santa Cruz Biotechnology) for T18 fragment detection.
4. Rabbit polyclonal antiserum directed against purified *B. pertussis* CyaA protein (serum L24023, DL unpublished) for T25 fragment detection.

3 Methods

3.1 General Methodology

The general methodology to analyze interactions between two proteins of interest with the BACTH system is diagrammed in Fig. 2.

- In a first step, clone the genes encoding the two proteins of interest (e.g., X and Y) into the two sets of BACTH vectors (pKT25 or pKNT25 and pUT18C or pUT18) using standard molecular biology techniques [18] or with the Gateway® recombineering technique [19].
- In a second step, cotransform the recombinant plasmids encoding the T25-X (or X-T25) and T18-Y (or Y-T18) hybrid proteins into competent BACTH cells (DHM1, DHT1, or BTH101), and plate the transformed cells on indicator plates (i.e., LB-X-Gal or MacConkey media supplemented with maltose) or on selective plates (synthetic medium supplemented with maltose as unique carbon source) [3, 4, 10, 20, 21] (*see* **Note 1**). Complementation is usually detected within 1–3 days of incubation at 30 °C (or 37 °C, although it is usually less efficient at this temperature). If no interaction occurs, colonies will remain colorless on indicator plates or will not grow on selective plates.

3.2 Construction of BACTH Plasmids Encoding Hybrid Proteins

This section assumes that the reader has background knowledge in basic molecular biology techniques. Additional protocols for molecular cloning, PCR, DNA analysis, and transformation can be found in many textbooks (e.g., [18]) or on the Internet.

3.2.1 Standard Cloning of Genes Encoding Proteins of Interest into BACTH Vectors

1. Design specific primers to amplify the genes encoding the proteins of interest. The primers should include restriction sites (e.g., *Bam*HI on 5′ primer and *Kpn*I on 3′ primer) for allowing oriented cloning of the amplified genes into the BACTH vectors. Be careful to correctly position these restriction sites so that the genes of interest will be in frame with the T25 and T18 open reading frames.
2. PCR amplify the genes encoding the proteins of interest using a standard protocol [18].
3. Purify the PCR-amplified DNA fragments using a standard PCR purification kit (available from various companies) and

digest them with the appropriate restriction enzymes (e.g., *Bam*HI and *Kpn*I or others depending on the restriction sites introduced into the primers). Digest the BACTH vectors with the same restriction enzymes.

4. Ligate the digested fragments and vectors with T4 DNA ligase [18]. Transform the ligation mixtures into competent XL1-Blue cells (Stratagene) and plate transformants on LB plates supplemented with appropriate antibiotics. Incubate plates at 30 °C for 24–36 h.
5. Pick 6–12 colonies for each cloning experiment and grow them overnight at 30 °C in 4 mL LB medium plus antibiotics (*see* **Note 5**). Purifiy plasmid DNA using standard protocol or commercial kit (e.g., QIAprep Spin Miniprep Kit from Qiagen). Check recombinant plasmids by restriction analysis and DNA sequencing to verify that no mutation was introduced during PCR amplification.

3.2.2 Gateway™ Cloning of Genes Encoding Proteins of Interest to $BACTH_{GW}$ Vectors

The Gateway® cloning technology (Life Technologies, Thermo Fisher Scientific) is used to transfer genes of interest to the BACTH-Gateway destination vectors, pST25-DEST, pSNT25-DEST, and pUT18C-DEST [17]. For detailed descriptions of the Gateway® cloning techniques, the reader may refer to the manufacturer's guidelines.

1. PCR-amplify the genes of interest (from genomic DNA or from other appropriate sources) using appropriate primers that also contain specific attB sites (*see* **Note 6**) and purify the PCR products as described earlier.
2. Mix the purified PCR products with the pDONR™221 plasmid [19], add the BP Clonase™ II enzyme and incubate 2 h at room temperature to allow the BP recombination reaction. Add 2 μg proteinase K to terminate recombination reaction and transform mixture into *E. coli* XL1 competent cells. Select transformants on LB plates supplemented with 50 μg/mL kanamycin [17].
3. Purify plasmid DNA from 3 to 4 independent clones from each cloning as described earlier, and check the recombinant plasmids by restriction analysis and DNA sequencing.

   In the resulting plasmids (pDONR™-gene X) the genes of interest are flanked by attL recombination sites and can be easily transferred into Gateway® destination vectors by a so-called LR reaction [19].

4. Mix the entry pDONR™-gene X plasmid with the appropriate $BACTH_{GW}$ destination vectors, pST25-DEST, pSNT25-DEST, or pUT18C-DEST. Add the LR Clonase™ enzyme mix

(see manufacturer's guidelines) and incubate 1 h at 25 °C. Longer incubation times may be necessary for larger inserts. Add proteinase K as described earlier to terminate reaction.

5. Transform mixture in *E. coli* XL1 competent cells. Select transformants on LB plates supplemented with appropriate antibiotic (spectinomycin or ampicillin).
6. Purify plasmid DNA from two to three independent clones from each cloning as earlier and check recombinant plasmids by restriction analysis or DNA sequencing.

The resulting plasmids encode the T25 or T18 fragments fused in frame to the gene of interest (gene X).

3.3 Analysis of Interactions by Screening Procedure on Indicator Plates

1. Prepare chemically competent or electro-competent DHT1, DHM1, or BTH101 cells by using standard procedures ([18]; *see* **Note 7**).
2. Cotransform the BACTH competent cells with one of the recombinant plasmids encoding the T25-fusions (pKT25, pKNT25, pST25-DEST, or pSNT25-DEST derivatives) and one of the recombinant plasmids encoding the T18-fusions (pUT18, pUT18C, or pUT18C-DEST derivatives).
3. In parallel, cotransform a separate aliquot of cells with plasmids pKT25 and pUT18C (coding for the unfused T25 and T18 fragments) to serve as a negative control. For a positive control, cotransform another aliquot of cells with plasmids pKT25-zip and pUT18C-zip (encoding the T25 and T18 fragments fused to a leucine-zipper dimerization motif).
4. Plate different amounts of the transformation mixture (in order to have no more than 2–500 colonies per plate) on LB-X-Gal or MacConkey-maltose indicator plates (plus antibiotics) and incubate at 30 °C for 24–48 h.

The results of typical phenotypic assays on LB-X-Gal or MacConkey-maltose plates are shown in Fig. 2. DHM1 (or other BACTH strains) transformants, expressing the T25-zip and T18-zip hybrid proteins that can heterodimerize through their leucine zipper motif, form blue colonies on LB X-Gal medium and red colonies on MacConkey/maltose, while cells expressing the unfused T25 and T18 remain colorless.

3.4 BACTH Screening of Interacting Partners: Selection Procedure on Minimal Medium

The BACTH system can be used to screen libraries to isolate partners of a protein of interest (e.g., protein X, classically designated as "bait") as follows:

1. Construct a library of genomic DNA (or cDNA) fragments in one of the BACTH vectors, e.g., pKT25, using standard procedures [18]. Obviously the quality (i.e., complexity) of the

library is critical for the success in isolating putative partners. A brief summary of a procedure used in our laboratory to construct a library of genomic *E. coli* chromosomal DNA fragments in the pKT25 vector is provided in **Note 8** (further experimental details can be found in [20, 21]). Clone the gene encoding protein X into one of the BACTH vectors (e.g., pUT18C) to generate so-called bait plasmid pUT18C-*X* that codes for the T18-X hybrid protein.

2. Transform pUT18C-*X* into a BACTH reporter strain (e.g., DHM1).
3. Prepare electrocompetent cells from the resulting transformants DHM1/pUT18C-*X* (*see* **Note 9**).
4. Transform electrocompetent DHM1/pUT18C-*X* cells with 50–100 ng of DNA from the BACTH DNA library constructed in plasmid pKT25. Add 1 mL LB medium and incubate 90 min at 30 °C. Collect the cells by centrifugation, wash them four to five times with M63 medium, and plate them (approximately 1×10^6 transformants/plate) on M63 minimal medium agar supplemented with maltose (0.2%), as the sole carbon source, kanamycin, ampicillin, IPTG, and X-Gal (to facilitate the detection of Cya^+ clones that are Mal^+ and Lac^+).
5. Incubate plates at 30 °C for 4–8 days until appearance of blue Cya^+ colonies. Reisolate these colonies on fresh plates, purify their pKT25 plasmids, and further characterize the DNA inserts by sequencing.

This procedure (and related ones) has been used in our laboratory to isolate several novel components of the *E. coli* cell division machinery [20, 21].

3.5 Quantification of Functional Complementation Between Hybrid Proteins by β-Galactosidase Assays

Quantification of the functional complementation mediated by interaction between the different hybrid proteins is performed by measuring β-galactosidase activities in bacterial liquid cultures [3, 10]. These β-galactosidase activity assays are conveniently carried out in 96-well microtiter plate format as it allows performing many assays in parallel [17, 21]. Other methods for β-galactosidase assays can be found elsewhere [16, 18, 22].

1. Pick eight independent colonies from each set of transformation (i.e., expressing a given couple of T25 and T18 hybrid proteins), and use them to inoculate 300–400 μL sterile LB broth supplemented with 0.5 mM IPTG and appropriate antibiotics and distributed to individual wells of a 96-well microtiter plate (2.2 mL 96-well storage plate or deep-well storage plate). Seal the plate with a microporous tape sheet to allow gas exchange and incubate overnight at 30 °C on a rotary shaker.

2. Dilute the cultures fivefold by adding appropriate volume of M63 medium to the same microplate.
3. Transfer 175 μL of the diluted cultures into a flat-bottom microtiter plate and record the $OD_{595\,nm}$ absorbance data with a microplate reader.
4. Transfer 200 μL of the diluted bacterial suspensions into a new microtiter plate (1.2 mL polypropylene 96-well storage block) and add 7 μL 0.05% SDS and 10 μL chloroform to permeabilize the cells. Mix vigorously and then leave the plate under a fume hood at room temperature for 30–40 min to allow chloroform evaporation.
5. In a new microtiter plate, distribute 105 μL/well PM2 reaction buffer containing 100 mM β-mercaptoethanol, and 0.1% o-nitrophenol-β-galactoside (ONPG). Start the enzymatic reactions by adding 20 μL aliquots of the permeabilized cells and incubate the plate at room temperature for 20–30 min or until sufficient yellow color has developed. In parallel, perform control assays with 20 μL of M63 medium instead of cells.
6. Stop the reaction by adding 50 μL 1 M Na_2CO_3 and record the OD_{405} absorbance data with a microplate reader.
7. Analyze data with an appropriate software (e.g., Microsoft Excel or other spreadsheet program). For each well, calculate the enzymatic activity, A (in relative units), according to

$$A = 1000 \times \left(OD_{405} - OD_{405}\text{in control wells}\right) / \left(OD_{595} - OD_{595}\text{in control wells}\right) / t\left(\text{min}\right)\text{of incubation.}$$

Results are given in relative units (RUs) of β-galactosidase activity. It is important to include in the assay negative and positive controls, i.e., bacteria that express noninteracting (e.g., T25 and T18 only or fused to proteins that do not interact or should not interact with the protein of interest) and interacting (e.g., T25-zip and T18-zip) hybrid proteins, respectively. Under routine conditions, the β-galactosidase activities measured with the positive controls (T25-zip/T18-zip) is defined as 100% activity, while the β-galactosidase activities of the negative controls (T25/T18) should be below 2–3% of positive control activity. The β-galactosidase activities in cells expressing the hybrid proteins of interest should be at least four to five times higher than the background level to demonstrate a positive interaction in the BACTH assay [17, 20, 21, 23].

3.6 Characterization of Hybrid Proteins by Western Blot

In many cases, it is important to characterize immunologically or biochemically the hybrid proteins and eventually to quantify their level of expression in the complementing cells. For this, western blot analysis of hybrid proteins can be carried out using standard procedures [18]. The T25 fragment can be detected with a rabbit polyclonal antiserum directed against the purified *B. pertussis* CyaA protein (serum L24023, DL unpublished) while the T18 fragment is revealed by an anti-CyaA monoclonal antibody (3D1, sc-13582) that reacts specifically with the C-terminal region of T18 [20, 24]. Alternatively, it is also possible to append to the T25 or T18 fragments different epitope tags that can be detected with specific monoclonal antibodies (e.g., myc, HA, or T7 tags) or a 6× histidine tag that permits purification of the complex of hybrid proteins by chromatography on Ni-NTA-agarose resin [25]. These modified fragments can be used to perform immunoprecipitation experiments or pull-down assays to demonstrate by direct biochemical means the physical association of the hybrid proteins [18].

4 Notes

1. Two types of indicator plates are commonly used to reveal protein interaction with the BACTH assay:

LB-X-Galplates: In *E. coli*, expression of the *lacZ* gene encoding β-galactosidase is positively controlled by cAMP/CAP. Hence, bacteria expressing interacting hybrid proteins form blue colonies on rich LB medium in the presence of the chromogenic substrate X-Gal (*see* Fig. 2), while cells expressing noninteracting proteins remain white (pale blue).

MacConkeymedium: *E. coli cya* bacteria are unable to ferment lactose or maltose [15, 18]; they form white (or pale pink) colonies on MacConkey indicator media containing maltose (*see* Fig. 2). In contrast, Cya^+ bacteria form red colonies on the same media (fermentation of the sugar results in the acidification of the medium and induces a color change of the phenol red dye). Note that not all MacConkey agar base media are of equal quality. MacConkey from Difco Laboratories (cat # 216830) is strongly recommended.

Cells expressing interacting proteins can be selected by plating transformants on a selective medium consisting of a synthetic minimal medium supplemented with maltose as a unique carbon source [4, 20, 21]: as the *mal* regulon (involved in maltose catabolism) expression is under a strict cAMP/CAP dependency, only Cya^+ bacteria can utilize maltose as a carbon source. Hence, only the cells that express interacting hybrid proteins will be able to grow on this minimal medium (Fig. 2). X-Gal and IPTG are also commonly added to the selective medium to facilitate the early visualization of the Cya^+ colonies (these cells should also be Lac^+ and therefore exhibit a blue phenotype on X-Gal). Note that when using the DHT1 as a reporter strain [23, 26], casamino acids should be added to the minimal medium/maltose plates to allow growth, as this strain is *ilv*⁻ (i.e., unable to synthetize isoleucine and valine).

2. β-mercaptoethanol is considered toxic, causing irritation to the skin and respiratory tract upon inhalation and should be manipulated under a fume hood. In fact, it can be easily omitted from the PM2 buffer: the β-galactosidase activities will be reduced by a factor of 2, which is not problematic because only relative enzymatic activities will be considered.
3. Several adenylate-cyclase-deficient *(cya) E. coli* reporter strains, DHT1, DHM1, and BTH101 (see genotypes below), can be used as host organisms for the detection of protein–protein interactions in a BACTH assay [4, 13, 27]. Other *E. coli cya* strains (see *E. coli* strain collection at http://cgsc.biology.yale.edu) may also be used. The different genetic backgrounds of these strains provide different complementation efficiencies and different reporter gene stringencies. DHT1 [F^-, *cya-854, ilv* 691::Tn10, *recA1, endA1, gyrA96(Nal$_r$), thil, hsdR17, spoT1, rfbD1, glnV44(AS)*] is a *recA* strain that displays a high BACTH complementation efficiency and fast growth, but it requires casamino acid supplementation for growth on minimal medium as it carries an *ilv* mutation. DHM1 [F^-, *cya-854, recA1, endA1, gyrA96*(Nal$_r$), *thil, hsdR17, spoT1, rfbD1, glnV44(AS)*] is an *ilv* + DHT1 derivative able to grow on minimal media plus sugars, but it displays a lower complementation efficiency and slower growth than the parental DHT1. BTH101 [F^-, *cya-99, araD139, galE15, galK16, rpsL1* (Str^r), *hsdR2, mcrA1, mcrB1*] also displays a good BACTH efficiency and fast growth, but some instability of plasmids may be observed owing to the Rec$^+$ character of the strain. The frequencies of spontaneous Lac$^+$ revertants (owing to cAMP/CAP independent promoter mutations) of these different strains range from 10^{-7} to 10^{-8}, while frequencies of spontaneous Mal$^+$ revertants are below the detection threshold (i.e., $<10^{-10}$).
4. The BACTH technology requires coexpression of two hybrid proteins within the same recipient *cya* bacteria. For this, two sets of compatible vectors allowing genetic fusions of the proteins of interest at either the N- or the C-terminus of the T25 fragment (pKT25 and pKNT25) or of the T18 fragment (pUT18 and pUT18C) are available: their schematic maps are shown in Fig. 3a, and their nucleotide sequences are available upon request [4, 13].

Plasmid pKT25 expresses under a *lac* promoter control, the T25 fragment (corresponding to the first 224 amino acids of CyaA). It is a low-copy-number pSU40 plasmid derivative harboring a kanamycin resistance selectable marker. It contains a multicloning site sequence (MCS) at the 3′ end of T25 to allow construction of in-frame fusions at the C-terminal end of the T25 polypeptide. Plasmid pKNT25 is similar to pKT25 except that MCS is located at the 5′ end of T25 coding region to allow fusions of proteins to the N-terminus of T25.

Plasmid pUT18 is an ampicillin-resistant pUC19 derivative that expresses the T18 fragment (amino acids 225 to 399 of CyaA) under the transcriptional control of the *lac* promoter. The T18 open reading frame is located downstream of the pUC19 MCS, and therefore pUT18 is used to express chimeric proteins in which the polypeptide of interest is fused to the N-terminal end of T18. In plasmid pUT18C, the same MCS is located at the 3′ end of the T18 open reading frame to allow fusions of proteins to the C-terminus of T18.

In addition two plasmids, pKT25-zip and pUT18C-zip, are commonly used as positive controls for BACTH complementation. They are derivatives of pKT25 and pUT18C, respectively, that code for the T25 and T18 fragments fused to the leucine zipper of GCN4 [10, 11].

Another set of vectors was recently designed to be compatible with the Gateway® recombineering technique (Thermo Fisher Scientific). The Gateway® technique allows for facile recombinase-mediated transfer of an open reading frame (ORF), flanked by recombination sites, from an "entry" vector into a wide variety of "destination" vectors [19]. The Gateway®-compatible destination vectors were constructed by the insertion of a recombination cassette encoding a chloramphenicol resistance marker and the toxin CcdB and flanked by *attR* bacteriophage lambda recombination sites (Fig. 3b) in the BACTH vectors. The resulting plasmids pST25-DEST and pUT18C-DEST are suitable for fusing proteins of interest to the C-terminus of the T25 and T18 fragments, respectively, while pSNT25-DEST is used for fusing ORFs to the N-terminus of the T25 fragment [17]. Importantly, the pST25-DEST and pSNT25-DEST $BACTH_{GW}$ vectors contain a spectinomycin-resistant (instead of kanamycin-resistant) marker to be compatible with the popular Gateway® entry vector pDONR221 that harbors a kanamycin resistance gene. The pST25-DEST, pSNT25-DEST, and pUT18C-DEST plasmids must be propagated in *E. coli* strains that are resistant to the lethal effect of the CcdB toxin, such as the DB3.1™ *E. coli* strain (harboring the CcdB-resistant, *gyrA462* gyrase mutation; see manufacturer's guidelines).

5. Vectors and recombinant plasmids are commonly propagated at 30 °C in standard *E. coli K12 recA* strains (such as XL1-Blue). To avoid any problems during construction of the plasmids, it is wise to grow the cells in LB medium containing 0.2% glucose or to use an *E. coli* host strain that overproduces the LacI repressor to prevent expression of the hybrid proteins (e.g., XL1-Blue). Plasmid DNA is routinely purified with commercial kits used for minipreparation of DNA, according to the manufacturer's instructions.

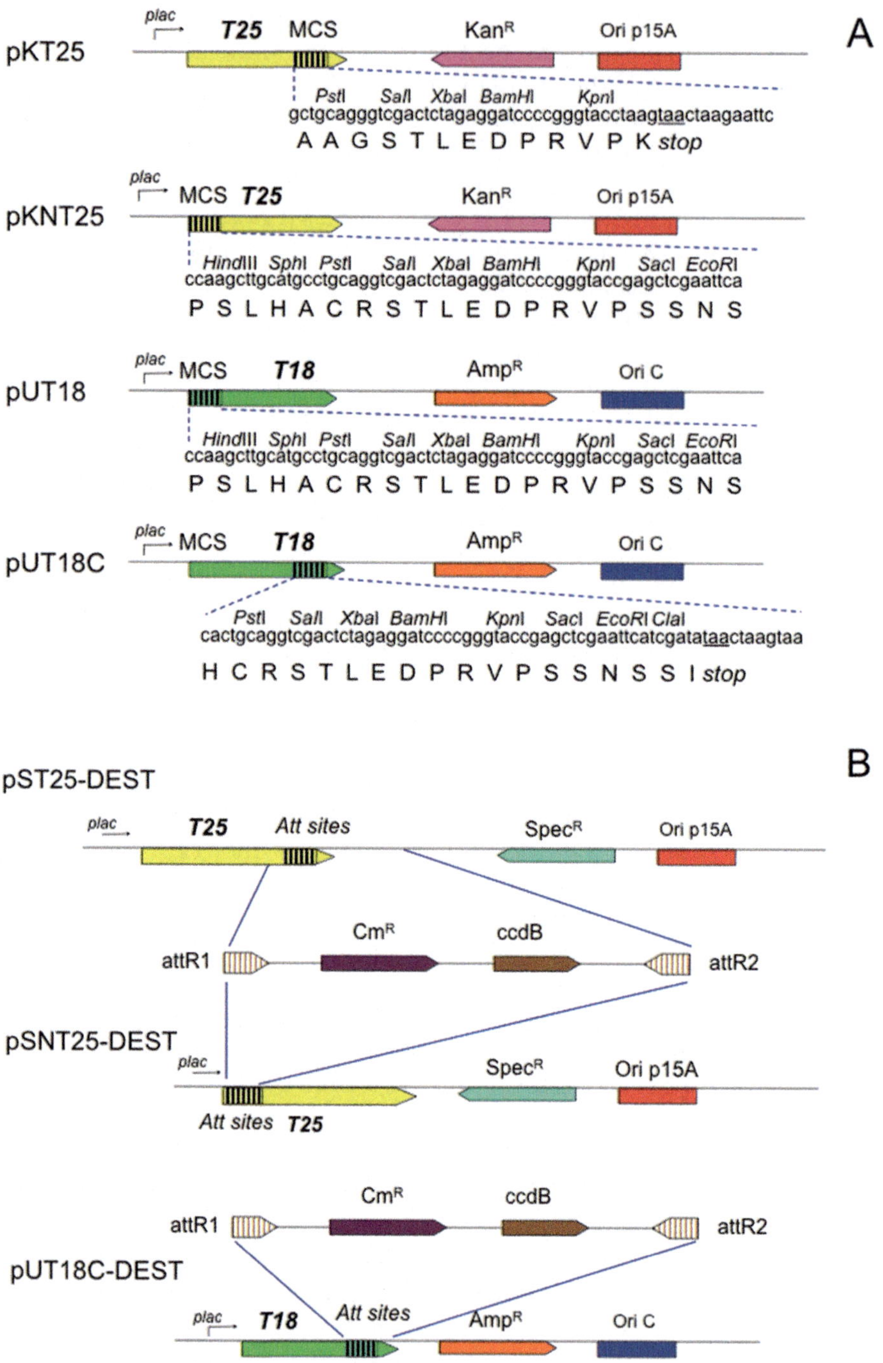

Fig. 3 Schematic representation of BACTH plasmids. (**a**) Standard BACTH plasmids. *Yellow* and *green rectangles* represent the open reading frames of T25 and T18 fragments, respectively, under the control of the *lac* promoter (small arrow). *Pink* and *orange arrows* indicate the antibiotic selectable markers and the direction of transcription. The *red* and *blue boxes* indicate the plasmid origin of replication. The *hatched boxes* represent the multicloning sequences (MCS) that allow insertion of foreign genes: some unique restriction sites are displayed above the nucleotide sequence, and the encoded polypeptide sequences are shown below. (**b**) Gateway™-compatible, BACTH$_{GW}$ plasmids

6. During the design of primers to amplify the genes of interest, the following sequences (red, underlined) corresponding to the specific *attB* sites (necessary for recombination reaction) should be appended to the gene-specific sequences (indicated by **XXX**...):

Direct primer (the bold ATG corresponds to the initiation codon of the open reading frame): 5′-GCCGCACAAGTTTGTACAAAAAAGCAGGCTTT**ATGXXXXXXXXX**

Reverse primer: (the stop codon can be removed if desired): 5′-GCGGACCACTTTGTACAAGAAAGCTGGGTT**XXXXXXXX**

Refer to the manufacturer's guidelines for more precise explanations regarding the design of primers for Gateway® cloning.

7. Before use, the strains (DHT1, DHM1, or BTH101) from the LB-DMSO stock should be restreaked on either MacConkey/maltose or LB/X-Gal/IPTG plates and grown overnight at 37 °C. White colonies (i.e., *cya*) should be picked up to start the overnight liquid preculture. Any red (on MacConkey/maltose) or blue colonies (on LB/X-Gal/IPTG) that may appear should be avoided (they likely correspond to Lac^+ or Mal^+ revertants or contaminants). If too many contaminants are present upon restreaking of the stock, a selective antibiotic may be added to the MacConkey/maltose or the LB/X-Gal/IPTG plates: DHT1 and DHM1 are resistant to nalidix acid (30 μg/mL), whereas BTH101 is resistant to streptomycin (100 μg/mL).

DHT1, DHM1, or BTH101 competent cells can be prepared by the classical $CaCl_2$ technique [18], which yields a competency level (>10^6 cfu/μg) sufficient for most routine transformations. Briefly, freshly reisolated cells are grown in 1 L LB medium at 37 °C to OD 0.25–0.3, cooled on ice, and pelleted by centrifugation. Cells are washed twice in 100 mL ice-cold 0.1 M $CaCl_2$ solution. Cells are finally resuspended in 30–40 mL ice-cold 0.1 M $CaCl_2$ and incubated overnight at 4 °C (it is critical to keep cells, buffers, and vessels well chilled at all stages of the process).

For transformation, 50 μL of chemically competent DHM1 cells are mixed in a chilled microcentrifuge tube with 5–10 ng of each plasmid, incubated 30 min at 4 °C, and then heat-shocked at 42 °C for 2 min. Then 1 mL LB is added, and the cell suspension is further incubated at 30 °C for 60–90 min before being plated. Different volumes of the transformation mixture should be plated to obtain about 100–200 colonies per plate. It is important that the number of colonies not exceed 500; otherwise, the detection of positive clones might be difficult. It should be noted that after prolonged incubation (4–5 days), negative colonies (i.e., cya^-) will show a weak red (on MacConkey-maltose) or blue spot (on LB-X-Gal) in the center but will remain colorless at the periphery. It might also be worth testing the complementation at 37 °C, although in many cases complementation at 37 °C is less efficient than at 30 °C.

8. The genomic DNA (≈50 μg) from a Δ*cya* derivative of the *E. coli* strain MG1655 was randomly fragmented by sonication (size range of 500–1500 bp). The fragments were end-repaired by Mung Bean nuclease and treated with a mixture of T4 DNA polymerase and Klenow fragment (with dNTP). In parallel, the pKT25 vector (10 μg) was digested with *Sma*I and dephosphorylated with shrimp alkaline phosphatase, and the linearized vector was gel purified. The blunt-ended DNA fragments were then ligated with the *Sma*I-digested pKT25 vector and transformed into electrocompetent ElectroMAX DH10B cells (Thermo Fisher Scientific). About 5 × 105 independent clones were thus obtained. All these colonies were pooled and their plasmid DNA was purified and used as a stock for the BACTH DNA library [21].
9. Efficient (>10^8 cfu/μg) electrocompetent DHM1/pUT18C-*X* cells can be prepared as follows [18]: freshly reisolated cells are grown at 37 °C in 1 L LB containing 100 μg/mL ampicillin until OD_{600} of 0.5–0.7. Cells are chilled on ice and pelleted by centrifugation at 4 °C. Cells are washed at least three times with ice-cold water and resuspended in 10 mL 10% glycerol (in water). For transformation, 50 μL are transferred into an electroporation cuvette (1 mm wide) previously equilibrated on ice, and 50–100 ng DNA from the BACTH plasmid DNA library are added. After mixing and a few minutes of incubation at 4 °C, the cuvette is placed in an electroporator (e.g., BioRad) set at 2.5 KV, 100 Ohms capacitance, and electroporation is carried out. One milliliter of LB media is immediately added to the cuvette, and cells are further incubated at 30 °C for 60–90 min. Cells are then collected by centrifugation (5 min at 6000 rpm, or 4500 × *g*) and washed several times with M63 medium (to remove all nutrients from the rich medium) before being plated on M63 minimal medium agar (approximately 1 × 10^6 transformants/plate).

Acknowledgments

This work was supported by Institut Pasteur and the Centre National de la Recherche Scientifique (CNRS UMR 3528, Biologie Structurale et Agents Infectieux). E.G. was supported by Ph.D. funding from the Université Paris Diderot, Sorbonne Paris Cité, Cellule Pasteur, Paris, France.

References

1. Fields S, Song O (1989) A novel genetic system to detect protein-protein interactions. Nature 340:245–246
2. Stynen B, Tournu H, Tavernier J, Van Dijck P (2012) Diversity in genetic in vivo methods for protein-protein interaction studies: from the yeast two-hybrid system to the mammalian split-luciferase system. Microbiol Mol Biol Rev 76:331–382
3. Karimova G, Pidoux J, Ullmann A, Ladant D (1998) A bacterial two-hybrid system based on a reconstituted signal transduction pathway. Proc Natl Acad Sci U S A 95:5752–5756
4. Karimova G, Dautin N, Ladant D (2005) Interaction network among *Escherichia coli* membrane proteins involved in cell division as revealed by bacterial two-hybrid analysis. J Bacteriol 187:2233–2243
5. Jack RL, Buchanan G, Dubini A, Hatzixanthis K, Palmer T, Sargent F (2004) Coordinating assembly and export of complex bacterial proteins. EMBO J 23:3962–3972
6. Paschos A, den Hartigh A, Smith MA, Atluri VL, Sivanesan D, Tsolis RM, Baron C (2011) An in vivo high-throughput screening approach targeting the type IV secretion system component VirB8 identified inhibitors of Brucella abortus 2308 proliferation. Infect Immun 79:1033–1043
7. Cisneros DA, Bond PJ, Pugsley AP, Campos M, Francetic O (2012) Minor pseudopilin self-assembly primes type II secretion pseudopilus elongation. EMBO J 31:1041–1053
8. Georgiadou M, Castagnini M, Karimova G, Ladant D, Pelicic V (2012) Large-scale study of the interactions between proteins involved in type IV pilus biology in Neisseria meningitidis: characterization of a subcomplex involved in pilus assembly. Mol Microbiol 84:857–873
9. Zoued A, Durand E, Brunet YR, Spinelli S, Douzi B, Guzzo M, Flaugnatti N, Legrand P, Journet L, Fronzes R et al (2016) Priming and polymerization of a bacterial contractile tail structure. Nature 531:59–63
10. Karimova G, Ullmann A, Ladant D (2000) A bacterial two-hybrid system that exploits a cAMP signaling cascade in *Escherichia coli*. Methods Enzymol 328:59–73
11. Ladant D, Ullmann A (1999) Bordatella pertussis adenylate cyclase: a toxin with multiple talents. Trends Microbiol 7:172–176
12. Lawson CL, Swigon D, Murakami KS, Darst SA, Berman HM, Ebright RH (2004) Catabolite activator protein: DNA binding and transcription activation. Curr Opin Struct Biol 14:10–20
13. Karimova G, Ullmann A, Ladant D (2001) Protein-protein interaction between *Bacillus stearothermophilus* tyrosyl-tRNA synthetase subdomains revealed by a bacterial two-hybrid system. J Mol Microbiol Biotechnol 3(1): 73–82
14. Fransen M, Brees C, Ghys K, Amery L, Mannaerts GP, Ladant D, Van Veldhoven PP (2002) Analysis of mammalian peroxin interactions using a non-transcription-based bacterial two-hybrid assay. Mol Cell Proteomics 1: 243–252
15. Dautin N, Karimova G, Ladant D (2003) Human immunodeficiency virus (HIV) type 1 transframe protein can restore activity to a dimerization-deficient HIV protease variant. J Virol 77:8216–8226
16. Battesti A, Bouveret E (2012) The bacterial two-hybrid system based on adenylate cyclase reconstitution in *Escherichia coli*. Methods 58:325–334
17. Ouellette SP, Gauliard E, Antosova Z, Ladant D (2014) A Gateway((R)) -compatible bacterial adenylate cyclase-based two-hybrid system. Environ Microbiol Rep 6:259–267
18. Sambrook J, Russell DW (2006) The condensed protocols from molecular cloning: a laboratory manual. Cold Spring Harbor Laboratory Press, Cold Spring Harbor, NY
19. Hartley JL, Temple GF, Brasch MA (2000) DNA cloning using in vitro site-specific recombination. Genome Res 10:1788–1795
20. Karimova G, Robichon C, Ladant D (2009) Characterization of YmgF, a 72-residue inner membrane protein that associates with the *Escherichia coli* cell division machinery. J Bacteriol 191:333–346
21. Karimova G, Davi M, Ladant D (2012) The beta-lactam resistance protein Blr, a small membrane polypeptide, is a component of the *Escherichia coli* cell division machinery. J Bacteriol 194:5576–5588
22. Griffith KL, Wolf REJ (2002) Measuring beta-galactosidase activity in bacteria: cell growth, permeabilization, and enzyme assays in 96-well arrays. Biochem Biophys Res Commun 290:397–402
23. Ouellette SP, Rueden KJ, Gauliard E, Persons L, de Boer PA, Ladant D (2014) Analysis of MreB interactors in Chlamydia reveals a RodZ homolog but fails to detect an interaction with MraY. Front Microbiol 5:279

24. Robichon C, Karimova G, Beckwith J, Ladant D (2011) Role of leucine zipper motifs in association of the *Escherichia coli* cell division proteins FtsL and FtsB. J Bacteriol 193:4988–4992
25. Battesti A, Bouveret E (2008) Improvement of bacterial two-hybrid vectors for detection of fusion proteins and transfer to pBAD-tandem affinity purification, calmodulin binding peptide, or 6-histidine tag vectors. Proteomics 8:4768–4771
26. Ouellette SP, Karimova G, Subtil A, Ladant D (2012) Chlamydia co-opts the rod shape-determining proteins MreB and Pbp2 for cell division. Mol Microbiol 85:164–178
27. Dautin N, Karimova G, Ullmann A, Ladant D (2000) Sensitive genetic screen for protease activity based on a cyclic AMP signaling cascade in *Escherichia coli*. J Bacteriol 182:7060–7066

Chapter 14

Protein–Protein Interactions: Yeast Two-Hybrid System

Jer-Sheng Lin and Erh-Min Lai

Abstract

The yeast two-hybrid system is a powerful and commonly used genetic tool to investigate interactions between artificial fusion proteins inside the nucleus of yeast. Here we describe how to use the Matchmaker GAL4-based yeast two-hybrid system to detect the interaction of the *Agrobacterium* type VI secretion system (T6SS) sheath components TssB and $TssC_{41}$. The bait and prey gene are expressed as a fusion to the GAL4 DNA-binding domain (DNA-BD) and GAL4 activation domain (AD, prey/library fusion protein) respectively. When bait and prey fusion proteins interact in yeast nucleus, the DNA-BD and AD are brought into proximity, thereby activating the transcription of reporter genes. This technology can be widely used to identify interacting partners, confirm suspected interactions, and define interacting domains.

Key words Protein–protein interaction, Yeast two-hybrid, Gal4 transcriptional activation domain (AD), Gal4 DNA-binding domain (BD), *Saccharomyces cerevisiae* AH109, Type VI secretion system, TssB, TssC

1 Introduction

The yeast two-hybrid system (Y2H) was first developed in 1989 and revolutionized the process of searching for and identifying interacting proteins [1]. To date, the Y2H system has proven to be a useful and sensitive method for detecting not only stable interacting proteins but also weak and transient protein interactions [2]. Because Y2H is performed in vivo, the great advantage of the system is that the testing proteins are more likely to be in their native conformations, which may lead to increased sensitivity and accuracy of detection [1, 3, 4]. Importantly, the Y2H system is complementary to biochemical methods such as co-immunoprecipitation/pulldown followed by western blotting or mass spectrometry analysis to increase accuracy and dynamics for a more complete and reliable map of interactions [2]. Notably, the Y2H method has been modified and improved greatly in recent years, including applications in protein–DNA interactions and yeast three-hybrid, and has proven to be amenable to interaction studies of membrane proteins, DNA

Laure Journet and Eric Cascales (eds.), *Bacterial Protein Secretion Systems: Methods and Protocols*, Methods in Molecular Biology, vol. 1615, DOI 10.1007/978-1-4939-7033-9_14, © Springer Science+Business Media LLC 2017

binding proteins, and RNA binding proteins [5–8]. Using the Matchmaker GAL4-based Y2H system (Clontech, Mountain View, CA) as an example, the principle of the Y2H system is illustrated in Fig. 1a. Based on the properties of the yeast GAL4 transcription factor it consists of separable domains responsible for DNA binding and transcriptional activation [3]. Bait proteins are expressed as a fusion to the GAL4 DNA binding domain (DNA-BD), while prey proteins are expressed as fusions to the GAL4 activation domain (AD). When bait and prey fusion proteins interact in yeast nucleus, the DNA-BD and AD are brought into proximity to restore to a functional GAL4 transcriptional activator, which binds onto an upstream activating sequence (UAS) of reporter genes (such as

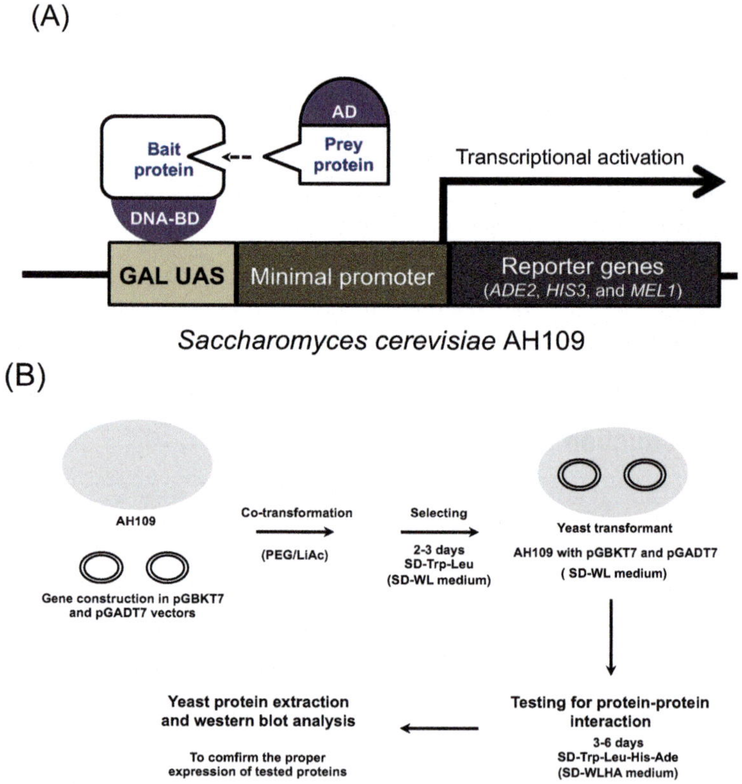

Fig. 1 Principle and experimental flowchart of yeast two-hybrid system. (**a**) Schematic diagram of principle of Y2H system. Two testing proteins are each fused with two different Gal4 domains, with the bait protein fused to the Gal4 DNA-binding domain (DNA-BD, 1-147 a.a.) and the prey protein fused to the Gal4 transcriptional activation domain (AD, 768-881 a.a.). In yeast strain AH109, transcriptional activation of the reporters (*ADE2*, *HIS3*, and *MEL1*) only occurs in a cell that bait protein interact with prey protein to restore functional Gal4 transcription factor binding to the Gal4-responsive promoter GAL UAS [3]. (**b**) Experimental flowchart of Y2H system. Cotransformation was performed using PEG/LiAc-mediated transformation method. SD-WL medium is representative of synthetic defined (SD) minimal medium lacking tryptophan (Trp) and leucine (Leu). SD-WLHA medium is representative of synthetic dextrose minimal medium lacking Trp, Leu, adenine (Ade), and histidine (His)

ADE2 and *HIS3*) for transcriptional activation. The Y2H system has been widely used to detect the interactions of a wide range of proteins from yeast, bacteria, animal, and plant systems. Y2H has been successfully used to study the interaction of proteins involved in bacterial protein secretion from type IV [9–12] and type VI secretion systems (T4SS, T6SS) in *Agrobacterium tumefaciens* [13, 14].

Here, the Y2H protocol describes the use of Matchmaker Y2H system to detect the interaction of the *Agrobacterium* T6SS sheath components TssB and $TssC_{41}$ according to the instructions of the user manual (Clontech, Mountain View, CA), with minor modifications. TssB and TssC interact to form a cogwheel-like tubular structure, which is analogous to the outer sheath structure of a contractile phage, and wraps around the T6SS tail tube to propel the tail tube toward the target cell interior upon infection [15, 16]. In *A. tumefaciens*, we showed the interaction of the T6SS sheath components TssB and $TssC_{41}$ by Y2H assay, copurification in *E. coli*, and co-IP in *A. tumefaciens* [14]. For the Y2H assay, each bait and prey plasmid pair was cotransformed into the *Saccharomyces cerevisiae* strain AH109. The transformants were selected by their growth on synthetic dextrose (SD) minimal medium lacking tryptophan (Trp) and leucine (Leu) (SD-WL medium), which are the nutritional selection markers for pGBKT7 and pGADT7, respectively. The positive interaction of expressed fusion proteins was then determined by their growth on SD lacking Trp, Leu, adenine (Ade), and histidine (His) (SD-WLHA medium) at 30 °C for at least 3 days (Fig. 1b). The positive interactions wwere observed only for plasmid pairs expressing TssB and $TssC_{41}$ but not when each of them coexpressed with vector only, suggesting the specific interactions of TssB and $TssC_{41}$ (Fig. 2) [14].

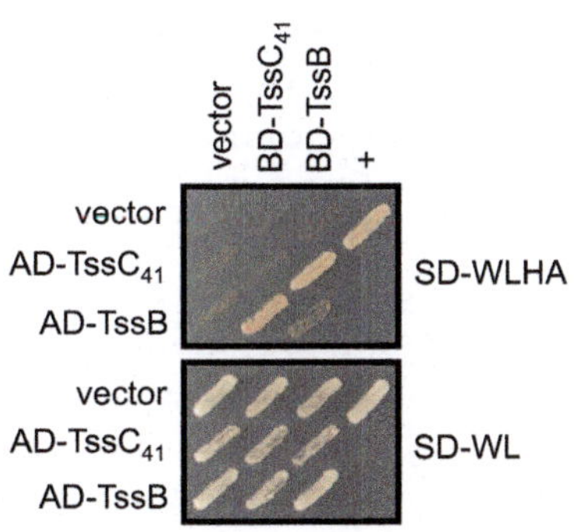

Fig. 2 TssB and $TssC_{41}$ interact with each other in yeast strain AH109. SD-WL medium (SD minimal medium lacking Trp and Leu) was used for the selection of plasmids. SD-WLHA medium (SD minimal medium lacking Trp, Leu, His, and Ade) was used for the auxotrophic selection of bait and prey protein interactions. The positive interaction was determined by the growth on SD-WLHA medium at 30 °C for at least 2 days. The positive control (+) showing interactions of SV40 large T-antigen and murine p53 and negative control (vector) are indicated (reproduced from [14]; no permission is required for reuse of the content published from Public Library of Science, PLoS)

2 Materials

All growth media and solutions are prepared using Milli-Q purified water and analytical or molecular biology grade reagents.

2.1 Yeast Strain and Vectors (the Following Information Is According to [3])

1. The yeast *Saccharomyces cerevisiae* strain AH109: The complete genotype of AH109 is provided in what follows.

   MATa, *trp1–901, leu2–3, 112, ura3–52, his3–200, gal4Δ, gal80Δ*, LYS2::GAL1$_{UAS}$-GAL1$_{TATA}$-HIS3, GAL2$_{UAS}$-GAL2$_{TATA}$-ADE2, URA3:: MEL1$_{UAS}$-MEL1$_{TATA}$-*lacZ*.

   AH109 strain is *gal4*$^{-}$ and *gal80*$^{-}$; this prevents interference of native regulatory proteins with the regulatory elements in the two-hybrid system. AH109 features three reporters, *ADE2*, *HIS3*, and *MEL1* (or *lacZ*), under the control of distinct GAL4 upstream activating sequences (UASs) and TATA boxes.
2. pGBKT7 vector: The pGBKT7 vector contains a multiple cloning site (MCS) for cloning to express proteins with N-terminal fusion to amino acids 1–147 of the GAL4 DNA binding domain (DNA-BD). In yeast, fusion proteins are expressed at high levels from the constitutive *ADH1* promoter (P_{ADH1}). Transcription is terminated by the T7 and ADH1 transcription termination signals (T_{ADH1}). The pGBKT7 vector can replicate autonomously in both *E. coli* and *S. cerevisiae* from the pUC and 2 μ ori, respectively. The vector carries the kanamycin-resistant gene for selection in *E. coli* and the *TRP1* nutritional marker for selection in yeast. In addition, pGBKT7 also contains the T7 promoter and a c-Myc epitope tag for in vitro transcription and translation of the c-Myc-tagged fusion protein without GAL4 DNA-BD.
3. The pGADT7 vector: The pGADT7 vector contains MCSs for cloning to express protein with *N*-terminal fusion to amino acids 768–881 of the GAL4 activation domain (AD). In yeast, fusion proteins are expressed at high levels from the constitutive *ADH1* promoter (P_{ADH1}). Transcription is terminated at the ADH1 transcription termination signal (T_{ADH1}). The fusion protein is targeted to the yeast nucleus by the SV40 nuclear localization sequences that have been added to the activation domain sequence. pGADT7 also contains the T7 promoter and an HA epitope tag for in vitro transcription and translation of the HA-tagged fusion protein without GAL4 AD. The pGADT7 vector can replicate autonomously in both *E. coli* and *S. cerevisiae* from the pUC and 2 μ ori, respectively. The vector carries the ampicillin-resistant gene for selection in *E. coli* and the *LEU2* nutritional marker for selection in yeast.

2.2 Yeast Cultures and Yeast Transformation [17]

1. Yeast peptone dextrose adenine (YPDA) medium: 20 g Bacto peptone, 10 g yeast extract, 20 g glucose, 40 mg adenine, 15 g agar (For plate use only), add water to 1 L, autoclave.
2. Minimal synthetic defined (SD) plate: 1.675 g yeast nitrogen base without amino acid, 5 g glucose, 3.75 g Agar, add water to 250 mL, autoclave. Dropout (DO) supplements (such as -Trp-Leu or -Trp-Leu-Ade-His) can be added to the Minimal SD Base to make a synthetic, defined medium lacking the specified nutrients (*see* **Note 1**).
3. Carrier DNA: 10 mg/mL salmon sperm DNA (ssDNA) (UltraPure™ Salmon Sperm DNA Solution, ThermoFisher), store at −20 °C (*see* **Note 2**).
4. 10× LiAc: 1 M Lithium acetate, pH 7.5 (*see* **Note 3**), autoclave and store at room temperature (RT).
5. 40% polyethylene glycol (PEG) solution: 22 g Polyethylene glycol (Molecular weight is 6000 or 3350 Da), add 31 mL of water, autoclave and store at room temperature.
6. Plasmid DNA: ~200 ng per plasmid for co-transformation (*see* **Note 4**).
7. Laminar flow.

2.3 Selective Media

1. Selective medium for transformants: Minimal synthetic defined (SD) plate with -Leu/-Trp DO supplement (containing every essential amino acids except for leucine and tryptophan) (*see* **Note 5**).
2. Selective medium for Protein–Protein interactions: Minimal synthetic defined (SD) plate with -Leu/-Trp/-His/-Ade DO supplement (containing all essential amino acids except for leucine, tryptophan, histidine, and adenine).

2.4 Preparation of Yeast Cultures for Protein Extraction and Western Blot

1. Yeast peptone dextrose adenine (YPDA) medium: *See* **item 1** in Subheading 2.2.
2. 2× minimal synthetic defined (SD) medium: 40 g Bacto peptone, 20 g yeast extract, 20 g glucose, 80 mg adenine, with 2X-Leu/-Trp dropout (DO) supplement (containing every essential amino acids except for leucine and tryptophan), add water to 1 L, autoclave.

2.5 Preparation of Yeast Protein Extracts

1. 1.5 mL microtube.
2. Acid-washed glass beads (425–600 μm).
3. Protease inhibitor cocktail solution.
4. Phenylmethylsulfonyl fluoride (PMSF) stock solution: 0.1 M
5. Yeast protein extraction buffer (*see* **Note 6**): 0.1% NP-40, 250 mM NaCl, 50 mM Tris–HCl, pH 7.5, 5 mM ethylenediaminetetraacetic acid (from a 0.5 M, pH 8.0 stock solution),

mix well, and place on ice. Before use, add 1 mM dithiothreitol (DTT), 2× protease inhibitor cocktails (stock: 50×), 4 mM PMSF, mix well, then ready for use.

3 Methods

3.1 Gene Construction in pGBKT7 and pGADT7 Vectors

The constructs used for Y2H analysis are generated based on the information of the vector map and MCS provided by the user manual (Clontech) [3]. Briefly, bait and prey coding sequence (without stop codon) are PCR-amplified with appropriate primers, digested with appropriate enzymes, and cloned to the same sites of pGBKT7 or pGADT7 [14].

3.2 Preparation of Yeast Cultures for Yeast Transformation [17]

1. Inoculate 3 mL YPDA with a colony of AH109 (*see* **Note 7**) and incubate at 30 °C overnight (>16 h) with shaking (250 rpm) to the stationary phase (*see* **Note 8**).
2. Subculture by adding 1 mL AH109 overnight culture into 50 mL fresh YPDA medium.
3. Incubate at 30 °C for 4 h with shaking (250 rpm) (*see* **Note 9**).
4. Pour cells into 50 mL tubes and pellet cells at 450 × *g* for 3 min at 4 °C or RT (*see* **Note 10**).
5. Discard supernatant and resuspend cell pellet with 10 mL sterile water by vortexing, and repellet cells at 450 × *g* for 3 min at 4 °C or RT (*see* **Note 11**).
6. Resuspend cell pellet in 100 μL 10× LiAc and 900 μL sterile water (final concentration is 1× LiAc) (*see* **Note 12**). Incubate cell suspension at 30 °C for 1 h with gentle shaking (150 rpm) (*see* **Note 13**).
7. The suspended yeast competent cells are ready to use for transformation (*see* **Note 14**).

3.3 PEG/LiAc-Mediated Transformation of Yeast (Small-Scale Transformation of Bait and Prey Plasmids) (see Note 15)

1. Pretreat ssDNA by heating at 100 °C for 10 min and then put on ice for 5–10 min before use (*see* **Note 16**).
2. Add 80 μL heat-treated ssDNA (10 μg/μL) to 1 mL yeast competent cell (final concentration ~ 0.8 mg/mL) and mix well (*see* **Note 17**).
3. Aliquot 100 μL of cell mixture into 1.5 mL microtube, and add an approximate amount of plasmid DNA (about 3–5 μL) (*see* **Note 18**). Mix well by vortexing and incubate at 30 °C for 30 min (*see* **Note 19**).
4. Freshly prepare LiAc-PEG solution (10× LiAc: 40% PEG = 1:10, mix 1 mL 10× LiAc with 10 mL 40% PEG) and add 700 μL LiAc-PEG solution to cell mixture after 30 min

incubation (*see* **Note 20**). Resuspend cell mixture immediately by vortexing (*see* **Note 21**) before incubation at 30 °C for 1 h.

5. Heat shock at 42 °C for 5 min (*see* **Note 22**).
6. Pellet cells by centrifugation at 14,500 × *g* for 1 min at RT (*see* **Note 23**).
7. Discard supernatant as much as possible to remove PEG. Resuspend cells in 300 μL sterile water (*see* **Note 24**).

3.4 Selection of Transformants

1. Streak cells on SD/-Trp-Leu selective plate and incubate at 30 °C for 2–3 days (*see* **Note 25**).
2. Patch single colony on SD/-Trp-Leu selection plate and incubate at 30 °C for 2 days (*see* **Note 26**).

3.5 Testing for Protein–Protein Interactions

1. Patch cells on both SD/-Trp-Leu (control) and SD/-Trp-Leu-His-Ade selection plates for 3–6 days (*see* **Note 27**).
2. Photograph the plate to record the final protein–protein interaction results (For example, TssB and $TssC_{41}$ can interact strongly in yeast, Fig. 2) [14].

3.6 Preparation of Yeast Cultures for Protein Extraction (see Note 28)

1. Grow culture from a single colony (*see* **Note 29**) in 3 mL YPDA or 2× SD selection medium (*see* **Note 30**) overnight at 30 °C.
2. Add 100 μL of overnight culture (OD_{600} should reach 1.5) in fresh 5 mL 2× SD selection medium. Incubate at 30 °C with shaking (about 250 rpm) until OD_{600} reaches 0.4–0.6. Depending on tested proteins, it may take 4–5 h to reach desired cell number (*see* **Note 31**).
3. Place cells in 15 mL tubes and pellet cells at 1000 × *g* for 5 min at 4 °C.
4. Discard supernatant and resuspend cell pellet with 10 mL sterile water by vortexing (*see* **Note 32**).
5. Repellet cells at 1000 × *g* for 5 min at 4 °C.
6. Repeat **steps 4** and **5**.
7. Discard supernatant. You can continue to extract yeast protein or immediately freeze cell pellet by placing tube in liquid nitrogen, then store cells at −80 °C until western blot analysis.

3.7 Preparation of Yeast Protein Extracts and Western Blot Analysis

1. Keep protein samples on ice, add 100 μL freshly prepared yeast protein extraction buffer to tube followed by addition of 50 μL acid-washed glass beads (*see* **Note 33**).
2. Vortex tubes at maximum speed for 30 s, then place tubes on ice for 30 s (*see* **Note 34**). Repeat this step six times.
3. Transfer supernatant above settled glass beads to a new 1.5 mL microtube using P200 PIPETMAN, and place tubes on ice. The supernatant is the first cell extract.

4. Add 50 μL yeast protein extraction buffer to tube containing glass beads and vortex tubes at highest speed for 30 s, then transfer supernatant (second cell extract) above settled glass beads to 1.5 mL microtube containing first cell extract.
5. Centrifuge combined cell extract at 14,500 × *g* for 5 min at 4 °C (*see* **Note 35**).
6. Transfer supernatant to new 1.5 mL microtube and measure protein concentration (*see* **Note 36**), and prepare protein samples for western blot analysis using appropriate antibodies.

4 Notes

1. Yeast nitrogen base without amino acid and DO supplements is very hygroscopic for curdling. Both them need to be stored in a humidity cabinet.
2. Aliquot stock carrier DNA in 100 μL working stock aliquots to avoid quality changes by repeated heating.
3. The pH of 1 M lithium acetate must be adjusted to 7.5 by acetic acid.
4. In general, we routinely obtain 100–200 colonies for a successful cotransformation using 200 ng per plasmid DNA.
5. It is not necessary to prepare the stock solution for DO supplement. Add DO supplement directly to minimal SD medium before autoclaving.
6. Yeast protein extraction buffer must be freshly prepared prior to use.
7. To grow overnight culture of yeast strain AH109, use colonies freshly streaked out (less than 2 months old). We also recommend refreshing yeast AH109 strain on YPDA plate every 2 months.
8. Grow yeast AH109 strain to stationary phase, which corresponds to $OD_{600} > 1.5$.
9. After 4 h incubation, the value of OD_{600} is about 0.3–0.4. Note that yeast cells may be precipitated. It is recommended to take out the flask and shake the culture a few times during incubation.
10. Centrifugation at 4 °C or RT does not significantly affect the transformation efficiency.
11. All steps are carried out in a laminar flow under aseptic conditions.
12. Discard supernatant as much as possible. Add 900 μL sterile water, then add 100 μL of 10× LiAc.

13. This is a very critical step as the excessive speed may cause yeast cell breakage and reduce the transformation efficiency.
14. Yeast competent cells must be freshly prepared to maintain high transformation efficiency.
15. All steps of PEG/LiAc-mediated transformation of yeast should be carried out in a laminar flow under aseptic conditions.
16. We recommend pretreating ssDNA at 100 °C for only 10–15 min. Prolonged heating may cause instability of ssDNA.
17. This step should be carried out on ice to maintain low temperature.
18. We recommend using 200 ng/plasmid for PEG/LiAc-mediated cotransformation of yeast.
19. We recommend vortexing mixture for only 1 s before incubation in incubator at 30 °C for 30 min. LiAc-PEG solution can be prepared during incubation time.
20. LiAc-PEG solution is very sticky, so it is better to use blunt-end tip by cutting the end of a regular pipette tip to draw the solution. This step is very critical and must be done within 2 min. Otherwise, the following resuspending step will be difficult to perform. Therefore, avoid handling more than ten samples at the same time.
21. Resuspend cell mixture immediately by vortexing using maximum speed for 2–3 s.
22. Before heat shock, we recommend gently shaking microtube several times to mix cell mixture well.
23. Pellet cells directly by centrifugation at 14,500 × *g* for 1 min. It is not necessary to incubate cells on ice before centrifugation.
24. Final cell suspension can be stored overnight at 4 °C before use.
25. In general, we recommend picking six to eight single colonies for further analysis. If possible, choose relatively large colonies, which usually correlate with high protein expression levels.
26. We highly recommend using a flat toothpick (750 flat toothpicks, Diamond Brands) to patch single colony on selection plate. The use of a sharp toothpick will often cause the breakage of the agar surface. The yeast cells are ready for further protein–protein interactions test when the yeast cells have grown nicely after 2–3 days of incubation.
27. Use a flat toothpick to patch cells on selection plate. In general, it is recommended to patch three individual colonies for

protein–protein interaction analysis in each tested interacting pair. In many cases, the growth rate of colonies on selection plate correlates with the binding strength of the two tested proteins.

28. It is highly recommended to perform western blot analysis to confirm the proper expression of tested proteins using commercially available antibody for tagged epitope of fusion proteins.
29. To prepare the yeast protein extraction, use yeast cells freshly grown on plates within a week.
30. It is recommended to use appropriate SD minimal medium with selection to maintain extrachromosomal plasmid(s) of transformants. The use of 2× SD minimal medium with more nutrients can facilitate faster and better growth of transformants.
31. During late log phase, the *ADH1* promoter shuts down and the expression level of endogenous yeast proteases is increased. Therefore, do not grow yeast cells beyond saturation.
32. Resuspend and wash cell pellets by vortexing at maximum speed for 2–3 s after adding 10 mL sterile water.
33. Because it is very difficult to take accurate amount of acid-washed glass beads using a pipet, we recommend to use a small spatula instead.
34. We use the Vortex-Genie 2 mixer (Scientific Industries, Inc.) to vortex the tubes at maximum speed for 30 s. Wear thick gloves to prevent your hands from being temporarily paralyzed during vortex.
35. The purpose of this step is to remove cell debris and glass beads by centrifugation.
36. To minimize the amount of protein extract used to determine protein concentration, it is recommended to use the NanoDrop 1000 Spectrophotometer (Thermo Fisher Scientific Inc.) for the measurement.

Acknowledgements

This work was supported by a research grant from the Ministry of Science and Technology (MOST 104-2311-B-001-025 -MY3) to E.M. Lai. J. S. Lin is the recipient of postdoctoral fellowships from Academia Sinica.

References

1. Fields S, Song O (1989) A novel genetic system to detect protein–protein interactions. Nature 340:245–246
2. Stasi M, De Luca M, Bucci C (2015) Two-hybrid-based systems: powerful tools for investigation of membrane traffic machineries. J Biotechnol 202:105–117
3. Clontech (2007) Matchmaker™ GAL4 two-hybrid system 3 & libraries user manual. http://www.clontech.com/images/pt/PT3247-1.PDF
4. Chien CT, Bartel PL, Sternglanz R, Fields S (1991) The two-hybrid system: a method to identify and clone genes for proteins that interact with a protein of interest. Proc Natl Acad Sci U S A 88:9578–9582
5. Causier B, Davies B (2002) Analysing protein–protein interactions with the yeast two-hybrid system. Plant Mol Biol 50:855–870
6. Petschnigg J, Groisman B, Kotlyar M, Taipale M, Zheng Y et al (2014) The mammalian-membrane two-hybrid assay (MaMTH) for probing membrane–protein interactions in human cells. Nat Methods 11:585–592
7. Reece-Hoyes JS, Barutcu AR, McCord RP, Jeong JS, Jiang L et al (2011) Yeast one-hybrid assays for gene-centered human gene regulatory network mapping. Nat Methods 8:1050–1052
8. Reece-Hoyes JS, Marian Walhout AJ (2012) Yeast one-hybrid assays: a historical and technical perspective. Methods 57:441–447
9. Tsai YL, Chiang YR, Narberhaus F, Baron C, Lai EM (2010) The small heat-shock protein HspL is a VirB8 chaperone promoting type IV secretion-mediated DNA transfer. J Biol Chem 285:19757–19766
10. Baron C, Thorstenson YR, Zambryski PC (1997) The lipoprotein VirB7 interacts with VirB9 in the membranes of Agrobacterium tumefaciens. J Bacteriol 179:1211–1218
11. Das A, Anderson LB, Xie YH (1997) Delineation of the interaction domains of Agrobacterium tumefaciens VirB7 and VirB9 by use of the yeast two-hybrid assay. J Bacteriol 179:3404–3409
12. Das A, Xie YH (2000) The Agrobacterium T-DNA transport pore proteins VirB8, VirB9, and VirB10 interact with one another. J Bacteriol 182:758–763
13. Ma LS, Lin JS, Lai EM (2009) An IcmF family protein, ImpLM, is an integral inner membrane protein interacting with ImpKL, and its walker a motif is required for type VI secretion system-mediated Hcp secretion in Agrobacterium tumefaciens. J Bacteriol 191:4316–4329
14. Lin JS, Ma LS, Lai EM (2013) Systematic dissection of the Agrobacterium type VI secretion system reveals machinery and secreted components for subcomplex formation. PLoS One 8:e67647
15. Bonemann G, Pietrosiuk A, Diemand A, Zentgraf H, Mogk A (2009) Remodelling of VipA/VipB tubules by ClpV-mediated threading is crucial for type VI protein secretion. EMBO J 28:315–325
16. Lossi NS, Manoli E, Forster A, Dajani R, Pape T et al (2013) The HsiB1C1 (TssB-TssC) complex of the Pseudomonas aeruginosa type VI secretion system forms a bacteriophage tail sheathlike structure. J Biol Chem 288:7536–7548
17. Ito H, Fukuda Y, Murata K, Kimura A (1983) Transformation of intact yeast cells treated with alkali cations. J Bacteriol 153:163–168

Chapter 15

Protein–Protein Interactions: Cytology Two-Hybrid

Krishnamohan Atmakuri

Abstract

Identifying protein–protein interactions between the machine components of bacterial secretion systems and their cognate substrates is essential. Establishing which component and substrate interactions are direct or indirect further facilitates (1) advancing the architecture and assembly of the machines and (2) understanding the substrates' translocation mechanistics. Currently, though biochemical means exist for identifying such direct interactions, they primarily remain in vitro and are quite labor intensive. Thus, adopting genetic approaches to help visualize these interactions in vivo is quick and advantageous. Here I describe bimolecular fluorescence complementation and cytology-based two-hybrid assays that could easily be adopted to understand the bacterial secretions systems.

Key words Bimolecular fluorescence complementation (BiFC), Cytology-based two-hybrid (C2H), Nonfluorescing halves, Protein–protein interactions, Retargeting fluorescence

1 Introduction

Evaluating protein–protein interactions among machine components of protein secretion systems and their cognate substrates is necessary and essential. To visualize them directly in bacterial cells, in the last decade, *bi*molecular *f*luorescence *c*omplementation (BiFC) and *c*ytology-based *two-hy*brid (C2H) assays have gained significant momentum [1–4].

Bimolecular fluorescence complementation. BiFC primarily requires the functional reconstitution of a fluorescent protein, such as green fluorescent protein (GFP) and yellow fluorescent protein (YFP). When complementary but nonfluorescing portions of a fluorescent protein that are genetically fused to putative protein partner pairs are brought together to restore fluorescence, the protein partners under evaluation are said to interact [5]. In recent times, improved versions of BiFC have evolved to help identify protein–protein interactions in varied environments and diverse systems [6–10]. Since BiFC operates through reconstitution, it is largely immune to influence by cellular conditions. Consequently, it

Laure Journet and Eric Cascales (eds.), *Bacterial Protein Secretion Systems: Methods and Protocols*, Methods in Molecular Biology, vol. 1615, DOI 10.1007/978-1-4939-7033-9_15, © Springer Science+Business Media LLC 2017

supersedes *f*luorescence *r*esonance *e*nergy *t*ransfer (FRET)-based in vivo visualization of protein–protein interactions [11], which is hypersensitive to interference from the same cellular conditions [5]. Further, BiFC assays for a given protein partner complex is not hampered by other interacting proteins because such interactions remain largely invisible [5, 8]. Finally, though several reporter proteins, such as ubiquitin, β-galactosidase, and dihydrofolate reductase, have also been fragmented and assembled back together through interacting partners, the reconstitution of fluorescence with fluorescent fragments eliminates requirements such as, for example, additional stains and reagents or stoichiometric protein-level constraints to visualize protein complexes.

Despite several advantages of BiFC, under some conditions and in a few experimental model systems, without being fused to interacting partners, the nonfluorescent halves intrinsically come together to fluoresce. Therefore, before setting up several BiFC assays, it is always useful to first test, with a couple of fluorescent proteins' halves, for the specificity of fluorescence complementation (in the system under study) and then move forward.

Cytology-based two-hybrid: In contrast to BiFC, C2H primarily involves the targeting of interacting protein partners by cell division proteins such as DivIVA (from *Bacillus subtilis*) or FtsZ (from *Escherichia coli*) to their native localizing sites, i.e., poles and midcell, respectively. Thus, when either of the cell division proteins fused to a bait protein target the interacting GFP/YFP-fused prey protein to the midcell/poles, the protein partners in question are said to interact [2]. While BiFC efficiently establishes interactions between soluble machine components [1, 4], C2H helps explore interactions between soluble and membrane-associated machine components of any secretion system [1, 2]. Recent advances in BiFC also make it possible to detect such interactions and determine membrane protein topology [5]. Since C2H involves the targeting of partner proteins to the midcell or poles, this assay works best when prey and bait proteins per se do not exhibit similar localization patterns. FtsZ fused to a bait protein can sometimes lead to cell filamentation, especially when the fusion protein dominates native FtsZ during cell division.

2 Materials

2.1 Bimolecular Fluorescence Complementation

1. Plasmid vectors (either regular or Gateway-based) for the expression of potential interacting partners as fusions to non-fluorescing halves. The plasmids must be of (i) different incompatibility groups, (ii) different antibiotic markers, and (iii) similar copy numbers and (iv) preferably harbor identical promoters for expression.

2. DNA encoding of either full length or N- and C-terminal nonfluorescing halves of fluorescent protein under consideration. (Also refer to Table 2 in ref. [12] for fragment lengths of each half of fluorescent protein(s) under consideration).
3. DNA encoding proteins of interest, i.e., the interacting partners.
4. Appropriate cloning primers (with required restriction sites designed into them) for polymerase chain reaction (PCR)-based amplification followed by directed cloning of the required halves and interacting proteins. Alternatively, Gateway-based pDONR vectors could be used for cloning and then the potential interacting proteins moved to appropriate pDEStination vectors harboring the two nonfluorescing halves.
5. DNA encoding mutated proteins of interest OR site-directed mutagenesis kit for generating mutant proteins to test that indeed the interaction of protein partners is driving nonfluorescent halves to interact.
6. Competent cells of *E. coli* and other bacterial systems under consideration (if any). If using Gateway technology, use *E. coli* DH5α for selection and *E. coli* DB3.1 for cloning and maintaining pDONR and pDEStination vectors.
7. Appropriate growth media for in vitro growing of bacteria under consideration.
8. Additional reagents to confirm fusions by immunoblotting.
9. Electroporator (for electroporating constructs into competent cells) OR a water bath or dry heat block (for heat shock-for the transformation of constructs into chemical-based competent cells).
10. Shaking incubator.
11. Fluorescence microscope equipped with 20× to 100× objectives, a 100× oil-immersion phase-contrast objective, a charge-coupled device (CCD) camera, appropriate filters to help visualize fluorescent proteins [13], and accompanying software for image capture, image analysis, and instrument control.

2.2 Cytology-Based Two-Hybrid

1. Plasmid vectors (either regular or Gateway-based) for the expression of potential interacting partners as fusions to either the cell division protein or fluorescent reporter. Again, the plasmids must be of (i) different incompatibility groups, (ii) different antibiotic markers, and (iii) similar copy numbers and (iv) harbor an identical promoter for expression.
2. DNA encoding fluorescent protein and cell division proteins.
3. DNA encoding proteins of interest, i.e., the interacting partners under study.

4. Appropriate cloning primers (with required restriction sites designed into them) for PCR-based amplification followed by directed cloning of interacting proteins, cell division proteins, and fluorescent reporters. Alternatively, Gateway-based pDONR vectors could be used for cloning and then the potential interacting proteins moved to appropriate pDEStination vectors harboring fluorescing reporter or cell division proteins.
5. DNA encoding mutated proteins of interest OR site-directed mutagenesis kit for generating mutant proteins to test that indeed the interaction of protein partners is driving the fluorescent reporter to midcell/poles in cells.
6. Competent cells of *E. coli* and other bacterial systems under consideration (if any). If using Gateway technology, use *E. coli* DH5α for selection and *E. coli* DB3.1 for cloning and maintaining pDONR and pDEStination vectors.
7. Appropriate growth media for in vitro growing of bacteria under consideration.
8. Additional reagents to confirm fusions by immunoblotting.
9. Electroporator (for electroporating constructs into competent cells) OR a water bath or dry heat block (for heat shock-for the transformation of constructs into chemical-based competent cells).
10. Shaking incubator.
11. Fluorescence microscope equipped with 20× to 100× objectives, a 100× oil immersion phase-contrast objective, a CCD camera, appropriate filters to help visualize fluorescent proteins [13], and accompanying software for image capture, image analysis, and instrument control.

3 Methods

Unless otherwise noted, all steps may be performed at room temperature.

3.1 Bimolecular Fluorescence Complementation

1. Select fluorescent protein, its halves, and appropriate fusion sites (*see* **Notes 1** and **2**).
2. Select appropriate controls (*see* **Note 3**).
3. Amplify required DNA fragment by PCR (*see* **Note 4**).
4. Use standard restriction-ligation procedures to clone PCR amplicons into appropriate expression vectors (*see* **Note 5**).
5. Transform ligation mixture into electro- or chemically competent bacteria (*see* **Notes 6** and 7).

6. Inoculate four to five colonies from freshly transformed bacteria separately into required growth media and grow under required conditions to approximately an OD (optical density at $A_{600\ nm}$) of 0.1.
7. Induce the expression of the proteins under study with appropriate inducers for required duration (*see* **Note 8**).
8. Wash a few hundred cells with fresh medium (to stop induction).
9. Observe cells under fluorescent microscope (*see* **Notes 9–11**).
10. Perform image analyses using ImageJ or commercially available software that comes with the fluorescent microscopes of most companies (*see* **Note 12**).

3.2 Cytology-Based Two-Hybrid

1. Select fluorescent protein, its halves, and appropriate fusion sites (*see* **Notes 13** and **14**).
2. Select appropriate controls (*see* **Note 15**).
3. Amplify the required DNA fragment by PCR (*see* **Note 4).**
4. Use standard restriction-ligation procedures to clone PCR amplicons into appropriate expression vectors (*see* **Note 5**).
5. Transform ligation mixture into electro- or chemically competent bacteria (*see* **Notes 6** and 7).
6. Inoculate four to five colonies from freshly transformed bacteria separately into required growth media and grow under required conditions to approximately an OD (optical density at $A_{600\ nm}$) of 0.1.
7. Induce expression of proteins under study with appropriate inducers for required duration (*see* **Note 8**).
8. Wash a few hundred cells with fresh media (to stop induction).
9. Observe cells under fluorescent microscope (*see* **Notes 9–11**).
10. Perform image analyses using ImageJ or commercially available software that comes with the fluorescent microscopes of most companies (*see* **Note 16**).

4 Notes

1. For interaction studies in *Agrobacterium*, we have used GFP halves, N'GFP (1–154 amino-acid residues), and GFP'C (153—end) [1, 4]. YFP and CFP halves can also be evaluated [12]. While generating fusion proteins, no linkers/spacers were used while connecting protein partners to nonfluorescing halves. However, recently, linkers/spacers of 5–17 amino acid residues have been used with better results in eukaryotic model systems [12].

2. The success of BiFC also largely relies on the end of the protein partners to which (N- or C-terminus) the nonfluorescent fragments gets fused. Primarily, we have fused the N-terminus end of the N'GFP half to the C-terminal end of a protein partner and the C-terminus end of the C'GFP half to the N-terminal end of another protein partner. However, it is important to evaluate other fusion ends (eight combinations in total—either ends of the partner proteins and nonfluorescing halves) to narrow down the suitable fusion ends. It is also important to ensure that the fused protein partners do not show altered localization patterns (e.g., cytosolic to membrane bound and vice versa). All fused protein partners need to be evaluated for their expression kinetics and accumulation by western blot analysis.
3. To make sure that BiFC works in a given model system under study, appropriate controls are to be evaluated first: (i) clone and express nonfluorescing halves alone (under the same promoter used for experimental study) to confirm that the halves per se do not interact. If they do, then switch to test several other fluorescent protein halves [12]; (ii) clone and express noninteracting protein partners (from established studies) fused to the same nonfluorescing halves to make sure that the noninteracting partners do not bring the nonfluorescing halves together to exhibit fluorescence; (iii) clone and express nonfluorescing halves fused to protein partner (under study) mutants (point mutants—at their site of interaction); this also helps to confirm/evaluate various residues at interaction sites. However, if the interacting protein pair is fairly uninvestigated or their site of interaction is undetermined, this negative control could easily be skipped. Alternatively, BiFC could be performed with these constructs to determine the site of interaction; (iv) clone and express either protein partners fused to a nonfluorescing half together with another construct expressing the other nonfluorescing half alone; this eliminates spurious interactions driving fluorescence; (v) clone and express positively interacting protein partners (from established studies) fused to same nonfluorescing halves to make sure that the assay works under the conditions being used.
4. Any high-fidelity proofreading polymerase could be used for PCR-based amplification of the required DNA fragments.
5. Standard cloning techniques could be applied to move fragments of interest to expression vectors chosen for the study. For Gateway-based cloning, the required kits are available with Life Technologies (now associated with Thermo Fisher Scientific).
6. Commercially available electro- or chemically competent bacteria or those generated by standard methods can be employed for transformations. We generally transform with 10–25 ng plasmid DNA to obtain several hundred colonies.

7. Better fluorescence levels are usually obtained with freshly transformed cells than when using cells stored at 4 °C or stocked at −80 °C.
8. Proper gene induction requires standardization. This depends on the model system under study, the type of inducer, the copy number of plasmids, and toxicity issues.
9. Generally, good images can be obtained by observing cells under a 100× oil-immersion phase-contrast objective.
10. As a standard practice, it is recommended to perform immunoblotting to check the level of expression and monitor the fusion proteins in the bacterial cultures under study.
11. If nonfluorescing halves of either YFP or CFP (BiFC) or YFP and CFP (C2H) are used, the cells might need to be briefly incubated at 30 °C for the fluorescent proteins to mature and generate fluorescence of high intensity.
12. If the two nonfluorescing halves interact in the model system under results study, alternative fluorescent protein halves must be evaluated. If point mutations at interaction sites of protein partners do not abolish fluorescence, then complementation of the nonfluorescing halves could be nonspecific. At this juncture, it might be important to determine whether higher protein levels would affect the outcome. If so, the concentration or time of induction could be modified. If not, an alternate promoter could be evaluated. Otherwise, alternatives to BiFC have been explored (e.g., C2H). However, if after evaluating all controls (*see* **Note 3**) fluorescence is observed only when both protein partners' fusions are available, then the two proteins under study are said to interact.
13. For C2H-based protein–protein interaction studies in *Agrobacterium* and *E. coli*, for retargeting we have used two cell division proteins DivIVA (from *B. subtilis*) and FtsZ (from *E. coli*), and for fluorescence GFP [1, 2, 4]. YFP and CFP can also be explored as alternate fluorescent proteins. While generating fusion proteins, no linkers/spacers were used and seem unnecessary. However, for the success of C2H, it is important to evaluate all possible fusion sites, i.e., the fusion of cell division proteins and fluorescent protein to either ends of protein partners under study.
14. We have fused the C-terminal end of FtsZ or DivIVA to the N-terminal end of protein partners [2, 4]. As a standard practice, it is recommended to routinely perform immunoblotting to evaluate the levels of expression and monitor protein fusion stability in the bacterial cultures under study.
15. To make sure C2H works in model systems under study, appropriate controls must be evaluated first: (i) clone and express interacting partners alone to confirm that neither of them localizes to the midcell and poles. If one of them does

localize to these locations, while it must be fused to either of the cell division proteins, the other partner in question must be fused to the fluorescent protein under consideration. If, however, both potential interacting partners localize to the midcell/poles, C2H cannot be utilized as the method of study for interactions; (ii) clone and express GFP/YFP/CFP fused protein partners (under study) to confirm that neither fusion localizes to the midcell/poles. If any fusions localize to the poles/midcell, it is important to determine whether the localization is natural or merely an artifact of inclusion bodies. Thus, expression and localization could be evaluated with alternate promoter(s), by altering the inducer concentrations, or by modifying the inducing time or temperature; (iii) clone and express fluorescent protein/cell division proteins fused to protein partner (under study) mutants (point mutants—at their site of interaction); this also helps confirm/evaluate various residues at interaction sites. However, if the interacting protein pair is fairly uninvestigated or its site of interaction undetermined, this negative control could easily be skipped; (iv) clone and express either protein partners fused to a fluorescent protein together with another construct expressing either cell division protein alone-this eliminates spurious interactions driving retargeting; (v) clone and express positively interacting protein partners (from established studies) fused to retargeting cell division proteins/fluorescent protein to make sure that the assay works under the conditions being used.

16. If in the model system under study the two protein partners localize to the midcell/poles, C2H cannot be employed. If point mutations at the interaction sites of the protein partners do not abolish retargeting, then localization could be nonspecific or a consequence of inclusion bodies. At this juncture, it might be important to determine whether higher protein levels would affect the outcome. If so, the concentration or time of induction could be modified. If not, an alternate promoter could be evaluated. Otherwise, alternatives to C2H must be explored (e.g., BiFC). However, if after evaluating all controls (*see* **Note 15**) fluorescence is observed only at the midcell/poles, then the two proteins under study are said to interact.

Acknowledgments

I thank Peter (Prof. Peter J. Christie) and Bill (Prof. William Margolin), both from the Department of Microbiology and Molecular Genetics, University of Texas Health Science Center, Houston, Texas, USA. Peter was instrumental in mentoring and providing an excellent opportunity to train as a post-doc in his lab and Bill provided a lot of help, guidance, and training in fluorescent microscopy.

References

1. Ding Z, Atmakuri K, Christie PJ (2003) The outs and ins of bacterial type IV secretion substrates. Trends Microbiol 11:527–535
2. Ding Z, Zhao Z, Jakubowski SJ, Atmakuri K, Margolin W, Christie PJ (2002) A novel cytology-based, two-hybrid screen for bacteria applied to protein–protein interaction studies of a type IV secretion system. J Bacteriol 184:5572–5582
3. Taylor KW, Kim JG, Su XB, Aakre CD, Roden JA, Adama CM, Mudgett MB (2012) Tomato TFT1 is required for PAPMP-triggered immunity and mutations that prevent T3S effector XopN from binding to TFT1 attenuate *Xanthomonas* virulence. PLoS Pathog 8:e1002768
4. Atmakuri K, Ding Z, Christie PJ (2003) VirE2, a type IV secretion substrate, interacts with the VirD4 transfer protein at cell poles of *Agrobacterium tumefaciens*. Mol Microbiol 49:1699–1713
5. Kerppola TK (2006) Visualization of molecular interactions by fluorescence complementation. Nat Rev Mol Cell Biol 7:449–456
6. Zhang XE, Cui Z, Wang D (2016) Sensing of biomolecular interactions using fluorescence complementing systems in living cells. Biosens Bioelectron 76:243–250
7. Kodama Y, Hu CD (2012) Biomolecular fluorescence complementation (BiFC): a 5-year update and future perspectives. BioTechniques 53:285–298
8. Kerppola TK (2008) Bimolecular Fluorescence Complementation (BiFC) analysis as a probe of protein interactions in living cells. Annu Rev Biophys 37:465–487
9. Hu CD, Chinenov Y, Kerppola TK (2002) Visualization of interactions among bZIP and Rel proteins in living cells using bimolecular fluorescence complementation. Mol Cell 9:789–798
10. Hu CD, Kerppola TK (2003) Simultaneous visualization of multiple protein interactions in living cells using multicolor fluorescence complementation analysis. Nat Biotechnol 21: 539–545
11. Jares-Erijman EA, Jovin TM (2003) FRET imaging. Nat Biotechnol 21:1387–1395
12. Kerppola TK (2006) Design and Implementation of Bimolecular Fluorescence Complementation (BiFC) assays for the visualization of protein interactions in living cells. Nat Protoc 1:1278–1286
13. Shaner NC, Steinbach PA, Tsein RY (2005) A guide to choosing fluorescent proteins. Nat Methods 2:905–909

Chapter 16

Fusion Reporter Approaches to Monitoring Transmembrane Helix Interactions in Bacterial Membranes

Laureen Logger, Abdelrahim Zoued, and Eric Cascales

Abstract

In transenvelope multiprotein machines such as bacterial secretion systems, protein–protein interactions not only occur between soluble domains but might also be mediated by helix–helix contacts in the inner membrane. Here we describe genetic assays commonly used to test interactions between transmembrane α-helices in their native membrane environment. These assays are based on the reconstitution of dimeric regulators allowing the control of expression of reporter genes. We provide detailed protocols for the TOXCAT and GALLEX assays used to monitor homotypic and heterotypic transmembrane helix–helix interactions.

Key words Membrane protein, Protein-protein interaction, Transmembrane segment, Helix–helix interaction, One-hybrid, Two-hybrid, cI repressor, TOXCAT, GALLEX

1 Introduction

The proper assembly of multiprotein complexes such as bacterial secretion systems requires specific interactions between the different subunits. While most of the interactions involve contacts between soluble domains of these subunits, the transmembrane helices (TMHs) of inner membrane proteins are also key players in membrane protein complex formation. For examples, the Type II secretion (T2SS)-associated GspC, GspL, and GspM proteins interact with each other via their TMHs [1]. A similar situation has been evidenced for the Type VI secretion system (T6SS) TssLM complex [2–4]. The TMH could be involved in homotypic interaction, i.e., participate in the formation of dimers such as the Type IV secretion (T4SS) and T6SS-associated VirB10 and TssL inner membrane proteins [4, 5] or in heterotypic interactions with other subunits [1–3]. Monitoring interactions between TMHs is not an easy task because mutations within or swapping of the TMH could interfere with the conformation of the soluble domains and therefore may indirectly affect protein–protein interactions. Thus, genetic one- or two-hybrid approaches based on fusion to transcriptional

Laure Journet and Eric Cascales (eds.), *Bacterial Protein Secretion Systems: Methods and Protocols*, Methods in Molecular Biology, vol. 1615, DOI 10.1007/978-1-4939-7033-9_16, © Springer Science+Business Media LLC 2017

reporters, such as the λcI repressor, TOXCAT, GALLEX, and Bacterial Adenylate Cyclase-Based Two-Hybrid (BACTH) assays, have been developed. While cI-repressor and TOXCAT can only be used for testing homotypic interactions, the GALLEX and BACTH approaches can also be used to monitor interactions between different TMHs. This chapter provides protocols to monitor homotypic and heterotypic transmembrane helix–helix interactions using TOXCAT and GALLEX. We refer the reader to excellent reviews summarizing the forces exerted to catalyze TMH folding and insertion, as well as the different methods to analyze TMH interactions in bacteria [6, 7].

1.1 Monitoring TMH Homotypic Interactions

Methods of testing the homodimerization of TMHs, such as the λ cI repressor and TOXCAT assays, are based on the one-hybrid reporter fusion approach.

The cI transcriptional regulator represses the expression of early promoters of the bacteriophage λ genome. Repression only occurs when cI dimerizes, a behavior conferred by the C-terminal domain. The λcI repressor assay is therefore based on the reconstitution of a dimeric λcI repressor by two interacting fragments [8–10]. The construct consists of a fusion between the monomeric N-terminal DNA-binding domain of λcI (called cI') with the TMH (Fig. 1a). TMH-mediated cI' dimerization induces binding of cI to its operator sequence, allowing repression of phage λ early genes, hence conferring protection against superinfection by phage λ (Fig. 1a). The cI repressor assay has been successfully used to demonstrate that the T2SS XcpR and T4SS VirB4 and VirB11 proteins oligomerize [11–13].

The TOXCAT assay is based on the characteristics of the *Vibrio cholerae* ToxR regulator: a strict dimerization-dependent transcriptional activator consisting of an N-terminal helix-turn-helix DNA-binding domain and a C-terminal dimerization domain. The construct consists of a fusion in which the TMH is inserted between the monomeric ToxR DNA-binding domain and the MalE periplasmic protein (Fig. 1b). By supporting growth on maltose-minimal media, MalE makes it possible to verify that the TMH is properly inserted. TMH-mediated ToxR dimerization induces binding of ToxR on its operator sequence, allowing transcription of a reporter gene. In the initial ToxR system, the reporter gene is *lacZ* [14], while the TOXCAT assay uses the *cat* gene [15] (Fig. 1b). Hence, dimerization of the TMH could then be assessed by measuring the β-galactosidase and chloramphenicol acetyltransferase (resistance to chloramphenicol) activities [16]. The TOXCAT assay has been successfully used to provide evidence that the TMH of the T4SS VirB10 subunit oligomerizes [5]. Further improvements of the ToxR and TOXCAT assays have been published [17–19].

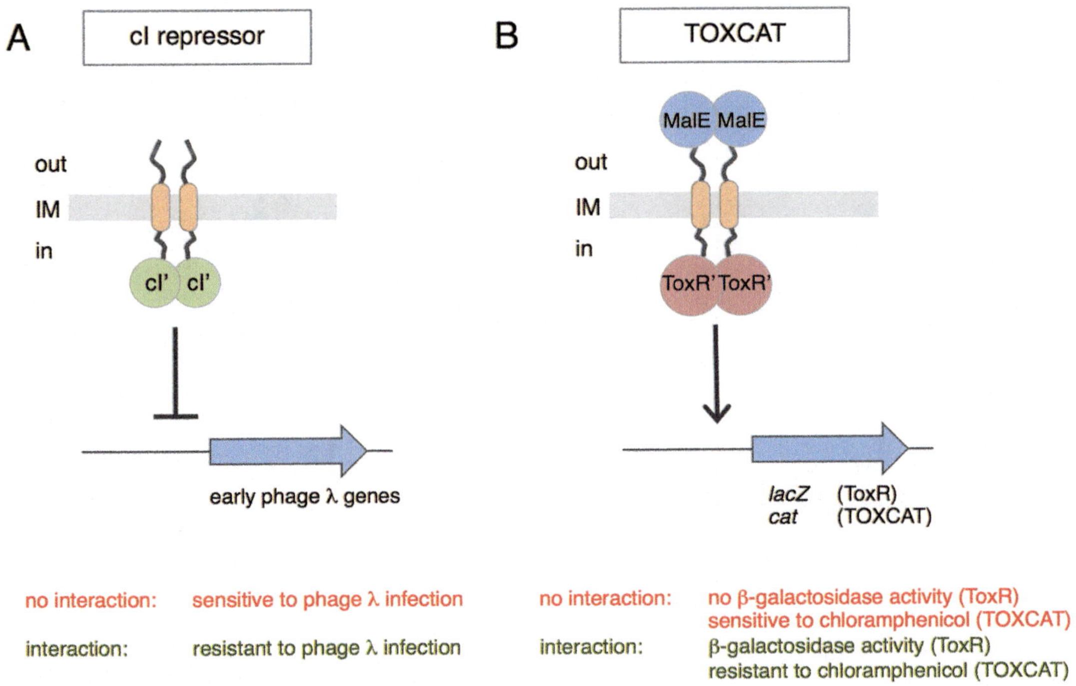

Fig. 1 Schematic representation of the assays for monitoring TMH homotypic interactions. (**a**) cI repressor assay. The TMH of interest (*orange*) is fused to the cI DNA-binding domain (*green*). Reconstitution of the cI dimer results in the repression of early phage λ genes (*blue*). Repression of phage λ genes confers protection against phage λ infection. (**b**) ToxR and TOXCAT assays. The TMH of interest (*orange*) is fused between the *Vibrio cholerae* ToxR DNA-binding domain (*red*) and the MalE protein (*blue*). Reconstitution of the ToxR dimer results in the expression of the reporter genes (*blue*, *lacZ* for the ToxR assay, *cat* for the TOXCAT assay)

1.2 Monitoring TMH Heterotypic Interactions

Methods of testing the heterodimerization of TMHs, such as the GALLEX and BACTH assays, are based on the two-hybrid reporter fusion approach.

The GALLEX assay is based on the reconstitution of a dimeric LexA transcriptional repressor by two interacting TMHs. The construct consists of a fusion in which each TMH is inserted between the monomeric LexA N-terminal DNA-binding domain and the MalE periplasmic protein. The elegant improvement is that one of the two TMHs is fused to the wild-type LexA N-terminal domain ($LexA^{WT}$), whereas the second TMH is fused to a LexA N-terminal domain variant bearing a mutation in the DNA-binding motif ($LexA^{408}$), allowing recognition of a different operator sequence (op^{408}). Formation of helix heterodimers induces binding of $LexA/LexA^{408}$ on a dual operator sequence (op^{WT}/op^{408}), allowing repression of a reporter gene [20–22] (Fig. 2a).

The BACTH assay is based on the reconstitution of the adenylate cyclase activity conferred by the T18 and T25 domains of the *Bordetella pertussis* Cya protein [23–25] (Fig. 2b). Widely

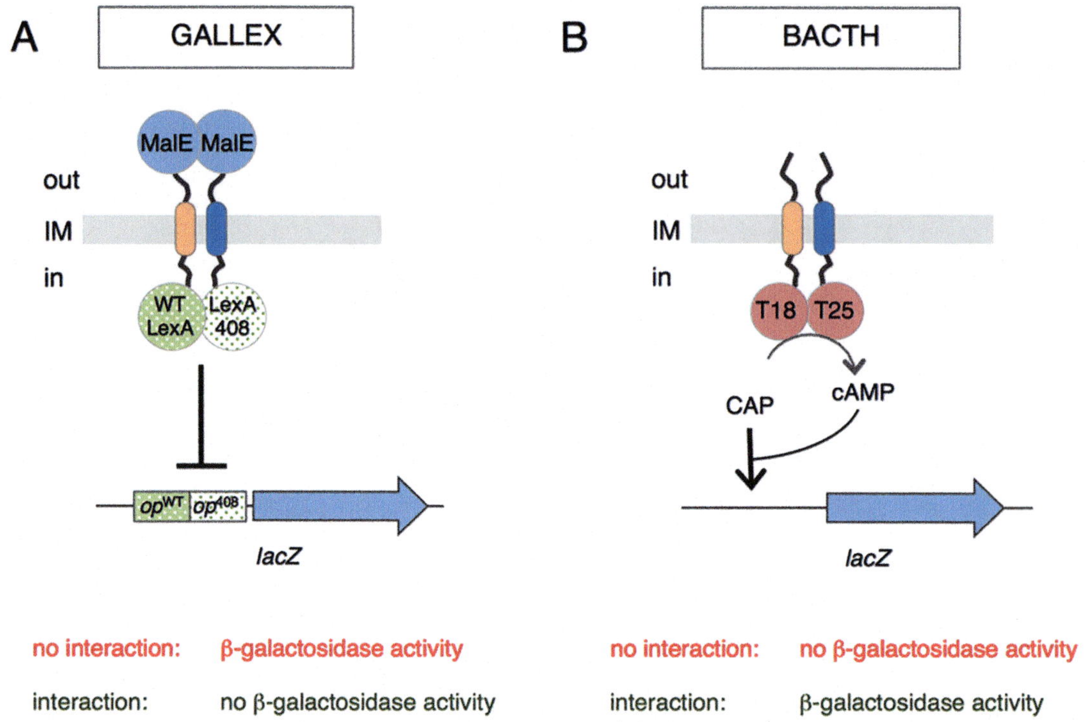

Fig. 2 Schematic representation of assays for monitoring TMH heterotypic interactions. (**a**) GALLEX assay. The first TMH of interest (*orange*) is fused between the wild-type LexA DNA-binding domain (WT LexA) and MalE, whereas the second TMH (*blue*) is fused between the $LexA^{408}$ variant (LexA 408) and MalE. Reconstitution of the $LexA^{WT}/LexA^{408}$ dimer results in the repression of the reporter gene (*blue*). (**b**) BACTH assays. The first TMH of interest (*orange*) is fused to the T18 domain of the *B. pertussis* adenylate cyclase, whereas the second TMH (*blue*) is fused to the T25 of adenylate cyclase. Reconstitution of the T18/T25 adenylate cyclase results in the production of cyclic adenosine monophosphate (cAMP). Binding of cAMP to the catabolite activator protein (CAP) induces the expression of the reporter gene (*blue*)

used for testing interactions between soluble domains or proteins in multiprotein complexes such as the divisome or secretion systems [26–34], it has only rarely been used for the study of transmembrane helix–helix interactions [35–37]. A detailed protocol for the bacterial two-hybrid assay is described in Chapter 13. In this chapter, we provide protocols for the TOXCAT and GALLEX assays.

2 Material

2.1 Monitoring TMH Homotypic Interactions: The TOXCAT Assay

1. pcckan vector [15] (*see* **Note 1**).
2. *Escherichia coli* NT326 or MM39 bacterial strains [15] (*see* **Note 2**).
3. Lysogeny broth (LB) medium: Dissolve 10 g tryptone, 5 g yeast extracts, and 10 g NaCl in 1 L distilled water. Autoclave for 15 min at 121 °C. For LB agar plates, add 15 g bacto-agar prior to autoclaving.

4. M9-maltose medium: Dissolve 0.6 g $Na_2HPO_4 \cdot 12H_2O$, 0.3 g $KH_2PO_4 \cdot H_2O$, 50 mg NaCl, 100 mg NH_4Cl, and 1.5 g bacto-agar in 90 mL distilled water. Autoclave. Add 100 mg casamino acids, 400 mg maltose, 25 mg $MgSO_4 \cdot 7H_2O$, and 1 mg $CaCl_2$.
5. Ampicillin stock solution (250×): 25 mg/mL ampicillin. Dissolve 250 mg ampicillin in 10 mL distilled water. Filter to sterilize. Store at 4 °C.
6. Chloramphenicol stock solution: Dissolve 90 mg chloramphenicol in 1 mL absolute ethanol. Store at −20 °C.
7. 2.5 mM chloramphenicol solution: Dissolve 8.1 mg chloramphenicol in 10 mL ethanol.
8. Sodium dodecyl sulfate (SDS)-polyacrylamide gel electrophoresis (PAGE) loading buffer: 60 mM Tris–HCl, pH 6.8, 2% SDS, 10% glycerol, 5% β-mercaptoethanol, 0.01% bromophenol blue.
9. Lysis buffer: 25 mM Tris–HCl, 2 mM EDTA, pH 8.0. Dissolve 303 mg Tris(hydroxymethyl) aminomethane and 58 mg ethylenediaminetetraacetic acid (EDTA, disodium salt) in 100 mL sterile distilled water. Adjust pH to 8.0.
10. Reaction buffer: 100 mM Tris–HCl, pH 7.8, 0.1 mM acetyl-CoA, 0.4 mg/mL 5,5'-dithiobis-(2 nitrobenzoic acid) (dTNB). Dissolve 121 mg Tris, 0.81 mg acetyl-CoA, and 4 mg dTNB in 10 mL sterile distilled water. Adjust pH to 7.8 with HCl.
11. 10 mm filter paper disk.
12. 96-well microplates.
13. Anti-maltose-binding protein (MBP) antibodies for MalE immunodetection.
14. Incubator.
15. Spectrophotometer.
16. Benchtop centrifuge.
17. Water bath at 96 °C.
18. Mini-gel caster system and SDS-PAGE apparatus.
19. Protein blotting apparatus.
20. Sonifier.
21. Microplate reader.

2.2 Monitoring TMH Heterotypic Interactions: The GALLEX Assay

1. pALM148 and pBML100 vectors [20].
2. *E. coli* NT326 or MM39 bacterial cells [15] (*see* **Note 2**).
3. *E. coli* SU202 bacterial strain [20, 38] (*see* **Note 3**).
4. LB medium: *see* Subheading 2.1.
5. M9-maltose medium: *see* Subheading 2.1.

6. Ampicillin stock solution (250×): *see* Subheading 2.1.
7. Tetracyclin stock solution (1000×): 12 mg/mL tetracyclin. Dissolve 120 mg tetracyclin in 10 mL ethanol. Filter to sterilize. Store at 4 °C.
8. Isopropyl-β-D-thiogalactopyranoside (IPTG) stock solution (500×). 0.1 M IPTG. Dissolve 238 mg IPTG in 10 mL sterile distilled water. Filter to sterilize. Store at 4 °C.
9. X-Gal stock solution (1000×): 40 mg/mL X-Gal. Dissolve 40 mg 5-Bromo-4-chloro-3-indolyl β-D-galactopyranoside (X-Gal) in 1 mL dimethylformamide. Prepare fresh and do not store.
10. 0.1% SDS: Dissolve 50 mg SDS in 50 mL distilled water.
11. Chloroform.
12. Ortho-nitrophenyl-β-D-galactopyranoside (ONPG) stock solution. 4 mg/mL ONPG: Dissolve 20 mg ONPG in 5 mL buffer Z.
13. SDS-PAGE loading buffer: 60 mM Tris–HCl, pH 6.8, 2% SDS, 10% glycerol, 5% β-mercaptoethanol, 0.01% bromophenol blue.
14. Buffer Z: Dissolve 2.15 g $Na_2HPO_4 \cdot 12\ H_2O$, 0.29 g $Na_2HPO_4 \cdot H_2O$, 75 mg KCl, and 25 mg $MgSO_4 \cdot 7H_2O$ in 100 mL distilled water. Adjust pH to 7.0. Add 270 μL β-mercaptoethanol. Prepare fresh and do not store.
15. Anti-MBP antibodies for MalE immunodetection.
16. 96-well microplates.
17. Incubator.
18. Spectrophotometer.
19. Benchtop centrifuge.
20. Water bath at 96 °C.
21. Mini-gel caster system and SDS-PAGE apparatus.
22. Protein blotting apparatus.
23. Microplate reader.

3 Methods

3.1 Monitoring TMH Homotypic Interactions: The TOXCAT Assay

1. Clone the DNA fragment corresponding to the TMH to be studied into the pcckan vector to yield a plasmid producing the ToxR'-TMH-MalE fusion protein. Before testing the homodimerization of the TMH, verify that your fusion protein is properly produced (**steps 3–8**) and inserted in the inner membrane (**steps 9** and **10**). The dimerization of the TMH is then assessed by the disk diffusion assay (**steps 11–16**) and quantified by measuring the chloramphenicol acetyltransferase activity (**steps 17–26**).

2. Transform the empty pcckan vector and your pcckan construct into NT326 or MM39 *E. coli* competent cells. Select on LB-ampicillin plates (*see* **Note 1**).
3. Pick a single colony of each transformation and grow cells in 20 mL LB medium supplemented with ampicillin (100 μg/mL) until an optical density at 600 nm (OD_{600}) of 0.8 is reached.
4. Harvest 2 mL of cells by centrifugation at 4000 × *g* for 5 min.
5. Discard supernatants and resuspend cell pellets into 20 μL SDS-PAGE loading buffer.
6. Boil samples for 10 min at 96 °C.
7. Separate proteins by SDS-PAGE and transfer onto nitrocellulose membrane using your favorite protocol.
8. Use western blotting to immunodetect your fusion protein using commercial anti-MalE (anti-MBP) antibodies.
9. Streak 20 μL of the bacterial culture obtained at **step 3** in Subheading 3.1 onto M9-maltose medium.
10. After incubation for 48 h at 37 °C, verify that your strain grew on M9-maltose medium.
11. Drop a 10 mm filter paper disk in center of LB-ampicillin plate (*see* **Note 4**).
12. Add 60 μL chloramphenicol stock solution (90 mg/mL) on filter paper disk.
13. Incubate LB plates with chloramphenicol disks for 6 h at 37 °C.
14. Remove disk.
15. Spread 2 mL of the culture obtained at **step 3** in Subheading 3.1 on the LB-ampicillin plate to make a lawn. Eliminate excess culture.
16. After incubation for 16 h at 37 °C, measure the halo of chloramphenicol sensitivity (*see* **Note 5**).
17. Centrifuge 3 mL of the culture obtained at **step 3** in Subheading 3.1 at 4000 × *g* for 5 min (in triplicate).
18. Discard supernatant and resuspend cell pellets in 500 μL lysis buffer. Vortex.
19. Lyse cells by sonication using a sonifier.
20. Clear lysate by centrifugation at 10,000 × *g* for 15 min.
21. In a 96-well microplate, mix 15 μL of the cleared lysate with 220 μL reaction buffer.
22. Measure absorbance at 412 nm (A_{412}) (*see* **Note 6**) and at 550 nm (A_{550}; cell debris) every 20 s for 4 min using a microplate reader.
23. Inject 15 μL 2.5 mM chloramphenicol in each well.

24. Measure absorbance at 412 nm (*see* **Note 6**) and at 550 nm (cell debris) every 20 s for 10 min using a microplate reader.
25. Divide each A_{412} value by the corresponding A_{550} value and plot these values against time.
26. Calculate the chloramphenicol acetyltransferase activity based on the slope in the linear part of the curve (initial rate).

3.2 Monitoring TMH Heterotypic Interactions: The GALLEX Assay

1. Clone the DNA fragment corresponding to the first TMH to be studied (TMH1) into the pBLM100 vector to yield a pBR322 derivative plasmid producing the $LexA_{WT}$'-TMH1-MalE fusion protein. Clone the DNA fragment corresponding to the second TMH to be studied (TMH2) into the pALM148 vector to yield a pACYC184 derivative plasmid producing the $LexA_{408}$'-TMH2-MalE fusion protein. Before testing the heterodimerization of the TMH, verify that your fusion protein is properly produced (**steps 3–8**) and inserted in the inner membrane (**steps 9** and **10**). The dimerization of the TMH is then assessed on LB-X-Gal plates (**steps 11–14**) and quantitated by measuring the β-galactosidase activity (**steps 15–22**).
2. Transform the empty pBLM100 and pALM148 vectors as well as the pBLM100-TMH1 and pALM148-TMH2 constructs into NT326 or MM39 *E. coli* competent cells. Select on LB plates supplemented with ampicillin (pBLM100 derivatives) or tetracyclin (pALM148 derivatives).
3. Pick a single colony of each transformation and grow cells in 3 mL LB medium supplemented with IPTG and ampicillin or tetracyclin until an OD_{600} of 0.8 is reached.
4. Harvest 2 mL of cells by centrifugation at 4000 × *g* for 5 min.
5. Discard supernatants and resuspend cell pellets into 20 μL of SDS-PAGE loading buffer.
6. Boil samples for 10 min at 96 °C.
7. Separate proteins by SDS-PAGE and transfer onto nitrocellulose membrane using your favorite protocol.
8. Use western blotting to immunodetect your fusion protein using commercial anti-MalE (anti-MBP) antibodies.
9. Streak 20 μL of the bacterial culture obtained at **step 3** in Subheading 3.2 onto M9-maltose medium.
10. After incubation for 48 h at 37 °C, verify that your strain grew on M9-maltose medium.
11. Cotransform pBLM100 and pBLM100-TMH1 vectors in combination with pALM148 and pALM148-TMH2 vectors into SU202 *E. coli* competent cells (*see* **Note 7**). Select on LB plates supplemented with ampicillin and tetracyclin.

12. Pick a single colony of each transformation and grow cells in 3 mL LB medium supplemented with IPTG, ampicillin, and tetracyclin until an OD_{600} of 0.8 is reached.
13. Drop 15 μL of the bacterial culture obtained at **step 12** in Subheading 3.2 onto LB plates supplemented with IPTG, ampicillin, tetracyclin, and X-Gal.
14. After 6, 14, and 24 h of incubation at 37 °C, observe the coloration of the spots. White spots correspond to strains with no β-galactosidase activity (*i.e.*, interaction between the two TMHs), whereas blue spots correspond to strains with β-galactosidase activity (*i.e.*, no interaction between the two TMHs) (*see* **Note 8**).
15. Mix 200 μL of the bacterial culture obtained at **step 12** in Subheading 3.2 with 800 μL of buffer Z into a 1.5 mL Eppendorf tube. Vortex.
16. Add one drop of 0.1% SDS and two drops of chloroform to lyse cells. Vortex for 10 s.
17. In a 96-well microplate, mix 50 μL of the cleared lysate with 150 μL buffer Z.
18. Measure the absorbance at 420 nm (absorption wavelength of ortho-nitrophenol, the product of degradation of ONPG) and at 550 nm (A_{550}; cell debris) every 30 s for 2 min using a microplate reader.
19. Inject 40 μL ONPG solution in each well.
20. Measure the absorbance at 420 nm and at 550 nm every 30 s for 20 min using a microplate reader.
21. Divide each A_{420} value by the corresponding A_{550} value and plot these values against time.
22. Calculate the β-galactosidase activity based on the slope in the linear part of the curve (initial rate).

4 Notes

1. pcckan is a vector comprising the sequence corresponding to the ToxR N-terminal domain and that corresponding to MalE separated by a multiple cloning site allowing insertion of the sequence corresponding to the TMH of interest. Positive and negative controls have been developed by Russ and Engelman corresponding to the wild-type and mutated TMH of the glycophorin A, respectively [15].
2. NT326 and MM39 strains do not produce the MBP and therefore could be used as reporters to verify the proper insertion of the ToxR'-TMH-MalE and LexA-TMH-MalE fusions.

3. Strain SU202 is a reporter for the GALLEX assay. It has a chromosomally integrated fragment corresponding to a hybrid operator sequence (op^{WT}/op^{408}) controlling the expression of the *lacZ* reporter gene. Transformed SU202 cells are not stable and therefore transformations should be made fresh and colonies should not be stored at 4°C.
4. Use three LB-ampicillin plates per strain to be tested.
5. The diameter of the halo reflects the ability of the strain to resist chloramphenicol and therefore is directly and inversely linked to the expression of the *cat* gene that is induced by TMH dimerization. If the TMH dimerizes, the expression level of *cat* will be high and, hence, the diameter of the halo small.
6. The reaction catalyzed by the chloramphenicol acetyltransferase consists in the acetylation of the chloramphenicol and the release of free coenzyme A. Coenzyme A then reacts with the 5,5'-dithiobis-(2-nitrobenzoic acid), resulting in an increase of the absorbance at 412 nm.
7. You should obtain the combinations pBLM100 + pALM148, pBLM100 + pALM148-TMH2, pBLM100-TMH1 + pALM148, and pBLM100-TMH1 + pALM148-TMH2.
8. MacConkey/maltose could be used as reporter medium instead of LB-X-Gal plates. If MacConkey/maltose plates are used, the coloration of the spots differs: yellow spots correspond to strains with no β-galactosidase activity (*i.e.*, interaction between the two TMHs), whereas red spots correspond to strains with β-galactosidase activity (*i.e.*, no interaction between the two TMHs).

Acknowledgements

Work in the EC laboratory is supported by the Centre National de la Recherche Scientifique, the Aix-Marseille Université, and grants from the Agence Nationale de la Recherche (ANR-14-CE14-0006-02 and ANR-15-CE11-0019-01). LL and AZ are recipients of doctoral fellowships from the French Ministère de l'Enseignement Supérieur et de la Recherche and end-of-thesis fellowships from the Fondation pour la Recherche Médicale (FDT20160435498 and FDT20140931060).

References

1. Lallemand M, Login FH, Guschinskaya N, Pineau C, Effantin G, Robert X, Shevchik VE (2013) Dynamic interplay between the periplasmic and transmembrane domains of GspL and GspM in the type II secretion system. PLoS One 8:e79562
2. Ma LS, Lin JS, Lai EM (2009) An IcmF family protein, ImpLM, is an integral inner membrane protein interacting with ImpKL, and its walker a motif is required for type VI secretion system-mediated Hcp secretion in Agrobacterium tumefaciens. J Bacteriol 191:4316–4329

3. Aschtgen MS, Gavioli M, Dessen A, Lloubès R, Cascales E (2010) The SciZ protein anchors the enteroaggregative Escherichia coli Type VI secretion system to the cell wall. Mol Microbiol 75:886–899
4. Durand E, Zoued A, Spinelli S, Watson PJ, Aschtgen MS, Journet L, Cambillau C, Cascales E (2012) Structural characterization and oligomerization of the TssL protein, a component shared by bacterial type VI and type IVb secretion systems. J Biol Chem 287:14157–14168
5. Garza I, Christie PJ (2013) A putative transmembrane leucine zipper of agrobacterium VirB10 is essential for T-pilus biogenesis but not type IV secretion. J Bacteriol 195:3022–3034
6. Schneider D, Finger C, Prodöhl A, Volkmer T (2007) From interactions of single transmembrane helices to folding of alpha-helical membrane proteins: analyzing transmembrane helix-helix interactions in bacteria. Curr Protein Pept Sci 8:45–61
7. Fink A, Sal-Man N, Gerber D, Shai Y (2012) Transmembrane domains interactions within the membrane milieu: principles, advances and challenges. Biochim Biophys Acta 1818:974–983
8. Hu JC (1995) Repressor fusions as a tool to study protein-protein interactions. Structure 3:431–433
9. Leeds JA, Beckwith J (1998) Lambda repressor N-terminal DNA-binding domain as an assay for protein transmembrane segment interactions in vivo. J Mol Biol 280:799–810
10. Leeds JA, Beckwith J (2000) A gene fusion method for assaying interactions of protein transmembrane segments in vivo. Methods Enzymol 2327:165–175
11. Turner LR, Olson JW, Lory S (1997) The XcpR protein of Pseudomonas aeruginosa dimerizes via its N-terminus. Mol Microbiol 26:877–887
12. Dang TA, Zhou XR, Graf B, Christie PJ (1999) Dimerization of the agrobacterium tumefaciens VirB4 ATPase and the effect of ATP-binding cassette mutations on the assembly and function of the T-DNA transporter. Mol Microbiol 32:1239–1253
13. Rashkova S, Zhou XR, Chen J, Christie PJ (2000) Self-assembly of the Agrobacterium tumefaciens VirB11 traffic ATPase. J Bacteriol 182:4137–4145
14. Langosch D, Brosig B, Kolmar H, Fritz HJ (1996) Dimerisation of the glycophorin A transmembrane segment in membranes probed with the ToxR transcription activator. J Mol Biol 263:525–530
15. Russ WP, Engelman DM (1999) TOXCAT: a measure of transmembrane helix association in a biological membrane. Proc Natl Acad Sci U S A 96:863–868
16. Joce C, Wiener A, Yin H (2011) Transmembrane domain oligomerization propensity determined by ToxR assay. J Vis Exp 51
17. Lindner E, Langosch D (2006) A ToxR-based dominant-negative system to investigate heterotypic transmembrane domain interactions. Proteins 65:803–807
18. Lindner E, Unterreitmeier S, Ridder AN, Langosch D (2007) An extended ToxR POSSYCCAT system for positive and negative selection of self-interacting transmembrane domains. J Microbiol Methods 69:298–305
19. Lis M, Blumenthal K (2006) A modified, dual reporter TOXCAT system for monitoring homodimerization of transmembrane segments of proteins. Biochem Biophys Res Commun 339:321–324
20. Schneider D, Engelman DM (2003) GALLEX, a measurement of heterologous association of transmembrane helices in a biological membrane. J Biol Chem 278:3105–3111
21. Cymer F, Sanders CR, Schneider D (2013) Analyzing oligomerization of individual transmembrane helices and of entire membrane proteins in E. coli: a hitchhiker's guide to GALLEX. Methods Mol Biol 932:259–276
22. Tome L, Steindorf D, Schneider D (2013) Genetic systems for monitoring interactions of transmembrane domains in bacterial membranes. Methods Mol Biol 1063:57–91
23. Karimova G, Pidoux J, Ullmann A, Ladant D (1998) A bacterial two-hybrid system based on a reconstituted signal transduction pathway. Proc Natl Acad Sci U S A 95:5752–5756
24. Ladant D, Karimova G (2000) Genetic systems for analyzing protein-protein interactions in bacteria. Res Microbiol 151:711–720
25. Battesti A, Bouveret E (2012) The bacterial two-hybrid system based on adenylate cyclase reconstitution in Escherichia coli. Methods 58:325–334
26. Karimova G, Dautin N, Ladant D (2005) Interaction network among Escherichia coli membrane proteins involved in cell division as revealed by bacterial two-hybrid analysis. J Bacteriol 187:2233–2243
27. Sivanesan D, Hancock MA, Villamil Giraldo AM, Baron C (2010) Quantitative analysis of VirB8-VirB9-VirB10 interactions provides a dynamic model of type IV secretion system core complex assembly. Biochemistry 49:4483–4493

28. Cisneros DA, Bond PJ, Pugsley AP, Campos M, Francetic O (2012) Minor pseudopilin self-assembly primes type II secretion pseudopilus elongation. EMBO J 31:1041–1053
29. Georgiadou M, Castagnini M, Karimova G, Ladant D, Pelicic V (2012) Large-scale study of the interactions between proteins involved in type IV pilus biology in Neisseria meningitidis: characterization of a subcomplex involved in pilus assembly. Mol Microbiol 84:857–873
30. Zoued A, Durand E, Bebeacua C, Brunet YR, Douzi B, Cambillau C, Cascales E, Journet L (2013) TssK is a trimeric cytoplasmic protein interacting with components of both phage-like and membrane anchoring complexes of the type VI secretion system. J Biol Chem 288:27031–27041
31. Pais SV, Milho C, Almeida F, Mota LJ (2013) Identification of novel type III secretion chaperone-substrate complexes of Chlamydia trachomatis. PLoS One 8:e56292
32. Pineau C, Guschinskaya N, Robert X, Gouet P, Ballut L, Shevchik VE (2014) Substrate recognition by the bacterial type II secretion system: more than a simple interaction. Mol Microbiol 94:126–140
33. Brunet YR, Zoued A, Boyer F, Douzi B, Cascales E (2015) The type VI secretion TssEFGK-VgrG phage-like baseplate is recruited to the TssJLM membrane complex via multiple contacts and serves as assembly platform for tail tube/sheath polymerization. PLoS Genet 11:e1005545
34. Zoued A, Durand E, Brunet YR, Spinelli S, Douzi B, Guzzo M, Flaugnatti N, Legrand P, Journet L, Fronzes R, Mignot T, Cambillau C, Cascales E (2016) Priming and polymerization of a bacterial contractile tail structure. Nature 531:59–63
35. Llosa M, Zunzunegui S, de la Cruz F (2003) Conjugative coupling proteins interact with cognate and heterologous VirB10-like proteins while exhibiting specificity for cognate relaxosomes. Proc Natl Acad Sci U S A 100:10465–10470
36. Segura RL, Aguila-Arcos S, Ugarte-Uribe B, Vecino AJ, de la Cruz F, Goñi FM, Alkorta I (2013) The transmembrane domain of the T4SS coupling protein TrwB and its role in protein-protein interactions. Biochim Biophys Acta 1828:2015–2025
37. Sawma P, Roth L, Blanchard C, Bagnard D, Crémel G, Bouveret E, Duneau JP, Sturgis JN, Hubert P (2014) Evidence for new homotypic and heterotypic interactions between transmembrane helices of proteins involved in receptor tyrosine kinase and neuropilin signaling. J Mol Biol 426:4099–4111
38. Dimitrova M, Younès-Cauet G, Oertel-Buchheit P, Porte D, Schnarr M, Granger-Schnarr M (1998) A new LexA-based genetic system for monitoring and analyzing protein heterodimerization in Escherichia coli. Mol Gen Genet 257:205–212

Chapter 17

Protein–Protein Interactions: Co-Immunoprecipitation

Jer-Sheng Lin and Erh-Min Lai

Abstract

Proteins often do not function as single substances but rather as team players in a dynamic network. Growing evidence shows that protein–protein interactions are crucial in many biological processes in living cells. Genetic (such as yeast two-hybrid, Y2H) and biochemical (such as co-immunoprecipitation, co-IP) methods are the methods commonly used at the beginning of a study to identify the interacting proteins. Immunoprecipitation (IP), a method using a target protein-specific antibody in conjunction with Protein A/G affinity beads, is a powerful tool to identify molecules that interact with specific proteins. Therefore, co-IP is considered to be one of the standard methods of identifying or confirming the occurrence of protein–protein interaction events in vivo. Co-IP experiments can identify proteins via direct or indirect interactions or in a protein complex. Here, we use *Agrobacterium* type VI secretion system (T6SS) sheath components TssB-$TssC_{41}$ interaction as an example to describe the principle, procedure, and experimental problems of co-IP.

Key words Protein–protein interaction, Immunoprecipitation (IP), Co-immunoprecipitation (Co-IP), Immobilization, Protein A/G Sepharose, Physical interaction

1 Introduction

The earliest concept of immunoprecipitation (IP) was used to trace protein turnover by pulse labeling of total proteins during translation using radioactive amino acids added in the cell culture [1, 2]. The use of antibodies for IP can cause the spontaneous precipitation of antigen–antibody complexes formed by the interaction of certain polyclonal antibodies with their antigens. As a consequence, in the study discussed here, the antigen was purified from the protein mixture using a specific antibody immobilized on beads directly or precipitated by affinity beads conjugated by protein A/G, which binds the conserved region of the antibody. Purified antigens (proteins) were then visualized by sodium dodecyl sulfate (SDS)-polyacrylamide gel electrophoresis (PAGE) followed by autoradiography [3]. Co-IP adapts the concept of IP to identify interacting partners and has become one of the most popular methods for protein–protein interaction studies in recent years.

Laure Journet and Eric Cascales (eds.), *Bacterial Protein Secretion Systems: Methods and Protocols*, Methods in Molecular Biology, vol. 1615, DOI 10.1007/978-1-4939-7033-9_17, © Springer Science+Business Media LLC 2017

In a typical experiment, co-IP consists of several steps, including preparation of protein extract (usually a cell lysate), coupling a specific antibody to beads, purification of specific protein complexes, and analysis of the co-IP complexes (Fig. 1) [4]. The unbound proteins are washed away while the antibody, bait protein, and proteins associated to the bait are eluted. Purified protein complexes can then be identified by mass spectrometry or western blot analysis. Depending on the specificity and quality of antibody and experimental conditions, co-IP experiments may generate significant background noise owing to nonspecific binding to the antibody or beads. Thus, negative controls of samples without bait protein or antibody run in parallel are critical to identifying specific interacting proteins. The increased sensitivity of mass spectrometry instrumentation has also greatly reduced the quality and quantity

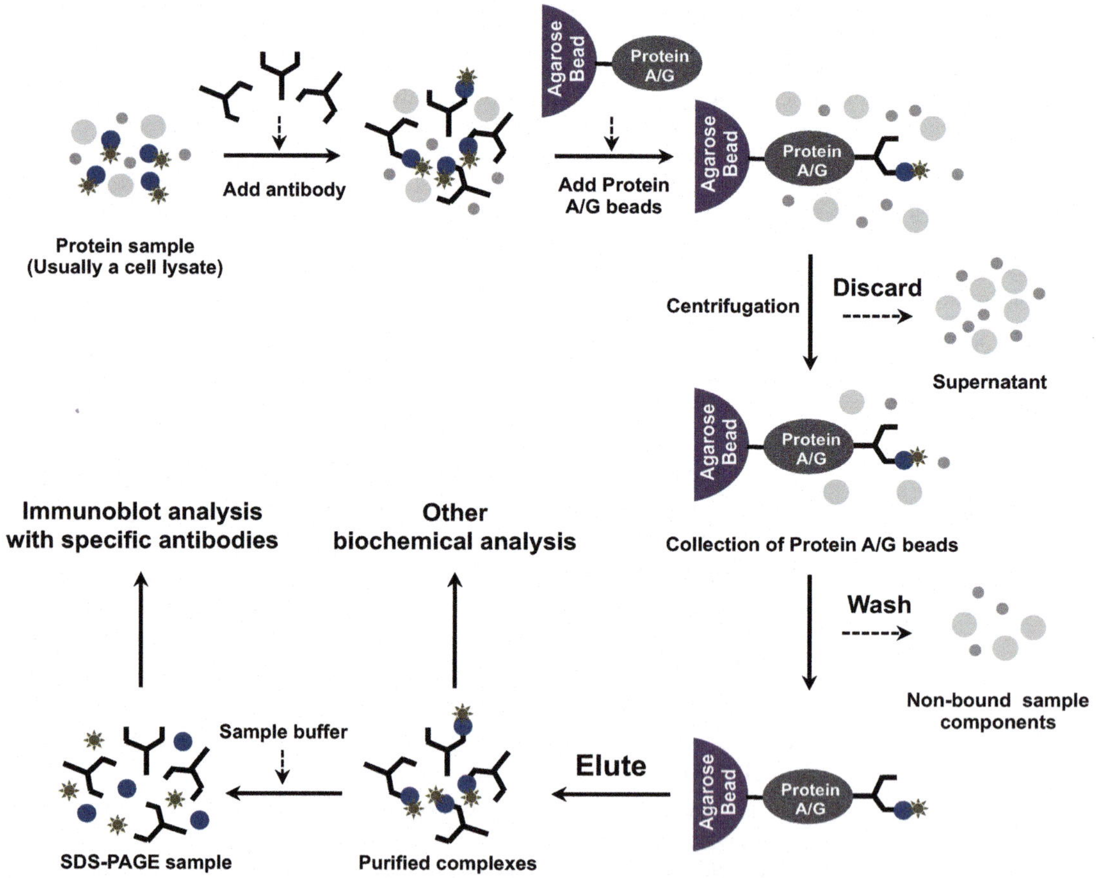

Fig. 1 Schematic diagram of principle of co-IP. Antigen-containing protein sample (usually a cell lysate), specific antibody, and affinity beads (usually protein A/G, which can specifically bind to conserved region of antibody) are added sequentially for binding reaction. The affinity beads with bound proteins are collected by centrifugation. The supernatant containing unbound proteins is discarded and further washed away during washing steps. Antibody and antigen are eluted with a buffer that dissociates proteins from affinity beads. Purified protein complexes can be further used for immunoblot or other biochemical analysis

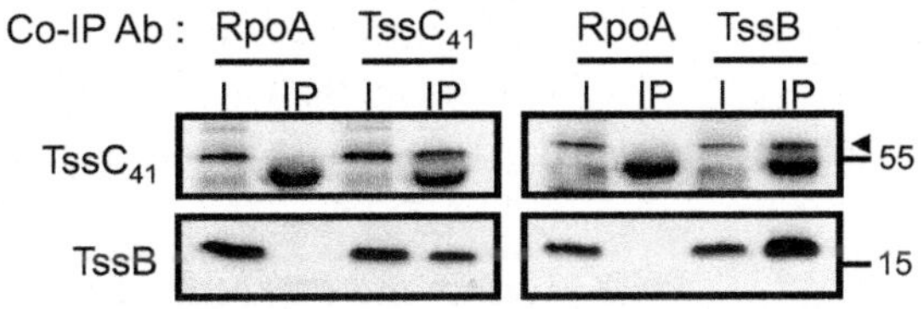

Fig. 2 Co-IP analysis of TssB and $TssC_{41}$ in *A. tumefaciens*. Total protein extracts isolated from *A. tumefaciens* wild-type strain C58 treated with DTBP cross-linker were solubilized by buffer containing 1% SDS, then diluted in Triton X-100-containing solution for IP. Coprecipitated proteins were identified by western blotting. Co-IP was also performed with antibody against RNA polymerase alpha subunit (RpoA) as a negative control. The proteins analyzed and sizes of molecular weight standards are indicated on the *left* and *right*, respectively, and by *arrows* when necessary (*I* input, *IP* immunoprecipitation) (reproduced from Ref. [10]; no permission is required for reuse of content published from Public Library of Science, PLOS

of starting protein samples required for successful protein identifications, which allows for a more complete and reliable map of interactions [5, 6].

Co-IP has been successfully used to study the interaction of proteins involved in bacterial type IV and type VI secretion systems (T4SS and T6SS) in *Agrobacterium tumefaciens* [7–11]. An example is presented in Fig. 2, which illustrates a protocol for a co-IP experiment adapted from "Immunoprecipitation (IP) technical guide and protocols" [3] and "Detection of Protein–Protein Interactions by Coprecipitation" [12], with minor modifications. The cleavable and membrane-permeable cross-linker dimethyl 3,3′-dithiobispropionimidate (DTBP) is used to cross-link interacting proteins before cell lysis to ensure the identification of both stable and weak interacting proteins of type VI secretion components, including the T6SS sheath components TssB and $TssC_{41}$ in *A. tumefaciens* [10]. Coprecipitated proteins are further identified by western blotting. Using this protocol, we found that TssB and $TssC_{41}$ were respectively coprecipitated with each other in *A. tumefaciens* (Fig. 2). In contrast, $TssC_{41}$ and TssB were not precipitated by control antibody (Fig. 2). Together with the interacting data obtained by yeast two-hybrid, copurification in *E. coli*, we concluded that TssB and $TssC_{41}$ could interact strongly to form a protein complex [10].

2 Materials

2.1 Cross-Linking of the Bacterial Cells (see Note 1)

1. Bacterial cells (*see* **Note 2**).
2. DTBP: 0.5 M (*see* **Note 3**).
3. Phosphate buffer: 20 mM sodium phosphate, pH 7.6, 20 mM sodium chloride.
4. 1 M Tris–HCl buffer, pH 7.6.

2.2 Preparation of Bacterial Cell Extracts (see Note 4)

1. TES buffer: 50 mM Tris–HCl, pH 6.8, 2 mM ethylenediaminetetraacetic acid (EDTA), 1% SDS.
2. NP1 buffer: 150 mM Tris–HCl, pH 8.0, 0.5 M sucrose, 10 mM EDTA.
3. Lysozyme (*see* **Note 5**).
4. Triton X-100.
5. Rotating wheel.
6. Protease inhibitor cocktail.

2.3 Protein Sample Preclearing

1. Protein A-Sepharose™ CL4B (GE Healthcare Life Sciences) or equivalent (*see* **Note 6**).
2. 2 mL microtube.
3. Rotating wheel.

2.4 Coupling of Antibody to Protein A-Sepharose Beads

1. Specific antibody and control antibody.
2. Protein A-Sepharose™ CL4B.
3. Rotating wheel.

2.5 Purification and Isolation of Protein Complexes

1. NP1 buffer supplemented with 1% Triton X-100 (*see* **Note 7**).
2. NP1 buffer supplemented with 0.1% Triton X-100.
3. Elution buffer: 0.1 M glycine–HCl, pH 2.5.
4. 2× SDS sample buffer: 100 mM Tris–HCl, pH 6.8, 4% SDS, 20% glycerol, 5% 2-mercaptoethanol, 2 mM EDTA, 0.1 mg/mL bromophenol blue (*see* **Note 8**).

2.6 TrueBlot for Protein Detection of Co-IP Complexes

1. Minigel caster system and SDS-PAGE apparatus.
2. Transblot apparatus for western blot transfer.
3. Rabbit TrueBlot®: Anti-Rabbit IgG horseradish peroxidase (HRP) conjugated secondary antibody, which enables unhindered detection of molecules (eBioscience Inc.).

3 Methods

All procedures are performed in a cold room or on ice unless otherwise indicated. For example, we perform the cross-linking and preclearing steps at room temperature.

3.1 Cross-Linking of Sample

1. Grow bacterial cells (ex: *A. tumefaciens*) under appropriate culture conditions.
2. Centrifuge bacterial cell culture at 6000 × *g* for 10 min at 4 °C, wash the cells by resuspending the cell pellets with 12 mL phosphate buffer, followed by centrifugation at 6000 × *g* for

10 min at 4 °C. Repeat this washing step twice, then the cell pellets in the same buffer are adjusted to OD_{600} 4 (*see* **Note 9**).

3. Add cross-linker DTBP in the cell suspension to a final concentration of 5 mM (*see* **Note 10**).
4. Incubate mixture at room temperature for 45 min (*see* **Note 11**).
5. Stop cross-linking reaction by adding Tris–HCl, pH 7.6, to a final concentration of 20 mM for 15 min (*see* **Note 12**).
6. Collect cells by centrifugation at 6000 × *g* for 10 min at 4 °C, and wash the cells twice by resuspending the cell pellets with 12 mL 50 mM Tris–HCl, pH 7.6, followed by centrifugation at 6000 × *g* for 10 min at 4 °C before cell lysis (*see* **Note 13**).

3.2 Preparation of Bacterial Cell Extracts

1. Pellet the cross-linked cells by centrifugation at 6000 × *g* for 10 min at 4 °C and resuspend the cell pellets in 4 mL TES buffer to OD_{600} 20 (*see* **Note 14**).
2. Incubate the cell resuspension for 30 min at 37 °C with shaking at 200 rpm.
3. Add 18 mL NP1 buffer supplemented with 1.5 mg/mL lysozyme (*see* **Note 15**) and incubate for 2 h on ice.
4. Incubate mixture for 30 min at 37 °C with shaking at 200 rpm.
5. Add Triton X-100 to a final concentration of 4% and incubate for 20 min at room temperature with rotation (*see* **Note 16**).
6. Add protease inhibitor cocktail to working concentration (1×) (*see* **Note 17**) and incubate for 15 min at 37 °C with shaking at 200 rpm.
7. Store the sample mixture for at least 3 h at 4 °C with rotation (*see* **Note 18**).
8. Add 64 mL NP1 buffer to the mixture (*see* **Note 19**); the final concentration of SDS and Triton X-100 are about 0.05% and 1% respectively. The insoluble material is removed by centrifugating twice for 15 min at 14,000 × *g*. The resulting supernatant is the detergent-solubilized solution (*see* **Note 20**).
9. The choice of detergents and the appropriate concentration used for co-IP is made according to the user's manual of protein A-Sepharose.

3.3 Protein Sample Preclearing and Coupling of Antibody to Protein A/G Beads

The preclearing step will reduce the background noise caused by the adhesion of some protein components to the protein A-Sepharose.

1. For each 2 mL of the detergent-solubilized solution, add a 60 μL bed volume of protein A-Sepharose and incubate for 60 min at room temperature with rotation (*see* **Note 21**).
2. Remove protein A-Sepharose with nonspecifically bound proteins by centrifugation for 5 min at 5000 × *g* at 4 °C (*see* **Note 22**).

3. After the preclearing step, the supernatant (protein sample) will serve as the "starting material" for co-IP. The supernatant (about 1.5 mL) is directly incubated with antibody with optimized titer (*see* **Note 23**) and protein A-Sepharose (about 60 μL) overnight at 4 °C with slow rotation (*see* **Note 24**).

3.4 Purification and Isolation of Protein Complexes

1. After overnight incubation, pellet the beads by centrifugation at 5000 × *g* at 4 °C. The supernatant is designated as the "flow through" for co-IP.
2. Wash the beads twice with 1 mL NP1 buffer supplemented with 1% Triton X-100 and once with 1 mL NP1 buffer supplemented with 0.1% Triton X-100 by centrifugation at 5000 × *g* at 4 °C. Discard each wash solution (supernatant) or collect them for SDS-PAGE analysis to examine the washing efficiency. Typically repeat this washing step four to eight times to remove nonspecific binding proteins. In general, the washing steps are carried out until no signal can be detected in the negative control. Then proceed to **step 3** or **step 4** to recover the co-IP complexes.
3. Sample-buffer elution: add 100 μL of 2× SDS sample buffer to the microtube containing beads and place the tube at 96 °C for 20 min. After centrifugation at 10,000 × *g* at room temperature for 5 min, the supernatant is ready to use for western blot analysis (*see* **Note 25**).
4. Low-pH elution: add 100 μL elution buffer to microtube containing beads and place tube at room temperature with low-speed rotation for 20 min to elute the proteins. After centrifugation at 10,000 × *g* at room temperature for 5 min, the co-IP proteins should be eluted in supernatant (*see* **Note 26**).

3.5 TrueBlot for Protein Detection of Co-IP Complexes

1. Analyze the samples by western blot.
2. When the antibody used for co-IP is generated from rabbit, we recommend using the Rabbit TrueBlot® as a secondary antibody with 5000× dilution to minimize interfering signals caused by immunoglobulin heavy and light chains (*see* **Note 27**).

4 Notes

1. The purpose of cross-linking of the bacterial cells is to fix the protein interactions before cell lysis and detergent treatment, especially for weak and dynamic interactions.
2. You must use fresh, live bacterial cells (without freezing) for cross-linking reaction.
3. Always freshly prepare DTBP before use. It is not feasible to dissolve DTBP directly in the chemical bottle by pipetting.

Thus, it is easier to handle by weighing DTBP powder on weight paper and transfer the powder to a microtube and then add a metered volume of buffer to slowly dissolve the powder by vortex.

4. Several methods can be used to prepare bacterial cell extracts (such as sonication or French press). Here, we use a lysozyme/detergent-solubilized method for our co-IP sample preparation [13].
5. We recommend freshly preparing the lysozyme stock solution in NP1 buffer at a concentration of 1 M before use.
6. Both protein A and protein G can associate to rabbit serum with high affinity [4]. Here we chose protein A-Sepharose for our co-IP experiment.
7. Because of the viscous property of Triton X-100, be sure to stir and mix the NP1 buffer with Triton X-100 well before use. We recommend preparing the 10% Triton X-100 stock solution prior to use.
8. 2-mercaptoethanol should be added freshly to 2× sample buffer prior to use.
9. The excess concentration of bacterial cells will cause poor cross-link efficiency.
10. DTBP should be freshly prepared as 0.5 M stock with buffer, in which phosphate buffer is used so as to be consistent with the cell suspension buffer.
11. Gently mix the mixture once every 10 min.
12. Directly add the 1 M Tris–HCl (pH 7.6) stock solution to the mixture until the final concentration of 20 mM is reached; then mix well to stop the cross-linking reaction.
13. The cross-linked cells can be frozen at −80 °C until use. However, we recommend conducting the experiments within 2 weeks after storage.
14. The concentration of cross-linked cells can be reduced. However, if the concentration of bacterial cells is higher than OD 20, the efficiency of cell lysis will become poor.
15. Use NP1 buffer to dissolve lysozyme first before use.
16. Use rotating wheel for this step.
17. The protease inhibitor cocktail is 100× stock.
18. Place the sample mixture on the rotating wheel in a cold room.
19. After adding NP1 buffer to the mixture, carefully mix the mixture well by vortex.
20. The detergent-solubilized solution can be directly used for the co-IP experiment.

21. Generally, the preclearing efficiency is better when incubation is performed at room temperature than in a cold room.
22. Transfer the supernatant to a new 2 mL microtube. The treated protein sample can be used directly for the subsequent co-IP experiment.
23. The titer of antibody for co-IP is based on the reference of antibody titer used for western blotting. In general, we use 5- to 10-fold higher titer than the titer used for western blotting.
24. Place the sample mixture on the rotating wheel in a cold room with gentle rotation (about 10–12 rpm). This is a critical step because an excessive rotation speed will affect the binding efficiency of the antibody and protein A-Sepharose.
25. The sample-buffer elution method is ideal for western blot analysis.
26. A low-pH elution is ideal for enzymatic or functional assays after the low pH is neutralized. In general, the efficiency of elution is lower using the low-pH elution method compared with the sample-buffer elution method.
27. Rabbit IgG TrueBlot® is a unique anti-rabbit IgG immunoblotting reagent (used as the secondary antibody). Rabbit IgG TrueBlot enables detection of immunoblotted target protein bands, with reduced interfering IP immunoglobulin heavy (55 kDa) and light (23 kDa) chains. The Rabbit TrueBlot®: Anti-Rabbit IgG HRP can be reused at least three to five times.

Acknowledgements

This work was supported by a research grant from the Taiwan Ministry of Science and Technology (MOST 104-2311-B-001-025 -MY3) to E.M. Lai. J.S. Lin is the recipient of postdoctoral fellowships from Academia Sinica.

References

1. Kessler SW (1975) Rapid isolation of antigens from cells with a staphylococcal protein A-antibody adsorbent: parameters of the interaction of antibody-antigen complexes with protein A. J Immunol 115:1617–1624
2. Kessler SW (1976) Cell membrane antigen isolation with the staphylococcal protein A-antibody adsorbent. J Immunol 117:1482–1490
3. ThermoFisher (2009) Immunoprecipitation (IP) technical guide and protocols. https://tools.thermofisher.com/content/sfs/brochures/TR0064-Immunoprecipitation-guide.pdf.
4. Lee C (2007) Coimmunoprecipitation assay. Methods Mol Biol 362:401–406
5. Aebersold R, Mann M (2003) Mass spectrometry-based proteomics. Nature 422:198–207
6. Gevaert K, Vandekerckhove J (2000) Protein identification methods in proteomics. Electrophoresis 21:1145–1154

7. Atmakuri K, Cascales E, Christie PJ (2004) Energetic components VirD4, VirB11 and VirB4 mediate early DNA transfer reactions required for bacterial type IV secretion. Mol Microbiol 54:1199–1211
8. Atmakuri K, Cascales E, Burton OT, Banta LM, Christie PJ (2007) Agrobacterium ParA/MinD-like VirC1 spatially coordinates early conjugative DNA transfer reactions. EMBO J 26:2540–2551
9. Anderson LB, Hertzel AV, Das A (1996) Agrobacterium tumefaciens VirB7 and VirB9 form a disulfide-linked protein complex. Proc Natl Acad Sci U S A 93:8889–8894
10. Lin JS, Ma LS, Lai EM (2013) Systematic dissection of the Agrobacterium type VI secretion system reveals machinery and secreted components for subcomplex formation. PLoS One 8:e67647
11. Ma LS, Narberhaus F, Lai EM (2012) IcmF family protein TssM exhibits ATPase activity and energizes type VI secretion. J Biol Chem 287:15610–15621
12. Elion EA (2007) Detection of protein-protein interactions by coprecipitation. Curr Protoc Immunol Chapter 8:Unit 8.7
13. Cascales E, Christie PJ (2004) Definition of a bacterial type IV secretion pathway for a DNA substrate. Science 304:1170–1173

Chapter 18

Protein–Protein Interaction: Tandem Affinity Purification in Bacteria

Julie P.M. Viala and Emmanuelle Bouveret

Abstract

The discovery of protein–protein interaction networks can lead to the unveiling of protein complex(es) forming cellular machinerie(s) or reveal component proteins of a specific cellular pathway. Deciphering protein–protein interaction networks therefore contributes to a deeper understanding of how cells function. Here we describe the protocol to perform tandem affinity purification (TAP) in bacteria, which enables the identification of the partners of a bait protein under native conditions. This method consists in two sequential steps of affinity purification using two different tags. For that purpose, the bait protein is translationally fused to the TAP tag, which consists of a calmodulin binding peptide (CBP) and two immunoglobulin G (IgG) binding domains of *Staphylococcus aureus* protein A (ProtA) that are separated by the tobacco etch virus (TEV) protease cleavage site. After the first round of purification based on the binding of ProtA to IgG coated beads, TEV protease cleavage releases CBP-tagged bait-protein along with its partners for a second round of purification on calmodulin affinity resin and leaves behind protein contaminants bound to IgG. Creating the TAP-tag translational fusion at the chromosomal locus allows detection of protein interactions occurring in physiological conditions.

Key words Protein–protein interaction, Protein complex, Affinity purification, Tandem affinity purification (TAP), Calmodulin binding peptide (CBP), ProtA, Tobacco etch virus (TEV), *Escherichia coli*, *Salmonella*

1 Introduction

At the end of the 1990s, mass spectrometry combined with genome sequencing rendered possible the rapid and systematic identification of all the proteins present in a purified sample. However, a protocol amenable to standardized and systematic purification of protein complexes without any prior knowledge was missing. In 1999, the laboratory of B. Séraphin at the European Molecular Biology Laboratory in Heidelberg, Germany, proposed such a generic procedure for the identification of protein complexes in yeast [1]. This permitted the subsequent description of the full interactome of yeast [2, 3]. This method has since been used in a variety of organisms. We first described its use in bacteria [4], and

Laure Journet and Eric Cascales (eds.), *Bacterial Protein Secretion Systems: Methods and Protocols*, Methods in Molecular Biology, vol. 1615, DOI 10.1007/978-1-4939-7033-9_18, © Springer Science+Business Media LLC 2017

soon after this, it was used to obtain the first interactome of *Escherichia coli* [5].

One general principle of the tandem affinity purification (TAP) method involves the use of two successive steps of affinity purification to lower as much as possible the amount of contaminants, together with an elution preserving the interactions (without significantly changing the buffer's chemical properties) between these two steps. Specifically, the original TAP tag consists of two repeats of the immunoglobulin G (IgG) binding domain of protein A (ProtA) from *Staphylococcus aureus* and a calmodulin binding peptide (CBP), separated by a TEVprotease cleavage site (Fig. 1). However, it must be noted that any combination of affinity tags is potentially usable. Published examples are the GS-TAP (protein G and Strep tag), the sequential peptide affinity (SPA) tag (CBP and 3Flag), the SF-TAP (Strep-tag II and Flag tag), and the HB tag (6Histidine and Biotin) (*see* [6] for specific references). The second general principle of the TAP procedure is to use physiological expression of the recombinant tagged protein. This needs to be adapted to each organism of interest. For *E. coli* and closely related bacteria, lambda Red-based recombination [7] combined with specific dedicated SPA and TAP cassettes [8], makes it very easy to introduce the tag at the 3′ extremity of the gene on the chromosome to obtain the physiological production of a recombinant protein tagged at its C-terminus (Fig. 2). If more convenient, however, TAP tag translational fusion can also be expressed from a plasmid (Fig. 3).

We present here the TAP protocol that has been successfully used in our institute to purify protein complexes of *E. coli*, *Salmonella*, and *Bacillus subtilis* [4, 9–11]. A detailed protocol for the SPA purification has been published [12]. To isolate a protein complex by TAP, a strain producing a recombinant bait protein tagged with the TAP tag must be constructed first (Fig. 1, step 1). Then a soluble extract is prepared from a sufficient volume of bacteria (about 500 mL). The complex is enriched by a first step of affinity chromatography on IgG beads (Fig. 1, step 5). After washes, TEVprotease is added, which cleaves the specific site located between the CBP and ProtA domains, resulting in the elution of the specifically bound material (Fig. 1, step 6). This material is purified a second time by affinity of the CBP tag with calmodulin beads (Fig. 1, step 7). After washes, the purified complex is eluted by adding ethylene glycol-bis(β-aminoethyl ether)-N,N,N',N'-tetraacetic acid (EGTA) that chelates the calcium required for the CBP/calmodulin interaction (Fig. 1, step 8). The totality of the purified material is analyzed on sodium dodecyl sulfate (SDS)-polyacrylamide gel electrophoresis (PAGE). Bands detected by coomassie blue or silver staining are cut from the gel and analyzed by mass spectrometry.

This is the basic TAP procedure. Note that the procedure can be amenable to adaptation or improvements depending on the

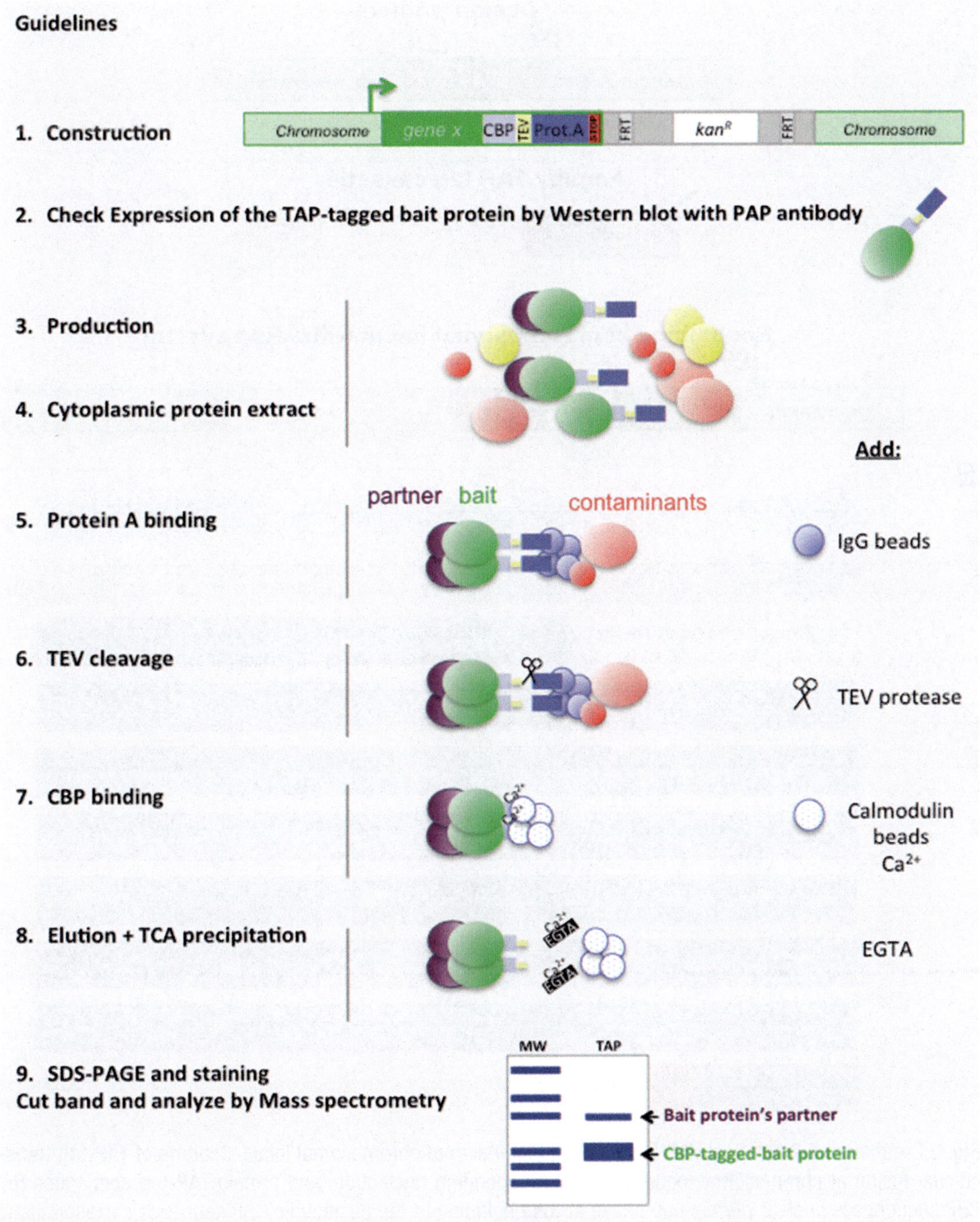

Fig. 1 Guidelines for overall TAP procedure

specific needs. For example, the extensive washes and the duration of the procedure only allow for the recovery of relatively stable complexes. For the detection of more transient or unstable interactions, a cross-linking procedure can be applied before purification [13].

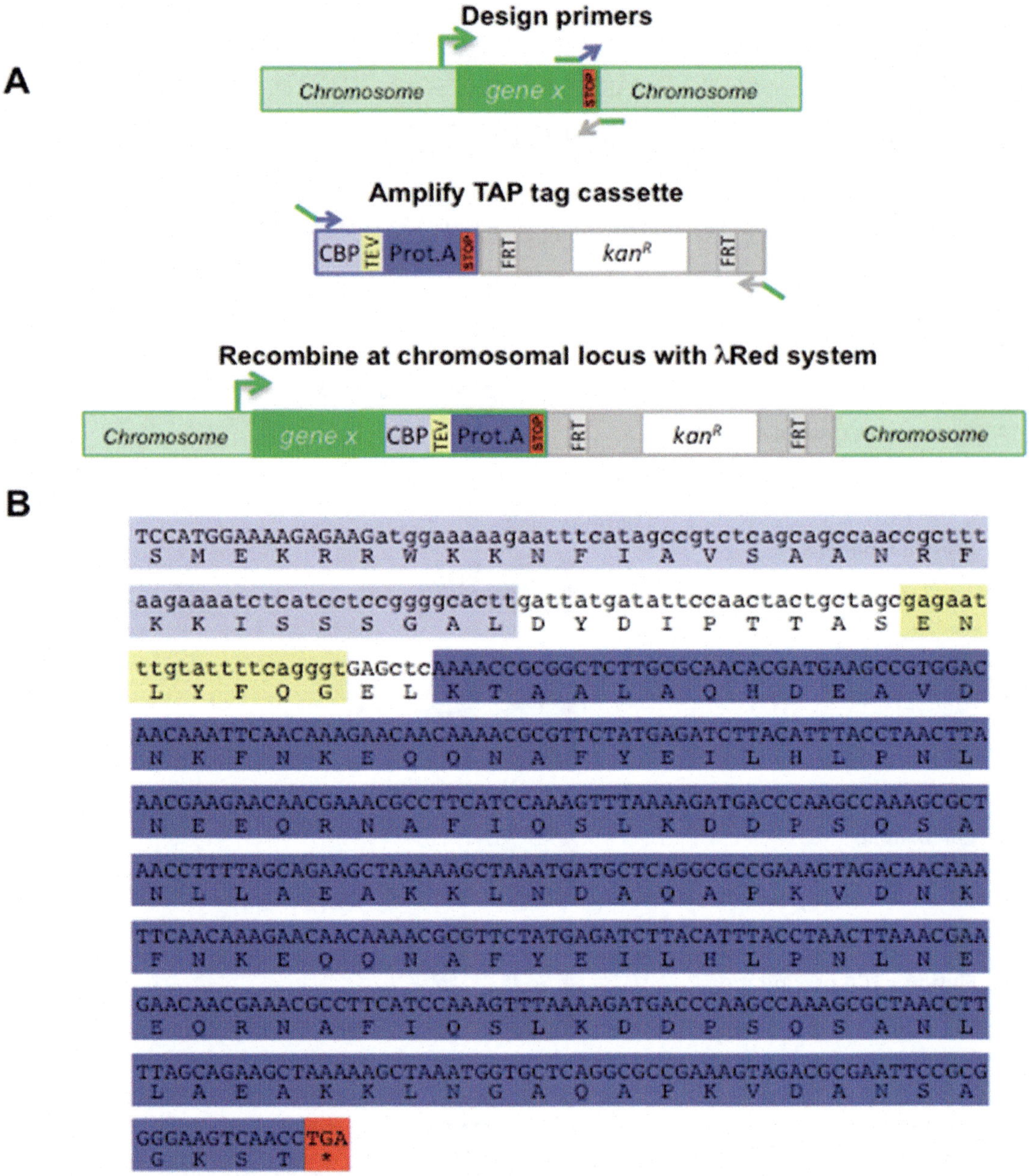

Fig. 2 Creation of a C-terminal TAP-tag translational fusion at chromosomal locus. Scheme of TAP-tag translational fusion at chromosomal locus (**a**) and corresponding nucleotidic and protein TAP-tag sequences (**b**). CBP sequence is in clear *purple*; uppercase at the beginning of the nucleotidic CBP sequence corresponds to the primer sequence mentioned in Subheading 3.1, TEVprotease cleavage site is in *yellow*, and ProtA sequence is in *dark purple*, with stop codon in *red*

This might be helpful also for the purification of membrane complexes, where modifications must be made in the protocol for the solubilization of membranes [3]. Finally, it is possible to play with the two tags to obtain information on the organization of the

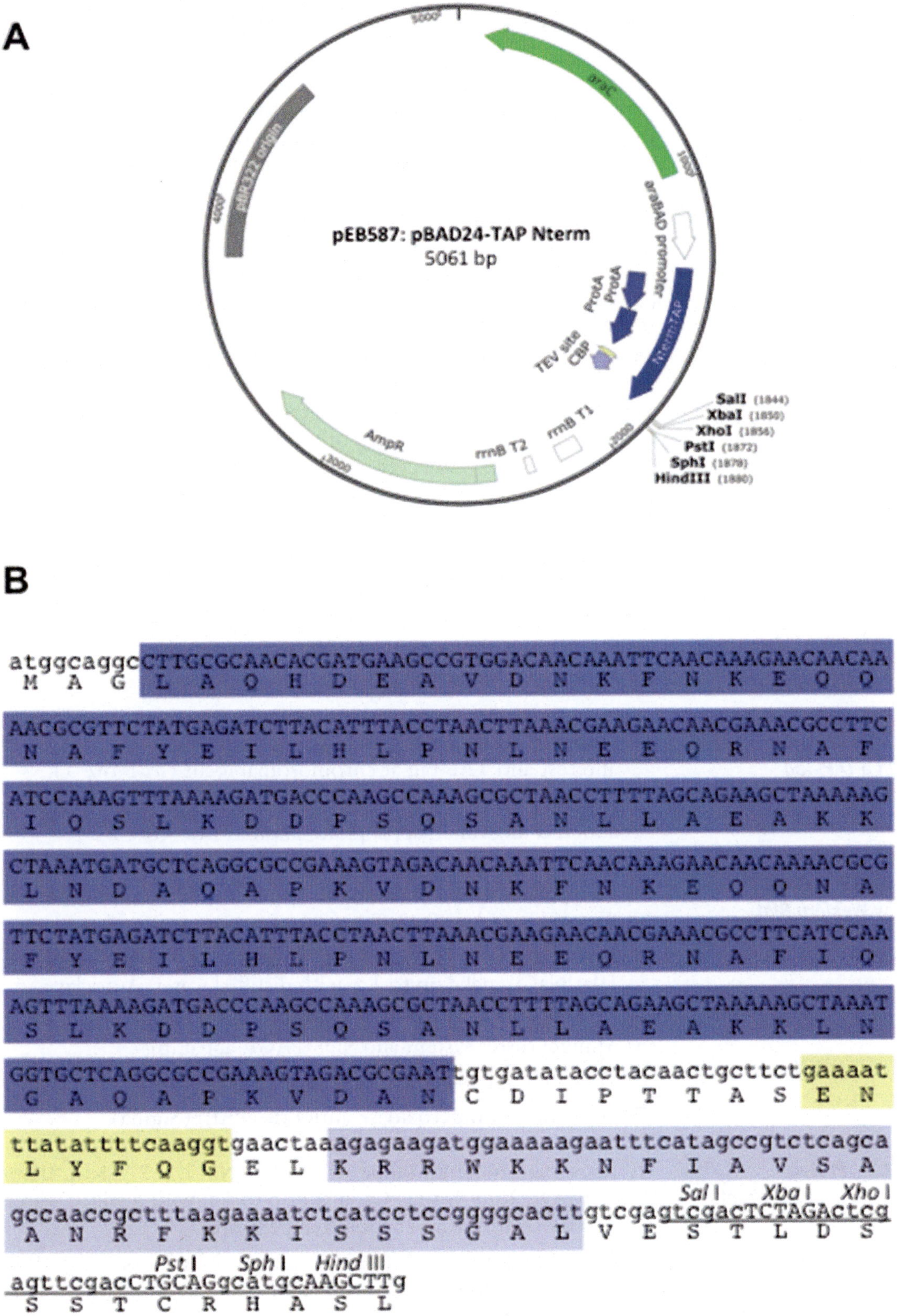

Fig. 3 Creation of plasmidic N-terminal TAP-tag translational fusion. Map of plasmid pEB587 used to create an N-terminal TAP-tag translational fusion under the control of the P_{BAD} arabinose-inducible promoter (**a**) and the corresponding nucleotidic and protein TAP-tag sequences (**b**). ProtA sequence is in *dark purple*, TEVprotease cleavage site is in *yellow*, CBP sequence is in *clear purple*, the multicloning site is *underlined*, and restriction enzyme sites are indicated

complexes. Indeed, in some cases, one bait protein might participate in the formation of several types of complexes. To purify one specific type of complex, it is therefore possible to put the two tags on two distinct proteins, both of which are members of the desired type of complex (split tag method [9, 14]). Alternatively, it is possible to perform the subtraction method that consists in eliminating the unwanted complex(es) during the first purification step by leaving it, for example, on IgG beads thanks to a partner protein of the bait that belongs to the unwanted complex and bears a noncleavable ProtA tag. The desired complex, made of untagged partner proteins, will elute with the bait after TEVprotease cleavage [14, 15].

To our knowledge, the TAP procedure has not been used very much for the characterization of secretion systems in bacteria, certainly owing to the difficulty of working with integral envelope components [11]. However, it has proved to be powerful in identifying the target of effectors of *Legionella* T4SS or *Pseudomonas* T6SS in eukaryotic host cells [16, 17]. In addition, as mentioned earlier, it is amenable to several improvements that might make it possible to identify unsuspected partners of the secretion machineries in the bacterium.

2 Materials

2.1 Creation of a TAP-Tag Translational Fusion and Verification of Production of Hybrid Protein by Western Blot

1. Bacterial strain with translational fusion between protein of interest and TAP tag at chromosomal locus (*see* **Note 1**).
2. Alternatively, translational fusion between protein of interest and TAP tag on an appropriate plasmid (*see* **Note 2**).
3. Yeast extract and tryptone media (2YT): 16 g yeast extract, 10 g tryptone, 10 g NaCl, make up to 1 L with distilled water. Autoclave and store at room temperature.
4. Lysogeny broth (LB): 5 g yeast extract, 10 g Bacto Tryptone, 10 g NaCl, make up to 1 L with distilled water. Autoclave and store at room temperature.
5. Minigel caster system and SDS-PAGE apparatus.
6. Transblot apparatus for western blot.
7. Peroxidase–antiperoxidase antibody (PAP) (Sigma).

2.2 Protein Cytoplasmic Extract

1. Phosphate buffered saline (PBS): 8 g NaCl, 0.2 g KCl, 0.2 g KH_2PO_4, 2.9 g Na_2HPO_4, make up to 1 L with distilled water. Autoclave and store at room temperature.
2. 10% Nonidet P-40 (NP-40 or Igepal): Mix 10 mL NP-40 in 90 mL distilled water, pass through a 0.2 μm filter, and store at room temperature (*see* **Note 3**).
3. ProtA binding buffer: 10 mM Tris–HCl, pH 8, 150 mM NaCl, 0.1% NP-40. Approximately 50 mL will be required per sample

per experiment. Prepare 500 mL containing 5 mL 1 M Tris–HCl, pH 8, 15 mL 5 M NaCl, 5 mL 10% NP-40, and 475 mL distilled water. Store at 4 °C.

4. 0.1 M phenylmethylsulfonyl fluoride (PMSF): Dissolve 87.1 mg PMSF in 5 mL isopropanol. Prepare 1 mL aliquots and store at −20 °C (*see* **Note 4**).
5. Liquid nitrogen.
6. Sonicator, French press, or cell disruptor.
7. Centrifuge tubes and rotor, compatible with spinning volumes of 250 mL, 10 mL, and 50 mL, at approximately 5000 × *g* and 25,000 × *g*, respectively.

2.3 Tandem Affinity Purification

1. IgG Sepharose 6 fast flow (GE Healthcare).
2. ProtA binding buffer: *See* Subheading 2.2.
3. 0.5 M EDTA ($C_{10}H_{14}N_2Na_2O_8{\cdot}2H_2O$): Dissolve 18.6 g EDTA in 80 mL distilled water, make up to 100 mL with distilled water once pH has been adjusted to 8 with 10 N NaOH (*see* **Note 5**). Autoclave and store at room temperature.
4. TEV cleavage buffer: 10 mM Tris–HCl, pH 8, 150 mM NaCl, 0.1% NP-40, 0.5 mM EDTA, 1 mM dithiothreitol (DTT) (*see* **Note 6**).
5. AcTEV™ protease (Invitrogen).
6. Calmodulin binding buffer: 10 mM Tris–HCl, pH 8, 150 mM NaCl, 0.1% NP-40, 1 mM magnesium acetate, 1 mM imidazole, 2 mM $CaCl_2$, 10 mM β-mercaptoethanol. Approximately 40 mL will be required per sample per experiment. Prepare 500 mL containing 5 mL 1 M Tris–HCl, pH 8, 15 mL 5 M NaCl, 5 mL 10% NP-40, 500 μL 1 M magnesium acetate, 500 μL 1 M imidazole, 1 mL 1 M $CaCl_2$, 348.5 μL 14.3 M β-mercaptoethanol (*see* **Note 7**), and 473 mL distilled water. Store at 4 °C.
7. 1 M $CaCl_2$: Dissolve 11.1 g in 100 mL distilled water. Autoclave and store at room temperature.
8. Calmodulin affinity resin (Agilent).
9. 1 M EGTA: Dissolve 19 g in 40 mL distilled water, make up to 50 mL with distilled water once pH has been adjusted to 8 with 10 N NaOH (*see* **Note 5**), 0.2 μm filter, and store at 4 °C.
10. Calmodulin elution buffer: 10 mM Tris–HCl, pH 8, 150 mM NaCl, 0.1% NP-40, 1 mM magnesium acetate, 1 mM imidazole, 2 mM EGTA, 10 mM β-mercaptoethanol. Approximately 1 mL will be required per sample per experiment. Prepare 100 mL containing 1 mL 1 M Tris–HCl, pH 8, 3 mL 5 M NaCl, 1 mL 10% NP-40, 100 μL 1 M magnesium acetate, 100 μL 1 M imidazole, 200 μL 1 M EGTA, 69.7 μL 14.3 M

β-mercaptoethanol (*see* **Note 7**), and 94.5 mL distilled water. Store at 4 °C.

11. Disposable chromatography columns of 10 mL with narrow bottom, for example, Poly-Prep chromatography column from Biorad.
12. Rotating wheel.
13. Centrifuge tubes and rotor compatible with spinning volumes of 10 mL at approximately 25,000 × *g*.

2.4 Trichloroacetic Acid Precipitation

1. 16 mg/mL sodium deoxycholate: Dissolve 160 mg sodium deoxycholate in 10 mL water. Pass through a 0.2 μm filter and store at room temperature.
2. Liquid trichloroacetic acid (TCA) (stock is 100%).
3. TCA washing buffer: Mix 70 mL acetone, 20 mL ethanol, 5 mL 1 M Tris–HCl, pH 8, and 5 mL distilled water. Store at 4 °C.
4. SDS-PAGE loading buffer.

3 Methods

A translational fusion between the protein of interest and TAP tag, either at the chromosomal locus or on an appropriate plasmid, must be constructed (*see* **Notes 1** and **2**). Translational fusion at the chromosomal locus will allow a physiological expression, while constructing the translational fusion on a plasmid may be more amenable.

3.1 Verification of Expression of TAP-Tag Translational Fusion by Western Blot

1. Prepare a cytoplasmic or a crude protein extract (*see* **Note 8**).
2. Load 10 μg of protein extract (or proteins corresponding to a bacterial sample of 0.3 OD_{600} unit) on a SDS-PAGE and proceed to transfer and western blot to verify production of hybrid protein (*see* **Note 9**).
3. Perform a one-step western blot using PAP antibody (*see* **Note 10**) and using an appropriate substrate to detect horseradish peroxidase activity (*see* **Note 11**).

3.2 Protein Cytoplasmic Extract

1. Day 1—Inoculate 10 mL 2YT media with a bacterial colony and grow overnight at 37 °C with shaking (*see* **Note 8**).
2. Day 2—Dilute culture 100-fold in 500 mL LB and grow for 5 h 30 min at 37 °C with shaking until $OD_{600} \approx 2$–3.
3. Pellet bacteria by centrifugation for 20 min, 5000 × *g* at 4 °C.
4. Wash once with cold PBS, transfer to 50 mL centrifuge tubes, centrifuge again for 10 min, 5000 × *g* at 4 °C, discard supernatant, and freeze bacterial pellets with liquid nitrogen.

Maintain frozen bacterial pellets at −80 °C until you are ready to prepare cytosolic protein extract and proceed to TAP.

5. Day 3—Resuspend frozen bacterial pellets with 10 mL ProtA binding buffer containing 0.5 mM PMSF (*see* **Note 4**).
6. Use sonication, French press, or cell disruptor to break bacterial cells (*see* **Note 12**).
7. Centrifuge for 30 min, 25,000 × *g* at 4 °C, and save supernatant, which is the cytoplasmic protein extract.

3.3 Tandem Affinity Purification

From here, carry out all procedures with gloves to avoid contamination of your sample(s) with keratine.

1. Put 200 μL IgG Sepharose beads in a disposable chromatography column and wash by gravity with 5 mL ProtA binding buffer.
2. *Binding of ProtA tag to IgG Sepharose beads.* After washing the beads, close the bottom of the chromatography column, and, using a pipette, transfer 9 mL of the cytoplasmic protein extract. Close the top of the column and put on a wheel for 2 h at 4 °C.
3. Remove first the top plug of the column and then the bottom one. Leave the unbound material flow by gravity and discard.
4. Wash three times the IgG beads with 10 mL of ProtA binding buffer.
5. *TEVprotease cleavage.* Close bottom of column, fill it with 1 mL TEV cleavage buffer and 100 units of AcTEV™ protease. Close top of column and put on wheel at room temperature for 1 h.
6. Remove top and bottom plugs and *recover* elution fraction by gravity. Add additional 200 μL TEV cleavage buffer in column to recover as much material as possible from sides of column.
7. Add 3 mL calmodulin binding buffer and 3 μL 1 M $CaCl_2$ (*see* **Note 13**) to elution fraction.
8. *Binding by the CBP tag part on calmodulin affinity resin.* Put 200 μL calmodulin affinity resin in a new disposable chromatography column, and wash it with 5 mL calmodulin binding buffer. Then close bottom of column.
9. Add the 4.2 mL of the elution fraction (obtained at **steps 6** and 7). Close top of column and put on wheel for 1 h at 4 °C.
10. Remove first the top plug of the column and then the bottom one. Leave unbound material to flow by gravity and discard.
11. Wash three times calmodulin affinity resin with 10 mL calmodulin binding buffer.
12. *Elution.* Elute with five times 200 μL of calmodulin elution buffer.
13. Pool fractions 2, 3, and 4 and proceed to TCA precipitation of elution fraction 1, pooled fractions 2, 3, and 4, and fraction 5.

3.4 TCA Precipitation

1. To each of the eluted protein samples add 1/100th of 16 mg/mL sodium deoxycholate. Vortex and leave on ice 30 min.
2. Add TCA to 10% final. Vortex and leave on ice 30 min.
3. Centrifuge 15 min, 15,000 × *g* at 4 °C.
4. Wash twice with TCA washing buffer.
5. Leave pellets to dry on bench and resuspend in 20 μL protein loading buffer 1×.

3.5 Analysis by SDS-PAGE and Mass Spectrometry

1. Load totality of samples on SDS-PAGE (*see* **Note 14**) and stain with coomassie blue.
2. Unstain and then rinse with distilled water.
3. Cut bands to identify partner proteins by mass spectrometry (*see* **Note 15**).

4 Notes

1. A C-terminal TAP tag translational fusion can be introduced at the chromosomal locus using the λ Red recombination system [7]. To prepare the appropriate polymerase chain reaction (PCR) product, use a pJL72 plasmid as template (this latter harbors a cassette made of the TAP tag and the kanamycin resistance gene, Fig. 2a) [8], design a forward primer that contains, in the 5′-end, the 45 nucleotides that are immediately upstream of the stop codon of the gene of interest, followed by the sequence 5′-TCCATGGAAAAGAGAAG-3′ (this sequence will form a hybrid to the CBP tag, Fig. 2b), and design a reverse primer that contains at its 5′-end the reverse complement of 45 nucleotides that are immediately downstream of the stop codon of the gene of interest, followed by the sequence 5′-CATATGAATATCCTCCTTAG-3′ (Fig. 2a).
2. Alternatively, the sequence corresponding to the open reading frame of the gene of interest can be cloned in the plasmid pEB587 [18] (Fig. 3a), which allows an N-terminal TAP tag translational fusion (Fig. 3b) under the control of the arabinose-inducible promoter P_{BAD}.
3. Gently agitate the solution for complete dissolution of NP-40 if necessary.
4. PMSF crystallizes at −20 °C, so heat the PMSF aliquot to 37 °C to redissolve the PMSF before use. We use PMSF as generic protease inhibitor, but a protease inhibitor cocktail can be used as well.
5. EDTA and EGTA may not be soluble until the pH is adjusted to 8 with 10 N NaOH.

6. Add DTT to the volume of buffer you will need when starting the experiment. DTT is necessary for TEV activity.
7. Add β-mercaptoethanol to the volume of buffer you will need when starting the experiment.
8. Also plan to prepare a protein extract of an untagged strain as the negative control of the experiment.
9. Translational TAP tag fusion adds 20 kDa to the mass of the protein of interest; 3 kDa corresponds to the CBP tag and 15 kDa to the ProtA tag.
10. Immunoglobulins will bind the ProtA fragment of the TAP tag.
11. In our experience, the detection of a tagged protein in crude extracts using this PAP antibody is mandatory for a successful TAP purification.
12. A French press or cell disruptor might be more gentle for preserving protein complexes.
13. The addition of extra $CaCl_2$ is required to quench the EDTA that was previously necessary for TEV protease activity.
14. Usually 12% SDS-PAGE allows visualization of low- and high-molecular-weight proteins.
15. Use one blade per band.

References

1. Rigaut G, Shevchenko A, Rutz B, Wilm M, Mann M et al (1999) A generic protein purification method for protein complex characterization and proteome exploration. Nat Biotechnol 17:1030–1032
2. Gavin AC, Bosche M, Krause R, Grandi P, Marzioch M et al (2002) Functional organization of the yeast proteome by systematic analysis of protein complexes. Nature 415:141–147
3. Gavin AC, Aloy P, Grandi P, Krause R, Boesche M et al (2006) Proteome survey reveals modularity of the yeast cell machinery. Nature 440:631–636
4. Gully D, Moinier D, Loiseau L, Bouveret E (2003) New partners of acyl carrier protein detected in Escherichia coli by tandem affinity purification. FEBS Lett 548:90–96
5. Butland G, Peregrin-Alvarez JM, Li J, Yang W, Yang X et al (2005) Interaction network containing conserved and essential protein complexes in Escherichia coli. Nature 433:531–537
6. Collins MO, Choudhary JS (2008) Mapping multiprotein complexes by affinity purification and mass spectrometry. Curr Opin Biotechnol 19:324–330
7. Datsenko KA, Wanner BL (2000) One-step inactivation of chromosomal genes in Escherichia coli K-12 using PCR products. Proc Natl Acad Sci U S A 97:6640–6645
8. Zeghouf M, Li J, Butland G, Borkowska A, Canadien V et al (2004) Sequential Peptide Affinity (SPA) system for the identification of mammalian and bacterial protein complexes. J Proteome Res 3:463–468
9. Gully D, Bouveret E (2006) A protein network for phospholipid synthesis uncovered by a variant of the tandem affinity purification method in Escherichia coli. Proteomics 6:282–293
10. Pompeo F, Luciano J, Galinier A (2007) Interaction of GapA with HPr and its homologue, Crh: novel levels of regulation of a key step of glycolysis in Bacillus subtilis? J Bacteriol 189:1154–1157
11. Viala JP, Prima V, Puppo R, Agrebi R, Canestrari MJ, Lignon S et al (2017) Acylation of the type 3 secretion system translocon using a dedicated acyl carrier protein. PLoS Genet 13(1):e1006556

12. Babu M, Butl G, Pogoutse O, Li J, Greenblatt JF et al (2009) Sequential peptide affinity purification system for the systematic isolation and identification of protein complexes from Escherichia coli. Methods Mol Biol 564:373–400
13. Stingl K, Schauer K, Ecobichon C, Labigne A, Lenormand P et al (2008) In vivo interactome of Helicobacter pylori urease revealed by tandem affinity purification. Mol Cell Proteomics 7:2429–2441
14. Puig O, Caspary F, Rigaut G, Rutz B, Bouveret E et al (2001) The tandem affinity purification (TAP) method: a general procedure of protein complex purification. Methods 24:218–229
15. Bouveret E, Rigaut G, Shevchenko A, Wilm M, Seraphin B (2000) A Sm-like protein complex that participates in mRNA degradation. EMBO J 19:1661–1671
16. So EC, Schroeder GN, Carson D, Mattheis C, Mousnier A et al (2016) The Rab-binding profiles of bacterial virulence factors during infection. J Biol Chem 291:5832–5843
17. Sana TG, Baumann C, Merdes A, Soscia C, Rattei T et al (2015) Internalization of Pseudomonas aeruginosa strain PAO1 into epithelial cells is promoted by interaction of a T6SS effector with the microtubule network. MBio 6:e00712
18. Battesti A, Bouveret E (2008) Improvement of bacterial two-hybrid vectors for detection of fusion proteins and transfer to pBAD-tandem affinity purification, calmodulin binding peptide, or 6-histidine tag vectors. Proteomics 8:4768–4771

Chapter 19

Site-Directed and Time-Resolved Photocrosslinking in Cells Metabolically Labeled with Radioisotopes

Raffaele Ieva

Abstract

To efficiently transport proteins into and across cellular membranes, specialized transport machineries engage in recognition events with different domains of their client proteins, forming sequential intermediate complexes. The short-lived nature of these interactions poses a big challenge in the identification of the key factors involved in transport reactions and their mechanism of action. Site-directed photocrosslinking is a powerful method for the detection and accurate mapping of interacting protein domains. This chapter describes a protocol that combines site-directed photocrosslinking to metabolic labeling of proteins and lipids as a method to characterize, with temporal and spatial resolution, the interactions of a secretory protein as it transverses the bacterial envelope.

Key words Site-directed photocrosslinking, Transient protein–protein interaction, Protein–lipid interaction, Protein secretion, *Escherichia coli*, Autotransporter, BAM complex

1 Introduction

Efficient and correct sorting of proteins, from the site of synthesis to the compartment where they function, is vital for cells. This task is particularly challenging for proteins that must cross one or more cellular membranes before reaching their destination, such as proteins secreted across the bacterial envelope. In bacteria, protein secretion is aided by specialized molecular machineries, which insert or transport their client proteins into and across lipid bilayers [1]. Capturing interactions between secretory proteins and their transport machineries "at work" is a key approach to elucidate the molecular events that govern these sophisticated reactions.

Several biochemical methods have been developed to describe the interactions of proteins within multisubunit complexes in their cellular native environment, including native complex isolation by affinity purification or immunoprecipitation, blue native polyacrylamide gel electrophoresis of assembly intermediates, and chemical crosslinking. Recognition events between secretory proteins and

Laure Journet and Eric Cascales (eds.), *Bacterial Protein Secretion Systems: Methods and Protocols*, Methods in Molecular Biology, vol. 1615, DOI 10.1007/978-1-4939-7033-9_19, © Springer Science+Business Media LLC 2017

their transport machineries, however, are often too short-lived to be detected by these methods. Different strategies can be designed to "trap" client proteins in a complex with their transport machineries. This can be achieved, for instance, by stabilizing structural motifs within the client protein or by fusing it to a bulkier protein moiety, generating an obstruction that prevents the completion of the transport reaction. Although such "trapping" strategy may help to capture a transient intermediate complex that forms at a specific stage of the transport reaction, it may not provide information on the sequence of molecular events that precede and follow the formation of the captured intermediate.

A further limitation of methods based on native complex isolation or chemical crosslinking concerns the mapping of protein–protein interactions. In order to map interaction sites, a method for the incorporation of unnatural photoactivatable amino acids into newly synthesized proteins was originally developed using an amber suppression approach in cellfree translation systems [2]. A subsequent modification of the method has made it possible to express proteins with site-specific photoprobes in vivo. To this end, cells that coexpress an engineered orthogonal aminoacyl tRNA synthetase, which loads a cognate amber suppressor tRNA with a photoreactive amino acid analog, such as *p*-benzoyl-L-phenylalanine (Bpa), are used [3]. The photoprobe can be introduced at an amber codon engineered at a specific position of the open reading frame of a gene of interest. Upon irradiation with ultraviolet (UV) light (350–365 nm), Bpa forms a highly reactive radical that can crosslink to C–H bonds in the vicinity [4]. Note that Bpa is a small amino acid of approximately 4 Å in diameter, thus it can crosslink only proteins that are in close proximity, allowing precise mapping of protein interaction sites.

The method described in this chapter exploits the in vivo site-directed photocrosslinking approach to resolve consecutive interactions of a client protein with its transport machinery. Site-directed photocrosslinking is combined with transient metabolic labeling of cells with radioactive amino acids. Labeling is performed using an in vivo pulse-chase methodology, thus the fate of radiolabeled proteins can be followed in time, helping to distinguish the order of the detected interactions. The example protocol described here analyzes the biogenesis of EspP, an autotransporter expressed by *Escherichia coli* O157:H7. Autotransporters are a class of virulence factors produced by Gram-negative bacteria. After transport across the inner membrane, the autotransporter carboxy-terminal "β domain" integrates into the outer membrane by folding into a β-barrel structure, while its amino-terminal "passenger domain" is ultimately translocated into the extracellular space. The mechanism by which the passenger domain is transported across the outer membrane has been debated for some time [5, 6]. In a series of studies, the described approach revealed that passenger domain secretion and outer membrane integration of the β domain are both mediated by the β-barrel assembly machinery (BAM complex) in a coordinated reaction. BAM is a multisubunit complex consisting of BamA, an integral outer membrane protein, and four lipoproteins, BamBCDE, that associate to the inner leaflet of the outer membrane lipid bilayer [7, 8]. Consecutive interactions of distinct EspP domains, first with periplasmic chaperones and then with subunits of the BAM machinery, were identified, leading to a detailed model of autotransporter biogenesis [9–11]. Finally, this chapter provides an example of how site-directed photocrosslinking can be combined with the metabolic labeling of cells using radioactive inorganic phosphate [12] to reveal the time of insertion of the EspP β domain into the lipid bilayer of the bacterial outer membrane [10, 11].

2 Materials

2.1 Plasmid Construction and Transformation of E. coli Cells

1. Site-directed mutagenesis kit or a similar set of reagents for high-fidelity polymerase chain reaction (PCR).
2. Chemically ultracompetent *E. coli* cells.
3. Plasmid miniprep kit.
4. Cuvettes for electroporation.
5. Electroporator.
6. Lysogenic broth (LB) agar plates containing selected antibiotics.

2.2 Expression of a EspP Variant Containing Bpa and Pulse-Chase Metabolic Labeling of Cells Using ^{35}S-Labeled Amino Acids

1. 10× M9 salt (67.8 g/L Na_2HPO_4, 30 g/L KH_2PO_4, 5 g/L NaCl, 10 g/L NH_4Cl).
2. M9 complete medium containing 1× M9 salt, 1 mM $MgSO_4$, 0.1 mM $CaCl_2$, 0.2% w/v glycerol, 40 μg/mL L-amino acids (except methionine and cysteine).
3. One disposable 125 mL Erlenmeyer flask.
4. A high-specific activity mixture of ^{35}S-cysteine and ^{35}S-methionine (1075–1175 Ci/mmol).
5. A stock solution of non-radiolabeled methionine and cysteine (100 mM methionine, 100 mM cysteine).
6. 100 mM isopropyl-β-D-thio-galactoside (IPTG) stock solution.
7. Bpa (Bachem).
8. Water bath shaker.

2.3 Metabolic Labeling with ^{32}P-Labeled Inorganic Phosphate and Expression of an EspP Variant Containing Bpa

1. One disposable 125 mL Erlenmeyer flask.
2. Modified G56 medium (45 mM MES, pH 7.0, 10 mM KCl, 10 mM $MgCl_2$, 15 mM $(NH_4)_2SO_4$, 5 μg/L thiamine, 0.2% glycerol, 40 μg/mL L-amino acids).
3. A stock solution of 0.5 M KH_2PO_4.
4. Radioactive $KH_2{}^{32}PO_4$ (900–1100 mCi/mmol).

5. 100 mM IPTG stock solution.
6. Bpa (Bachem).

2.4 Photocross linking

1. High-intensity UV lamp such as Spectroline SB-100P (Spectronics Corporation).
2. Six-well tissue culture plates.
3. Disposable pipettes.

2.5 Immunoprecipita tion and Sodium Dodecyl Sulfate (SDS)-Polyacrylamide Gel Electrophoresis (PAGE) Analysis of Photocrosslinked Proteins

1. Antisera specific for EspP, BamA, and BamB or other proteins of interest.
2. Protein solubilization buffer (15% glycerol, 200 mM Tris base, 15 mM EDTA, 4% SDS, 2 mM phenylmethylsulfonyl fluoride (PMSF)).
3. Radioimmunoprecipitation assay (RIPA) buffer (50 mM Tris–HCl, pH 8, 150 mM NaCl, 1% Igepal CA-630, 0.5% sodium deoxycholate, 0.1% SDS).
4. *Staphylococcus aureus* protein A-Sepharose beads.
5. SDS-polyacrylamide gels.
6. 4× SDS-PAGE sample buffer (8% SDS, 40% glycerol, 240 mM Tris–HCl, pH 6.8, 1.6% β-mercaptoethanol, 0.04% bromophenol blue).
7. Gel drying system.
8. Storage phosphor screen.

3 Method

3.1 Strategy Design and Plasmid Construction

1. Clone a construct encoding the selected protein of interest under the control of an inducible promoter in an expression plasmid that can be propagated in the desired bacterial model organism. In this example, RI23, an RB11-derived plasmid, harbors a construct encoding EspP(586TEV) (*see* **Note 1**) under the control of the IPTG-inducible *lac* promoter. The selected model organism is the laboratory *E. coli* strain AD202, an MC4100-derived strain that lacks the gene encoding the outer membrane protease OmpT [13].
2. Use a PCR-based site-specific mutagenesis approach to incorporate an amber codon at a specific position of the protein of interest. In this example, the mutagenesis approach is conducted on pRI23 to replace with an amber codon the EspP Trp 1149 codon, generating pRI23-1149Bpa (*see* **Note 2**).
3. Digest the parental DNA used as template in the PCR with the restriction enzyme DpnI.
4. Transform ultracompetent *E. coli* cells using a small aliquot of the amplified DNA product.

5. Select single colonies growing on LB containing 100 μg/mL of ampicillin, extract plasmidic DNA, and verify the correct incorporation of the amber codon by plasmid sequencing.
6. Cotransform by electroporation the *E. coli* strain AD202 with pDULE-*p*Bpa, a plasmid encoding an engineered amino acyl tRNA synthetase and a cognate amber suppressor tRNA of *Methanococcus jannaschii* (*see* **Note 3**), and the generated plasmid (pRI23-1149Bpa). Plate cells on LB-agar containing 100 μg/mL ampicillin and 5 μg/mL tetracycline.
7. Select a single colony of transformed cells.

3.2 Preparation of Cell Cultures for EspP Expression and Metabolic Labeling

Two methodologies of metabolic labeling are used (Fig. 1). In one culture ^{35}S labeling is conducted following a "pulse-chase" procedure, in which all newly synthesized proteins become radiolabeled by exposing cells for a short time period to ^{35}S-methionine and ^{35}S-cysteine (pulse phase). Subsequently, the addition of an excess of non-radiolabeled ("cold") methionine and cysteine prevents further incorporation of radiolabeled amino acids into proteins, thereby ending the pulse phase. The fate of radiolabeled proteins can be chased over time (chase phase). By combining ^{35}S-labeling with site-specific

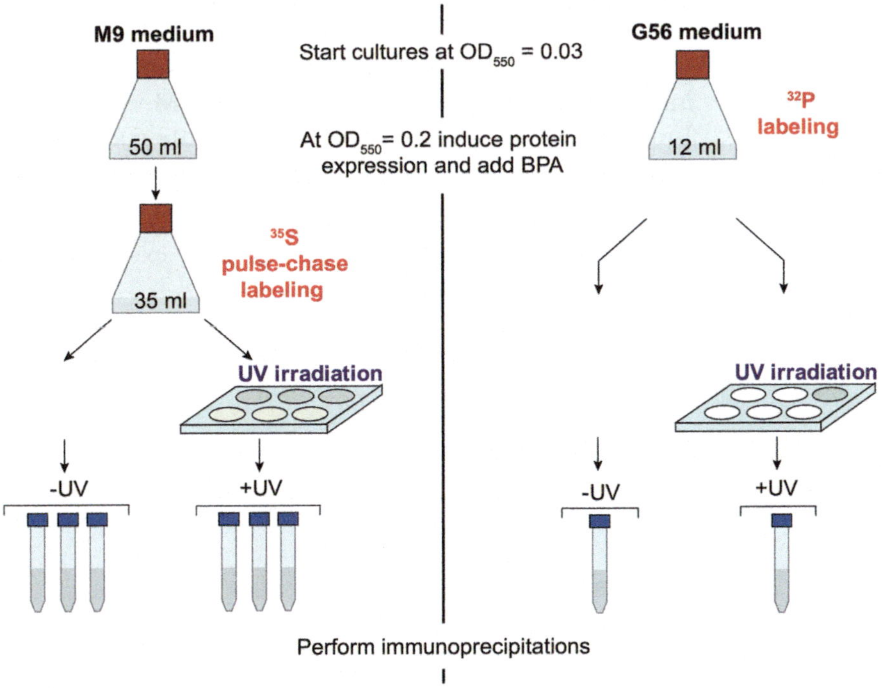

Fig. 1 Flow diagram of the illustrated experimental procedure. Two cultures are conducted in parallel. One culture is subjected to pulse-chase labeling using ^{35}S-methionine and ^{35}S-cysteine in order to label all newly synthesized proteins (*left*). The other culture is subjected to labeling with ^{32}P-inorganic phosphate in order to label cellular phospholipids (*right*)

photocrosslinking and protein immunoprecipitation, the interactions of EspP at sequential steps of its biogenesis can be monitored over time. In a parallel culture, ^{32}P-labeled inorganic phosphate is incorporated into phospholipids and lipopolysaccharide (LPS) and allows for the identification of interactions between the EspP β domain and lipids of the outer membrane.

1. Start two 5 mL overnight cultures of AD202 cells transformed with pRI23-1149Bpa and pDULE-*p*Bpa, one in M9 medium and one in G56 medium containing 0.13 mM KH_2PO_4. Supplement both cultures with 100 μg/mL ampicillin and 5 μg/mL tetracycline.
2. The following day, centrifuge cells from both cultures and wash them once with fresh medium (*see* **Note 4**). Finally resuspend cells in 2 mL fresh M9 and 2 mL fresh G56 media, respectively, and measure the optical density of the cell suspension at a wavelength of 550 nm (OD_{550}) in a spectrophotometer.
3. Use M9 resupended cells to inoculate a 50 mL culture in M9 medium.
4. Use G56 resuspended cells to inoculate a 12 mL culture in G56 medium w/o phosphate in a disposable flask. The starting OD_{550} of both cultures is set to 0.03 (Fig. 1).

3.3 Expression of Photoprobed EspP and Preparation of Cells for ^{35}S-Pulse-Chase Labeling

When the 50 mL culture in M9 medium reaches an OD_{550} of 0.2, induce EspP expression by adding 200 μM IPTG (*see* **Note 5**). Immediately after, add 1 mM Bpa (*see* **Note 6**). Incubate the culture for an additional 30 min at 37 °C (Fig. 1).

1. During this incubation time, label six 15 mL tubes as follows (this step is preparatory for the next experimental phase of metabolic labeling):
   - ^{35}S Time 1 min −UV;
   - ^{35}S Time 7 min −UV;
   - ^{35}S Time 15 min −UV;
   - ^{35}S Time 1 min +UV;
   - ^{35}S Time 7 min +UV;
   - ^{35}S Time 15 min +UV.
2. Label three wells of a six-well plate as follows:
   - ^{35}S Time 1 min +UV;
   - ^{35}S Time 7 min +UV;
   - ^{35}S Time 15 min +UV.
3. Fill the tubes with ice chips, approximately to the 5 mL volume line. Place the ice collected in the +UV tube into the corresponding labeled wells of the tissue culture plate. Keep the −UV tubes in one ice bucket and the cell culture plate in another ice bucket.

4. After 30 min from IPTG addition, perform pulse-chase labeling with ^{35}S-methionine and ^{35}S-cysteine, as described in Subheading 3.5.

3.4 Cell Labeling with ^{32}P-Inorganic Phosphate, Expression of Photoprobed EspP, and Photocrosslinking

1. Immediately after inoculation, supplement the 12 mL culture in G56 medium with 133 μCi/mL $KH_2{}^{32}PO_4$ (*see* **Note 7**).
2. When the culture reaches an OD_{550} of 0.2, induce expression of EspP by adding 200 μM IPTG (*see* **Note 5**). Immediately after, add 1 mM Bpa (*see* **Note 6**). Incubate the culture for an additional 30 min at 37 °C (Fig. 1).
3. During this incubation time, label two 15 mL tubes as follows:
   - ^{32}P –UV;
   - ^{32}P +UV.
4. Fill both tubes with ice chips, approximately up to the 5 mL volume mark line. Place the ice collected in the "^{32}P +UV" tube into a well of a six-well tissue culture plate. Label the well accordingly. Keep the "^{32}P –UV" tube in an ice bucket and the cell culture plate in another ice bucket.
5. After 45 min from IPTG induction place 5 mL of the culture in the "^{32}P –UV"-labeled tube and 5 mL in the "^{32}P +UV"-labeled plate well containing ice chips.
6. Immediately move the multiwell plate under the lamp to expose the "^{32}P +UV" sample to UV light for 5 min (*see* **Notes 8** and **9**). After this time, place the sample into the corresponding 15 mL tube previously labeled "^{32}P +UV" (*see* **Note 10**).
7. Collect cells in both 15 mL tubes by centrifugation at 2000 × *g* for 5 min at 4 °C.
8. Resuspend the cells pelleted in each tube in 1 mL M9 salt, place each sample in 1.5 mL Eppendorf tubes. Precipitate the whole protein content in both samples by adding trichloracetic acid.
9. Collect precipitated proteins by centrifugation at 16,000 × *g* for 15 min.
10. Wash samples with ice-cold acetone. Dry protein pellets.

3.5 ^{35}S-Pulse-Chase Labeling of Cells and Photocrosslinking

1. Move 35 mL of the M9 culture into a disposable 125 mL flask.
2. Place the flask in the water bath shaker.
3. Start a timer counting up (*see* **Note 11**).
4. Time 30″: add 11 μCi/mL of ^{35}S-Met/^{35}S-Cys protein labeling mix. Close the flask and swirl rapidly by hand. Place the flask in the water bath to prevent excessive cooling of the culturing medium.
5. Time 1′00″: Add 350 μL of cold methionine and cysteine stock solution. Close flask, swirl rapidly by hand. Place the flask in the water bath.

6. Time 2′00″: Remove 10 mL using a disposable pipette. Place 5 mL into the well labeled "^{35}S Time 1 min +UV" containing ice chips and 5 mL into the 15 mL "^{35}S Time 1 min −UV" tube containing ice chips. Close the flask and place it in the water bath.
7. Immediately move the multiwell plate under the lamp to expose sample "^{35}S Time 1 min +UV" to UV light for 5 min (*see* **Notes 8** and **9**). After this time, place the sample in the corresponding 15 mL tube labeled "^{35}S Time 1 min +UV" (*see* **Note 10**).
8. Time 8′00″: Repeat **steps 5** and **6** using the well labeled "^{35}S Time 7 min +UV" and tubes labeled "^{35}S Time 7 min −UV" and "^{35}S Time 7 min +UV."
9. Time 16′00″: Repeat **steps 5** and **6** using the well labeled "^{35}S Time 15 min +UV" and tubes labeled "^{35}S Time 15 min −UV" and "^{35}S Time 15 min +UV."
10. Place all 15 mL tubes in a centrifuge and collect cells at 2000 × *g* for 5 min at 4 °C.
11. Resuspend the cells pelleted in each tube using 1 mL M9 salt, place each sample in 1.5 mL Eppendorf tubes. Precipitate the whole protein content in each sample by adding trichlor-acetic acid.
12. Collect precipitated proteins by centrifugation at 16,000 × *g* for 15 min at 4 °C.
13. Wash samples with ice-cold acetone. Dry protein pellets.

3.6 Immuno-precipitation of EspP and Analysis of Photocrosslinking Products

1. Solubilize precipitated proteins by adding 50 μL of protein solubilization buffer (15% glycerol, 200 mM Tris base, 15 mM EDTA, 4% SDS, 2 mM PMSF) to each sample.
2. Place tubes in a thermoblock at 95 °C with agitation set to 1000 rpm.
3. Add 1 mL RIPA buffer. Subject samples to a clarifying spin at 16,000 × *g* for 10 min at 95 °C to pellet nonsolubilized proteins.
4. Move three 300 μL aliquots of supernatant from ^{35}S-labeled samples into new Eppendorf tubes and supplement them with 1 μL of anti-EspP, anti-BamA, or anti-BamB antisera, respectively. Move a single 300 μL aliquot from each ^{32}P-labeled sample into a new tube and supplement it with 1 μL anti-EspP antiserum.
5. Mix and incubate samples on ice for 2 h.
6. Add *S. aureus* protein A-Sepharose beads to sediment antibodies.
7. Wash beads with high-salt RIPA buffer (containing 500 mM NaCl) at least twice (*see* **Note 12**).

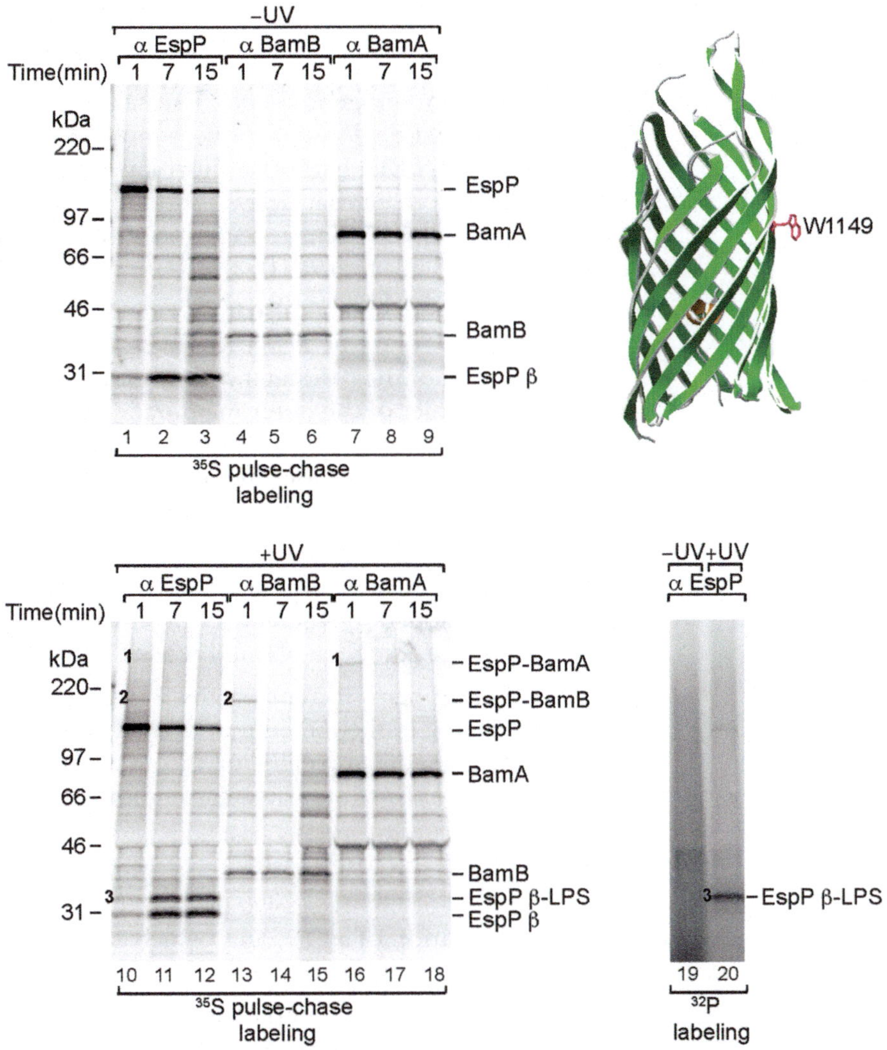

Fig. 2 Transient and stable interactions detected at a specific position of the EspP β domain. AD202 cells transformed with plasmids RI23-1149Bpa and pDULE-*p*Bpa are subjected either to pulse-chase labeling with radioactive amino acids (lanes 1–18) or to labeling with radioactive phosphate (lanes 19, 20). Each sample is divided into two equal aliquots, one of which is subjected to UV irradiation (lanes 10–18 and 20). All samples are subjected to immunoprecipitation using the indicated antisera. The crystal structure of EspP (PDB: 2QOM) is shown to highlight the position of W1149

8. Elute proteins using SDS-PAGE loading buffer.
9. Analyze eluted proteins by SDS-PAGE (*see* **Note 13**). Dry gels and expose them for autoradiography using storage phosphor screens. Phosphorimager-acquired autoradiography images are shown in Fig. 2 (*see* **Notes 14** and **15** for data analysis and interpretation).

4 Notes

1. To slow down EspP biogenesis and increase the temporal resolution of sequential EspP interactions, a modified EspP variant is used. EspP(586TEV) harbors a short linker insertion in the EspP passenger domain that delays (but does not permanently block) both passenger domain secretion and β domain insertion into the lipid bilayer [9, 10].
2. When available, use structural data to guide the strategy for site-specific incorporation of photoprobes. Trp 1149 is situated in a β-strand of the β domain with its side chain projecting into the outer leaflet of the outer membrane lipid bilayer [14]. Thus, a photoprobe at position 1149 would be predicted to be in close proximity to the outer membrane LPS upon completion of EspP biogenesis.
3. A series of plasmids for efficient incorporation of unnatural amino acids into proteins of interest using amber suppression have been developed by Schultz and coworkers [15] and deposited into the Addgene plasmid repository.
4. This washing step helps to remove any β-lactamase that might have been released in the medium of overnight cultures. Thus, the concentration of ampicillin in the new culture remains optimal for plasmid maintenance throughout the entire time of cell growth.
5. The level of induction of a given protein of interest must be determined empirically. Excessive protein overexpression may lead to the formation of aggregates, which can generate undesired photocrosslinking reactions.
6. Bpa is highly insoluble at a neutral pH, thus a 1000× stock solution of 1 M Bpa is prepared in 1 M NaOH. To facilitate rapid dilution of Bpa and prevent precipitation when added to the culture, slowly dispense the Bpa stock solution in the culture medium using a micropipette while gently swirling the culture flask.
7. It is recommended to start a parallel identical culture in G56 containing 0.13 mM KH_2PO_4 (instead of ^{32}P-labeled inorganic phosphate) to monitor cell growth. This will limit the risk of contaminating equipment with ^{32}P while measuring the culture optical density to assess bacterial growth.
8. Turn on the UV lamp 10 min before sample irradiation. After this warming-up time the lamp irradiates with maximal power. Wear protective goggles.
9. The positioning of the UV lamp with respect to the samples will depend on its wattage power. The described protocol is conducted using a 100 W mercury lamp. The lamp is positioned approximately 4 cm from the samples. To prevent excessive

heating of samples, it is important to adjust the position of the lamp so that the added ice chips do not melt completely during UV irradiation.

10. Do not switch off the UV lamp in between irradiations of different samples, unless the waiting time is longer than 20 min. Once switched off, it takes about 15 min before it can be reactivated.
11. To successfully conduct the pulse-chase technique, it is critical that the several described steps are conducted in parallel and coordinated in a timely fashion. Thus, the labeling and preparation of tubes where different samples will be collected (described in the previous section) must be conducted before proceeding with pulse-chase labeling. In addition, UV irradiation of samples is performed in the interval times of the chase phase. To minimize the time of handling and preparation of the samples, make sure that the following reagents have been thawed in advance and are ready for use on the bench: an aliquot of the ^{35}S-Met/^{35}S-Cys protein labeling mix, cold methionine and cysteine, an automatic pipettor and disposable 10 mL pipettes, a P100 micropipette set to 35 μL, and a P1000 micropipette set to 350 μL.
12. With samples labeled with ^{32}P-inorganic phosphate, four washing steps with high-salt RIPA buffer should be conducted to reduce the background of radioactivity detected after SDS-PAGE.
13. The type of gel and the concentration of polyacrylamide to choose will depend on the sizes of the protein that contains the photocrosslinker and the generated crosslinking products. In an initial experiment, it might be necessary to analyze samples with different gels of varying polyacrylamide concentrations to increase the chance of resolving crosslinking products.
14. Analysis of pulse-chase labeled samples not irradiated with UV light reveals that an approximately 135 kDa band, corresponding to EspP, is progressively converted to an approximately 30 kDa band (Fig. 2, lanes 1–3). This conversion is the result of EspP passenger domain cleavage by an autoproteolytic reaction occurring in the membrane-integrated β domain [16]. Upon exposure to UV irradiation, two high-molecular-weight products are immunoprecipitated with EspP antibodies (Fig. 2, lanes 10–12, products labeled "1" and "2"). In parallel reactions, an antiserum against BamA precipitates product 1 (Fig. 2, lane 16), while an antiserum against BamB precipitates product 2 (Fig. 2, lane 13). The signal intensities of EspP-BamA and EspP-BamB crosslinking products are maximal after 1 min chase and decrease over time (lanes 10–18), indicating that EspP position 1149 interacts with BamA and BamB at an early stage of biogenesis prior to cleavage of the secreted passenger domain. Another crosslink product detected in samples exposed to UV irradiation

migrates 2–4 kDa higher than the EspP β domain (Fig. 2, lanes 10–12, product labeled "3"). This crosslinking product is generated with high efficiency, and its amount is proportional to the amount of mature EspP β domain that forms over time (lanes 10–12). Thus crosslinking product 3 results from a stable interaction of the mature EspP β domain with a low-molecular-weight factor. In a parallel experiment, cellular phospholipids are labeled with ^{32}P. Upon UV irradiation, EspP-specific antibodies precipitate a ^{32}P-labeled product that runs with the same apparent molecular weight of crosslinking product 3 (Fig. 2, lane 20). Furthermore, this crosslinking product can be detected using antibodies specific for *E. coli* LPS [10, 11]. Thus, product 3 reveals an interaction of EspP with lipids of the outer leaflet of the outer membrane.

15. This protocol detects two types of interactions for EspP amino acid position 1149: (i) transient interactions with BamA and BamB at an early stage of EspP biogenesis, prior to passenger domain cleavage; (ii) an interaction of the mature EspP β domain with the LPS of the outer membrane, following secretion and cleavage of the passenger domain, that remains stable over time. Together with analyses of the interactions occurring at other amino acids of EspP, the site-directed photocrosslinking approach in metabolically labeled cells has helped to describe consecutive intermediate steps of the autotransporter assembly reaction in the bacterial outer membrane [9–11]. In brief, distinct autotransporter segments interact with periplasmic chaperons such as Skp and SurA at an early stage of its transport through the cellular envelope. At a later time, transient interactions with one of three subunits of the BAM complex (BamA, BamB, and BamD) are mapped at amino acid positions that are located approximately 120° from each other at the periplasmic side of the folded EspP β-barrel, suggesting the formation of an intermediate where the assembling EspP β domain is at the center of the BAM complex. On the circumference of the folded barrel, amino acid 1149 is positioned between the BamA and the BamB interacting sites. At a subsequent stage, a stable interaction of position 1149 with the outer membrane LPS indicates the release of the EspP β domain by the BAM complex into the lipid bilayer.

Acknowledgements

This protocol was originally developed in the laboratory of Dr. Harris Bernstein (National Institutes of Health, Bethesda, MD, USA). Plasmid pDULE-*p*Bpa was kindly provided by Dr. Peter Schultz (Scripps Research Institute, La Jolla, CA, USA). R.I. is supported by the CNRS-Inserm ATIP-Avenir program.

References

1. Holland IB (2010) The extraordinary diversity of bacterial protein secretion mechanisms. Methods Mol Biol 619:1–20
2. Ellman J, Mendel D, Anthony-Cahill S et al (1991) Biosynthetic method for introducing unnatural amino acids site-specifically into proteins. Methods Enzymol 202:301–336
3. Chin JW, Martin AB, King DS et al (2002) Addition of a photocrosslinking amino acid to the genetic code of *Escherichia coli*. Proc Natl Acad Sci U S A 99:11020–11024
4. Dormán G, Prestwich GD (1994) Benzophenone photophores in biochemistry. Biochemistry 33:5661–5673
5. Dautin N, Bernstein HD (2007) Protein secretion in gram-negative bacteria via the autotransporter pathway. Annu Rev Microbiol 61:89–112
6. Leyton DL, Rossiter AE, Henderson IR (2012) From self sufficiency to dependence: mechanisms and factors important for autotransporter biogenesis. Nat Rev Microbiol 10:213–225
7. Hagan CL, Silhavy TJ, Kahne D (2011) β-Barrel membrane protein assembly by the Bam complex. Annu Rev Biochem 80:189–210
8. Noinaj N, Rollauer SE, Buchanan SK (2015) The β-barrel membrane protein insertase machinery from Gram-negative bacteria. Curr Opin Struct Biol 31:35–42
9. Ieva R, Bernstein HD (2009) Interaction of an autotransporter passenger domain with BamA during its translocation across the bacterial outer membrane. Proc Natl Acad Sci U S A 106:19120–19125
10. Ieva R, Tian P, Peterson JH et al (2011) Sequential and spatially restricted interactions of assembly factors with an autotransporter beta domain. Proc Natl Acad Sci U S A 108:E383–E391
11. Pavlova O, Peterson JH, Ieva R et al (2013) Mechanistic link between β barrel assembly and the initiation of autotransporter secretion. Proc Natl Acad Sci U S A 110:E938–E947
12. Ganong BR, Leonard JM, Raetz CR (1980) Phosphatidic acid accumulation in the membranes of *Escherichia coli* mutants defective in CDP-diglyceride synthetase. J Biol Chem 255:1623–1629
13. Akiyama Y, Ito K (1990) SecY protein, a membrane-embedded secretion factor of *E coli*, is cleaved by the ompT protease *in vitro*. Biochem Biophys Res Commun 167:711–715
14. Barnard TJ, Dautin N, Lukacik P et al (2007) Autotransporter structure reveals intra-barrel cleavage followed by conformational changes. Nat Struct Mol Biol 14:1214–1220
15. Young TS, Ahmad I, Yin JA et al (2010) An enhanced system for unnatural amino acid mutagenesis in E coli. J Mol Biol 395:361–374
16. Dautin N, Barnard TJ, Anderson DE et al (2007) Cleavage of a bacterial autotransporter by an evolutionarily convergent autocatalytic mechanism. EMBO J 26:1942–1952

Chapter 20

Protein–Protein Interactions: Pull-Down Assays

Arthur Louche, Suzana P. Salcedo, and Sarah Bigot

Abstract

Determining protein partners is an essential step toward understanding protein function and identifying relevant biological pathways. Many methods exist for investigating protein–protein interactions. The pull-down assay is an in vitro technique used to detect physical interactions between two or more proteins and an invaluable tool for confirming a predicted protein–protein interaction or identifying novel interacting partners. This method typically involves the use of affinity purification with various wash and elution steps. In this chapter, we describe how an interaction between two purified bacterial proteins or between bacterial and eukaryotic proteins can be detected by pull-down experiments.

Key words Pull-down, Protein–protein interactions, Tagged protein, Affinity purification

1 Introduction

Pathogenic bacteria produce virulence factors that usually help the pathogen to survive in an environmental niche, to promote colonization and invasion of host tissues, or to modulate the immune system. Virulence factors are toxins or effector proteins than can be transported by diverse secretion machineries in bacteria [1, 2]. Once secreted, these proteins can be assembled on the bacterial cell surface, released in the extracellular space, or secreted directly into a host cell or a neighboring bacterium. Once in host cells, effectors often target key proteins to hijack the host cellular machinery to remodel signaling cascades. The yeast two-hybrid system is often used to screen a large number of host proteins that potentially interact with bacterial effectors [3]. Regarding the mechanism of the secretion systems, a bacterial two-hybrid system is frequently employed to identify interaction networks between components of the secretory apparatus, as well as interaction between effectors and proteins of the machinery [4]. However, protein–protein interactions that have been determined by two-hybrid assay must be confirmed by other methods [5].

Laure Journet and Eric Cascales (eds.), *Bacterial Protein Secretion Systems: Methods and Protocols*, Methods in Molecular Biology, vol. 1615, DOI 10.1007/978-1-4939-7033-9_20, © Springer Science+Business Media LLC 2017

Pull-down is an in vitro method widely used to detect or confirm interactions among multiple proteins. This assay is similar in methodology to co-immunoprecipitation experiments in its use of an affinity ligand to capture interacting proteins. The difference between these two methods is that while co-immunoprecipitation uses immobilized antibodies to capture protein complexes, the pull-down approach uses a purified and tagged protein as a "bait" to bind any interacting proteins. The method consists of first immobilizing the tagged protein (bait) on an affinity ligand specific to the tag, creating an affinity support to capture and purify other proteins (prey) that interact with the bait. The bait and prey proteins can be obtained from multiple sources, such as cell lysates, purified proteins, expression systems, and in vitro transcription/translation systems. Once the prey proteins have been incubated with an immobilized bait protein, interacting complexes are eluted using an eluting buffer depending on the affinity ligand. Each experiment needs proper controls to demonstrate that characterized interactions are not an artifact. For example, a positive control consisting of an immobilized bait protein alone is necessary to verify proper attachment of the tagged bait protein to the affinity support. To identify and eliminate false positives caused by nonspecific binding of prey proteins to the affinity support, cell lysates or purified proteins can be analyzed after being passed through a minus bait support. Following a pull-down experiment, protein fractions are resolved by sodium dodecyl sulfate-polyacrylamide gel electrophoresis (SDS-PAGE) and then visualized by gel staining or western-blotting detection.

In this chapter, we describe detailed pull-down assay procedures that allow the identification of interacting proteins. First, we focus on how to perform a pull-down experiment to identify an interaction between a bacterial bait protein and eukaryotic prey proteins expressed in host cells (Subheadings 3.1 and 3.2). Next, we present how the interaction between two purified proteins can be visualized by a pull-down assay (Subheading 3.3). In these procedures, pull-down experiments have been performed using specific bait proteins fused to a 6× histidine tag. As a consequence, we selected Ni-NTA agarose beads as the affinity support used to immobilize these recombinant proteins.

2 Materials

Prepare all solutions with distilled water at room temperature and keep them at the indicated temperatures.

2.1 Preparation of Cell Lysate

1. Eukaryotic cells.
2. Cell culture dish, treated for optimal cell attachment, with growth surface area around 55 cm^2, sterile.

3. Plasmid containing the gene of interest fused to a specific tag (obtained from a EndoFree maxipreparation).
4. Transfection reagent.
5. Phosphate buffered saline (PBS): Prepare a 10× solution with bidistilled water (18.2 MΩ cm) containing 10.6 mM KH_2PO_4, 30 mM Na_2HPO_4, $2H_2O$, and 1.54 M NaCl, and sterilize with a 0.2 μm filter. The 1× solution obtained following dilution with bidistilled water will have a pH of around 7.4.
6. Radioimmunoprecipitation assay (RIPA) buffer: Ready-to-use solution containing 150 mM NaCl, 1.0% IGEPAL® CA-630, 0.5% sodium deoxycholate, 0.1% SDS, 50 mM Tris, pH 8.0.
7. Antiprotease cocktail: Mix 1% (v/v) of protease inhibitor cocktail (Sigma-Aldrich), phosphatase inhibitor cocktail 2 (Sigma-Aldrich), phosphatase inhibitor cocktail 3 (Sigma-Aldrich), and phenylmethylsulfonyl fluoride (PMSF).

2.2 Pull-Down Assays

1. 1 M Tris–HCl, pH 7.5 stock solution. Weigh 121.1 g Tris base and transfer to a 1 L graduated cylinder. Add water to 800 mL, mix, adjust pH with HCl, and make up to 1 L with water. Store at room temperature (*see* **Note 1**).
2. 5 M NaCl stock solution. Weigh 292.2 g NaCl and transfer to a 1 L graduated cylinder. Add water to 800 mL, stir, and adjust volume to 1 L with water (*see* **Note 1**).
3. Equilibrium buffer (*see* **Note 2**): 20 mM Tris–HCl, pH 7.5, 250 mM NaCl. Mix 1 mL 1 M Tris–HCl, pH 7.5 stock solution with 2.5 mL 5 M NaCl stock solution in a 50 mL centrifuge tube, and add water to a volume of 50 mL. Keep at 4 °C (*see* **Note 3**).
4. Elution buffer (*see* **Note 2**): 20 mM Tris–HCl, pH 7.5, 250 mM NaCl, 500 mM imidazole. Weigh 1.7 g imidazole in 50 mL solution of equilibrium buffer. Keep at 4 °C (*see* **Note 3**).
5. Purified His-tagged protein (bait).
6. Ni-NTA agarose beads: 6% beaded agarose (cross-linked), precharged with Ni^{2+} (Protino® Ni-NTA Agarose, Macherey Nagel, or equivalent). Store at 4 °C (*see* **Note 4**).
7. 0.8 mL empty columns for gravity flow (Pierce™ Centrifuge Columns, Thermo Fisher Scientific, or equivalent).
8. Refrigerated microcentrifuge.

2.3 Sodium Dodecyl Sulfate (SDS) Polyacrylamide Gel Components

1. Resolving gel: 1.5 M Tris–HCl, pH 8.8. Weigh 90.8 g, transfer to 500 mL graduated cylinder, and add 300 mL water. Adjust pH with HCl and fill with water to 500 mL. Store at room temperature.
2. Stacking gel buffer: 0.5 M Tris–HCl, pH 6.8. Weigh 30.275 g, transfer to 500 mL graduated cylinder, and add 300 mL water.

Adjust pH with HCl and fill with water to 500 mL. Store at room temperature.

3. 30% acrylamide/Bis solution (37.5:1 acrylamide:Bis). Store at 4 °C.
4. Ammonium persulfate (APS): 20% solution in water. Store at −20 °C (*see* **Note 5**).
5. *N,N,N′,N′*-tétraméthyléthylènediamine (TEMED). Store at room temperature.
6. SDS-PAGE running buffer: 25 mM Tris–HCl, 192 mM glycine, 0.1% SDS. Prepare 10× running buffer solution: Weigh 30 g Tris base, 144 g glycine, and 10 g SDS and add distilled water to 1 L. Store at room temperature. Prepare fresh 1× solution before gel electrophoresis.
7. Laemmli lysis buffer [6], 4× concentrate: 62.5 mM Tris–HCl pH 6.8, 2% SDS, 10% glycerol, 0.01% bromophenol blue, 5% β-mercaptoethanol. Store at −20 °C (*see* **Note 6**).
8. Protein ladder.

3 Methods

3.1 Preparation of Cell Lysate

1. Seed eukaryotic cells at 5.10^5 in 10 cm cell culture dish (*see* **Note 7**) and incubate overnight at 37 °C in CO_2.
2. Transfect cells with plasmid containing gene of interest fused to a specific tag with appropriate transfection reagent for time necessary for optimal expression of protein (16–24 h is usually a good range).
3. Cool cells by placing plates on ice, wash cells with 1× PBS. Add 2 mL cold PBS and harvest cells using cell scraper.
4. Centrifuge 5 min at 80 × *g* at 4 °C.
5. Resuspend cells with 200 μL RIPA buffer supplemented with antiprotease cocktail.
6. Incubate on ice 20 min and mix gently every 5 min with a P200 micropipette.
7. Stock prepared cells at −80 °C (*see* **Note 8**).
8. Right before pull-down experiment, thaw prepared cell extract. Centrifuge at 17,000 × *g* at 4 °C for 20 min. Use the supernatant as prey by following **step 9** in Subheading 3.2 (*see* **Note 9**).

3.2 Pull-Down Assay Using Cell Lysate as Prey (See Notes 10 and 11)

1. Transfer 120 μL Ni-NTA agarose beads to gravity flow column (*see* **Note 12**).
2. Centrifuge column for 1 min at 1000 × *g* at 4 °C. Discard flow-through.

3. Add 400 μL distilled water to column (*see* **Note 13**).
4. Centrifuge column for 1 min at 1000 × *g* at 4 °C. Discard flow-through.
5. Mix carefully 50 μg His-tagged protein (bait) with 400 μL equilibrium buffer and load onto column (*see* **Notes 14** and **15**).
6. Incubate 1 h (*see* **Note 16**) with agitation at 4 °C (*see* **Note 17**) and 10 min on ice without agitation (*see* **Note 18**).
7. Centrifuge column for 1 min at 1000 × *g* at 4 °C and keep flow-through.
8. Load flow-through to column, and centrifuge column for 1 min at 1000 × *g* at 4 °C (*see* **Note 19**). Keep flow-through at 4 °C for analysis.
9. Mix 200 μL cell extract (*see* **Note 20**) with 200 μL equilibrium buffer and load onto column (*see* **Note 21**).
10. Incubate 1 h at 4 °C under agitation (*see* **Note 22**) then 10 min on ice without agitation (*see* **Note 18**).
11. Centrifuge column for 1 min at 1000 × *g* at 4 °C. Keep flow-through for analysis.
12. Wash column by adding to column 400 μL equilibrium buffer.
13. Centrifuge column for 1 min at 1000 × *g* at 4 °C. Discard flow-through.
14. Wash column by adding to column 400 μL equilibrium buffer containing 50 mM imidazole. Keep the first washing for analysis.
15. Centrifuge column for 1 min at 1000 × *g* at 4 °C. Discard flow-through.
16. Repeat **steps 14** and **15** three times and go to **step 17**. Keep last washing fraction at 4 °C for analysis.
17. Elute by loading 80 μL elution buffer to column and incubate 10 min at 4 °C (*see* **Note 18**).
18. Centrifuge column for 1 min at 1000 × *g* at 4 °C and keep eluted fraction.
19. Repeat **steps 17** and **18** with eluted fraction (*see* **Note 22**). Keep eluted fraction at 4 °C for analysis.

3.3 Pull-Down Assay Using Purified Protein as Prey (See Note 11)

1. Incubate 50 μg His-tagged bait protein with 50 μg purified prey protein in total volume of 400 μL equilibrium buffer (*see* **Note 23**) 2 h 30 min at 4 °C under agitation (*see* **Notes 17** and **24**).
2. Add 80 μL Ni-NTA agarose beads to gravity flow column and follow **steps 1–4** of Subheading 3.2.

3. Equilibrate column by adding 400 μL equilibrium buffer supplemented with 20 mM imidazole.
4. Centrifuge column for 1 min at 1000 × *g* at 4 °C. Discard flow-through.
5. Load 400 μL incubated bait and prey proteins onto column. Incubate 10 min on ice without agitation (*see* **Note 18**).
6. Centrifuge column for 1 min at 1000 × *g* at 4 °C. Keep flow-through at 4 °C for analysis.
7. Wash by adding to column 400 μL equilibrium buffer supplemented with 20 mM imidazole.
8. Centrifuge column for 1 min at 1000 × *g* at 4 °C. Save first washing at 4 °C for analysis.
9. Repeat washing **steps 7** and **8** four times and keep last washing fraction at 4 °C for analysis.
10. Add 200 μL elution buffer to column and incubate on ice 10 min.
11. Centrifuge column for 1 min at 1000 × *g* at 4 °C. Keep eluted fraction at 4 °C for analysis.

3.4 SDS-PAGE and Analysis of Protein Fractions

1. To 15 μL protein fraction add 5 μL Laemmli lysis buffer, 4× concentrate. Heat for 3 min at 100 °C and centrifuge 30 s using a microcentrifuge to bring down condensate.
2. Load 10 μL protein fraction and 5 μL protein ladder on SDS-polyacrylamide gel.
3. Electrophorese proteins in running buffer at 100 V for 15 min then 180 V until dye front has reached bottom of gel.
4. Identify interacting proteins by immunodetection or blue coomassie coloration (*see* **Note 25**).

4 Notes

1. We prefer not to use the solutions after 6 months of storage.
2. A different buffer, such as HEPES (4-(2-hydroxyethyl)-1-piperazineethanesulfonic acid), MES (2-(*N*-morpholino) ethanesulfonic acid), or phosphate buffers, may be required for your specific protein–protein interaction. Additionally, different pH values may be tested as these are specific and dependent on the interaction between proteins.
3. We found that pull-down experiments work better with fresh equilibrium and elution buffers.
4. The bait proteins used in this protocol are tagged with 6× His that bind the nickel agarose affinity support. The choice of the matrix-associated antibody depends on the fusion tag. The His

tag is composed of a peptide motif that consists of six histidine residues with a high affinity towards metals like nickel that composes the used Ni-NTA agarose but also the Ni-NDA, Ni-TED, or Ni-TALON resins. The 6× His tag is very small (~1 kDa), which renders it less immunogenic than other larger tags, is shown not to affect the native conformation of bait proteins, and maintains its partner binding activity. Few naturally occurring proteins also bind to Ni-NTA matrices, making this tag the most commonly used affinity tag. In pull-down assays, the choice of the matrix-associated antibody depends on the fusion tag. What follow are some examples of tags with their advantages and disadvantages. The FLAG tag is an octapeptide that is likely located on the surface of the fusion protein due to the hydrophilic nature of amino acid residues and has affinity to anti-FLAG resin. Like the His tag, the FLAG tag is small, but a disadvantage is that the monoclonal antibody matrix is not as stable as Ni-NTA. Glutathione S-transferase (GST) tag binds to glutathione-associated support with high affinity and specificity. This tag has the advantage that GST isoforms are not normally found in bacteria, so purified bacterial prey proteins normally do not have affinity with glutathione resin. However, GST tag is large (26 kDa), exists as a dimer, is prone to nonspecific interaction, is expensive, and affinity to its support depends on certain reagents. The maltose-binding protein (MBP) tag from an *Escherichia coli* periplasmic protein has affinity for matrix consisting of sugars or anti-MBP. This tag is used for the purposes of overcoming problems associated with the expression and purification of recombinant proteins [7]. However, the disadvantage of the MBP tag is its large size, its immunogenicity, and the mild elution of MBP-tagged proteins, which complicate pull-down experiments.

5. Make an aliquot of 1 mL before −20 °C storage. This will prevent the degradation caused by repeated thawing.
6. Make an aliquot of 500 μL before −20 °C storage. The used Laemmli lysis buffer can be kept at 4 °C for 1 month.
7. As negative control, prepare a cell lysate without expressing bait protein (negative cell lysate). This will eliminate false positives resulting from nonspecific interactions of cell lysate proteins with the Ni-NTA agarose beads. Additional negative controls can include an irrelevant protein with the same tag or expression of the tag alone, as in the case of the GFP.
8. Before stocking the cells, remove an aliquot and control by western blot the production of the prey protein.
9. Whole-cell lysate instead of the supernatant fraction can also be used to test whether the prey protein of interest localizes in the pellet fraction.

10. Pull-down experiments using cell lysates will not demonstrate that interaction between the bait and prey proteins is direct but only determine that they are part of the same complex. To prove a direct interaction, the prey protein must be purified and used in pull-down experiments as described in Subheading 3.3.
11. Try to work mostly on ice or at 4 °C to prevent the degradation or the denaturation of the proteins.
12. Break the end cap of the gravity flow column and place it on a 1.5 mL Eppendorf tube. Thoroughly resuspend the Ni-NTA resin by inverting the bottle several times to obtain a uniform suspension. Pipette tips must be cut to allow the Ni-NTA agarose beads to get into.
13. This step eliminates the left 30% ethanol present in the Ni-NTA resin.
14. Before loading the bait protein, plug the gravity flow column using a piece of parafilm before replacing it on a 2 mL Eppendorf tube.
15. Prepare a supplementary column by mixing 50 μg of a known noninteracting bait fused to 6× His tag with 400 μL equilibrium buffer to an empty column. Additionally, prepare a column by adding 400 μL equilibrium buffer to an empty column. These negative bait columns will be used in combination with cell lysates to eliminate false positives resulting from nonspecific interactions.
16. The incubation time can be increased from several hours to overnight at 4 °C under agitation depending on the strength of the interaction between bait and prey proteins.
17. Rotate on roller or rotating platform.
18. The column should stand straight on the ice. This step allows the resin to flow by gravity before centrifugation.
19. We found that loading two times the flow-through increases the capacity of the binding.
20. The volume is dependent on the protein concentration of the cell extract. As a guide, 125–150 μg of protein of a cell extract is usually incubated per microgram of bait protein. Alternatively, cell extract samples can be normalized by visualization of transfected proteins to ensure equivalent expression of the prey and the relevant controls (*see* **Note 7**).
21. Several controls should be added at this step. Load 400 μL equilibrium buffer without prey protein to analyze the efficiency of the immobilization of the bait protein. As negative controls, load onto the negative column (*see* **Note 12**) 200 μL cell lysate containing the prey protein or the negative cell lysate (*see* **Note 7**) mixed with 200 μL equilibrium buffer. Additionally,

load 200 μL negative cell lysate mixed with 200 μL equilibrium buffer onto the column associated with the bait protein.

22. We found that loading two times the eluted fraction increased its quantity.
23. As negative control, incubate 50 μg bait protein (minus prey) or prey protein alone (minus bait) in 400 μL equilibrium buffer. The minus prey control will ensure that the Ni-NTA agarose resin will correctly capture the His-tagged bait protein alone. The minus bait control will eliminate false positives resulting from an interaction between affinity support and prey protein.
24. A different incubation temperature and time may be required for your specific protein–protein interaction.
25. A prey protein that interacts with the bait protein will be found in the eluted fraction. In contrast, a noninteracting protein will not be retained by the bait protein, will pass through the column, and will be found in the flow-through protein fraction.

References

1. Costa TRD, Felisberto-Rodrigues C, Meir A, Prevost MS, Redzej A, Trokter M, Waksman G (2015) Secretion systems in Gram-negative bacteria: structural and mechanistic insights. Nat Rev Microbiol 13:343–359
2. McBride MJ, Nakane D (2015) *Flavobacterium* gliding motility and the type IX secretion system. Curr Opin Microbiol 28:72–77
3. Rodríguez-Negrete E, Bejarano ER, Castillo AG (2014) Using the yeast two-hybrid system to identify protein-protein interactions. Methods Mol Biol 1072:241–258
4. Zoued A, Brunet YR, Durand E, Aschtgen M-S, Logger L, Douzi B, Journet L, Cambillau C, Cascales E (2014) Architecture and assembly of the Type VI secretion system. Biochim Biophys Acta 1843:1664–1673
5. Boucrot E, Henry T, Borg J-P, Gorvel J-P, Méresse S (2005) The intracellular fate of *Salmonella* depends on the recruitment of kinesin. Science 308:1174–1178
6. Laemmli UK (1970) Cleavage of structural proteins during the assembly of the head of bacteriophage T4. Nature 227:680–685
7. Di Guan C, Li P, Riggs PD, Inouye H (1988) Vectors that facilitate the expression and purification of foreign peptides in *Escherichia coli* by fusion to maltose-binding protein. Gene 67:21–30

Chapter 21

Protein–Protein Interactions: Surface Plasmon Resonance

Badreddine Douzi

Abstract

Surface plasmon resonance (SPR) is one of the most commonly used techniques to study protein–protein interactions. The main advantage of SPR is it gives on the ability to measure the binding affinities and association/dissociation kinetics of complexes in real time, in a label-free environment, and using relatively small quantities of materials. The method is based on the immobilization of one of the binding partners, called the ligand, on a dedicated sensor surface. Immobilization is followed by the injection of the other partner, called the analyte, over the surface containing the ligand. The binding is monitored by subsequent changes in the refractive index of the medium close to the sensor surface upon injection of the analyte. During the last 10 years, SPR has been intensively used in the study of secretion systems because of its ability to detect highly dynamic complexes that are difficult to investigate using other techniques. This chapter will guide users in the setup of SPR experiments in order to identify protein complexes and to assess their binding affinity or kinetics. It will include detailed protocols for (i) the immobilization of proteins with the amine coupling capture method, (ii) analyte-binding analysis, (iii) affinity/kinetic measurements, and (iv) data analysis.

Key words Surface Plasmon Resonance, Protein–protein interaction, Analyte, Ligand, Affinity, Kinetics, BIAcore T200

1 Introduction

Secretion systems are multiprotein complexes allowing the transport of a large number of effectors from the inside to the outside of bacterial cells. The assembly of these supramolecular machineries is ensured by the formation of protein complexes with extremely different times of stability, from transitory to stable interactions. To understand the function of these machineries as well as their modes of association, it is important to study their building blocks by identifying the different interacting partners and assessing their relative affinities and association/dissociation kinetics. For that purpose, scientists combine genetic, biochemical, and biophysical tools. During the last decade, the use of surface plasmon resonance (SPR) in the study of secretion systems has increased spectacularly [1–12]. This in vitro approach is the method of choice to study

Laure Journet and Eric Cascales (eds.), *Bacterial Protein Secretion Systems: Methods and Protocols*, Methods in Molecular Biology, vol. 1615, DOI 10.1007/978-1-4939-7033-9_21, © Springer Science+Business Media LLC 2017

such dynamic systems owing to its ability to detect both weak and strong interactions ranging from the millimolar to the nanomolar range [13, 14]. SPR can be used as a primary tool to screen interacting partners or as a validation tool for interactions previously identified by other methods (e.g., bacterial two-hybrid, co-immunoprecipitation, chemical crosslinking). The determination of the affinity or kinetics of an interaction, as can be done by SPR, is fundamental to understanding the nature of binding at the cellular level.

In SPR terminology, the immobilized biomolecule is called the *ligand* and the binding partner present in solution is called the *analyte*. SPR uses an optical method based on the detection of very tiny modifications of the refractive index of the medium in close proximity to a metal surface. The binding of the analyte to the ligand induces a change in the mass concentration at the metal surface and, consequently, in the refractive index, which is converted to resonance or response units (RUs). The metal surface typically consists of a thin gold layer and contains flow cells with very small volume (less than 100 nL). In a typical SPR experiment, the first step consists in the immobilization of the ligand in one flow cell by covalent or noncovalent capture methods. In the second step, the analyte is injected into the flow cell containing the immobilized ligand. The presence of a reference flow cell, which can either be empty or contain a protein irrelevant for the studied complex, is important to monitor the true binding of the analyte to the ligand. The binding of the analyte on the ligand induces a change on the mass concentration at the surface of the flow cell. The signal induced by the binding of the analyte to the metal surface in the reference flow cell is subtracted from the signal obtained from the binding of the analyte to the ligand. Consequently, upon binding, the curve that we observe, called a *sensogram*, is divided into three distinct parts (Fig. 1). The first phase is the association phase, during which the analyte molecules bind to the binding sites of the ligand, leading to the increase in RUs. The curve obtained can be used to calculate the rate of association (k_{on}). The next phase is a steady state, during which the association/dissociation events are equal. The level of RUs obtained in this equilibrium phase is called the response at equilibrium, R_{eq}, and can be used to calculate the binding affinity (K_D). Upon stopping the injection of the analyte over the flow cell, the dissociation phase takes place: the ligand dissociates from the surface as the running buffer flows over the chip. The dissociation of the analyte from the ligand binding site induces a decrease in RUs. The corresponding curve can be used to measure the rate of dissociation (k_{off}). The full removal of the analyte from the ligand binding sites takes place during the regeneration step, which is determinant for keeping the surface intact and ready for new injection cycles.

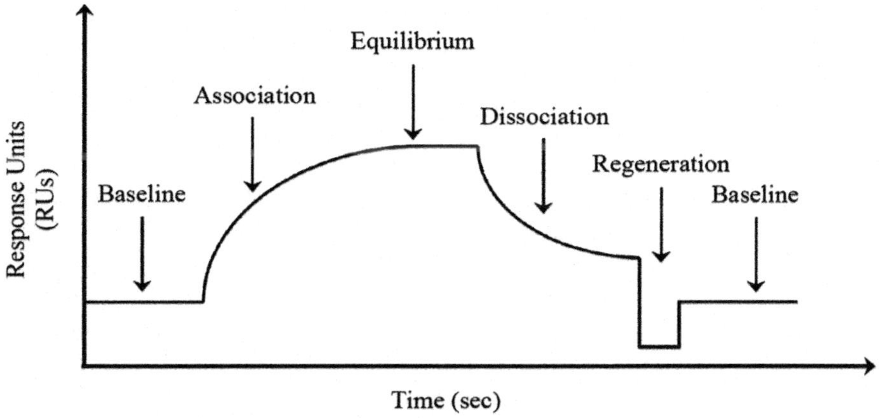

Fig. 1 Different phases of an SPR sensogram. Initial baseline is obtained after the immobilization of the ligand. Following injection of analyte over the ligand, the increase in the RUs corresponds to the association phase. When the number of association and dissociation events is equal, the equilibrium level is maintained. The dissociation phase starts when stopping the analyte injection: the decrease in the RUs corresponds to the flowing out of the analyte from the ligand surface. The regeneration step aims at the full removal of the analyte and at reaching the baseline level

Several SPR-based systems have been developed by different manufacturers. The most widely used system is the BIAcore developed by GE Healthcare. Despite the high cost of the machines and related products, this system offers many advantages, which has encouraged scientists to acquire it. The most important advantages are the capacity to detect weak interactions and the reproducibility of the obtained results.

This chapter illustrates the setup procedures routinely used for studying protein–protein interactions by SPR on a BIAcore T200. It describes the amine coupling immobilization method as well as analyte binding analysis, affinity/kinetic measurements, and data analysis.

2 Materials

All buffers are prepared with ultrapure water and analytical-grade reagents. All prepared buffers should be stored at 4 °C unless specifically required otherwise.

- Series S sensor chip CM5 from GE Healthcare (*see* **Note 1**).
- BIAcore T200 system from GE Healthcare Life Sciences (*see* **Note 2**).
- Amine coupling kit: the kit contains 750 mg 1-ethyl-3-(3-dimethylaminopropyl) carbodiimide hydrochloride (EDC), 115 mg *N*-hydroxysuccinimide (NHS), and 10.5 mL 1.0 M ethanolamine–HCl, pH 8.5.

- Immobilization buffers: 10 mM sodium acetate buffers at different pHs: pH 4, pH 4.5, pH 5, and pH 5.5.
- HBS-EP buffer: 0.01 M HEPES, 0.15 M NaCl, 3 mM EDTA, 0.05% v/v P20, pH 7.4.
- Regeneration scouting kit or equivalent homemade solutions: 11 mL ethylene glycol (100 wt%), 11 mL 10 mM glycine–HCl, pH 1.5, 11 mL 10 mM glycine–HCl, pH 2.0, 11 mL 10 mM glycine–HCl, pH 2.5, 11 mL 10 mM glycine–HCl, pH 3.0, 11 mL 4.0 M magnesium chloride, 11 mL 0.2 M sodium hydroxide, 11 mL 0.5% sodium dodecyl sulphate (SDS), 11 mL 5.0 M sodium chloride, 20 mL surfactant P20 (10% (v/v) solution of polysorbate 20).
- Vials: 0.8 mL rounded polypropylene microvials.
- Vial caps: Penetrable cap made of kraton G (SEBS).
- Ligand and analyte (*see* **Note 3**).
- Control ligand, e.g., Thioredoxin from *Escherichia coli.*
- Tabletop centrifuge.

3 Methods

In a typical protein–protein interaction study using an SPR-based system, many tasks should be undertaken in the following order:

1. Choice of protein to immobilize and protein to use as analyte (*see* **Note 4**).
2. Choice of immobilization type and sensor surface (*see* **Note 5**).
3. Choice of immobilization level (*see* **Note 6**).
4. Preparation of ligand and analyte.
5. Preparation of material and buffers (*see* Subheading 2).
6. pH scouting.
7. Immobilization of ligand.
8. Immobilization of control ligand.
9. Analyte binding analysis.
10. Regeneration optimization.
11. Affinity and kinetic measurements.
12. Data analysis.

These steps will be developed in subsequent sections in more detail.

3.1 Ligand and Analyte Preparation

1. The purified proteins must be dialyzed ON (overnight) at 4 °C against the running buffer to be used in the SPR experiment (*see* **Notes 7** and **8**).

2. Check the purity and the stability of the proteins at least 1 day before doing the experiment. This can be done by sodium dodecyl sulfate (SDS)-polyacrylamide gel electrophoresis (*PAGE*) and coomassie blue staining (*see* **Note 9**).
3. Thaw proteins on ice for 30 min.
4. Centrifuge proteins at 16,000 × *g* for 20 min at 4 °C.
5. Collect supernatant and keep it on ice on a new 1.5 mL Eppendorf tube.
6. Measure concentration using NanoDrop or alternative spectroscopic method (Bradford or bicinchoninic acid assays). Try to use the same method during the SPR experiment.

3.2 Material and Buffer Preparation

1. Turn on BIAcore T200 system and prime system with filtered ultrapure water (*see* **Note 10**).
2. Take sensor chip CM5 from 4 °C and keep it at room temperature (RT) for at least 1 h before experiment.
3. Prepare 1 L running buffer by diluting stock solution ten times.
4. Filter running buffer using 0.22 μM membrane filter.
5. Prime system using running buffer at least three times to be sure that all system tubing is properly flushed with running buffer.
6. Insert CM5 sensor chip on sensor chip port and close sensor port.
7. Prime system three times.
8. When priming is finished, the system will be automatically shifted to a continuous standby flow.
9. Start a manual run by setting flow path 2 and flow rate of 30 μL min^{-1}.
10. Prepare two vials, one containing 400 μL running buffer and the other with 400 μL 50 mM NaOH regeneration solution.
11. Eject rack tray and place vials in rack.
12. The rack will be inserted after 1 min.
13. Inject running buffer three times for 1 min.
14. Inject regeneration solution for 30 s after each running buffer injection until baseline becomes stable.

3.3 pH Scouting (See Note 11)

1. Open Biacore T200 control software and go to File/Open/New wizard template/Immobilization scouting wizard/New.
2. The next steps consist in setting the different parameters:
   - Specify different solutions used for your experiments and corresponding pH.
   - Set protein injection and dissociation times; injection time can be fixed at 2 min, dissociation time at 1 min.

- Specify regeneration solution, usually 50 mM NaOH (*see* **Note 12**).
- Set flow rate used for experiment at 10 μL min^{-1}.

3. Dilute protein at different pHs (4, 4.5, 5, and 5.5) using immobilization buffers (*see* Subheading 2) (*see* **Notes 13** and **14**).
4. Final protein concentration after dilution on immobilization buffers should be in range of 5–200 μg mL^{-1}.

3.4 Immobilization of Ligand Using Amine Coupling

3.4.1 Wizard Template Method

1. In the Biacore T200 control software dialogue window, go to File/Open/New wizard template/Immobilization/New.
2. Specify the sensor surface (CM5), the flow path (flow cell 2) and the amine coupling method.
3. The wizard template offers the possibility to choose between specifying the aim of your immobilization level on RUs or the contact time of your ligand and the flow rate.
   - If you choose to specify the aim of your immobilization level, you have to fix a target level. The system will inject 10 μL of the ligand at 5 μL min^{-1}. This short injection aims to estimate the rate of the preconcentration based on the level reached and the slope of the sensogram. This step is followed by the injection of the regeneration solution (50 mM NaOH to regenerate the surface before activation).
   - If you choose to specify the contact time and the flow rate, the concentration of your protein and the contact time should be estimated to obtain the desired level of Ri (amount of the immobilized ligand in RUs). It is recommended to use a low flow rate (5 μL min^{-1}) to maximize the ligand contact time to the surface.
4. Thaw EDC, NHS, and ethanolamine solutions from −20 °C at RT for 10 min (solutions are provided in amine coupling kit) (*see* **Notes 15** and **16**).
5. Thaw ligand and dilute it on corresponding pH solution at a concentration between 10 and 50 μg mL^{-1} (the concentration depends on the desired Ri and on the level of RUs obtained on the pH scouting).
6. Eject the rack tray.
7. Place vials in right positions specified when you set up experiment.
8. Insert rack.
9. Run immobilization program.

3.4.2 Manual Method

1. In Biacore T200 control software, open manual run.
2. Set flow rate to 10 μL min^{-1}.
3. Select flow path 2 (for flow cell 2).

4. Select start.
5. Dilute your protein in 100 μL 10 mM sodium acetate with suitable pH and at desired concentration (10–50 μg mL^{-1}).
6. Mix 120 μL EDC (0.4 M in water) and 120 μL NHS (0.1 M in water).
7. Prepare 120 μL ethanolamine (1 M ethanolamine–HCl at pH 8.5).
8. Inject EDC/NHS mixture between 6 and 10 min over surface.
9. After activation step, set flow rate to 5 μL min^{-1}.
10. Inject ligand (*see* **Note 17**).
11. Inject ethanolamine for 5 min.
12. To calculate the binding capacity, subtract RUs obtained after activation from RUs obtained after ethanolamine deactivation.
13. Wait until signal is stabilized to perform analyte-binding analysis (*see* **Note 18**).

3.5 Immobilization of Control Ligand (See Note 19)

1. Dissolve 1 mg Thioredoxin powder in 10 mM sodium acetate at pH 4 in 1.5 mL Eppendorf tube.
2. Centrifuge protein at 16,000 × *g* for 20 min at 4 °C.
3. Make aliquots of 80 μL and freeze using liquid nitrogen and conserve tube at −80 °C.
4. In Biacore T200 control software, open manual run.
5. Set flow rate to 10 μL min^{-1}.
6. Select flow cell number 1 (reference flow cell).
7. Perform tasks 4–13 from Subheading 3.4.2. The target level of the Thioredoxin immobilized on the reference flow cell should be the same as the Ri of the ligand.

3.6 Analyte Binding Analysis (See Note 20)

Despite the availability of wizard methods in the Biacore T200 control software (wizard template), it is recommended to perform a manual run (*see* **Note 21**).

1. In Biacore T200 control software, open manual run.
2. Set flow rate to 30 μL min^{-1}.
3. Select flow path 2-1 (analyte will be injected over flow cells 1 and 2).
4. Select start.
5. Thaw analyte on ice for 30 min.
6. Measure analyte concentration (*see* Subheading 3.1).
7. Make dilution of analyte on running buffer. If you estimate that the affinity is in the micromolar range, prepare 50 μM of analyte.

8. Eject rack tray.
9. Place sample in selected position.
10. Insert rack.
11. Inject analyte for 1 min (*see* **Note 22**).

3.7 Regeneration Optimization (See Note 23)

If analyte-binding analysis is performed using manual run, it is recommended to test the regeneration solutions in the same cycle.

1. After analyte injection (**step 11** in Subheading 3.6), estimate strength of interaction.
2. Prepare 100 μL 1 M NaCl, 2 M $MgCl_2$, 10 mM Glycine–HCl (pH 3), 10 mM HCl (pH 2), and 10 mM HEPES-NaOH (pH 9).
3. Inject 30 μL 1 M NaCl. If no dissociation is observed, go to next step.
4. Inject 30 μL 2 M $MgCl_2$. If no dissociation is observed, go to next step.
5. Inject 30 μL 10 mM Glycine–HCl. If no dissociation is observed, go to next step.
6. Inject 30 μL 10 mM HCl, If no dissociation is observed, go to next step.
7. Inject 30 μL 10 mM HEPES-NaOH.
8. After each injection step, measure amount of analyte remaining bound (*see* **Note 24**).

3.8 Affinity Measurements (See Notes 25–27)

1. Go to T200 Biacore software control dialog interface and open File/Open/New wizard template/Kinetics-affinity.
2. Specify flow path (2-1) if ligand is immobilized.
3. Specify chip type (CM5).
4. Select regeneration, if needed.
5. Select three startup cycles with buffer to stabilize baseline signal before starting experiment.
6. Specify contact time, flow rate, and dissociation time (*see* **Note 28**).
7. Specify contact time and flow rate of regeneration solution.
8. Specify stabilization period (100 s).
9. Fill sample identity and specify serial twofold dilutions of analyte used for experiment.
10. Prepare analyte (*see* Subheading 3.1).
11. Prepare ten twofold dilutions of analyte starting from $10 \times K_D$ as specified in wizard template (*see* **Note 29**). Dilutions must be prepared on vials (*see* **Note 30**).
12. Place vials in corresponding positions.
13. Start experiment.

3.9 Affinity Data Analysis

1. Open Biacore evaluation software.
2. Open Surface-bound kinetics/affinity from Kinetics/affinity at bottom of toolbar.
3. Select data to analyze. All sensograms are shown in different colors except for blanks (light gray) (*see* **Note 31**).
4. If you select Next, blank curves will be subtracted from other sensograms automatically.
5. Select Affinity for steady-state evaluation. The top panel shows a plot of the response at equilibrium (R_{eq}) against analyte concentrations (C^A) based on the average response of the selected region on the plateau of each sensogram (*see* **Note 32**).
6. Select Next.
7. Select the binding model (1:1 binding).
8. Select Fit.

The calculated K_D is shown as a vertical line in the curve plot. The K_D corresponds to the concentration of the analyte at half the R_{max}.

3.10 Kinetics Measurements (See Notes 33–35)

Perform **steps 3–13** as described in Subheading 3.8.

3.11 Kinetics Data Analysis (See Note 36)

1. Follow **steps 1–4** as described in Subheading 3.9.
2. Select kinetics.
3. Select 1:1 binding model and then select Fit.
4. Results are displayed as fitted curves in black over original sensograms.
5. The Quality control interface gives you the statistic quantifying the quality of your fitting.
6. For access to the calculated values, go to Report window.

4 Notes

1. XanTec bioanalytics GmbH (www.Xantec.com) offers a large number of different types of sensor surfaces. These chips are cheaper and give the same results as the Biacore chips.
2. For more details, read "Getting Started BIAcore T200" and BIAcore T200 instrument handbook, available for download from the GE Healthcare website.
3. To obtain good data, pure and stable protein samples are critical. The purity of the ligand is very important to ensure binding specificity and capacity. Impurities can be immobilized

with the ligand and therefore lead to a nonspecific binding or alter the accurate determination of the affinity or kinetic measurements. The purity of the analyte is important for the determination of the affinity and kinetic parameters. In fact, injections of impurities with the analyte over the ligand can give false constants or rates owing to an incorrect estimation of analyte concentrations. Analyte purity must exceed 95% and be analyzed by SDS-PAGE. It is important to check the quality of the ligand and the analyte at least 1 day before experiments. It is essential to apply rigorous purification and conservation protocols of protein samples.

4. When studying protein–protein interactions, generally each of the two protein candidates should be used as ligand in one experiment and as analyte in another experiment to validate the interaction. However, many factors must be taken into account in the choice of protein to be immobilized: the amount, size, stability, solubility, and valency of the proteins. For example, if you have a limitation in one of your proteins, you can use it as ligand because immobilization requires a very low amount of proteins (5–200 ng). If the protein is unstable, it should be used as analyte and to immobilize the stable one. If one of your proteins is multimeric in solution, it should be immobilized and monomeric proteins should be used as analytes.
5. For protein coupling, two approaches can be used:
   - Covalent immobilization. This strategy uses amine, thiol, or aldehyde functional groups on proteins. Covalent coupling methods use free carboxymethyl groups exposed on the surface of the sensor chip [15].
   - Noncovalent immobilization. Three capture approaches are frequently used: capture of biotinylated molecules on streptavidin sensor chip [16], nickel-chelated nitrilotriacetic acid (NTA) groups for His-tagged proteins [17], and capture by immobilized specific antibodies [18].

Amine coupling is the more frequently used method. It includes three-step reactions: EDC/NHS activation, followed by ligand immobilization by its primary amine (Lysine), and finally deactivation of the remaining free ester groups by ethanolamine.

The choice of immobilization method is critical in an SPR experiment. Generally, the amine coupling method is tested first. However, this approach has many limitations, and alternative immobilization techniques, such as noncovalent immobilization, must be adopted in some cases. Among these limitations, amine coupling can induce ligand precipitation or aggregation owing to the low pH used during immobilization, or the immobilization can hold the ligand in inactive conformation.

6. The immobilization is optimized when you have on the surface just enough ligand in active form and homogeneously oriented. High-density surfaces (immobilizing a high concentration of ligand) must be avoided because they can introduce mass transport issues during the association phase and a rebinding effect of analyte to ligand during the dissociation phase. Thus, a low level of ligand immobilization and the use of a higher flow rate are recommended in the case of kinetic measurements. A simple way to determine the level of the ligand to be immobilized is to calculate a theoretical R_{max} of the interaction to be studied. The R_{max} is the maximum analyte response capacity of the surface expressed on RUs. It depends on the molecular weight of both the ligand and the analyte and the stoichiometry of the reaction. This theoretical R_{max} is calculated assuming that 100% of the ligand molecules will be active once immobilized (Fig. 2).

For example, to study the interaction between GspH (16 kDa) and GspJ (26 kDa) from the Type II secretion system of *Pseudomonas aeruginosa*, 180 RUs of GspJ should be immobilized to obtain an R_{max} of 300 assuming that GspJ has one binding site on GspH ($S = 1$):

$$R_i = (MW \text{ ligand} / MW \text{ analyte}) \times R_{max} \times (1/S).$$

$$R_i = (16/26) \times 300 \times (1/1) \Rightarrow R_i = 180\, RUs.$$

The experimental R_{max} is always lower than the theoretical calculated R_{max}. In fact, it is difficult to obtain 100% active ligand on the surface (having the same orientation, immobilized at the same interface, where all binding sites are available and all molecules are correctly folded). To avoid this heterogeneity, it is recommended that between 20 and 50% of additional ligand be immobilized.

In general, kinetic measurements require a small amount of ligand immobilization (R_{max} between 50 and 150 RUs). For affinity studies, the binding capacity can vary from low to moderate levels (R_{max} between 100 and 800 RUs). The important factor in this last case is that the analyte should saturate the surface during contact time.

$$\text{Rmax} = \frac{\text{MW analyte}}{\text{MW ligand}} \text{ x Ri x S}$$

MW analyte: Molecular weight of the analyte.
MW ligand: molecular weight of the ligand.
Ri : amount of immobilized ligand in RUs.
S : the stoichiometry of the reaction.

Fig. 2 Determination of theoretical R_{max}

7. If your proteins are not stable in the running buffer (HBS-EP), a running buffer with a different composition, like phosphate buffer saline (PBS), should be used.
8. After purification, it is important to characterize the multimerization state of your proteins. This can be done using size exclusion chromatography (SEC), dynamic light scattering (DLS), or multiple angle light scattering coupled to SEC (SEC-MALS). If your protein forms multimers, it is recommended to use it as ligand.
9. If the proteins are already purified in the running buffer and conserved at −80 °C, check their quality after thawing by concentration quantification. Compare the concentration of the protein sample before freezing and after thawing 1 day before the experiment to be sure that thawing the protein did not lead to precipitation.
10. Be sure that the maintenance procedures are properly followed before starting the experiments.
11. The aim of the pH scouting experiment is to test the preconcentration of the ligand on the sensor surface at different pHs, and so to determine the pH at which the ligand can be adsorbed at highest concentration on the dextran matrix by electrostatic interaction. Typically, the optimal pH for preconcentration should be at 1 pH unit below the isoelectric point (pI) of your protein. At low pH ($3.5 < pH < 5.5$), the carboxylated dextran is negatively charged and the ligand is positively charged ($pI > 6$). The choice of the pH depends on the "real" pI of your protein. A low pH may be not suitable for many proteins. To avoid protein precipitation or denaturation, dilute the protein before performing the experiment. pH scouting is performed only in one flow cell. It is recommended that the flow cell planned for immobilization be used. The BIAcore T200 control software contains an immobilization wizard method to help users find the optimum pH for ligand immobilizing. Nevertheless, you can easily set up a manual run in which you inject the ligand diluted at different pHs.
12. The use of 50 mM NaOH solution for regeneration is sometimes sufficient. Nevertheless, the injection time may not be sufficient to dissociate the ligand from the sensor surface. In this case, it is recommended to use multiple injections with short contact times (e.g., three injections with 30 s of contact time).
13. If the theoretical pI of the ligand is around 5, start with a pH range between 3.5 and 4.5. If your ligand is an acidic protein ($pI < 4$), you can use the covalent immobilization using thiol groups or a noncovalent immobilization method.

14. Once the protein injection is performed, the choice of the optimal immobilization pH is based on the pattern of the sensograms. The ligand should bind rapidly to the sensor surface and completely dissociate after the end of the injection. If this condition is satisfied by all pHs used, use the highest pH because it is the least offensive to your protein. For example, in the case of the immobilization of TssE from the Type VI secretion system from enteroaggregative *E. coli*, and based on the pH scouting sensograms (Fig. 3), the optimum pH use for its immobilization is pH 5.
15. If you use the immobilization wizard template from the Biacore control software, do not mix the EDC and NHS solution. The software dialog box will ask you to prepare an empty vial to mix the two solutions.
16. The solutions should be prepared just before performing the experiments.
17. The injection of the ligand on the sensor surface is an irreversible step. Be careful when you inject the ligand; do not immobilize over the desired level. To minimize this risk, perform a short injection (5–10 μL of your ligand to estimate the immobilizing level). Once the relationship between the injection time and the reached RU values is established, perform a second injection to reach the final R_i. Ligand contact should be completed within 15 min after activation of the surface with EDC/NHS to ensure coupling.

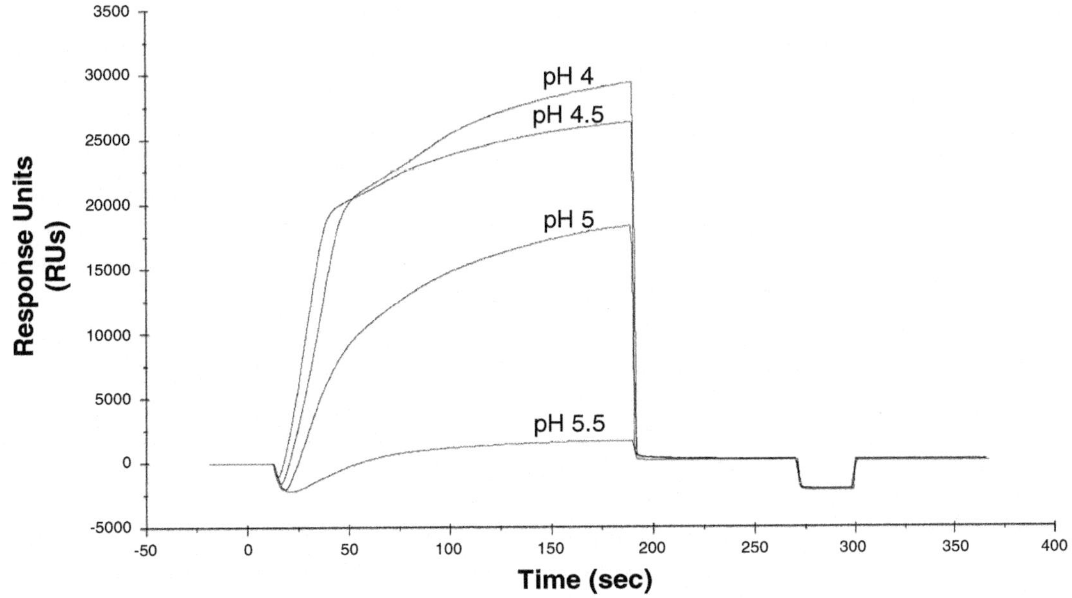

Fig. 3 pH scouting of TssE protein from Type VI secretion system over CM5 sensor surface. The pH of each solution tested is indicated at the top of the corresponding sensogram

18. The baseline stability is an indication of the "good" quality of the ligand after immobilization. It is important to wait until the baseline becomes stable and no decrease in RUs is observed.
19. The nonspecific binding of the analyte to the sensor surface is a common problem faced by SPR users. In some cases, the activation/deactivation of the reference cell is sufficient to remove nonspecific binding. However, frequently activation/ deactivation does not help and immobilization of an irrelevant protein on the sensor surface is recommended. A large number of proteins can be used to this end: for example Thioredoxin, maltose-binding protein (MBP), or a homemade protein not related to the ligand.
20. This step is performed after ligand immobilization and baseline stability. The choice of the analyte concentration depends on the strength of the interaction. If you are testing a new interaction, it is recommended to start with micromolar concentration (10–50 μM). If this leads to a high signal with a very slow dissociation, decrease your analyte concentration to the nanomolar range. In the case of antibody–antigen interactions, it is known that they are tight and you can start with a nanomolar concentration (100 nM). The function of your proteins can help you to estimate the strength of the interaction. For example, in the case of haemolysin-coregulated protein Hcp and the tail-sheath component TssB, in the T6SS, Hcp hexamers are able to pack and form a tail tube wrapped by the tail sheath composed of two proteins, TssB and TssC [19]. The contraction of the tail sheath leads to the expulsion of the Hcp tail tube to the extracellular milieu. Based on the dynamic of such a complex, you can estimate a weak interaction between Hcp and TssB. Consequently, we started with TssB as analyte in the micromolar range.
21. The use of the wizard template method is recommended if the regeneration buffer is known or described in the literature. If not, manual runs with different regeneration buffers should be carried out to optimize the regeneration step after analyte binding.
22. When examining the binding of the analyte to the ligand, the subtraction of the signal of the reference flow cell from that of the flow cell containing the ligand should give a typical sensogram of protein–protein interaction (Fig. 1). If the interaction is weak (micromolar range) or transient, low RU values are obtained following injection of the analyte (e.g., less than 20 RUs) and the sensograms have fast association and dissociation phases. In this case, it is recommended to perform a binding analysis with increasing concentrations of the analyte.

By contrast, if the interaction is strong (nanomolar range), higher RU values are obtained and the association and dissociation phases are slow. In this case, after the regeneration step, try a lower analyte concentration.

If no change in the RU value is observed, either the analyte did not bind to the ligand or the ligand immobilized to the sensor surface was inactivated. In this case, the immobilization of the ligand by the noncovalent method (*see* **Note 5**) may help solve the problem. Another possibility may be to immobilize the analyte (if the protein is suitable for immobilization) and test the opposite interaction by injecting the ligand.

23. Once the binding of the analyte on the ligand is confirmed, the regeneration step can be started. The goal of this step is to totally dissociate the analyte from the ligand binding sites without affecting the activity of the immobilized ligand. To this end, many regeneration solutions can be tested depending on the type of interaction and the nature of the ligand. If the interaction is reversible and fast dissociation is observed, washing with low-ionic-strength solutions (1 M NaCl, 2 M $MgCl_2$) or ethylene glycol (from 10 to 100%) for hydrophobic interactions may accelerate the dissociation. In the case of high-affinity interactions, stronger solutions may be necessary (low- or high-pH solutions or highly hydrophobic solutions). If the ligand is an antibody, it is recommended to use a strong acid solution (10 mM phosphoric acid).
24. If the decrease in the signal is below 30%, move to the next regeneration solution.
    - If the amount of the decrease is more than 30% but below 90%, try to repeat the injection.
    - If this procedure is not sufficient to remove more than 90% of the analyte, try a higher concentration of the regeneration solution.
    - If the decrease in the signal is significant (more than 100%), the regeneration solution is not suitable and may affect the ligand activity. If so, try to inject the analyte at the same concentration used in **step 3** in Subheading 3.6.
    - If the same binding level is obtained, use the same regeneration solution at a lower concentration.
    - If the binding level is lower compared to the first analyte injection, try another type of regeneration buffer.
    - If the regeneration fails, try to combine two solutions among those giving more than 30% surface regeneration.
    - If the residual activity of the ligand is less than 90% following regeneration, the flow cell is no longer suitable for binding analysis. It is recommended that the ligand be

immobilized by amine coupling in a different flow cell or that a noncovalent immobilization method be used.

– If an efficient regeneration buffer is not found, immobilize the ligand in a new flow cell. Kinetic measurement could be undertaken using single cycle kinetics (SCK) with no regeneration between injections. In an SCK experiment, increased concentrations of the analyte are injected sequentially in the same cycle.

25. Observation of the pattern of the sensogram resulting from ligand immobilization and the analyte binding gives you information on the strength of the interaction. If the association and the dissociation are fast (1–2 min), it is difficult to estimate the interaction rates. This suggests that the experiment will allow only for the estimation of the binding affinity of the complex. On the other hand, if the sensogram shows slow association/dissociation phases (5 min for association and 10–60 min or more for dissociation), it will be possible to estimate the kinetics of the interaction.

26. The affinity of molecule A to molecule B is described by the dissociation constant K_D (Fig. 4). K_D is expressed in molars (M). The K_D can be calculated using SPR data by the equilibrium binding analysis. The steady-state binding level is related to the concentration of the analyte (Fig. 5).

27. To perform the experiment, the analyte concentrations must be varied from $0.1 \times K_D$ to $10–100 \times K_D$.

28. The association and the dissociation times are determined based on the sensogram obtained after the analyte-binding analysis. The dissociation time should be sufficient to go back to the original level of the baseline. If it is time consuming, a regeneration step can be added with a soft regeneration solution to avoid ligand inactivation.

$$A + B \underset{k_{off}}{\overset{k_{on}}{\longleftrightarrow}} AB$$

$$K_D = \frac{C^A \text{ x } C^B}{C^{AB}} = \frac{k_{off}}{k_{on}}$$

C^A : Concentration of the molecule A.
C^B : Concentration of the molecule B.
C^{AB} : Concentration of the complex AB.
k_{on} : Association rate constant of the complex AB.
k_{off} : Dissociation rate constant of the complex AB.

Fig. 4 Representation of different binding parameters of an interaction between two molecules A and B

$$Req = \frac{C^A \times Rmax}{C^A + K_D}$$

Req: response at the equilibrium.
Rmax: maximum binding capacity of the analyte.

Fig. 5 Determination of K_D using response units at equilibrium and concentrations of analyte

$$R = \left(\frac{C^A \times Rmax}{C^A + K_D}\right)\left(1 - \frac{1}{e^{((k_{on} \times C^A) + k_{off}) \times t}}\right)$$

Fig. 6 Determination of rate of association k_{on} of complex AB

$$R = R_0 \times e^{(- k_{off} \times dt)}$$

R_0: the Response Unit at the end of the association.

Fig. 7 Determination of rate of dissociation k_{off} of complex AB

29. The volume of the analyte dilution solution depends on the contact times and the flow rate. These parameters are estimated during the analyte-binding analysis test.
30. When making twofold dilutions, avoid forming bubbles when mixing your analyte on the buffer. If bubbles are present, centrifuge your sample and put it in a new vial.
31. The sensograms are adjusted to zero at the start of the analyte injection on both the response and the time axis.
32. Adjust the region used for the calculation of R_{eq} by selecting the setting bottom.
33. Using SPR you can determine the association and the dissociation rate constants. The first phase in the sensogram, corresponding to the injection of the analyte over the ligand and the binding, allows for the determination of the rate of formation of the complex k_{on} according to the equation described in Fig. 6. The unit of k_{on} is M^{-1} s^{-1}.

The dissociation phase, when the analyte is removed from the flow cell till a zero concentration is reached, makes it possible to calculate the rate of dissociation k_{off} using the equation presented in Fig. 7. The unit of k_{off} is s^{-1}.

34. Mass transport is one of the more widely known limitations for the determination of the association rate constant. Mass transport takes place when the rate of analyte binding is higher than the rate of diffusion of the analyte. By contrast, a limitation on the determination of the dissociation rate constant comes from the rebinding of the analyte to the ligand owing to the inefficient effusion of the free analyte from the ligand surface. These issues can be avoided by the immobilization of a low amount of ligand (50–150 RUs) and performing the binding analysis at a high flow rate (30–100 μL min^{-1}).
35. To perform the experiment, the analyte concentrations must be varied from 0.1 × K_D to 10 × K_D.
36. The best way to analyze the kinetic data is to use the Biacore T200 evaluation software. Kinetics data are interpreted in terms of an interaction model, and the kinetic constants obtained from the SPR data analysis are apparent constants, which are valid in the context of the binding model adopted. A simple model used to determine the rates of an interaction is the Langmuir model, in which it is assumed that molecule A binds B with a 1:1 stoichiometry and the binding events are independent and equivalent.

Acknowledgments

I am grateful to Dr. Romé Voulhoux and Dr. Mariella Tegoni for their constant training, encouragement, and support and to Dr. Sawsan Amara and John Young for their careful reading of the manuscript.

References

1. Barison N, Lambers J, Hurwitz R, Kolbe M (2012) Interaction of MxiG with the cytosolic complex of the type III secretion system controls Shigella virulence. FASEB J 26:1717–1726
2. Benabdelhak H, Kiontke S, Horn C, Ernst R, Blight MA, Holland IB, Schmitt L (2003) A specific interaction between the NBD of the ABC-transporter HlyB and a C-terminal fragment of its transport substrate haemolysin A. J Mol Biol 327:1169–1179
3. Douzi B, Ball G, Cambillau C, Tegoni M, Voulhoux R (2011) Deciphering the Xcp *Pseudomonas aeruginosa* type II secretion machinery through multiple interactions with substrates. J Biol Chem 286:40792–40801
4. Douzi B, Durand E, Bernard C, Alphonse S, Cambillau C, Filloux A, Tegoni M, Voulhoux R (2009) The XcpV/GspI pseudopilin has a central role in the assembly of a quaternary complex within the T2SS pseudopilus. J Biol Chem 284:34580–34589
5. Douzi B, Spinelli S, Blangy S, Roussel A, Durand E, Brunet YR, Cascales E, Cambillau C (2014) Crystal structure and self-interaction of the type VI secretion tail-tube protein from enteroaggregative *Escherichia coli*. PLoS One 9:e86918
6. Felisberto-Rodrigues C, Durand E, Aschtgen MS, Blangy S, Ortiz-Lombardia M, Douzi B, Cambillau C, Cascales E (2011) Towards a structural comprehension of bacterial type VI secretion systems: characterization of the TssJ-TssM complex of an *Escherichia coli* pathovar. PLoS Pathog 7:e1002386
7. Girard V, Cote JP, Charbonneau ME, Campos M, Berthiaume F, Hancock MA, Siddiqui N, Mourez M (2010) Conformation change in a

self-recognizing autotransporter modulates bacterial cell–cell interaction. J Biol Chem 285:10616–10626

8. Schroder G, Lanka E (2003) TraG-like proteins of type IV secretion systems: functional dissection of the multiple activities of TraG (RP4) and TrwB (R388). J Bacteriol 185:4371–4381
9. Swietnicki W, O'Brien S, Holman K, Cherry S, Brueggemann E, Tropea JE, Hines HB, Waugh DS, Ulrich RG (2004) Novel protein-protein interactions of the *Yersinia pestis* type III secretion system elucidated with a matrix analysis by surface plasmon resonance and mass spectrometry. J Biol Chem 279:38693–38700
10. Zoued A, Durand E, Bebeacua C, Brunet YR, Douzi B, Cambillau C, Cascales E, Journet L (2013) TssK is a trimeric cytoplasmic protein interacting with components of both phage-like and membrane anchoring complexes of the type VI secretion system. J Biol Chem 288:27031–27041
11. Zoued A, Durand E, Brunet YR, Spinelli S, Douzi B, Guzzo M, Flaugnatti N, Legrand P, Journet L, Fronzes R, Mignot T, Cambillau C, Cascales E (2016) Priming and polymerization of a bacterial contractile tail structure. Nature 531(7592):59–63
12. Pineau C, Guschinskaya N, Robert X, Gouet P, Ballut L, Shevchik VE (2014) Substrate recognition by the bacterial type II secretion system: more than a simple interaction. Mol Microbiol 94:126–140
13. Ohlson S, Strandh M, Nilshans H (1997) Detection and characterization of weak affinity antibody antigen recognition with biomolecular interaction analysis. J Mol Recognit 10: 135–138
14. Peess C, von Proff L, Goller S, Andersson K, Gerg M, Malmqvist M, Bossenmaier B, Schraml M (2015) Deciphering the stepwise binding mode of HRG1beta to HER3 by surface plasmon resonance and interaction map. PLoS One 10:e0116870
15. Fischer MJ (2010) Amine coupling through EDC/NHS: a practical approach. Methods Mol Biol 627:55–73
16. Hutsell SQ, Kimple RJ, Siderovski DP, Willard FS, Kimple AJ (2010) High-affinity immobilization of proteins using biotin- and GST-based coupling strategies. Methods Mol Biol 627:75–90
17. Khan F, He M, Taussig MJ (2006) Double-hexahistidine tag with high-affinity binding for protein immobilization, purification, and detection on ni-nitrilotriacetic acid surfaces. Anal Chem 78:3072–3079
18. Della Pia EA, Martinez KL (2015) Single domain antibodies as a powerful tool for high quality surface plasmon resonance studies. PLoS One 10:e0124303
19. Cianfanelli FR, Monlezun L, Coulthurst SJ (2015) Aim, load, fire: the type VI secretion system, a bacterial nanoweapon. Trends Microbiol 24(1):51–62

Chapter 22

Assessing Energy-Dependent Protein Conformational Changes in the TonB System

Ray A. Larsen

Abstract

Changes in conformation can alter a protein's vulnerability to proteolysis. Thus, in vivo differential proteinase sensitivity provides a means for identifying conformational changes that mark discrete states in the activity cycle of a protein. The ability to detect a specific conformational state allows for experiments to address specific protein–protein interactions and other physiological components that potentially contribute to the function of the protein. This chapter presents the application of this technique to the TonB-dependent energy transduction system of Gram-negative bacteria, a strategy that has refined our understanding of how the TonB protein is coupled to the ion electrochemical gradient of the cytoplasmic membrane.

Key words Protein conformation, Ion electrochemical gradient, Proteinase sensitivity, TonB, Spheroplast, Energy transduction

1 Introduction

Many proteins contain regions that are intrinsically disordered, unable to achieve stable structures on their own. Rather than a detriment, this conformational flexibility is an essential feature in many protein-mediated processes, particularly those involving protein–protein interactions [1]. A feature common to most proteins with intrinsically disordered regions is an enhanced sensitivity to proteolysis [2]. One such protein is the Gram-negative bacterial cell envelope protein TonB [3].

Anchored to the cytoplasmic membrane by a single N-terminal transmembrane domain, the bulk of TonB occurs in the periplasmic space, which TonB spans to interact with outer membrane transporters. TonB functions as an energy transducer, coupling the ion electrochemical gradient of the cytoplasmic membrane to drive active transport processes at the outer membrane (reviewed in [4]). In this capacity, TonB appears to cycle through several distinct conformations, as evidenced in vivo by differential sensitivity to

Laure Journet and Eric Cascales (eds.), *Bacterial Protein Secretion Systems: Methods and Protocols*, Methods in Molecular Biology, vol. 1615, DOI 10.1007/978-1-4939-7033-9_22, © Springer Science+Business Media LLC 2017

both intrinsic and extrinsic proteinases [5]. We have proposed that these conformational changes in TonB involve regions of intrinsic disorder [6]. Interactions with known proteins appear to drive at least some of these conformational changes, and in doing so support the energy transduction function of TonB [7, 8]. These and other studies suggest a current working model of the TonB energy transduction cycle, wherein the conformational flexibility of intrinsically disordered regions is a key feature [9].

It had long been recognized that intrinsic proteolysis was important in TonB and that this sensitivity correlated with apparent interactions between TonB and other proteins [10]. The discovery that TonB was differentially sensitive to exogenous proteinases, and the correlation of this sensitivity to the energy state of the cytoplasmic membrane was accidental. We were characterizing a set of mutations in the signal anchor of TonB that rendered TonB inactive [11]. Because these mutations occurred in a region of TonB involved in the trafficking of TonB to the cytoplasmic membrane, a trivial explanation for the TonB negative phenotype could have been that the protein was simply not delivered to the cytoplasmic membrane. To determine whether the TonB derivatives were properly oriented in the cytoplasmic membrane, we modified a method initially developed to examine the delivery of maltose-binding protein to the periplasmic space [12]. Here, the outer membrane is permeabilized and the peptidoglycan layer disrupted, converting the cells into spheroplasts. This allows large molecules, in this case proteinase K, access to periplasmically exposed proteins. Using the same general strategy, we found that both the wild-type and the mutant TonB derivatives were readily degraded by proteinase K in spheroplasts, but not in intact cells, indicating that the mutant derivatives were properly trafficked to and oriented in the cytoplasmic membrane.

In the experiments just described we included a control where the spheroplasts were osmotically lysed to allow proteinase K access to the cytoplasmic contents. Thus, if the mutant derivatives were not susceptible to proteolysis in the intact spheroplasts, we could confirm that the problem involved trafficking (and not simple resistance to proteinase K). To our great surprise, we found that in lysed spheroplasts, wild-type TonB was not fully degraded but rather reduced to a smaller fragment. Further studies suggested that this fragment represented the ~130 residue N-terminal domain of TonB present in the cytoplasmic membrane. Interestingly, the ability to form this proteinase K-resistant fragment was dependent upon the presence of the cytoplasmic membrane protein ExbB, with which TonB was known to associate. Also of interest, the mutant derivatives did not form the proteinase K-resistant fragment in either the presence or absence of ExbB, suggesting the mutations disrupted that particular interaction. Together, these data suggested that ExbB influences the conformation of TonB [11].

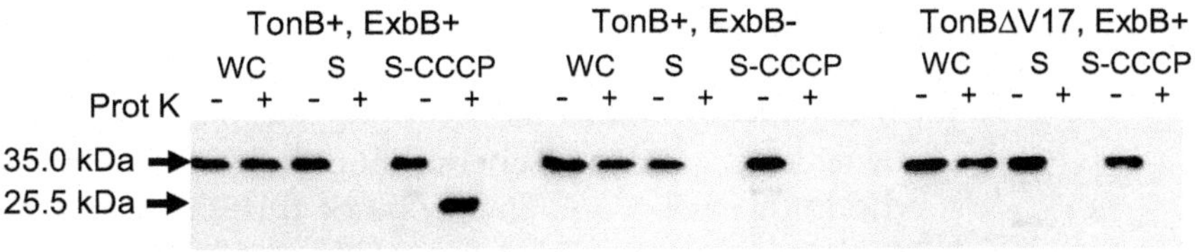

Fig. 1 Identification of an ion electrochemical responsive TonB conformation by differential proteinase K susceptibility. Western blots of whole-cell (WC), spheroplast (S), and CCCP-treated spheroplast (S-CCCP) samples, treated with (+) or without (–) proteinase K ("Prot K"), are shown. The sample set on the left was generated from the wild-type strain W3110 and is wild type for TonB, ExbB, and ExbD. The sample set in the middle was generated from the W3110 derivative and is wild type for TonB but does not express ExbB and ExbD. The sample set on the right was generated from a W3110 derivative bearing a *tonB* allele that encodes an inactive TonB with a deletion of the valine residue from position 17 and is wild type for ExbB and ExbD. Samples were developed on 11% SDS-polyacrylamide gels, transferred to polyvinylidene fluoride membranes and probed with the monoclonal antibody 4H4 (specific for TonB residues 79–84). The positions and apparent molecular masses of intact TonB and the proteinase K-resistant fragment are indicated on the left of the panel. Reproduced in part from Larsen et al. [5] with permission from John Wiley & Sons

While it seems obvious in retrospect, it was initially unclear to us how lysing the spheroplast could alter TonB conformation. Eventually it occurred to us that spheroplast lysis collapses the cytoplasmic membrane ion electrochemical gradient. We tested this hypothesis by treating spheroplasts with protonophores and determined that indeed the proteinase K-resistant fragment represented a conformation that energized TonB can become trapped in. Similar to the results obtained with lysed spheroplasts, only functional TonB in the presence ExbB could achieve this conformation and only when the ion electrochemical gradient was collapsed (Fig. 1). This was the first direct evidence that TonB conformation was coupled to the ion electrochemical potential of the cytoplasmic membrane [5]. Soon thereafter it was demonstrated that the ability of the TonB analog TolA to interact with an outer membrane lipoprotein was dependent upon the cytoplasmic membrane proton gradient [13]. Subsequently, our strategy was used to demonstrate that TolA undergoes conformational changes in response to the ion electrochemical gradient similar to that occurring in TonB [14].

This general strategy provides a means to examine conformational changes in periplasmically exposed proteins in response to varied stimuli as well as through protein–protein interactions. We have focused on the ion electrochemical gradient, but, depending upon the protein and the system, the influence of a wide range of physical and chemical conditions on protein conformation and, consequently, their influence on function could be addressed. Likewise, we have relied upon proteinase K as a probe; a myriad of other proteases with a diversity of mechanisms and specificities have potential as informative probes.

Ultimately, the readout of this assay requires visualization of the protein of interest. Here, samples are developed on sodium dodecyl sulfate (SDS) 11% polyacrylamide gels, transferred by electrophoretic elution to polyvinylidine difluoride membranes, probed with a monoclonal antibody specific for the protein, and visualized by enhanced chemiluminescence. Description of this methodology is beyond the scope of this chapter; it is presumed that laboratories interested in adapting a strategy that uses differential sensitivity to proteolysis to detect conformational changes will have a similar visualization technique optimized for their protein of interest (*see* **Note 1**).

2 Materials

Prepare solutions and culture media with double-distilled water and reagent-grade materials. The procedures described here use the laboratory-adapted *Escherichia coli* K-12 strain W3110 [15] and derivatives thereof (*see* **Note 2**).

2.1 Spheroplast Production

1. Luria–Bertani (LB) agar medium ([16]) (*see* **Note 3**).
2. Supplemented M9 medium: Prepare a 10× stock solution containing 60 g Na_2HPO_4, 30 g KH_2PO_4, 5 g NaCl, and 10 g NH_4Cl per liter. Sterilize the 10× solution by autoclaving and store at room temperature. To make the culture medium, add 10 mL of the 10× solution to 85.6 mL sterile distilled water, then supplement this mixture with 2 mL of a filter-sterilized 20% (w/v) carbohydrate source (*see* **Note 4**), 1 mL 20% (w/v) vitamin-free casamino acids (autoclaved), 1 mL filter-sterilized 0.4% (w/v) tryptophan, 200 μL filter-sterilized 0.2% (w/v) thiamin (*see* **Note 5**), 100 μL 1 M $MgSO_4$, 100 μL 0.5 M $CaCl_2$, and 50 μL 0.1 (w/v) $FeCl_3 \cdot 6H_2O$ (*see* **Notes 6** and **7**).
3. Isopropyl β-D-1-thiogalactopyranoside (IPTG): 1 M (*see* **Note 8**).
4. Buffer 1: 200 mM Tris–acetate, pH 8.2, 500 mM sucrose, 0.5 mM ethylenediaminetetraacetic acid (EDTA). Keep solution at 4 °C.
5. Buffer 2: 200 mM Tris–acetate, pH 8.2. Keep solution at 4 °C.
6. Buffer 3: 200 mM Tris–acetate, pH 8.2, 250 mM sucrose, 20 mM $MgSO_4$ (*see* **Note 9**). Keep solution at 4 °C.
7. Lysozyme: Stock solution of 2 mg lysozyme in 1 mL sterile water (*see* **Note 10**). Keep solution at 4 °C.
8. Incubator.
9. Spectrophotometer (*see* **Note 11**).
10. Microcentrifuge and microfuge tubes.
11. Micropipette or a pulled Pasteur pipet and a rubber bulb.

2.2 Proteinase Accessibility

1. Proteinase K: Stock solution of 2 mg proteinase K in 1 mL sterile water (*see* **Note 10**).
2. Phenylmethylsulfonyl fluoride (PMSF): Stock solution of 100 mM in dimethyl sulfoxide (DMSO) (*see* **Note 12**).
3. Carbonylcyanide *m*-chlorophenylhydrazone (CCCP): Stock solution of 15 mM in DMSO (*see* **Note 12**).
4. Trichloroacetic acid (TCA): Stock solution of 10% (w/v) (*see* **Note 13**).
5. 100 mM Tris–HCl, pH 6.8.
6. 1× Laemmli sample buffer: 63 mM Tris–HCl, pH 6.8, 2% (w/v) sodium dodecyl sulfate, 10% (v/v) glycerol, 0.1% (v/v) β-mercaptoethanol, and 0.0005% (w/v) bromophenol blue.

3 Methods

Bacterial strains are grown at 37 °C with shaking to provide aeration; all subsequent steps are performed at 4 °C, except for β-galactosidase assays and solubilization of samples for electrophoresis.

3.1 Production of Spheroplasts

1. Select an isolated colony from an LB agar stock plate (*see* **Note 3**) and inoculate 5 mL of supplemented M9 medium in a standard culture tube. Incubate at 37 °C overnight with shaking (~200 rpm).
2. Inoculate a fresh culture tube containing 5 mL of supplemented M9 medium with 25 μL of the overnight culture and incubate as described earlier, with periodic spectrophotometric monitoring of growth (*see* **Notes 11** and **14**).
3. At or near an A_{550} of 0.2, add 5 μL 1 M IPTG (1 mM final concentration) to each culture to induce β-galactosidase expression for subsequent measurement of spheroplast integrity.
4. At an A_{550} of 0.4, harvest cells as six 500 μL samples into prechilled 1.5 mL Eppendorf microfuge tubes, and centrifuge at ~20,000 × *g* for 5 min at 4 °C (*see* **Note 15**).
5. Remove supernatant by manual aspiration using either a micropipette with a disposable tip or a pulled Pasteur pipet and a rubber bulb. Discard the supernatant and place tubes on ice.
6. Four of the six tubes are used to generate spheroplasts. To each of these tubes, add 250 μL chilled (4 °C) buffer 1. Suspend pellets by gently pipetting up and down with a micropipette. When fully suspended, add 20 μL of the 2 mg/mL lysozyme solution. Mix by gently flicking each tube, then immediately add 250 μL chilled (4 °C) buffer 2. Mix gently as above, then incubate on ice for 5 min (*see* **Note 16**). Proceed to **step 7** while incubating these tubes, then proceed to **step 8**.

7. The two remaining pellets are used as whole-cell controls. To each of these add 500 μL chilled (4 °C) buffer 3. Suspend pellets by gently pipetting up and down with a micropipette, then store on ice.
8. After 5 min on ice, add 10 μL 1 M $MgSO_4$ to each of the four tubes used to produce spheroplasts in **step 6** (*see* **Note 17**). Mix gently.
9. Centrifuge all six tubes (prepared in **steps 7** and **8**) at ~20,000 × *g* for 5 min at 4 °C. Remove the supernatant by manual aspiration and discard, then add 500 μL buffer 3 to each tube and gently suspend each pellet using a micropipette (*see* **Note 18**).

3.2 Proteinase Accessibility

Preparations are divided into pairs for analysis. The whole-cell control pair was produced separately at the same time the four tubes of spheroplasts were produced. The spheroplast tubes are now divided into two groups: The first is used to evaluate baseline proteinase susceptibility, and the second pair of tubes is used to examine the effect of collapsing the ion electrochemical gradient of the cytoplasmic (i.e., spheroplast) membrane on the conformation of TonB. Take care to maintain all tubes at 4 °C and to handle tubes gently through the following steps.

1. Add the protonophore CCCP to two spheroplast tubes as 3.4 μL of a 15 mM solution of CCCP in DMSO, for a final concentration of ~50 μM (*see* **Note 19**). These two tubes become the second pair. The two remaining spheroplast tubes (pair 1) and the two whole-cell tubes receive 3.4 μL of DMSO alone.
2. Add 6.3 μL of the 2.0 mg/mL solution of proteinase K to one tube of the whole-cell pair and one tube of each spheroplast pair (for a final concentration of 25 μg/mL). Add 6.3 μL water to the other tube in each pair. Incubate all tubes for 15 min at 4 °C, with each tube gently flicked every several minutes to maintain cells and spheroplasts in suspension (*see* **Note 20**).
3. Add 5 μL 100 mM PMSF to each tube (for a final concentration of 1 mM), mix gently, and incubate for 2 min at 4 °C to inactivate proteinase K.
4. Optional: Remove two 100 μL aliquots from each tube for the determination of β-galactosidase activity. Store on ice while **step 5** is performed, then proceed to **step 6**.
5. Add 500 μL (300 μL if **step 4** performed) of cold (4 °C) 10% (w/v) TCA to both whole-cell tubes and to each tube in spheroplast pairs 1 and 2. Mix by inverting several times, then incubate the samples on ice for 15 min to precipitate total protein (*see* **Note 21**). Proceed to **step 6** while incubating these tubes.

6. Optional: Centrifuge one of the two aliquots taken from each tube at ~20,000 × *g* for 5 min at 4 °C. Transfer the supernatant from each to a fresh tube. These samples can then be assayed for the presence of β-galactosidase to determine spheroplast membrane integrity (*see* **Note 22**).
7. Centrifuge the six TCA precipitation tubes at ~20,000 × *g* for 5 min at 4 °C (*see* **Note 23**). Remove supernatant by hand aspiration, then gently rinse the tube with 200 μL 100 mM Tris–HCl, pH 6.8. Add 50 μL 1× Laemmli sample buffer, then incubate at 96 °C for 5 min to denature sample for subsequent electrophoretic analysis.

4 Notes

1. Our system uses a monoclonal antibody [17] that recognizes a linear epitope present in the proteinase K-resistant fragment of TonB. Other fragments might occur that would be missed using this antibody as a probe. In our case, no additional or alternative proteinase K-resistant fragments were identified using several other monoclonal antibodies specific for epitopes located elsewhere in TonB. In the absence of a good mono-specific antibody, expression of the protein of interest modified to include an epitope tag or a small binding domain for a detectable ligand might prove useful.
2. The strain W3110 is a commonly used so-called wild-type *E. coli* K-12 strain, with a complete genome sequence available in Genbank (NC_007779.1). The methods described in this chapter should work well with other wild-type K-12 strains and most of their mutant derivatives. Successful application to non-laboratory-adapted strains with more robust outer membranes may require some modifications, particularly in creating spheroplasts.
3. Several different formulations of LB agar are available; there is nothing particularly special about the Miller formulation, but it is the one we first used long ago, and so we continue to use it for the sake of consistency. Other formulations are not likely to alter the results of the methods described here, nor should any other rich medium commonly used to grow laboratory *E. coli* stains.
4. Normally glucose is used; however, many of our studies involve products of arabinose-regulated genes that are subject to catabolite repression; therefore, we substitute glycerol as the major carbon source. The solution is filter-sterilized to avoid caramelization.
5. Thiamin and tryptophan are both heat-labile, so the stock solution should be sterilized by filtration and stored at 4 °C.

Tryptophan is also light sensitive and should be stored in an opaque container.

6. Autoclaving of phosphates with ionic metals (such as magnesium and calcium) can result in precipitation. The metal cations will preferentially precipitate with only one ionic form of phosphate, thereby altering the pH of the medium.
7. For strains carrying plasmids, the medium is supplemented with the appropriate amount of antibiotic required to provide selection for a given plasmid. Additional supplementation with arabinose is made for cells carrying plasmids with arabinose-regulated genes of interest; the working concentration for a given strain is that which has been predetermined to provide for the expression of the gene product at a normal physiologic level.
8. IPTG stock solution is made at 1 M in water and should be stored at −20 °C.
9. Buffers 1–3 are made fresh as 10 mL volumes from stocks of 1.0 M Tris–acetate, pH 8.2, 2.0 M sucrose, 1.0 M $MgSO_4$, and 500 mM EDTA. For buffer 1: to 2.0 mL 1.0 M Tris–acetate, pH 8.2, add 2.5 mL 2.0 M sucrose, 10 μL 500 mM EDTA, and add water to 10 mL final volume. For buffer 2: to 2.0 mL 1.0 M Tris–acetate, pH 8.2, add water to 10 mL final volume. For buffer 3: to 2.0 mL 1.0 M Tris–acetate, pH 8.2, add 1.25 mL 2.0 M sucrose, 200 μL 1.0 M $MgSO_4$, and add water to 10 mL final volume.
10. Store enzyme solutions at −20 °C.
11. Cultures are monitored using a Spectronic 20 spectrophotometer fitted with an adaptor to receive our culture tubes, which have an internal diameter of 1.5 cm. Culture tubes must contain at least 5 mL of medium to insure that the light path goes through the medium below the meniscus. Absorbance of light at 550 nm is measured. Because the path length of light is 1.5 cm, we refer to our measure as an "A_{550}," as the standard term optical density (OD_{550}) is specifically defined for a 1.0 cm path length.
12. PMSF is unstable in water, and CCCP has a low solubility in water, hence the use of DMSO as a carrier. Both compounds are toxic and should be handled with caution; a mask should be worn to protect against inhalation when weighing the solid, and gloves should be worn at all times, as the carrier solvent (DMSO) facilitates absorption through the skin.
13. 10% (w/v) TCA is prepared by dilution in water from a 100% (w/v) stock solution, made by adding 227 mL water to a new reagent bottle containing 500 g TCA.
14. For W3110 grown under the conditions described, cells experience an initial lag phase of ~45 min, transitioning into

exponential growth with a generation time of 45–50 min. After four to five generations (about 4–4.5 h) cells reach an A_{550} of 0.4, at which point they will be harvested. Grown under these conditions, an A_{550} of 0.4 corresponds to ~1.0×10^8 colony forming units (CFU)/mL.

15. It is challenging to catch cells at exactly at $A_{550} = 0.4$, and when dealing with more than one independently inoculated tube, it is unlikely that they will all reach $A_{550} = 0.4$ at the same time. What is important is to harvest an equivalent number of cells for each culture. One approach is to allow each culture to grow past $A_{550} = 0.4$ and then dilute back with fresh medium, good in theory but logistically complex in practice. A more practical solution is to vary the volume harvested relative to the A_{550} value. To determine how much to sample, we consider this in terms of A_{550} equivalents (A_{550}eq), where A_{550}eq = A_{550} × volume (in milliliters). Here, the A_{550}eq = 0.4 × 0.5 mL = 0.2 A_{550}eq. Rearranged, A_{550}eq/A_{550} = volume (mL) harvested. For example, for a culture at $A_{550} = 0.43$: 0.2/0.43 = 0.465; harvest 465 μL. For a second culture at $A_{550} = 0.36$: 0.2/0.36 = 0.556; harvest 556 μL. This approach does introduce some error; the A_{550} value is a measure of light absorbance, and as culture densities increase, the probability of any single bacterium's being in the shadow of another bacterium increases. Thus, the correlation of the measured A_{550} value and the CFU is not linear. However, working in the range of A_{550} = 0.35–0.45, the amount of error introduced is lower than the range of error in determining CFU.
16. The 500 mM sucrose in buffer 1 is hypertonic, creating turgor pressure on the outer membrane. The EDTA chelates divalent cations that stabilize the anionic inner core of lipopolysaccharide. The addition of buffer 2 renders the solution relatively isotonic. This rapid shift of pressure on the cation-depleted outer membrane permeabilizes the barrier, allowing lysozyme access to the periplasmic side, where it then catalyzes degradation of the peptidoglycan layer.
17. In the absence of peptidoglycan the cytoplasmic membrane becomes very fragile (which is why we are keeping everything cold and treating the spheroplasts gently). The addition of magnesium cations partially compensates for the absence of peptidoglycan, stabilizing the cytoplasmic membrane of the spheroplasts by ionic interactions with anionic phosphoryl lipid head groups.
18. Cells are now ready for the in vivo assay of differential proteinase K susceptibility. The centrifugation and suspension of the two whole-cell controls seems superfluous because they were already in suspension in the final buffer, but they are included in this step to minimize any handling-based differences between samples.

19. We have also used the protonophore dinitrophenol (DNP), with similar results. However, the effective working concentration of DNP is 10 mM, vs. 50 μM for CCCP. We therefore chose not to use DNP because such higher concentrations might allow other, unidentified physiological perturbations.
20. The amount of proteinase K used corresponds to roughly one unit per reaction. Given the small sample size and length of incubation, this seems excessive; however, at 4 °C the reaction is occurring at well below the optimal of 37 °C for this enzyme. Initial experiments using several different concentrations of proteinase should be performed to optimize the concentration for the system and the protein of interest.
21. Precipitation with TCA is routinely used to denature proteins, including proteinases, to protect samples as they are prepared for electrophoresis.
22. The enzyme β-galactosidase is a readily assayed cytoplasmic protein, providing a means to evaluate spheroplast membrane integrity. We determine β-galactosidase activity for 100 μL of supernatant and uncentrifuged samples. The percentage of lysed spheroplasts is calculated as = 100 × (supernatant)/(uncentrifuged). We routinely find that the amount of β-galactosidase activity in the supernatant samples is less than 15% of that found in the uncentrifuged samples. Because some spheroplast lysis occurs with centrifugation, the percentage of spheroplasts damaged during the proteolysis phase of the experiment is less. We measure β-galactosidase activity using the assay described by Miller [16]. This is a classic, widely used assay whose description is beyond the scope of this chapter.
23. The pellet formed in a TCA precipitation is not always tightly formed and in a fixed-angle rotor will form as a streak along the upward-facing side of the tube. It is important to note tube orientation in the centrifuge so that one can avoid inadvertent loss of sample when aspirating the supernatant.

References

1. Oldfield CJ, Dunker AK (2014) Intrinsically disordered proteins and intrinsically disordered protein regions. Annu Rev Biochem 83: 553–584
2. Johnson DE, Xue B, Sickmeier MD, Meng J, Cortese MS, Oldfield CJ, Gall TL, Dunker AK, Uversky VN (2012) High-throughput characterization of intrinsic disorder in proteins from the protein structure initiative. J Struct Biol 180:201–215
3. Peacock RS, Weijie AM, Howard SP, Price FD, Vogel HJ (2005) The solution structure of the C-terminal domain of TonB and interaction studies with TonB box peptides. J Mol Biol 345:1185–1197
4. Postle K, Larsen RA (2007) TonB dependent energy transduction between outer and cytoplasmic membranes. Biometals 20:453–465
5. Larsen RA, Thomas MG, Postle K (1999) Protonmotive force, ExbB and ligand-bound FepA drive conformational changes in TonB. Mol Microbiol 31:1809–1824
6. Larsen RA, Deckert G, Kastead K, Devanathan S, Keller KL, Postle K (2007) His_{20} provides

the sole functionally significant side chain in the essential TonB transmembrane domain. J Bacteriol 189:2825–2833

7. Ollis AA, Postle K (2012) Identification of functionally important TonB-ExbD periplasmic domain interactions in vivo. J Bacteriol 194:3078–3087
8. Ollis AA, Kumar A, Postle K (2012) The ExbD periplasmic domain contains distinct functional regions for two stages in TonB energization. J Bacteriol 194:3069–3077
9. Gresock MG, Kastead KA, Postle K (2015) From homodimer to heterodimer and back: elucidating the TonB energy transduction cycle. J Bacteriol 197:3433–3445
10. Fischer E, Günter K, Braun V (1989) Involvement of ExbB and TonB in transport across the outer membrane of *Escherichia coli*: phenotypic complementation of *exbB* mutants by overexpressed *tonB* and physical stabilization of TonB by ExbB. J Bacteriol 171: 5127–5134
11. Larsen RA, Thomas MG, Wood GE, Postle K (1994) Partial suppression of an *Escherichia coli* TonB transmembrane domain mutation (ΔV17) by a missense mutation in ExbB. Mol Microbiol 13:627–640
12. Randall LL, Hardy SJS (1986) Correlation of competence for export with lack of tertiary structure of the mature species: a study in vivo of maltose-binding protein in *E. coli*. Cell 46:921–928
13. Cascales E, Gavioli M, Sturgis JN, Lloubés R (2000) Proton motive force drives the interaction of the inner membrane TolA and outer membrane pal proteins in *Escherichia coli*. Mol Microbiol 38:904–915
14. Germon P, Ray MC, Vianney A, Lazzaroni JC (2001) Energy-dependent conformational change in the TolA protein of *Escherichia coli* involves its N-terminal domain, TolQ, and TolR. J Bacteriol 183:4110–41104
15. Hill CW, Harnish BW (1981) Inversions between ribosomal RNA genes of *Escherichia coli*. Proc Natl Acad Sci U S A 78:7069–7072
16. Miller JH (1972) Experiments in molecular genetics. Cold Spring Harbor Press, Cold Spring Harbor
17. Larsen RA, Myers PS, Skare JT, Seachord CL, Darveau RP, Postle K (1996) Identification of TonB homologs in the family *Enterobacteriaceae* and evidence for conservation of TonB-dependent energy transduction complexes. J Bacteriol 178:1363–1373

Chapter 23

Defining Assembly Pathways by Fluorescence Microscopy

Abdelrahim Zoued and Andreas Diepold

Abstract

Bacterial secretion systems are among the largest protein complexes in prokaryotes and display remarkably complex architectures. Their assembly often follows clearly defined pathways. Deciphering these pathways not only reveals how bacteria accomplish building these large functional complexes but can provide crucial information on the interactions and subcomplexes within secretion systems, their distribution within bacteria, and even functional insights. The emergence of fluorescent proteins has provided a new powerful tool for biological imaging, and the use of fluorescently labeled components presents an interesting method to accurately define the biogenesis of macromolecular complexes. Here we describe the use of this method to decipher the assembly pathway of bacterial secretion systems.

Key words Fluorescence microscopy, Biogenesis, Secretion systems, Fluorescently labeled proteins, Macromolecular complexes, Epistasis experiments, Subcellular localization

1 Introduction

Bacterial secretion systems are macromolecular machines that mediate the transport of proteins between bacteria or from bacteria to eukaryotic cells [1, 2]. These complexes incorporate one or multiple copies of a large number of different proteins that are recruited in a hierarchical order. Important insights into the assembly of secretion systems were obtained by the purification and visualization of assembly intermediates, either in strains lacking certain components of the system or upon overexpression of defined components [3–10]. However, the often low number of stable intermediates and the difficulties in obtaining and visualizing these have limited the utilization of this approach. Increasingly, the assembly of secretion systems is therefore deciphered based on the localization of fluorescently labeled subunits. Variations of this approach have been applied for a variety of secretion systems including the Tat system [11, 12], the type II secretion system (T2SS) [13, 14], the T3SS [15, 16], the T4SS [17], or, more recently, the T6SS [18]. The rationale is that a fluorescently labeled component forms

Laure Journet and Eric Cascales (eds.), *Bacterial Protein Secretion Systems: Methods and Protocols*, Methods in Molecular Biology, vol. 1615, DOI 10.1007/978-1-4939-7033-9_23, © Springer Science+Business Media LLC 2017

distinctive fluorescent foci where the secretion system resides, in wild-type cells and in strains lacking components that are not required for its assembly, but will have a diffuse fluorescent pattern when components responsible for its recruitment are missing. Prerequisites for this approach are the genetic amenability of the bacterium to create fluorescent fusion proteins and a specific distribution of the secretion system(s) within the bacterium (*see* **Note 1**). Beyond studying the kinetics of secretion system assembly in live bacteria, this method can be used to obtain a detailed description of the assembly pathway by visualizing the labeled subunits in strains lacking other components of the secretion system. In this chapter, we describe a generally applicable approach to deciphering the assembly pathway of secretion systems using fluorescently labeled proteins.

2 Materials

2.1 Strains

1. Strain(s) expressing a chromosomal fusion of a fluorescent protein to the target protein of interest, e.g., superfolder green fluorescent protein (sfGFP) fused either to the N- or C-terminus of the target protein (*see* **Notes 2** and **3**).
2. Additional deletions of other components of the secretion system in the strain background mentioned earlier to allow the investigation of the order of assembly.
3. Recommended: Untagged strain as a control for autofluorescence.
4. Recommended: Strain expressing fluorescent protein in cytosol from plasmid as a control (*see* **Note 4**).

2.2 Sample Preparation

1. Incubator shaker.
2. Spectrophotometer.
3. Culture medium: Standard culture medium and appropriate antibiotics for overnight incubation and growth of bacteria, e.g., lysogeny broth (LB) or M9 minimal medium (*see* **Note 5**).
4. Microscopy buffer: Nonfluorescent minimal medium or imaging buffer, e.g., phosphate buffered saline (PBS) (*see* **Note 6**).

2.3 Microscope Slide Preparation

1. Low-melting agarose (or agar), commercially available.
2. Microscopy slides and cover slips compatible with used microscope. Standard sizes include 75 × 25 × 1 mm glass slides and 22 × 22 mm cover slips, with No. 1 (0.13–0.16 mm) being the most commonly used thickness.
3. Microwave oven to prepare agarose solution.

2.4 Image Acquisition

1. Automated inverted epifluorescence microscope with 60× or 100× objective (*see* **Note 7**).
2. Optical filters for visualizing fluorescence, e.g., ET-GFP filter set (Chroma 49002) for visualizing GFP fluorescence and ET-mCherry filter set (Chroma 49008) for visualizing mCherry fluorescence.
3. Dichroic mirrors compatible with fluorophores and filter sets used.
4. Incubation chamber for microscope stage, where required.

2.5 Software for Image Processing

1. Proprietary software often preinstalled on microscopy controller, commercial software Like Adobe Photoshop, or open-source solutions like ImageJ, a widely used and adaptable open-source image-processing program [19].

3 Methods

Because the protocols for propagation of bacteria and induction of secretion systems vary greatly, we aim to provide a general protocol that can be adapted to the specificities of the studied secretion system.

All buffers and solutions should be prepared using ultrapure water at room temperature.

3.1 Preparation of Bacteria and Setup of Microscopy Equipment

1. Streak bacteria from −80 °C stock onto LB agar plates containing the required additives and antibiotics and incubate at the required temperature (e.g., 37 °C) until single colonies are visible (usually 12–36 h).
2. Inoculate overnight cultures of bacteria with single colonies from agar plates and grow in shaking incubator at required temperature and agitation.
3. On the next day, determine the optical density at 600 nm wavelength (OD_{600}) of the overnight culture and inoculate a main culture to OD_{600} that is suitable for the expression of the analyzed secretion system (a culture volume of 5 mL is sufficient, and an OD_{600} of 0.1 is a good starting point for many systems).
4. Incubate bacteria in a shaking incubator until they reach the early stationary phase (usually 1–2 h).
5. Induce secretion system according to standard conditions (e.g., temperature shift, addition of inducer).
6. In the meantime, prepare a 1.5% solution of low-fluorescence agarose in microscopy buffer (*see* **Note 8**). While less than 100 μL of agarose solution is required per strain, a higher volume

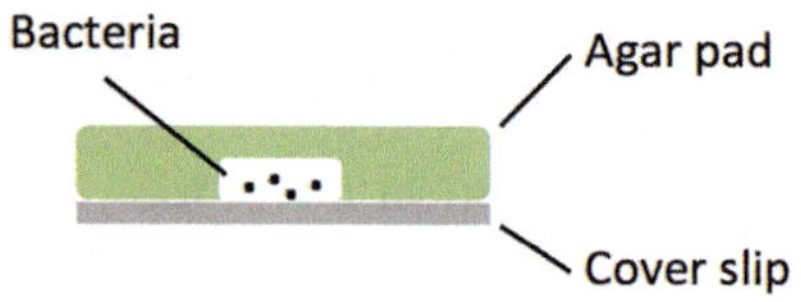

Fig. 1 Methods for the preparation of agarose pads. (**a**) Approximately 50 μL of agarose solution (*beige*) are transferred onto a microscope slide and quickly covered with a cover slip (*gray*), which is then gently and evenly pressed onto the agarose to form an evenly distributed patch. This simple method is sufficient for quick imaging but often leads to uneven or sloped agarose patches, which reduce image quality. (**b**) To ensure a more level surface of the agarose patch, the covering cover slip can be supported by two flanking cover slips. (**c**) Double-sided tape or commercially available adhesives (e.g. Gene Frame, Thermo Fisher) (rippled pattern) can be used to permanently adhere the cover slip to the sample. This prevents evaporation during imaging, which is very useful for longer experiments. However, the decreasing oxygenation of the sample should be kept in mind in this case

(20–50 mL) is easier to prepare, and the agarose concentration is less influenced by evaporation. Add the agarose to the buffer, which is then carefully brought to a boil in a microwave oven. Be aware of the possibility of delays in boiling and use precaution. Check that the agarose is completely dissolved and allow to cool to approximately 55 °C. After cooling, add any required additives (antibiotics, inducers) (*see* **Note 9**).

7. Prepare a thin pad of 1.5% agarose on a microscopy slide (Fig. 1) (*see* **Note 10**).
8. Harvest exponentially growing cells (OD_{600} ~ 0.8–1) by centrifugation (2400 × *g*, 4 min; these values depend on the bacterium). Resuspend in imaging buffer and recentrifuge once, using the same settings, before resuspending in imaging buffer to an OD_{600} ~ 2 (*see* **Note 11**).
9. Remove cover slip from agarose patch prior to spotting the bacteria, and wait until no more liquid areas are visible on the surface of the agarose patch (usually 1–5 min).
10. Spot bacteria by either of the following two methods:
    (a) Pipette 1–2 μL of resuspended bacteria into the center of the agarose or agar patch without damaging the patch itself, let dry for about 1–2 min (*see* **Note 12**), and carefully cover with a cover slip.
    (b) Spot 1–2 μL of resuspended bacteria onto a cover slip and carefully cover with an agarose or agar pad (Fig. 2).

3.2 Microscopy

1. Place a drop of immersion oil onto the center of the cover slip, and flip the slide so that the cover slip faces the objective. Carefully insert the slide into the microscope and bring the lens into contact with the immersion oil.

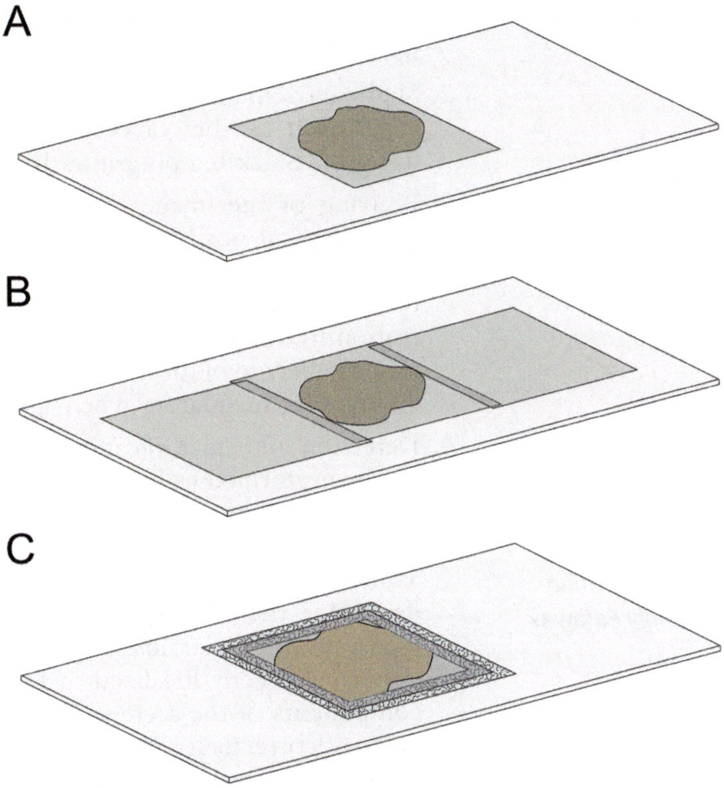

Fig. 2 Schematic representation of bacteria loaded onto cover slip and covered with agar pad

2. In phase contrast or differential interference contrast (DIC) mode, slowly decrease the distance between the lens and the cover slip until bacteria are visible (*see* **Note 12** in case large numbers of detached or swimming bacteria are visible).
3. Adjust Koehler illumination of microscope for optimal phase contrast or DIC images [20].
4. To determine which phase contrast/DIC plane corresponds to the best fluorescence plane, run an automated *z* stack of phase contrast/DIC and fluorescence images (*see* **Note 13**). For bacteria with a diameter of about 1 μm, *z* stacks containing 10–20 planes with Δz = 100 nm yield sufficient coverage.
5. Capture phase contrast and fluorescence micrographs:
   (a) For kinetic studies: Every 30 s with a minimal exposure time to minimize bleaching and phototoxicity effects (*see* **Note 14**).
   (b) For the determination of the assembly pathway: Single micrograph or *z* stack image (*see* **Note 15**) in wild-type and mutant strains.

3.3 Image Processing

1. Phase contrast and fluorescence images can be adjusted and merged using ImageJ or equivalent software (*see* **Note 16**).
2. Slight movements of the whole field during the time of the experiment can be corrected by registering individual frames using the StackReg plug-in in ImageJ [19].
3. Blurring of the image can be reduced by deconvolution, a mathematical postimaging process that removes or reassigns the fraction of detected photons caused by out-of-focus structures. This is especially useful when three-dimensional information from *z* stacks is available. Care should be taken not to mistake deconvolution artifacts for clustering, and negative controls are mandatory when applying deconvolution.
4. Detection and quantification of bacteria and fluorescent foci can be performed in ImageJ or using the Oufti software package (formerly known as MicrobeTracker [21]).

3.4 Determination of Assembly Pathway

1. To build an assembly pathway from the aforementioned data, determine the presence and number of foci per cell in wild-type and mutant strains. Nucleating component(s) of the system are correctly localized in foci in strains lacking all other components of the secretion system. Subsequently assembled proteins correctly localize in foci in strains expressing all earlier assembled proteins. Finally, proteins that require the presence of all the other components of the system are recruited at the end of the assembly. Proteins that are recruited at the same time or that interact before being recruited to the system will show the same localization in mutant strains and usually require each other to form foci.

4 Notes

Notes 1–4 correspond to the prerequisites for the applicability of this approach.

1. Most bacterial secretion systems have a sufficiently distinct localization within the bacterium that differs from the distribution of the free component. Should this not be the case, standard fluorescence microscopy cannot distinguish between the free and assembled state of a labeled component, and more sophisticated methods, such as diffusion-based fluorescence correlation spectroscopy (FCS) or interaction-based Förster resonance energy transfer microscopy (FRET), must be applied.
2. While most fluorescent proteins are relatively inert to interactions, their size of 25–30 kDa [22] can lead to cleavage or degradation of the fusion protein and impede the assembly or

functionality of the tagged protein. Smaller alternatives, such as tetracysteine tags [23, 24], require additional manipulation [25] and may also disturb the functioning of the protein (own unpublished observations). It is therefore essential to test the expression level and stability of the fusion protein (by immunoblot), as well as the functionality of the secretion system in the corresponding strains (by a functional assay). While fusions that influence functioning may be perfectly fine tools for deciphering the assembly, this must be corroborated by independent experiments. To maximize the chances of obtaining a functional fusion protein, both termini of the protein as well as internal flexible loops should be considered. Flexible linkers between the fluorophore and the secretion system component (e.g., a stretch of 6–15 amino acids with a high glycine content) have been shown to preserve the functionality of the fusion protein. Chromosomal fusion proteins are preferred to avoid mislocalization owing to overproduction of the protein or wrong timing and order of the expression of subunits. Moreover, chromosomal fusions make it possible to analyze the secretion system under close-to-wild-type conditions. However, especially for C-terminal fusions, care must be taken not to disturb the expression of downstream genes in the same operon, and it has proven helpful to repeat the genetic region upstream of the following gene.

3. Concerning the choice of fluorescent protein, many variants of GFP have been produced that vary in spectral properties, degrees of multimerization, folding rates, and functionality in oxidizing environments, so it can be helpful to try different fusion proteins [22]. Owing to their fast folding, low multimerization tendency, and proper folding in the periplasm, sfGFP and mCherry are good starting points. mCherry has additionally been observed to retain functionality of the T3SS in cases where GFP fusions were nonfunctional ([26] and unpublished results). Most bacteria also display considerably less autofluorescence in the red spectrum; however, mCherry is less photostable than GFP, which might make it less suitable for time course studies.

4. While these control strains are not absolutely required, especially in the presence of good controls for the protein of choice itself (i.e., deletion of a protein required for its localization), they are immensely valuable for setting up and testing the microscopy pipeline.

5. M9 and similar buffers have the advantage that, owing to their low autofluorescence, they can be directly used in microscopy. This ensures constant external conditions for the bacteria and may eliminate the need for the washing step described in **step 8**.

6. The choice of imaging buffer is crucial to obtain reproducible results because it will influence bacterial metabolism and possibly the state of the secretion system to be analyzed. While phosphate buffered saline (PBS) is a popular and easy-to-obtain imaging buffer, some bacteria show visible alterations in cell morphology in PBS within less than an hour. Preliminary experiments can reveal whether cell morphology and the distribution of secretion systems are affected in different imaging buffers.
7. The microscope must have a sufficient resolution and, most importantly, a high sensitivity to visualize and resolve the assembled proteins. A 100× objective is required for most distributions of secretion systems, although the formation of a polar spot or the distribution of few membrane-bound foci in large bacteria can be detected with a 60× objective. The sensitivity of the microscope is crucial, especially for low-stoichiometry components. The labeled protein must be present in multiple copies within the complex to be detectable. In our experience, sensitive wide-field microscopes can detect approximately ten molecules within a diffraction-limited spot over low background. For single-molecule detection, more sensitive methods, such as total internal reflection microscopy (TIRF) [27] or photoactivated localization microscopy/stochastic optical reconstruction microscopy (PALM/STORM) [28], must be applied.
8. The pad can also be done with agar (instead of agarose). This can be especially useful for longer time course experiments, where bacteria can be incubated for 1 h at optimal growth temperature prior to microscopy acquisition to allow cell division on a plane surface.
9. The agarose solution should be prepared or redissolved freshly before the experiment. The solution will stay liquid in a 55 °C water bath; small aliquots can be kept in a tabletop incubator shaker for 1.7 mL reaction tubes (vigorous shaking is required to prevent solidification of the agarose in this case).
10. The depth of the pad can vary; however, the surface should remain as smooth as possible. The pad can be prepared using a microscopy slide and a cover slip, with spacers, using commercially available systems (e.g. GeneFrame) (Fig. 1) or, alternatively, using two microscope slides or prewarmed protein gel chambers for larger patches. The pad should be bubble free to facilitate observation. Let the agarose solidify and dry at room temperature (>1 min; longer storage times are possible if the pad remains covered).
11. An OD of 2 leads to about 5% of the area being covered with bacteria (for *E. coli*; this obviously depends on the size of the bacterium). Increasing the OD will increase confluence, leading to more cell–cell contact.

12. Depending on their surface and the properties of the agarose patch, bacteria may take some time to settle at this point. If a large part of the bacteria are still moving after some minutes, the volume of bacterial resuspension should be reduced and drying times increased. Do not dry the agarose pad at 4 °C to avoid drifts during observation.
13. To avoid saturation, strong photobleaching, or phototoxicity effects, all fluorescence images should be acquired with the minimal exposure time required to reach a sufficient signal-to-noise ratio. The optimal exposure time must be determined for each protein; depending on the sensitivity of the microscope, exposure times of 20–100 ms for phase contrast or DIC and 100 ms to 2 s for fluorophores are good starting points. Narrow-band microscopy filters can also reduce photobleaching.
14. For time-lapse experiments, many microscopy systems allow one to define fields of view (x, y, z, focus offset) that are stored and then automatically accessed by a motorized stage. The fields of view should be sufficiently far apart to avoid cross-photobleaching (run a preliminary experiment with long exposures to determine the area of bleaching, if required). Ten fields of view at an OD of about 2 usually yield a sufficient number of bacteria for further analysis.
15. *Z* stacks allow a more complete coverage of the bacterium, ensuring images that comprise the region of interest (often the center of the bacterium). In addition, the three-dimensional data yield information about the spatial distribution of the secretion systems within the bacterium and allow better deconvolution of the images. However, imaging z stacks leads to stronger photobleaching and is therefore generally avoided in kinetic experiments. For kinetic experiments, keeping the focus is of particular importance, and hardware-based focusing systems or sealed plates can be advantageous.
16. To avoid the loss of raw data, the original image files should be preserved. Keep the low and high boundary below the background value and above the highest measured intensity, respectively, to prevent misinterpretation of the data. Within an experiment, these values should be kept constant after background correction.

References

1. Tseng T-T, Tyler B, Setubal J (2009) Protein secretion systems in bacterial-host associations, and their description in the Gene Ontology. BMC Microbiol 9:S2
2. Costa TRD, Felisberto-Rodrigues C, Meir A, Prevost MS, Redzej A, Trokter M, Waksman G (2015) Secretion systems in Gram-negative bacteria: structural and mechanistic insights. Nat Rev Microbiol 13:343–359
3. Kimbrough TG, Miller SI (2000) Contribution of Salmonella typhimurium type III secretion components to needle complex formation. Proc Natl Acad Sci U S A 97:11008–11013

4. Sukhan A, Kubori T, Wilson J, Galán JE (2001) Genetic analysis of assembly of the Salmonella enterica serovar Typhimurium type III secretion-associated needle complex. J Bacteriol 183:1159–1167
5. Kimbrough TG, Miller SI (2002) Assembly of the type III secretion needle complex of Salmonella typhimurium. Microbes Infect 4:75–82
6. Ogino T, Ohno R, Sekiya K, Kuwae A, Matsuzawa T, Nonaka T, Fukuda H, Imajoh-Ohmi S, Abe A (2006) Assembly of the type III secretion apparatus of enteropathogenic *Escherichia coli*. J Bacteriol 188:2801–2811
7. Fronzes R, Schäfer E, Wang L, Saibil H, Orlova E, Waksman G (2009) Structure of a type IV secretion system core complex. Science 323:266–268
8. Schraidt O, Lefebre MD, Brunner MJ, Schmied WH, Schmidt A, Radics J, Mechtler K, Galán JE, Marlovits TC (2010) Topology and organization of the Salmonella typhimurium type III secretion needle complex components. PLoS Pathog 6:e1000824
9. Reichow SL, Korotkov KV, Hol WGJ, Gonen T (2010) Structure of the cholera toxin secretion channel in its closed state. Nat Struct Mol Biol 17:1226–1232
10. Chandran Darbari V, Waksman G (2015) Structural biology of bacterial type IV secretion systems. Annu Rev Biochem 84:603–629
11. Rose P, Fröbel J, Graumann PL, Müller M (2013) Substrate-dependent assembly of the Tat translocase as observed in live *Escherichia coli* cells. PLoS One 8:e69488
12. Alcock F, Baker MAB, Greene NP, Palmer T, Wallace MI, Berks BC (2013) Live cell imaging shows reversible assembly of the TatA component of the twin-arginine protein transport system. Proc Natl Acad Sci U S A 110:3650–3659
13. Lybarger S, Johnson TL, Gray M, Sikora A, Sandkvist M (2009) Docking and assembly of the type II secretion complex of Vibrio cholerae. J Bacteriol 191:3149–3161
14. Johnson TL, Sikora AE, Zielke RA, Sandkvist M (2013) Fluorescence microscopy and proteomics to investigate subcellular localization, assembly, and function of the type II secretion system. Methods Mol Biol 966:157–172
15. Diepold A, Amstutz M, Abel S, Sorg I, Jenal U, Cornelis GR (2010) Deciphering the assembly of the Yersinia type III secretion injectisome. EMBO J 29:1928–1940
16. Diepold A, Wiesand U, Cornelis GR (2011) The assembly of the export apparatus (YscR,S,T,U,V) of the Yersinia type III secretion apparatus occurs independently of other structural components and involves the formation of an YscV oligomer. Mol Microbiol 82:502–514
17. Aguilar J, Zupan J, Cameron TA, Zambryski PC (2010) Agrobacterium type IV secretion system and its substrates form helical arrays around the circumference of virulence-induced cells. Proc Natl Acad Sci U S A 107:3758–3763
18. Durand E, Nguyen VS, Zoued A, Logger L, Péhau-Arnaudet G, Aschtgen M-S, Spinelli S, Desmyter A, Bardiaux B, Dujeancourt A, Roussel A, Cambillau C, Cascales E, Fronzes R (2015) Biogenesis and structure of a type VI secretion membrane core complex. Nature 523:555–560
19. Schneider CA, Rasband WS, Eliceiri KW (2012) NIH Image to ImageJ: 25 years of image analysis. Nat Methods 9:671–675
20. Köhler A (1893) Ein neues Beleuchtungsverfahren für mikrophotographische Zwecke. Z Wiss Mikrosk 10:433–440
21. Paintdakhi A, Parry B, Campos M, Irnov I, Elf J, Surovtsev I, Jacobs-Wagner C (2015) Oufti: an integrated software package for high-accuracy, high-throughput quantitative microscopy analysis. Mol Microbiol 99:767–777
22. Shaner NC, Steinbach PA, Tsien RY (2005) A guide to choosing fluorescent proteins. Nat Methods 2:905–909
23. Adams S, Campbell R, Gross L, Martin B, Walkup G, Yao Y, Llopis J, Tsien RY (2002) New biarsenical ligands and tetracysteine motifs for protein labeling in vitro and in vivo: synthesis and biological applications. J Am Chem Soc 124:6063–6076
24. Andresen M, Schmitz-Salue R, Jakobs S (2004) Short tetracysteine tags to beta-tubulin demonstrate the significance of small labels for live cell imaging. Mol Biol Cell 15:5616–5622
25. Enninga J, Mounier J, Sansonetti P, Tran Van Nhieu G, Van Nhieu GT (2005) Secretion of type III effectors into host cells in real time. Nat Methods 2:959–965
26. Diepold A, Kudryashev M, Delalez NJ, Berry RM, Armitage JP (2015) Composition, formation, and regulation of the cytosolic C-ring, a dynamic component of the type III secretion injectisome. PLoS Biol 13:e1002039
27. Poulter NS, Pitkeathly WTE, Smith PJ, Rappoport JZ (2015) In: Verveer PJ (ed) Advanced fluorescence microscopy. Springer, New York
28. MacDonald L, Baldini G, Storrie B (2015) Does super-resolution fluorescence microscopy obsolete previous microscopic approaches to protein co-localization? Methods Mol Biol 1270:255–275

Chapter 24

Large Complexes: Cloning Strategy, Production, and Purification

Eric Durand and Roland Lloubes

Abstract

Membrane proteins can assemble and form complexes in the cell envelope. In Gram-negative bacteria, a number of multiprotein complexes, including secretion systems, efflux pumps, molecular motors, and pilus assembly machines, comprise proteins from the inner and outer membranes. Besides the structures of isolated soluble domains, only a few atomic structures of these assembled molecular machines have been elucidated. To better understand the function and to solve the structure of protein complexes, it is thus necessary to design dedicated production and purification processes. Here we present cloning procedures to overproduce membrane proteins into *Escherichia coli* cells and describe the cloning and purification strategy for the Type VI secretion TssJLM membrane complex.

Key words Membrane protein complexes, *Escherichia coli*, T7 overexpression, Protein purification

1 Introduction

Protein overproduction results from cloning a gene of interest into a plasmid vector, downstream of a tightly regulated promoter, and from inducing its expression after plasmid transformation into a bacterial strain. For large protein complexes containing multiple subunits, the genes encoding the different subunits can be expressed under the control of an inducible promoter either from a single plasmid containing a cluster of genes or from different compatible plasmids harboring single or multiple genes.

1.1 Cloning Vectors

Several inducible promoters have been described and are available to overexpress a gene of interest. These promoters are usually cloned into vectors that also contain the gene encoding the cognate regulatory protein and a transcriptional terminator to prevent non-productive transcription from the downstream gene [1–7]. The *tac* and *trc* promoters that contain the −35 and −10 sequences from the *trp* and *lacUV5* promoters, respectively, have been optimized for high expression levels [1]. The most famous regulated *E. coli*

Laure Journet and Eric Cascales (eds.), *Bacterial Protein Secretion Systems: Methods and Protocols*, Methods in Molecular Biology, vol. 1615, DOI 10.1007/978-1-4939-7033-9_24, © Springer Science+Business Media LLC 2017

promoters are the tetA and araBAD promoters, regulated by the TetR repressor and the AraC activator, respectively [3–6]. In addition, heterologous combinations of promoter/regulator binding sequences can be used [2]. The last family of inducible promoters gathers sequences that are not recognized by the *E. coli* RNA polymerase but are rather recognized by phage RNA polymerases such as the SP6, T3, and T7 promoters. Vectors and strains that express the T7 RNA polymerase (T7RNAp) have been extensively developed. Three independent methods are used to regulate the T7 expression systems. First, the expression of the chromosomally encoded or plasmid-encoded T7 RNAp gene can be itself under the control of an inducible promoter (*see* **Note 1**). Second, the inhibition of the T7 RNAp basal activity can be controlled by producing the T7 lysozyme under constitutive or regulated conditions [8, 9]. Finally, the transcription by the T7 RNAp can be repressed by the Lac repressor, adding the *lac* operator sequence [10], the *lacI* gene being cloned on the expression vector.

Many cloning vectors are available and can be selected according to the cellular location, toxicity, stability and folding rate of the proteins to overproduce:

- Production of proteins in the cytoplasm, the periplasm, or the membrane of *E. coli* (using the addition of synthetic N-terminal signal sequences, such as that of the OmpA and PelB proteins);
- Compatible T7 expression vectors containing one or two T7 promoters to overproduce protein complexes with up to eight subunits (see Duet vectors from Novagen);
- Protein tagging sequences (6–10×His, *Strep*-Tag II…) or fusion partners (protein G, glutathione-S-transferase (GST), calmodulin-binding peptide (CBP), maltose-binding protein (MBP)) to increase protein solubility or to simplify purification using affinity chromatography techniques. In addition, these tag affinity sequences can be removed using specific proteases (the most commonly used proteases are Tobacco Etch Virus (TEV) protease, enterokinase, Thrombin, Xa Factor, PreScission). For this purpose, the corresponding protease recognition sequence is inserted either downstream (for N-terminal tagging) or upstream (for C-terminal tagging) of the tag sequence. After protease digestion, the purified protein complex is subjected to a new purification step to remove the protease, the uncleaved protein, and the tag peptide (e.g., affinity chromatography to remove the His-tag peptide and the His-tagged TEV).

Currently, cloning into these vectors is facilitated by polymerase chain reaction (PCR) techniques based on gene and plasmid amplifications, leading to restriction site/ligation-free cloning methods. These techniques have proven efficient to carry out rapid gene expression strategies. (*see* Subheading 3.1 and **Note 2**).

If none of the overproduction strategies is sufficient to produce and purify protein complexes, in vitro alternatives, such as cell-free transcription–translation systems using the T7RNAp and *E. coli* cell extracts, have been found efficient at producing milligram-range amounts of some membrane proteins [11].

1.2 Membrane Protein Complex Overproduction and Purification

Membrane protein complexes have been successfully overproduced and purified based on pBAD or T7 vectors. For example, the Tol proteins from the Tol cell envelope molecular motor were overproduced and specifically radiolabeled with ^{35}S–Methionine after cloning into a pT7-driven vector using restriction sites (RS) and ligation techniques (RSl) ([12, 13], *see* **Notes 3** and **4**). Subcomplexes from the Ton system [14–17], the PomA-PomB flagellar rotor [18], the AcrAB-TolC efflux pump [19], the Type IV and Type VI secretion systems (T4SS and T6SS, respectively) [20, 21], and the β-barrel assembly machinery [22] have also been successfully purified using similar approaches.

In addition to the cloning strategy and the production to levels compatible with purification, studying these multiprotein machines also requires extracting and solubilizing the protein complex without disrupting the contacts between the subunits. Finally, since the overproduction condition may induce stoichiometry artifacts, the tagging of the minor subunit or the specific tagging of different subunits should be performed to purify stable protein complexes (see subsequent discussion and Subheading 3.2).

Here we describe the cloning strategy as well as the production, extraction, and purification of the T6SS TssJLM membrane complex [21].

2 Materials

2.1 Cloning of TssJLM Complex

1. Desalted oligonucleotides of up to 35 bases, 5′ ends unphosphorylated.
2. Restriction enzymes (NdeI, XhoI, DpnI).
3. T4 DNA ligase.
4. Thermoblock (16–42 °C).
5. *E. coli* DH5α competent cells.
6. PCR thermocycler.
7. High-performance liquid chromatography purified Megaprimer pairs (≥50 bases), 5′ ends unphosphorylated.
8. Vector: pRSF-Duet1.
9. High-fidelity Taq DNA polymerase (Pfu turbo, Agilent).
10. dNTPs, 10 mM stock solution in water.
11. Agarose gel electrophoresis (AGE) system.

2.2 Production and Purification of TssJLM Complex

1. Isopropyl β-D-1-thiogalactopyranoside (IPTG) stock solution: 0.1 M in water.
2. BL21(DE3): *E. coli* strain B F⁻ *ompT gal dcm lon hsdSB(rB⁻mB⁻)* λ(DE3 [*lacI lacUV5-T7 gene 1 ind1 sam7 nin5*]) [*malB*⁺] (λ^S).
3. Lysozyme stock solution: 10 mg/mL in water.
4. DNase stock solution: 10 mg/mL in water.
5. Tris(2-carboxyethyl)phosphine hydrochloride (TCEP) extemporaneously added.
6. Ethylenediaminetetraacetic acid (EDTA) stock solution: 0.5 M in water.
7. $MgCl_2$ stock solution: 1 M in water.
8. Lysis buffer: 50 mM Tris–HCl, pH 8.0, 50 mM NaCl, 1 mM EDTA.
9. Solubilization buffer: 50 mM Tris–HCl, pH 8.0, 50 mM NaCl, 1 mM EDTA, 0.5% (w/v) n-dodecyl-β-D-maltopyranoside (DDM), 0.75% (w/v) decyl maltose neopentyl glycol (DM-NPG), 0.5% (w/v) digitonin (Sigma-Aldrich).
10. EDTA-free protease inhibitor tablets.
11. Affinity buffer: 50 mM Tris–HCl, pH 8.0, 50 mM NaCl, 0.05% (w/v) DM-NPG.
12. Imidazole-HCl, pH 8.0: Stock solution 4 M.
13. Incubation shakers (16–37 °C).
14. Centrifugation (from 5000 to 100,000 × *g*).
15. Glass potter.
16. Emulsiflex-C5 (Avestin).
17. 5-mL StrepTrap HP and 5-mL HisTrap HP columns (GE Healthcare).
18. Desthiobiotin.

3 Methods

3.1 Cloning TssJLM Membrane Complex

To overproduce the T6SS TssJLM membrane complex, the *tssJ*, *tssL*, and *tssM* genes were assembled in an artificial operon with individual optimized ribosome-binding sites (RBS). In addition, based on previous data regarding permissive positions, each subunit was tagged with a specific tag (*see* Fig. 1).

The pRSF–$TssJ^{Strep}$-$TssL^{FLAG_His6}$TssM plasmid was constructed by both RSI (*see* **Note 4**) and RSI-free (*see* **Note 2**, [23]) cloning methods (Fig. 1).

1. PCR amplify the *tssJ* gene (encoding the TssJ lipoprotein) using high-fidelity Taq DNA polymerase and primers containing the NdeI (5′ primer) and the XhoI RS (3′ primer) extensions.

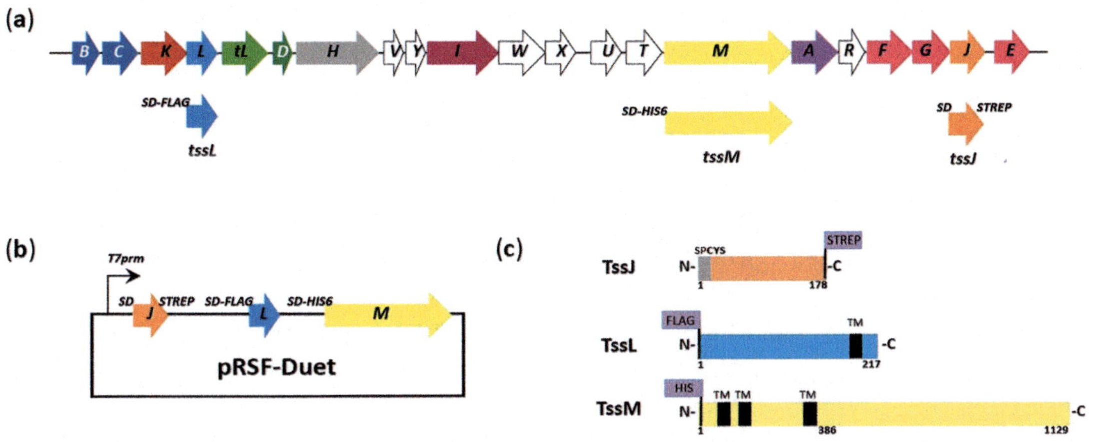

Fig. 1 Cloning of synthetic *tssJLM* operon encoding the T6SS membrane core complex. (**a**) The *tssL*, *tssM*, and *tssJ* genes were PCR amplified from the entero-aggregative *sci-1* T6SS operon. DNA sequences were added encoding an optimized RBS at the 5′ end of each gene and sequences encoding a C-terminal Strep-II tag (STREP) for TssJ and N-terminal 6×His and Flag tags for TssM and TssL, respectively. (**b**) The *tssJLM* artificial operon was cloned in the pRSF-Duet vector. (**c**) Schemes representing the protein constructs, the subdomain boundaries, and some indicated characteristics (*TM* transmembrane segments, *SP* signal peptide, and *CYS* acylated cysteine)

The *Strep*-tag II DNA sequence followed by a stop codon was introduced in the 3′ primer (*see* **Note 5**). Check the correct amplification by AGE.

2. Digest PCR product and pRSF-Duet vector with NdeI and XhoI restriction enzymes (sites present in pRSF-Duet MCS2).
3. Mix the digested PCR amplified fragment with about 50 ng of the digested pRSF-Duet vector, with a molar ratio of insert/vector of 2/1 to 5/1.
4. Add 1 U of T4 DNA ligase to the DNA mixture plus its specific buffer in a total volume of 15 μL and incubate at 16–20 °C for at least 2 h.
5. Transform *E. coli* DH5α competent cells with about 30% of the ligation mix, plate on lysogeny broth (LB) agar containing the appropriate antibiotic (in case of pRSF-Duet, use kanamycin 50 μg/mL), and incubate overnight at 37 °C.
6. Select positive clones containing the pRSF–TssJ^{Strep} plasmid by colony PCR screening using the same primers. Extract the plasmid DNA and check the accuracy of the cloned sequence by DNA sequencing.
7. PCR amplify the *tssL* gene using (1) a 5′ primer containing a 5′ extension (22 bp) corresponding to the *Strep*-TagII and the stop codon followed by the ribosome binding site (RBS) sequence (*see* **Note 6**) and the 5′-FLAG-*tssL* extension (63 bp), and (2) a 3′ primer containing the 3′-*tssL* gene extension

(35 bp) including the stop codon followed by RBS, the ATG start codon, and 5′-His6-*tssM* extension (14 bp). Check correct amplification by AGE.

8. PCR amplify the *tssM* gene using (1) the 5′ primer complementary to the 3′ end of *tssL* gene, RBS, His-tag (20 bp), and 5′ end of *tssM* (20 bp) and (2) the 3′ primer containing the 3′-*tssM* gene extension (35 bp) including the stop codon followed by the pRSF-Duet sequence extension (35 bp). Check correct amplification by AGE.
9. Mix the *flag-tssL*, *his-tssM* PCR products and the pRSF–TssJ^{Strep} plasmid, and PCR amplify with high-fidelity DNA polymerase in a single reaction, as described for the RSl-free technique (*see* **Note 2**).
10. Digest the mixture with *DpnI*. Transform into *E. coli* DH5α competent cells and plate on LB agar containing the appropriate antibiotic and incubate overnight at 37 °C.
11. Select positive clones containing the pRSF–TssJ^{Strep}-^{FLAG}TssL-^{6His}TssM plasmid by colony PCR screening using the same primers. Extract the plasmid DNA and check the accuracy of the cloned sequence by DNA sequencing.

3.2 TssJLM Membrane Complex, Extraction and Purification

1. Transform the expression vector (pRSF-TssJ^{Strep}-^{FLAG}TssL-^{6His}TssM) into the *E. coli* BL21(DE3) expression strain.
2. Grow cells at 37 °C in 8 L of LB to an optical density at 600 nm (OD_{600}) ~ 0.7. Induce the expression of the *tssJLM* genes with 1.0 mM IPTG for 16 h at 16 °C (*see* **Note 7**).
3. Pellet cells by centrifugation at 7000 × *g* for 20 min. Resuspend cell pellets in 300 mL of ice-cold lysis buffer supplemented with 1 mM TCEP, 100 μg/mL DNase I, 100 μg/mL lysozyme, and one tablet of EDTA-free protease inhibitor. Add $MgCl_2$ to the final concentration of 10 mM.
4. Break the cell suspension with an Emulsiflex-C5 by four passages at 15,000 psi (100 MPa). Pellet unbroken cells by centrifugation at 7000 × *g* (*see* **Note 8**).
5. Pellet membranes by ultracentrifugation at 98,000 × *g* for 45 min.
6. Resuspend membranes in 120 mL of solubilization buffer supplemented with 1 mM TCEP at 22 °C and homogenize membranes mechanically with a potter (duration about 45 min) (*see* **Note 9**).
7. Clarify the membrane suspension by centrifugation at 98,000 × *g* for 20 min.
8. Load the supernatant onto a 5 mL StrepTrap HP column and then wash with affinity buffer at 4 °C.

9. Elute the TssJLM core complex in affinity buffer supplemented with 2.5 mM desthiobiotin into a 5 mL HisTrap HP column.
10. Wash the HisTrap HP column in affinity buffer supplemented with 20 mM imidazole and proceed to the elution of the TssJLM core complex in the same buffer supplemented with 500 mM imidazole.
11. Pool the peak fractions and load onto a Superose 6 10/300 column equilibrated in 50 mM Tris–HCl, pH 8.0, 50 mM NaCl, 0.025% (w/v) DM-NPG (*see* **Note 10**). Elute the TssJLM complex as a single monodisperse peak close to the void volume of the column.

4 Notes

1. *E. coli* T7RNAp expression systems result from (1) plasmid (pGP1–2) expression: the T7RNAp under the control of the Lambda P_L promoter is regulated by the plasmid-encoded temperature-sensitive C1–857 repressor [24]; (2) chromosomal expression: the T7RNAp gene is under the control of the *lacUV5* [25] or the *ara* promoter [26]; (3) besides tight control of the chromosomal T7RNAp expression upon AraC control [26], infection by phages (M13 mGP1–2 or Lambda CE6) encoding the T7RNAp [27, 28] has proven to be efficient for the expression of toxic gene products.
2. The RSl-free cloning strategy allows the insertion of a gene at a precise position on a target plasmid but does not require the presence of RS. The restriction free cloning method consists in two sequential PCR amplifications using only two primers. These primers (with ≥25 bp homology extension) are designed to hybridize on the 5′ and 3′ ends of the gene of interest and contain additional extension to hybridize on the expression vector at a selected position downstream of the regulated promoter sequence. First, the gene of interest is amplified by PCR, then the amplified mega-primer pairs containing the gene are annealed to the vector of interest. A new amplification with high-fidelity polymerase produces a linear gene-vector amplification [29]. The gene is thus plasmid included in a nicked and circular DNA molecule. The PCR is treated by *DpnI* restriction enzyme to digest the unwanted Dam methylated plasmid templates. Competent *E. coli* cells are further transformed with the annealed DNA complex directly. An alternative method has been developed [30]. It uses a PCR-amplified gene containing at least 15 bp extensions that are homologous to each end of the linearized vector (PCR-amplified or RS-digested). PCR DNA and vector treated with T4 polymerase, to create 5′ overhangs, are annealed to form a recombinant plasmid.

3. Addition of a T7 promoter by an RS-dependent cloning strategy. RS present in the upstream sequence of the gene of interest can be used to insert a synthetic T7 promoter DNA fragment that is formed by two overlapping oligonucleotides containing the T7 consensus promoter sequence of 23 bases corresponding to 5′-taatacgactcactatagggaga-3′. The T7 promoter sequence is inserted upstream of the RBS of the gene of interest (it is also possible to insert the T7 promoter upstream of the natural promoter sequence). For this purpose, the synthetic DNA fragment contains additional 5′- and 3′-end extensions that are complementary to the sticky ends of the RS present in the plasmid (blunt-ended RSs can also be used but with lower ligation efficiency and random insertion). Two complementary desalted oligonucleotides (ONs) of about 35 bases, 5′ end unphosphorylated, are hybridized following heat denaturation in ultrapure water and further cooling at room temperature. The annealed DNA fragment is further ligated into an RS-digested plasmid. It is possible to favor the selection of positive clones using synthetic DNA that do not recreate the initial RS. Then the heat-inactivated ligation mix is digested with the RE corresponding to the RS that was destroyed upon DNA fragment ligation (e.g., DNA fragment insertion into EcoRI RS: after EcoRI digestion, the overhang EcoRI: 5′-AATTC... should be filled with the synthetic sequence 5′-AATTX..., where the X nucleotide does not correspond to the C nucleotide). It is notable that this fast technique does not require plasmid sequencing. It should be used for plasmids harboring resistance and regulatory genes in the opposite orientation from that of T7 regulated genes [31].
4. The general RSl-dependent strategy is often used for cloning genes of interest into the multiple cloning sites (MCS) of expression vectors. The genes can be either purified from RS digestion [32] or obtained from PCR amplification using primers that contain additional RS extensions. Then the PCR-amplified DNA is RE digested and inserted into the cognate RS of the MCS present in the vector [16].
5. Since lipoproteins undergo a post-translational modification at their N-termini [33], the *Strep*-Tag II affinity tag sequence was introduced at the C-terminal of TssJ. It is important to note that the positions of affinity tags (Strep, FLAG, and His tags) were rationally chosen to maintain functional proteins.
6. To optimize the production of the TssL and TssM proteins, the endogenous RBS-ATG 5′ sequence in front of the genes was replaced (in the 5′ primer sequence) by the consensual sequence "**AAGGAG**ATATACAT*ATG*" [34] (RBS and start codon are presented, bold and italic letters, respectively).
7. Growth and induction at 37 °C for 3 h led to very low biomass and protein yield. The culture was conducted at 37 °C until

OD_{600} ~ 0.7–0.9 before shifting the incubator at 16 °C prior to induction by IPTG. The final OD after the 16H induction was around 1.4–1.6. It is important not to induce around 0.4 OD since the cell would stop growing shortly thereafter.

8. It is important to note that whereas many cell-disrupting protocols have been tried, only lysis with Emulsiflex produced a stable and homogeneous sample of the TssJLM membrane core complex.
9. This specific combination of detergents (same recipe as that used for the purification of the T4SS [20]) gave a higher extraction yield of the TssJLM membrane complex. Other isolated detergents (Triton-X-100, n-Octyl-β-D-Glucoside, DDM, and DM-NPG) gave poor extraction yields. All detergent buffers were prepared the day of the purification. Digitonin tends to precipitate in high salt buffers and without other detergents. The addition of DDM and DM-NPG prevents digitonin precipitation.
10. DM-NPG was chosen because the solubilized TssJLM complex is stable and since this detergent has been used successfully for high-resolution structural biology studies, giving a very clear and reproducible electron microscopy background after negative staining. Using a low salt concentration (50 mM maximum) was key to preventing the aggregation of the TssJLM complex. The affinity purification steps were of paramount importance for isolating a stoichiometric complex. Indeed, a large excess of the TssJ lipoprotein was eliminated during the second His column.

Acknowledgements

We would like to thank M. Petiti, L. Houot, H. Célia, and D. Duché for their careful reading of the text. E.D. and R.L. are funded by the Centre National de la Recherche Scientifique, the Aix-Marseille Université, and two grants from the Agence Nationale de la Recherche (ANR-10-JCJC-1303-03 and ANR-14-CE09-0023, respectively). ED is supported by the Institut National de la Santé Et de la Recherche Médicale through a permanent research position.

References

1. Amann E, Brosius J, Ptashne M (1983) Vectors bearing a hybrid trp-lac promoter useful for regulated expression of cloned genes in *Escherichia coli*. Gene 25:167–178
2. Lutz R, Bujard H (1997) Independent and tight regulation of transcriptional units in *Escherichia coli* via the LacR/O, the TetR/O and AraC/I1-I2 regulatory elements. Nucleic Acids Res 25:1203–1210
3. Cagnon C, Valverde V, Masson JM (1991) A new family of sugar-inducible expression vectors for *Escherichia coli*. Protein Eng 4:843–847
4. Skerra A (1994) Use of the tetracycline promoter for the tightly regulated production of a murine antibody fragment in *Escherichia coli*. Gene 151:131–135
5. Guzman LM, Belin D, Carson MJ, Beckwith J (1995) Tight regulation, modulation, and

high-level expression by vectors containing the arabinose PBAD promoter. J Bacteriol 177: 4121–4130

6. Haldimann A, Daniels LL, Wanner BL (1998) Use of new methods for construction of tightly regulated arabinose and rhamnose promoter fusions in studies of the *Escherichia coli* phosphate regulon. J Bacteriol 180:1277–1286
7. Balzer S, Kucharova V, Megerle J, Lale R, Brautaset T, Valla S (2013) A comparative analysis of the properties of regulated promoter systems commonly used for recombinant gene expression in *Escherichia coli*. Microb Cell Fact 12:26
8. Studier FW (1991) Use of bacteriophage T7 lysozyme to improve an inducible T7 expression system. J Mol Biol 219:37–44
9. Schlegel S, Löfblom J, Lee C, Hjelm A, Klepsch M, Strous M, Drew D, Slotboom DJ, de Gier JW (2012) Optimizing membrane protein overexpression in the *Escherichia coli* strain Lemo21(DE3). J Mol Biol 423:648–659
10. Dubendorff JW, Studier FW (1991) Controlling basal expression in an inducible T7 expression system by blocking the target T7 promoter with lac repressor. J Mol Biol 219:45–59
11. Schwarz D, Junge F, Durst F, Frölich N, Schneider B, Reckel S, Sobhanifar S, Dötsch V, Bernhard F (2007) Preparative scale expression of membrane proteins in *Escherichia coli*-based continuous exchange cell-free systems. Nat Protoc 2:2945–2957
12. Guihard G, Boulanger P, Bénédetti H, Lloubes R, Besnard M, Letellier L (1994) Colicin A and the Tol proteins involved in its translocation are preferentially located in the contact sites between the inner and outer membranes of *Escherichia coli* cells. J Biol Chem 269:5874–5880
13. Cascales E, Lloubes R, Sturgis JN (2001) The TolQ-TolR proteins energize TolA and share homologies with the flagellar motor proteins MotA-MotB. Mol Microbiol 42:795–807
14. Celia H, Noinaj N, Zakharov SD, Bordignon E, Botos I, Santamaria M, Barnard TJ, Cramer WA, Lloubes R, Buchanan SK (2016) Structural insight into the role of the Ton complex in energy transduction. Nature 538:60–65
15. Pramanik A, Zhang F, Schwarz H, Schreiber F, Braun V (2010) ExbB protein in the cytoplasmic membrane of *Escherichia coli* forms a stable oligomer. Biochemistry 49:8721–8728
16. Pramanik A, Hauf W, Hoffmann J, Cernescu M, Brutschy B, Braun V (2011) Oligomeric structure of ExbB and ExbB-ExbD isolated from *Escherichia coli* as revealed by LILBID mass spectrometry. Biochemistry 50:8950–8956
17. Sverzhinsky A, Fabre L, Cottreau AL, Biot-Pelletier DM, Khalil S, Bostina M, Rouiller I, Coulton JW (2014) Coordinated rearrangements between cytoplasmic and periplasmic domains of the membrane protein complex ExbB-ExbD of *Escherichia coli*. Structure 22:791–797
18. Yonekura K, Maki-Yonekura S, Homma M (2011) Structure of the flagellar motor protein complex PomAB: implications for the torque-generating conformation. J Bacteriol 193: 3863–3870
19. Kim JS, Jeong H, Song S, Kim HY, Lee K, Hyun J, Ha NC (2015) Structure of the tripartite multidrug efflux pump AcrAB-TolC suggests an alternative assembly mode. Mol Cells 38:180–186
20. Low HH, Gubellini F, Rivera-Calzada A, Braun N, Connery S, Dujeancourt A, Lu F, Redzej A, Fronzes R, Orlova EV, Waksman G (2014) Structure of a type IV secretion system. Nature 508:550–553
21. Durand E, Nguyen VS, Zoued A, Logger L, Péhau-Arnaudet G, Aschtgen MS, Spinelli S, Desmyter A, Bardiaux B, Dujeancourt A, Roussel A, Cambillau C, Cascales E, Fronzes R (2015) Biogenesis and structure of a type VI secretion membrane core complex. Nature 523:555–560
22. Bakelar J, Buchanan SK, Noinaj N (2016) The structure of the β-barrel assembly machinery complex. Science 351:180–186
23. Unger T, Jacobovitch Y, Dantes A, Bernheim R, Peleg Y (2010) Applications of the Restriction Free (RF) cloning procedure for molecular manipulations and protein expression. J Struct Biol 172:34–44
24. Tabor S, Richardson CC (1985) A bacteriophage T7 RNA polymerase/promoter system for controlled exclusive expression of specific genes. Proc Natl Acad Sci U S A 82:1074–1078
25. Studier FW, Moffatt BA (1986) Use of bacteriophage T7 RNA polymerase to direct selective high-level expression of cloned genes. J Mol Biol 189:113–130
26. Narayanan A, Ridilla M, Yernool DA (2011) Restrained expression, a method to overproduce toxic membrane proteins by exploiting operator-repressor interactions. Protein Sci 20:51–61
27. Stuchlík S, Turna J (1998) Overexpression of the FNR protein of *Escherichia coli* with T7 expression system. Folia Microbiol (Praha) 43:601–604

28. Doherty AJ, Connolly BA, Worrall AF (1993) Overproduction of the toxic protein, bovine pancreatic DNaseI, in *Escherichia coli* using a tightly controlled T7-promoter-based vector. Gene 136:337–340
29. van den Ent F, Löwe J (2006) RF cloning: a restriction-free method for inserting target genes into plasmids. J Biochem Biophys Methods 67:67–74
30. Jeong J, Yim H, Ryu J, Lee H, Lee J, Seen D, Kang S (2012) One-step sequence- and ligation-independent cloning as a rapid and versatile cloning method for functional genomics studies. Appl Environ Microbiol 78: 5440–5443
31. Bénédetti H, Lazdunski C, Lloubes R (1991) Protein import into *Escherichia coli*: colicins A and E1 interact with a component of their translocation system. EMBO J 10:1989–1995
32. Derouiche R, Bénédetti H, Lazzaroni JC, Lazdunski C, Lloubes R (1995) Protein complex within *Escherichia coli* inner membrane. TolA N-terminal domain interacts with TolQ and TolR proteins. J Biol Chem 270: 11078–11084
33. Konovalova A, Silhavy TJ (2015) Outer membrane lipoprotein biogenesis: Lol is not the end. Philos Trans R Soc Lond B Biol Sci 370:1679
34. Ringquist S, Shinedling S, Barrick D, Green L, Binkley J, Stormo GD, Gold L (1992) Translation initiation in *Escherichia coli:* sequences within the ribosome-binding site. Mol Microbiol 6:1219–1229

Chapter 25

Shearing and Enrichment of Extracellular Type IV Pili

Alba Katiria Gonzalez Rivera and Katrina T. Forest

Abstract

Pili are widespread among bacteria. Type IVa pili (T4aP) are associated with a variety of bacterial functions, including adhesion, motility, natural transformation, biofilm formation, and force-dependent signaling. In pathogenic bacteria, T4aP play a crucial role during infection and have been the subject of hundreds of studies. Methods for the isolation and purification of T4aP were first described in the 1970s. Purified pili have been used for studies of filament protein content, morphology, immunogenicity, post-translational modifications, and X-ray crystallography. We detail a tried-and-true method of isolating large amounts of native T4aP from bacterial surfaces. The method requires supplies and equipment that are available in most microbiology labs.

Key words T4P, Fimbriae, Pili, Filament, Shear, Isolation

1 Introduction

Pili are hairlike appendages displayed by many pathogenic and environmental bacteria. One well-studied class is the Type IV pili (T4P). At approximately 6 nm in diameter, T4P filaments are thinner than flagella and can reach many micrometers in length [1]. They are expressed in some species at cell poles (e.g., in *Myxococcus xanthus* or *Pseudomonas aeruginosa*) [2, 3] and in others peritrichously (e.g., *in Neisseria gonorrhoeae or Deinococcus geothermalis*) [4, 5]. Found in diverse Gram-negative and Gram-positive species [6, 7], T4P are involved in various bacterial functions, including adherence, microcolony formation, biofilm initiation, and long-range electron transfer [8]. One distinctive feature of these filaments is the capacity of some to retract, a property that has been associated with phage sensitivity, force generation, twitching motility, natural transformation, and virulence [8].

T4P from Gram-negative organisms can be subdivided into two families; subtype a (T4aP) and subtype b (T4bP), with notable differences both in gene organization and biochemical properties. At the genome level, genes for T4aP subunits and assembly machinery are located in multiple operons dispersed around the

Laure Journet and Eric Cascales (eds.), *Bacterial Protein Secretion Systems: Methods and Protocols*, Methods in Molecular Biology, vol. 1615, DOI 10.1007/978-1-4939-7033-9_25, © Springer Science+Business Media LLC 2017

chromosome, whereas those for T4bP cluster in a single genomic region [9]. We describe a method for the isolation of T4aP that capitalizes on their properties in two kinds of buffers [10]. T4aP filaments can be isolated by disaggregation in high-pH, low salt buffer and aggregation (due to bundling) in buffer with near-neutral pH and physiological salt concentration. Cycles of these two steps can increase the purity of a preparation because contaminants that do not share these solubility properties are differentially lost. This method is not ideal for the more hydrophobic T4bP, which in some cases are successfully purified using an ammonium sulfate precipitation procedure [11].

Depending on the application, pili may be further purified. For example, to generate anti-pilus antibodies, it may be desirable to remove lipopolysaccharides (LPS). This can be accomplished by incubating the purified pili with polymixin B agarose. For high-resolution crystallographic studies of the full-length pilin monomer, dissociation of filaments in nondenaturing detergent followed by filtration is appropriate [12–15].

2 Materials

2.1 Growth and Harvest of Piliated Bacteria

1. Bacterial strain: *P. aeruginosa* PAK/2Pfs or PAK Δ*pilT* (*see* **Notes 1** and **2**).
2. 60 Tryptic soy agar (TSA) plates, 1.5% agar (*see* **Notes 3** and **4**).
3. Tryptic soy broth (TSB) (*see* **Note 5**).
4. 30 °C incubator (*see* **Note 6**).
5. Stereo microscope.
6. Glass spreader.
7. Inoculating turntable.
8. 50 mL disposable conical tube.
9. 70% v/v ethanol solution.
10. Bunsen burner.

2.2 Collection of Pili

1. High-speed floor model centrifuge and appropriate rotor (e.g., Beckman JA25.50).
2. Oak Ridge centrifuge tubes with O-ring screw caps (maximum relative centrifugal force (rcf) > 17,500 × *g*). These are needed to handle Biosafety Level-2 (BSL-2) organisms.
3. 250-mL beaker.
4. Magnetic stir bar and magnetic stir plate.
5. Parafilm.
6. 25 and 5 mL serological pipets.

7. Plastic transfer pipets.
8. Pilus disaggregation buffer (PDB): 1 mM dithiothreitol (DTT), 150 mM ethanolamine, pH 10.5 (4 °C). DTT must be added shortly before use.

2.3 Purification of Pili

1. Centrifuge tubes (maximum rcf > 20,000 × *g*). 50 mL disposable conical centrifuge tubes are convenient since at this stage, BSL-2 safety containment is not required. Standard Oak Ridge tubes are fine, too.
2. Pilus bundling buffer (PBB): 150 mM NaCl, 0.02% NaN_3, 50 mM Tris–HCl, pH 7.5 (4 °C).
3. Dialysis tank or large flask (at least 4 L).
4. Dialysis tubing (molecular weight cut-off >3.5 kDa, 29 mm diameter).

2.4 Assessment of Results

1. 16% Tricine-SDS-PAGE Gel (*see* **Note 7**).

3 Methods

3.1 Growth and Harvest of Piliated Bacteria (see Note 1)

1. To isolate single colonies of *P. aeruginosa*, streak bacteria from frozen stock stored at −80 °C onto a TSA plate. Incubate plate for 24 h at 30 °C.
2. Identify single colonies with round morphology and smooth margins using a stereo microscope (Fig. 1, and *see* **Note 8**). Streak four colonies onto fresh TSA plates. Incubate for 24 h at 30 °C.
3. Select a patch of cells with a sterile swab and resuspend in TSB to reach an optical density at a wavelength of 600 nm (OD_{600}) of 8.0. If at this stage there is already sufficient volume for

Fig. 1 Colony morphologies. *P. aeruginosa* strain K Δ*pilT* is plated on TSA plates. Piliated colonies have a smooth, domed appearance (*black arrow*). Avoid flat, spreading colonies (*white arrow*)

spreading 50 μL of this bacterial suspension onto 50 plates of TSA (that is, 2.5 mL), skip **steps 4–6**.

4. Grow a robust lawn of bacterial cells by inoculating 100 μL of a cell suspension of OD_{600} = 8.0 onto five TSA plates. Grow bacteria for 24 h at 30 °C.
5. Using a glass spreader and a turntable, remove bacterial lawns from all five TSA plates.
6. Transfer bacteria into a 50 mL conical tube with 5 mL TSB. Use a sterile transfer pipet to scrape cells from the glass spreader into the tube as well as to gently but thoroughly resuspend bacteria. Adjust TSB volume so that the suspension has an OD_{600} of 8.0.
7. Inoculate 50 μL of this cell suspension onto each of the 50 TSA plates and spread using a glass spreader and a turntable. Incubate plates for 24 h at 30 °C.
8. Collect bacterial lawns as just described using glass spreader and turntable. Ideally these lawns should be confluent and sticky; they will come off the plate in a goopy clump. We find it convenient to use the same glass spreader for approximately four plates before transferring bacteria into a 250 mL beaker with 25 mL ice-cold PDB. Use transfer pipet to remove bacteria from glass spreader and to resuspend cells in PDB. Keep adding PDB to suspended bacteria from all 50 plates. A total of 75 mL PDB should be sufficient. This will be suspension 0 (S0) (Fig. 2). During the following procedures, keep bacterial suspension and pilus suspension on ice at all times.
9. To dissociate large bacterial clumps, use a transfer pipet and then a 10 mL serological pipet to gently pull cells up and down to achieve a uniform suspension without large clumps. Be patient as bacterial resuspension will increase pilus yield; however, overly vigorous mixing will lyse cells and contaminate the final pilus preparation with other proteins.
10. After dissociating large bacterial clumps, add a magnetic stir bar and cover the beaker with parafilm. Then stir the bacterial suspension S0 at medium-low speed in cold room for 1 h.

3.2 Collection of Pili

1. Transfer samples to centrifuge tubes with O-ring screw caps. Fill only half of the centrifugation tube to allow good vortexing action.
2. Shear pili from cells by vortexing the cell suspension three times in 1 min bursts at maximum strength. Cool on ice for 2 min between vortexing steps. Collect sample for sodium dodecyl sulfate (SDS)-polyacrylamide gel electrophoresis (PAGE) analysis. In choosing a volume to load into gel, keep in mind this solution will have a high protein concentration.

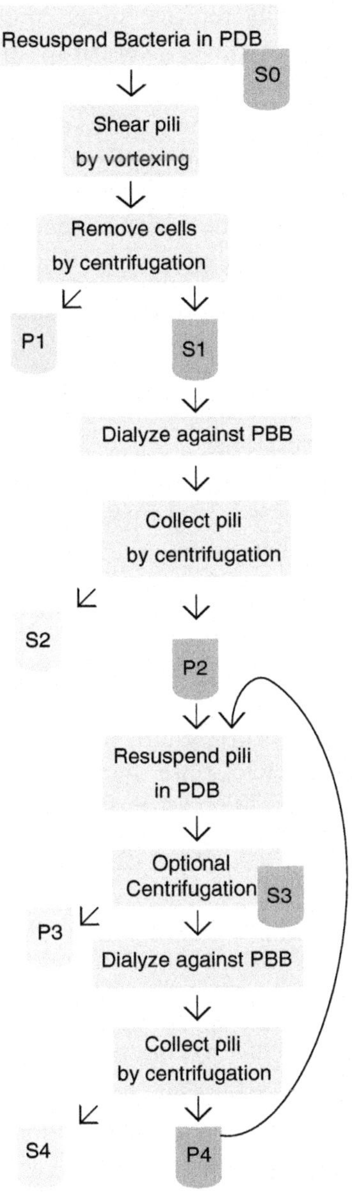

Fig. 2 Workflow for T4aP isolation. Bacteria are suspended in PDB and pili are sheared by vortexing (S0). Cell debris and bacteria are removed by centrifugation (P1). The remaining solubilized pili (S1) are aggregated by dialyzing against PBB and collected by centrifugation (P2). For purification purposes, pili in P2 are resuspended in PDB (S3) and aggregated by another dialysis against PBB (P4). Cycles of resuspension and aggregation can be repeated to improve pilus purity, although at a loss of overall yield

3. Sheared pili are now in suspension. To remove cells and cell debris, combine samples to fill and balance centrifuge tubes, and centrifuge samples at 15,000 × *g* for 20 min (*see* **Note 9**).

4. Carefully remove supernatant using a serological pipet and transfer to clean centrifuge tubes with O-ring screw caps. Do not disrupt pellet (P1).
5. Centrifuge the supernatant a second time to remove residual cells at 15,000 × *g* for 10 min (*see* **Note 9**). Transfer supernatant (S1) containing sheared pili to a clean disposable tube. Set aside a sample of the supernatant for analysis.
6. Prepare a 16% Tricine-SDS-PAGE gel following ref. 16.
7. Run samples to confirm successful separation of pili from cell fractions.

3.3 Purification of Pili

1. Prepare dialysis membrane according to manufacturer's instructions.
2. Load sheared pili sample S1 into dialysis tubing and dialyze against 4 L cold PBB for 4 h at 4 °C. Stir at low speed to facilitate buffer exchange.
3. Change dialysis buffer and repeat dialysis procedure until pH of sample reaches 7.5 (*see* **Note 10**).
4. Upon neutralization of pH during dialysis, the pilus filaments will aggregate by bundling. Carefully remove sample from dialysis tubing and transfer it to a centrifuge tube. For maximum yield, rinse the inside of the tubing with PBB and add to sample. Keep an aliquot for analysis.
5. To collect aggregated pili, centrifuge sample at 20,000 × *g* for 40 min.
6. Remove supernatant (S2) using serological pipet. Invert tubes and drain onto paper towel. At this step, the pellet (P2) contains sheared pili, and its appearance should be white. (If this pellet contains pink, it includes residual cells.) Resuspend pelleted pili (P2) in PDB, starting with ~3 mL. Add PDB as needed to make a nonviscous solution.
7. Optionally, centrifuge to remove contaminants not yet in suspension, which will pellet (P3) while pili remain in supernatant (S3).
8. Repeat dialysis procedure, now dialyzing S3 against PBB.
9. Collect purified pili by centrifugation; remove supernatant S4, and resuspend pellet (P4) in desired final buffer to reach appropriate concentration (e.g., 5 mL) (*see* **Notes 11** and **12**).

3.4 Assessment of Results

1. To monitor pilus purity, run pilus preparation samples in 16% Tricine-SDS-PAGE gel (Fig. 3a, Table 1) and visualize by Coomassie Brilliant Blue or silver staining. The latter also permits estimation of LPS contamination. Yields of ~0.2–0.5 mg of pili per plate at the S3 step can be achieved; although the overall yield of pilin falls with each cycle, the purity increases (Fig. 3a, Table 1).

Table 1
Purification yields at each step

	P3		S3		P4		S4	
Total protein (mg)	1	0.9	16[a]	30.5	10	11.7	6.4	12.6
Purity (%)	84	92	71	88	81	98	70	82
Pilin yield/loss (mg)	1	0.5	11	26.8	8	11.5	*4.5*	*14.5*

Approximate purity was estimated by image analysis of the Coomassie-stained gel in Fig. 3a using ImageJ [23]. Protein concentration was estimated using OD_{280} and a calculated extinction coefficient for pilin of 14,100/M/cm (as calculated using ExPasy ProtParam [24]). We present data for two representative examples of yield from the protocol described here. The left-hand columns are based on a purification by a relatively inexperienced researcher using 44 plates with volumes exactly as described in protocol (and shown in Fig. 3a). The right-hand columns are based on a purification by a more practiced investigator from 32 plates with correspondingly reduced volumes

[a]Note in this case the S3 concentration was estimated from the dialysate after S3 dialysis against PBB rather than in the initial S3. Thus, this value is an underestimate of the true yield at the S3 stage

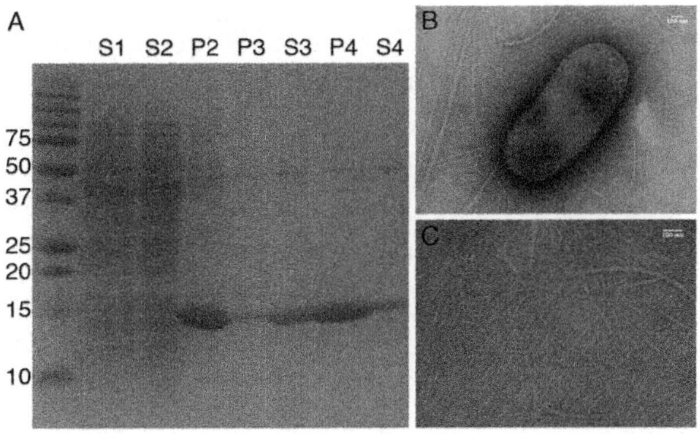

Fig. 3 Pili isolated from *P. aeruginosa* strain K Δ*pilT*. (**a**) Proteins were separated in 16% Tricine-SDS-PAGE gel (molecular mass markers are in kDa). (**b**) Whole cells and (**c**) purified pili (plus contaminating flagella) analyzed by negative staining and transmission electron microscopy (scale bars: 100 nm)

2. Negative staining can be used for the visualization of pilus filaments using transmission electron microscopy (Fig. 3b, c).

4 Notes

1. *P. aeruginosa* must be handled in a BSL-2 laboratory. These procedures will produce aerosols that could contain pathogens. This method will yield high amounts of biohazardous waste. All steps here should be carried out under BSL-2 conditions with appropriate personal protection and caution against aerosol release of bacteria.
2. PAK/2Pfs is a strain with nonretractile pili [17] caused by a mutation in the *pilT* gene [18]. PAK/2Pfs is available from

ATCC as strain 53,308 and is also known as PAK2.2 [19]. Strains that lack the PilT pilus retraction motor express higher amounts of pili than retraction-proficient strains and thus are commonly used for pilus purification, with yields up to 10 times higher than wild-type strains [17, 20]. We have achieved similar results with a PAK derivative carrying an in-frame deletion in the *pilT* gene (a gift of Dr. Stephen Lory, Harvard Medical School). In this chapter, data are presented for the PAK Δ*pilT* strain.

3. Prepare TSA following manufacturer's instructions or by mixing 15 g tryptone, 5 g soytone, 5 g NaCl, and 15 g agar per liter of purified water. Autoclave to sterilize. Pour 20–22 mL into 100 × 15 mm Petri dishes. Allow plates to dry (with lids on) at room temperature for 2–3 days.
4. In our experiments, TSA or LB in 1.5% agar plates yield equivalent amounts of pili based on assessment by SDS-PAGE.
5. Prepare TSB media as directed by manufacturer or dissolve and sterilize 17 g tryptone, 3 g soytone, 2.5 g D-glucose, 5.0 g NaCl, and 2.5 g dipotassium hydrogen phosphate per liter of purified water.
6. Avoid incubating plates in high humidity. If applicable, leave out water tray.
7. To prepare Tricine-SDS-PAGE gel follow method of Schagger [16]. We favor this recipe for separating low-molecular-weight proteins; however, any standard SDS-PAGE protocol will work.
8. This is true for the isolation of *P. aeruginosa* piliated but non-twitching mutant strains. Colony morphology for piliated strains varies among bacterial strains and species [21, 22].
9. The centrifugal force used for these cell-pelleting steps should be high enough to pellet cells but not high enough to lyse cells or pellet disaggregated pilus filaments. 15,000 × *g* is a good guide; however, lower rcf (8–10,000 × *g*) for a longer time (30 min) may be sufficient. It is worthwhile to follow pilin protein by SDS-PAGE to ensure pili are not spinning down with cells.
10. For best success, actually check the pH of the solution inside the dialysis bag. The lowering of salt together with pH neutralization is the fundamental basis of the method, and we have found that failure to reach equilibrium at this stage leads to very low pilus yield.
11. To achieve the desired purity, repeat resuspension and aggregation cycles. One modification to this procedure is at the last step, where the cold solution is brought to a final concentration

of 20% ammonium sulfate and stirred for 2 h to precipitate pili while leaving contaminating flagellar filaments in solution [20].

12. Depending on the application, one can remove residual LPS using polymixin B agarose beads (our unpublished result).

Acknowledgements

We are grateful to Dr. Nicole Koropatkin for purifying grams of pili and optimizing this protocol in the process, and to Dr. Lisa Craig for many years of collegial interactions and helpful suggestions for this chapter.

References

1. Hansen JK, Forest KT (2006) Type IV pilin structures: insights on shared architecture, fiber assembly, receptor binding and type II secretion. J Mol Microbiol Biotechnol 11: 192–207
2. MacRae TH, Dobson WJ, McCurdy HD (1977) Fimbriation in gliding bacteria. Can J Microbiol 23:1096–1108
3. Henrichsen J, Blom J (1975) Examination of fimbriation of some gram-negative rods with and without twitching and gliding motility. Acta Pathol Microbiol Scand B 83:161–170
4. Swanson J (1973) Studies on gonococcus infection. IV. Pili: their role in attachment of gonococci to tissue culture cells. J Exp Med 137:571–589
5. Saarimaa C, Peltola M, Raulio M, Neu TR, Salkinoja-Salonen MS, Neubauer P (2006) Characterization of adhesion threads of Deinococcus geothermalis as type IV pili. J Bacteriol 188:7016–7021
6. Imam S, Chen Z, Roos DS, Pohlschroder M (2011) Identification of surprisingly diverse type IV pili, across a broad range of gram-positive bacteria. PLoS One 6:e28919
7. Melville S, Craig L (2013) Type IV pili in Gram-positive bacteria. Microbiol Mol Biol Rev 77:323–341
8. Berry JL, Pelicic V (2015) Exceptionally widespread nanomachines composed of type IV pilins: the prokaryotic Swiss Army knives. FEMS Microbiol Rev 39:134–154
9. Pelicic V (2008) Type IV pili: e pluribus unum? Mol Microbiol 68:827–837
10. Brinton CC, Bryan J, Dillon J-A, Guerina N, Jen Jacobson L, Labik A, Lee S, McMichael J, Polen S, Rogers K, ACC T, SCM T (1978) Uses of pili in Gonorrhea control: role of bacterial pili in disease, purification and properties of Gonococcal pilus vaccine for Gonorrhea. In: Brooks GF (ed) Immunobiology of *Neisseria gonorrhoeae*: proceedings of a conference held in San Francisco, CA. American Society for Microbiology, Washington, DC
11. Li J, Lim MS, Li S, Brock M, Pique ME, Woods VL Jr, Craig L (2008) Vibrio cholerae toxin-coregulated pilus structure analyzed by hydrogen/deuterium exchange mass spectrometry. Structure 16:137–148
12. Forest KT, Dunham SA, Koomey M, Tainer JA (1999) Crystallographic structure reveals phosphorylated pilin from Neisseria: phosphoserine sites modify type IV pilus surface chemistry and fibre morphology. Mol Microbiol 31:743–752
13. Parge HE, Bernstein SL, Deal CD, McRee DE, Christensen D, Capozza MA, Kays BW, Fieser TM, Draper D, So M (1990) Biochemical purification and crystallographic characterization of the fiber-forming protein pilin from Neisseria gonorrhoeae. J Biol Chem 265: 2278–2285
14. Parge HE, Forest KT, Hickey MJ, Christensen DA, Getzoff ED, Tainer JA (1995) Structure of the fibre-forming protein pilin at 2.6 Å resolution. Nature 378:32–38
15. Craig L, Taylor RK, Pique ME, Adair BD, Arvai AS, Singh M, Lloyd SJ, Shin DS, Getzoff ED, Yeager M, Forest KT, Tainer JA (2003) Type IV pilin structure and assembly: X-ray and EM analyses of Vibrio cholerae toxin-coregulated pilus and Pseudomonas aeruginosa PAK pilin. Mol Cell 11:1139–1150
16. Schagger H (2006) Tricine-SDS-PAGE. Nat Protoc 1:16–22
17. Bradley DE (1974) The adsorption of Pseudomonas aeruginosa pilus-dependent bac-

teriophages to a host mutant with nonretractile pili. Virology 58:149–163

18. Whitchurch CB, Hobbs M, Livingston SP, Krishnapillai V, Mattick JS (1991) Characterisation of a Pseudomonas aeruginosa twitching motility gene and evidence for a specialised protein export system widespread in eubacteria. Gene 101:33–44
19. Bradley DE (1980) A function of Pseudomonas aeruginosa PAO polar pili: twitching motility. Can J Microbiol 26:146–154
20. Frost LS, Paranchych W (1977) Composition and molecular weight of pili purified from Pseudomonas aeruginosa K. J Bacteriol 131:259–269
21. Han X, Kennan RM, Davies JK, Reddacliff LA, Dhungyel OP, Whittington RJ, Turnbull L, Whitchurch CB, Rood JI (2008) Twitching motility is essential for virulence in Dichelobacter nodosus. J Bacteriol 190:3323–3335
22. Meng Y, Li Y, Galvani CD, Hao G, Turner JN, Burr TJ, Hoch HC (2005) Upstream migration of Xylella fastidiosa via pilus-driven twitching motility. J Bacteriol 187:5560–5567
23. Schneider CA, Rasband WS, Eliceiri KW (2012) NIH Image to ImageJ: 25 years of image analysis. Nat Methods 9:671–675
24. Artimo P, Jonnalagedda M, Arnold K, Baratin D, Csardi G, de Castro E, Duvaud S, Flegel V, Fortier A, Gasteiger E, Grosdidier A, Hernandez C, Ioannidis V, Kuznetsov D, Liechti R, Moretti S, Mostaguir K, Redaschi N, Rossier G, Xenarios I, Stockinger H (2012) ExPASy: SIB bioinformatics resource portal. Nucleic Acids Res 40:W597–W603

Chapter 26

Blue Native PAGE Analysis of Bacterial Secretion Complexes

Susann Zilkenat, Tobias Dietsche, Julia V. Monjarás Feria, Claudia E. Torres-Vargas, Mehari Tesfazgi Mebrhatu, and Samuel Wagner

Abstract

Bacterial protein secretion systems serve to translocate substrate proteins across up to three biological membranes, a task accomplished by hydrophobic, membrane-spanning macromolecular complexes. The overexpression, purification, and biochemical characterization of these complexes is often difficult, impeding progress in understanding the structure and function of these systems. Blue native (BN) polyacrylamide gel electrophoresis (PAGE) allows for the investigation of these transmembrane complexes right from their originating membranes, without the need for long preparative steps, and is amenable to the parallel characterization of a number of samples under near-native conditions. Here we present protocols for sample preparation, one-dimensional BN PAGE and two-dimensional BN/sodium dodecyl sulfate (SDS)-PAGE, as well as for downstream analysis by staining, immunoblotting, and mass spectrometry on the example of the type III secretion system encoded on *Salmonella* pathogenicity island 1.

Key words Bacterial secretion systems, Membrane proteins, Blue native polyacrylamide gel electrophoresis, Two-dimensional polyacrylamide gel electrophoresis, Sucrose gradients, Bacterial cell fractionation, Type III secretion systems, *Salmonella typhimurium*, *Escherichia coli*

1 Introduction

Blue native (BN) polyacrylamide gel electrophoresis (PAGE) is an electrophoretic method originally developed by Schägger and von Jagow to investigate the composition of protein complexes of the respiratory chain from isolated mitochondrial membranes [1]. Since its introduction, it has been widely adopted to characterize individual membrane protein complexes like the bacterial Sec and Tat translocons [2, 3] or mitochondrial and chloroplast import complexes [4, 5], as well as to assess the compositions of global complexomes in wild-type conditions and upon perturbation [6–10]. Lately it has also been recognized as a suitable tool for the elucidation of the composition and assembly of bacterial protein secretion systems: T4SS [11, 12], T3SS [13–17], and T7SS [18].

Laure Journet and Eric Cascales (eds.), *Bacterial Protein Secretion Systems: Methods and Protocols*, Methods in Molecular Biology, vol. 1615, DOI 10.1007/978-1-4939-7033-9_26,

BN PAGE relies on the mild extraction of membrane proteins and membrane protein complexes by nonionic detergents, which preserve the native protein conformation [19]. Charging and electrophoretic migration of the extracted proteins is facilitated by the adsorption of the anionic, water-soluble blue dye coomassie G to the hydrophobic regions of the extracted membrane proteins, which are then separated based on their complex size in gradient gels of a Tricine-based PAGE.

Here we explain the use of BN PAGE for the characterization of bacterial secretion systems on the example of the needle complex of the *Salmonella typhimurium* type III secretion system encoded on *Salmonella* pathogenicity island 1, a >180 component complex of 4.5 MDa [17, 20, 21]. We provide protocols for the preparation of samples from whole bacterial cells, crude and purified membranes, and immunoprecipitated material, for BN PAGE–based separation of complexes using one-dimensional BN PAGE or two-dimensional BN/ sodium dodecyl sulfate (SDS)-PAGE, and for the detection and analysis of separated complexes by coomassie and silver staining, immunoblotting, and mass spectrometry (MS). The individual protocols are provided as modules that can be combined freely according to individual needs.

2 Materials

2.1 Sample Preparation

2.1.1 Sample Preparation General Materials

1. Buffer K: 50 mM triethanolamin (TEA), 250 mM sucrose, 1 mM ethylenediaminetetraaceticacid (EDTA) (*see* **Note 1**), pH 7.5 (adjusted with acetic acid (HAc)). Store at 4 °C.
2. Lysozyme solution: 10 mg/mL in deionized distilled water (ddH_2O). Store aliquots of 100 μL at −20 °C.
3. Phosphate buffered saline (PBS): 137 mM NaCl, 2.7 mM KCl, 4.3 mM Na_2HPO_4, 1.47 mM KH_2PO_4, pH 7.4, adjusted with 1 M NaOH.
4. DNase solution: 10 mg/mL in 1× PBS. Store aliquots of 100 μL at −20 °C.
5. 1 M $MgSO_4$ in ddH_2O. Store at room temperature.
6. Complete protease inhibitor cocktail (without EDTA).
7. ACA750: 750 mM aminocaproic acid in ddH_2O.
8. Liquid nitrogen.
9. 10% (w/v) *n*-dodecyl-β-D-maltoside (DDM).
10. BN loading buffer: 5% Serva Blue G (*see* **Notes 2** and **3**), 250 mM aminocaproic acid, 25% glycerol in ddH_2O.

2.1.2 Extraction of Membrane Proteins from Crude Bacterial Membrane Preparations

1. Common material from Subheading 2.1.
2. Glass beads, acid washed, 150–212 μm.

Table 1
Sucrose solutions for two gradients

Percentage of sucrose (%) (w/w)	Sucrose (g)	Buffer 2× M (mL)	57% sucrose solution (mL)	Buffer 1× M (mL)
57	22.8	17.2		
32			12.8	10.0

Percentage of sucrose (%) (w/w)	Sucrose (g)	Buffer 2× M (mL)	55% sucrose solution (mL)	Buffer 1× M (mL)
55	19.25	13.5		
50			4.55	0.45
45			4.1	0.91
40			3.6	1.36
35			3.2	1.8
30			2.7	2.3

3. Homogenizer, e.g., SpeedMill Plus (Analytik Jena) or FastPrep-24 (MPBio).
4. Ultracentrifuge.

2.1.3 Membrane Fractionation by Sucrose Density Gradient Centrifugation

1. Common material from Subheadiing 2.1.
2. Buffer 2× M: 100 mM TEA, 2 mM EDTA (*see* **Note 1**), pH 7.5 adjusted with acetic acid (HAc).
3. Sucrose solutions (*see* **Notes 4** and **5**) for two gradients in 14 × 89 mm tubes; *see* Table 1.
4. Buffer L: 50 mM TEA, 250 mM sucrose (*see* **Note 1**), pH 7.5 (adjusted with HAc). Store at 4 °C.
5. French press.
6. SW 41 Ti rotor (Beckman) for ultracentrifuge.
7. Dounce homogenizer.
8. Seton 14 × 89 mm open-top polyclear ultracentrifuge tubes for SW 41 Ti rotor (*see* **Notes 6–8**) (optional; *see* Subheading 3.1.4).
9. Syringes and hypodermic needles (optional; *see* Subheading 3.1.4).
10. Gradient station (Biocomp Instruments), including accessories (optional; *see* Subheading 3.1.4).
11. Bicinchoninic acid (BCA) protein assay.

2.1.4 Immunoprecipitation of Membrane Protein Complexes

1. Anti-FLAG M2 affinity gel (*see* **Note 9**).
2. 3× FLAG peptide (*see* **Note 9**).
3. Rotating wheel.

2.2 Blue Native PAGE

2.2.1 One-Dimensional BN PAGE Using Precast Mini Gels

1. Precast native polyacrylamide gradient gels, e.g., Novex® NativePAGE™ Bis–Tris or SERVAGel™ N gel systems.
2. 10× BN anode buffer: 500 mM Bis-Tris-HCl, pH 7.0. Store at 4 °C.
3. 10× BN cathode buffer A: 500 mM Tricine, 150 mM Bis–Tris, 0.2% (w/v) Serva Blue G. Do not adjust pH and store at 4 °C.
4. 10× BN cathode buffer B: 500 mM Tricine, 150 mM Bis–Tris. Do not adjust pH and store at 4 °C.
5. NativeMark™ Unstained Protein Standard (Thermo Fisher).

2.2.2 Two-Dimensional BN/SDS PAGE

1. Hoefer SE600 or SE660 vertical electrophoresis unit.
2. GelBond PAG film (Lonza).
3. Double-sided Scotch tape (Tesa Photo® Film) (*see* **Note 10**).
4. Gel seal.
5. 3× BN gel buffer: 1.5 M ACA750, 150 mM Bis–Tris–HCl, pH 7.0. Store at 4 °C
6. Acrylamide 30% T, 3% C.
7. Ammonium persulfate (APS): 10% (w/v) in ddH_2O. Store at −20 °C.
8. *N*,*N*,*N*,*N*′-tetramethyl-ethylenediamine (TEMED). Store at 4 °C.
9. 80% (v/v) glycerol in ddH_2O. Autoclave and store at room temperature.
10. Gradient maker, e.g., Hoefer SG30 or SG50.
11. Peristaltic pump including tubing, e.g., GE Healtcare P-1.
12. 10× BN anode buffer: 500 mM Bis–Tris–HCl, pH 7.0. Store at 4 °C.
13. 10× BN cathode buffer A: 500 mM Tricine, 150 mM Bis–Tris, 0.2% (w/v) Serva Blue G. Do not adjust pH and store at 4 °C.
14. 10× BN cathode buffer B: 500 mM Tricine, 150 mM Bis–Tris. Do not adjust pH and store at 4 °C.
15. NativeMark™ Unstained Protein Standard (Thermo Fisher).
16. Acrylamide 30% T, 2.6% C.
17. 10% (w/v) SDS solution in ddH_2O. Store at room temperature.
18. SDS stacking gel buffer: 0.5 M Tris–HCl, pH 6.8.
19. SDS resolving gel buffer: 1.5 M Tris–HCl, pH 8.8.

20. 10× SDS-PAGE running buffer: 0.25 M Tris, 1.92 M glycine, 1% (w/v) SDS.
21. SDS equilibration buffer: 2% (w/v) SDS, 1% (v/v) β-mercaptoethanol, 50 mM Tris–HCl, pH 6.8.
22. Low melting agarose solution: 1% (w/v) low melting agarose, 0.5% (w/v) SDS, a few grains of bromphenol blue, 50 mM Tris–HCl, pH 6.8.
23. Razor blades, scissors.
24. Filter paper.
25. Protein standard.

2.3 Protein Detection, and Analysis

2.3.1 Colloidal Coomassie Staining

1. Fixing solution: 50% (v/v) ethanol, 3% (w/v) phosphoric acid in ddH_2O.
2. Serva Blue G (*see* **Note 2**).
3. ddH_2O.
4. Neuhoff's solution: 16% (w/v) ammonium sulfate, 25% (v/v) methanol, 5% (v/v) phosphoric acid, fill up to 100% with ddH_2O.
5. Storage solution: 5% (v/v) glacial acetic acid in ddH_2O.

2.3.2 Silver Staining

Because silver staining is extremely sensitive to trace impurities in water, it is strongly recommended to use high-quality ddH_2O in all recipes and protocol steps.

1. Fixing solution: 45% (v/v) methanol and 5% (v/v) glacial acetic acid in ddH_2O.
2. Sensitizing solution: 0.02% (w/v) sodium thiosulfate in ddH_2O. Always prepare fresh.
3. Silver nitrate solution: 0.1% (w/v) silver nitrate in ddH_2O. Always prepare fresh.
4. Developing solution: 2% (w/v) sodium carbonate and 0.04% (v/v) formalin in ddH_2O. Prepare this solution the day of use and add formalin to the developer at most 1 h before use.
5. Stop solution: 1% (v/v) glacial acetic acid in ddH_2O.
6. Plastic containers: Polyethylene trays are recommended (*see* **Note 11**).
7. Rocking table.
8. Special waste container for silver nitrate waste.

2.3.3 Immunoblotting Using Dual-Color Detection

1. 10× SDS-PAGE running buffer: 0.25 M Tris base, 1.92 M glycine, 1% (w/v) SDS.
2. 10× transfer buffer: 0.25 M Tris base, 1.92 M glycine, 0.05% SDS.
3. Polyvinylidene fluoride (PVDF) membrane.
4. Wet blot unit.

5. 100% methanol.
6. 10× Tris-buffered saline (TBS): 200 mM Tris-HCl, 1.5 M NaCl, pH 8.0.
7. TBS-T: TBS/0.05% Tween 20.
8. 0.1% Ponceau S, 5% acetic acid in ddH_2O.
9. Primary antibodies.
10. Secondary DyLight antibodies (Thermo Fisher).
11. LiCor Odyssey infrared scanner.
12. Image Studio software (LiCor) or Quantity One software (Bio-Rad).

2.3.4 Preparation of BN PAGE-Separated Complexes for Analysis by Mass Spectrometry

1. 5% HAc (v/v).
2. 100% ethanol.
3. Powder-free gloves (newly opened box).
4. Light table.
5. Razor blades.
6. Tweezers.

3 Methods

Methods are divided into sample preparation (Subheading 3.1), BN PAGE (Subheading 3.2), and protein detection and analysis (Subheading 3.3). Modules within these divisions can be combined as needed to yield a complete protocol. Each module contains a suggestion for suitable methodological combinations.

3.1 Sample Preparation

3.1.1 Extraction of Membrane Proteins from Whole Bacterial Cells

This protocol serves to extract membrane protein complexes from small amounts of whole bacterial cells. The extracted proteins are best run on BN mini gels and analyzed by immunoblotting as described previously [13].

1. All steps are to be performed at 4 °C or on ice unless otherwise stated.
2. Harvest 0.4–0.6 optical density units (ODU, *see* **Note 12**) of bacterial cells by centrifugation for 2 min at 5000 × *g* and 4 °C.
3. Aspirate supernatant.
4. Optional: Snap freeze cell pellet in liquid nitrogen and store cell pellet at −80 °C (*see* **Note 13**). Thaw pellet on ice to continue.
5. Resuspend bacterial cell pellet in 10 μL freshly prepared buffer K, supplemented with protease inhibitor cocktail, 10 μg/mL lysozyme, 10 μg/mL DNase, and 1 mM $MgSO_4$ (*see* **Note 14**).
6. Incubate for 30 min on ice to allow for cell wall digestion.

7. Add 70 μL buffer ACA750 and mix (*see* **Note 15**).
8. Freeze-thaw three times in liquid nitrogen 1 min/20 °C water 1 min.
9. Add 10 μL freshly prepared 10% DDM and mix to allow for extraction of membrane proteins (*see* **Note 16**).
10. Incubate for 1 h on ice with occasional mixing. Alternatively, place into shaker for 1.5 mL reaction tubes, cooled down to 4 °C, and mix at 1000 rpm.
11. Spin sample for 20 min at 20,000 × *g* and 4 °C to pellet unsolubilized material (*see* **Note 17**).
12. Transfer 45 μL of supernatant to new 1.5 mL tube containing 5 μL BN loading buffer and mix.
13. Load 25 μL of the suspension per well of a BN mini gel (*see* Subheading 3.2.1).

3.1.2 Extraction of Membrane Proteins from Crude Bacterial Membrane Preparations

This protocol describes the preparation of small amounts of crude bacterial membranes and the extraction of membrane protein complexes thereof (*see* **Note 18**). The extracted proteins are best run on BN mini gels and analyzed by immunoblotting as described previously [16]; however, we have also had good experience with the analysis of these preparations by two-dimensional BN/SDS-PAGE.

1. All steps are to be performed at 4 °C or on ice unless otherwise stated.
2. Harvest 2–10 ODU of bacterial cells by centrifugation for 2 min at 5000 × *g* and 4 °C (*see* **Notes 19** and **20**).
3. Aspirate supernatant.
4. Optional: Snap freeze cell pellet in liquid nitrogen and store cell pellet at −80 °C (*see* **Note 13**). Thaw pellet on ice to continue.
5. Resuspend bacterial cell pellet in 750 μL freshly prepared buffer K supplemented with protease inhibitor cocktail, 10 μg/mL lysozyme,10 μg/mL DNase (*see* **Note 14**).
6. Incubate for 30 min on ice to allow for cell wall digestion.
7. Meanwhile, prepare 2 mL screw-cap tubes with 500 μL glass beads. Label tubes on lid and on sides. Place tubes on ice for cooling (*see* **Notes 21** and **22**).
8. Add 0.8 μL 1 M $MgSO_4$ to each sample (final concentration).
9. Transfer cell suspension to a bead-containing tube and close properly.
10. Break up cells by bead milling (2 min continuous mode using SpeedMill Plus, 2 × 20 s at 4 m/s using FastPrep-24). For cooling of samples, cool down the SpeedMill Plus sample holder to −20 °C or place the samples on ice for 5 min after first 20 s cycle when using the FastPrep-24.

11. Pellet glass beads by centrifugation for 1 min at 1000 × *g* and 4 °C.
12. Transfer supernatant to a fresh 1.5 mL tube. Take care to pipet carefully and transfer as few glass beads as possible in this step.
13. Add 1 mL buffer K to glass beads and mix vigorously to wash off remaining material.
14. Pellet glass beads again by centrifugation for 1 min at 1000 × *g* and 4 °C.
15. Transfer supernatant to 1.5 mL tube containing supernatant of step 12. Take care to pipet carefully and transfer as few glass beads as possible in this step.
16. Pellet glass beads and cell debris by centrifugation for 10 min at 10,000 × *g* and 4 °C.
17. Transfer 1.3 mL of supernatant to 1.5 mL ultracentrifugation tubes (*see* **Note 23**). Take care to avoid the transfer of glass beads or cell debris.
18. Balance tubes to the same weight using buffer K for subsequent ultracentrifugation.
19. Pellet crude membranes by ultracentrifugation for 45 min at 55,000 rpm (135,520 × *g*) and 4 °C (Beckman TLA-55 rotor).
20. Discard supernatant.
21. Optional: Store membrane pellet at −80 °C. Thaw pellet on ice to continue.
22. Resuspend membrane pellet in ACA750 (*see* **Note 15**) by carefully pipetting up and down 40 times using 100–200 μL pipette tips. 8 μL buffer per ODU harvested bacterial cells is recommended, e.g., 24 μL for 3 ODU bacteria.
23. Add 1 μL freshly prepared 10% DDM per ODU harvested bacteria (3 μL for 3 ODU) to allow for extraction of membrane proteins (*see* **Note 16**).
24. Incubate for 1 h on ice with occasional mixing. Alternatively, place into shaker for 1.5 mL reaction tubes, cooled down to 4 °C, and mix at 1000 rpm.
25. Spin sample for 20 min at 20,000 × *g* and 4 °C to pellet unsolubilized material (*see* **Note 17**).
26. Transfer 18 μL supernatant to new 1.5 mL tube containing 2 μL BN loading buffer and mix.
27. Load 10–20 μL of the suspension per well of a BN mini gel (*see* **Note 24**). Larger amounts of up to 50 μL must be used for the analysis of membrane protein complexes by two-dimensional BN/SDS-PAGE (*see* dedicated protocol in Subheading 3.2.2).

3.1.3 Preparation of Crude Membranes for Sucrose Density Gradient Centrifugation

This protocol describes the large-scale preparation of crude membranes that are used for further membrane fractionation using sucrose density gradient centrifugation described in Subheadings 3.1.4 and 3.1.5 (*see* **Note 25**).

1. All steps are to be performed at 4 °C or on ice unless otherwise stated.
2. Harvest 500–2000 ODU of bacterial cells by centrifugation for 15 min at 6000 × *g* and 4 °C, e.g., in Beckman JLA-8.1000 (*see* **Note 26**).
3. Pour off supernatant.
4. Resuspend bacterial pellet in 35 mL cold PBS.
5. Transfer suspension to 50 mL Falcon tube.
6. Pellet bacteria by centrifugation for 10 min at 6000 × *g* and 4 °C.
7. Resuspend bacterial pellet in 10–15 mL buffer K supplemented with protease inhibitor cocktail, 1 mM EDTA, 10 μg/mL DNase, and 10 μg/mL lysozyme (all final concentrations).
8. Pass the cell suspension twice through a French press at 124 MPa (high position, 1000–1100 units).
9. Add $MgSO_4$ to a final concentration of 1 mM to activate the DNase. Mix.
10. Pellet cell debris by centrifugation for 20 min at 24,000 × *g* and 4 °C (*see* **Notes 27** and **28**).
11. Transfer supernatant to suitable ultracentrifugation tubes or bottles, e.g., for Beckman Type 45 Ti. Fill up bottles with buffer K.
12. Pellet crude membranes by centrifugation for 45 min at 234,000 × *g* and 4 °C (45,000 rpm in Beckman Type 45 Ti).
13. Discard supernatant. Wipe off residual supernatant with lint-free tissue.
14. Resuspend crude membranes in 500 μL buffer 1× M. Use a cut 1 mL pipette tip to detach membrane from tube wall. Transfer crude suspension to 1 mL dounce homogenizer. Homogenize with loose piston 15 times.
15. Store crude membrane suspension on ice until further use and proceed to alternatives (*see* Subheading 3.1.4 or 3.1.5).

3.1.4 Membrane Fractionation by Sucrose Density Gradient Centrifugation Using a Biocomp Gradient Station

This protocol describes the preparation of separated inner and outer membranes of Gram-negative bacteria using sucrose density gradient centrifugation and extraction of membrane protein complexes thereof. Sucrose gradient formation and fractionation are described in two ways: either supported by a Biocomp gradient station (this section, *see* **Note 29**) or manually without the need for specialized equipment (*see* Subheading 3.1.5).

The extracted proteins of inner or outer membrane fractions are suitable for any of the described downstream analyses. We previously described the use of purified inner membranes for the analysis of complete membrane complexomes by two-dimensional BN/SDS-PAGE [6–10], one-dimensional BN PAGE followed by quantitative immunoblotting [13], and immunoprecipitation followed by one-dimensional BN PAGE or two-dimensional BN/SDS-PAGE[13–15].

1. Prepare Seton 14 × 89 mm open-top polyclear centrifuge tubes for SW 41 Ti rotor. Mark at "half-full, long cap," as described in the gradient station manual.
2. Place tubes in tube holder of gradient station. Fill up to 2–3 mm above half-full mark with the 32% sucrose solution using a syringe and the needle provided with the gradient station (Fig. 1a).
3. Carefully underlay with 57% sucrose solution using a syringe and the needle provided with the gradient station (*see* **Note 30**). Bring up 57% sucrose solution exactly to the half-full mark. While underlaying, keep tip of syringe just below phase boundary. Strictly avoid having air bubbles come out of needle (Fig. 1b).
4. When removing the syringe needle, do so swiftly and make sure to fix the piston to prevent further leakage of the 57% solution during the movement.
5. Close the tubes with the long caps provided by the gradient station. Avoid entrapment of air bubbles by pushing the cap down at a slight angle so that air can leave through the ventilation whole in the cap (Fig. 1c).
6. Level gradient platform according to the gradient station manual.
7. Place tube holder at center of gradient platform.
8. Run gradient protocol SW41 LONG-SUCR-32-57%-w/w-2ST (step 1: 4:00 min, 70°, 30 rpm; step 2: 0:25 min, 85°, 25 rpm, check gradient station manual for programming instructions).
9. Remove excess sucrose solution from cap (Fig. 1d).
10. Carefully place gradients on ice until further use. Proceed to step 11 for the ultracentrifugation of sucrose gradients.
11. Carefully layer membrane suspension from **step 15** in Subheading 3.1.3 on top of continuous 32–57% (*see* **Note 31**, Fig. 1e, g).
12. Tare two opposing tubes with buffer 1× M to <0.01 g.
13. Centrifuge gradients for 14 h at 41,000 rpm and 4 °C in a Beckman SW41 Ti swing-out rotor (287,000 × *g*) (*see* **Note 32**). Set to slow acceleration and slow brake (*see* **Note 33**).

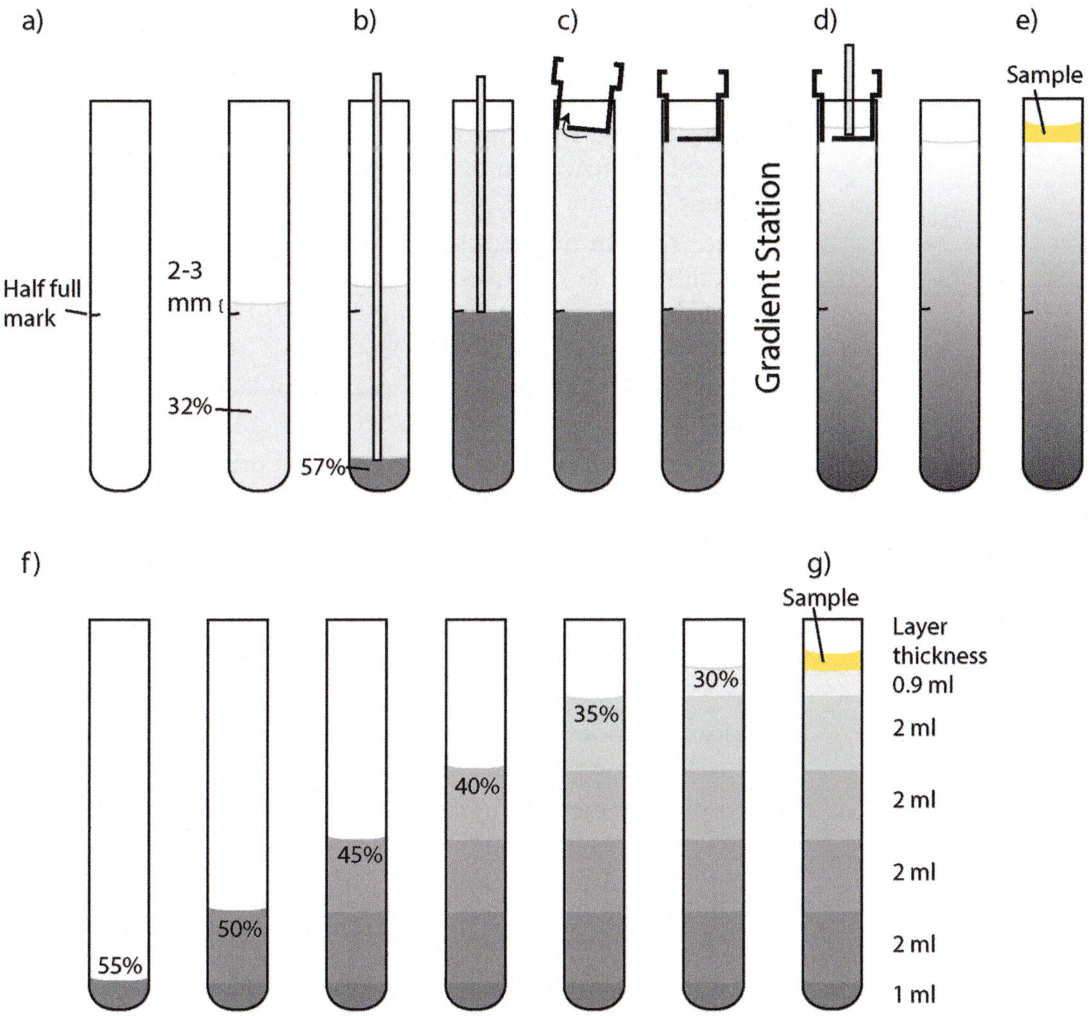

Fig. 1 Preparation of sucrose gradients (**a**–**e**) using Biocomp Instruments Gradient Station or (**f**–**g**) by hand. For details *see* Subheadings 3.1.4 and 3.1.5

14. Carefully place tubes on ice once the run is finished. Proceed to fractionation using Biocomp gradient station fractionator following **step 15**.
15. Label 1.5 mL tubes for collecting gradient fractions (13 per sample). Place tubes on ice.
16. Visually check if membranes are clearly separated and membrane bands are approximately at the same height. Make note if this is not the case (*see* **Note 34**). Take a photo of your gradient.
17. Program fractionation protocol: Speed 0.3 mm/s, distance 6.6 mm, number of fractions 12, rinse #0 (no automatic washing steps). Check manual for programming instructions.

18. Fill buffer reservoir with buffer 1× M.
19. Rinse tubing with buffer 1× M, then blow out remaining buffer from tubing.
20. Pour ice-cold water into tube holder of fractionator. Ice water should not reach rim of tube. Avoid ice clumps in holder that might dislocate tube.
21. Place tube in tube holder, lock lid, and place tube holder on fractionator as described in manual of gradient station. Take care that the correct piston tip (SW 41 Ti) is attached to the piston and its edge is undamaged.
22. Dim light of fractionator as low as possible to avoid heating up sample.
23. Move down position of piston until first drop appears at end of sample tubing. Reset position to zero (*see* **Note 35**).
24. Choose your program, put sample tubing into first collection tube, and start fractionation.
25. Collect twelve 6.6 mm fractions (*see* **Note 36**). Note which fractions contain the desired membrane bands.
26. Move up piston and collect remaining material (fraction 13).
27. Clean fractionator soon after completion of fractionation as dried sucrose solution will clog tubing of fractionator.
28. To prepare membrane fractions for downstream experiments, proceed to Subheading 3.1.6.

3.1.5 Membrane Fractionation by Sucrose Density Gradient Centrifugation Using a Manual Sucrose Step Gradient

1. Carefully pour gradient layers on top of each other, starting with 55% sucrose solution. Use 5 mL pipettes and Peleus ball or a cut 1 mL pipette tip (Fig. 1f) (*see* **Note 37**).
2. Carefully place gradients on ice until further use. Proceed to **step 3** for ultracentrifugation of sucrose gradients.
3. Carefully layer membrane suspension from **step 15** in Subheading 3.1.3 on top of continuous 30–55% step gradient (*see* **Note 31**, Fig. 1e, g).
4. Tare two opposing tubes with buffer 1× M to <0.01 g.
5. Centrifuge gradients for 14 h at 41,000 rpm and 4 °C in a Beckman SW41 Ti swing-out rotor (287,000 × *g*) (*see* **Note 32**). Set to slow acceleration and slow brake (*see* **Note 33**).
6. Carefully place tubes on ice once run is finished. Proceed to fractionation of sucrose gradient by hand following **step 7**.
7. Label 1.5 mL tubes for collecting gradient fractions (13 per sample). Place tubes on ice.
8. Visually check if membranes are clearly separated and membrane bands are approximately at same height. Make note if this is not the case. Take a photo of your gradient.

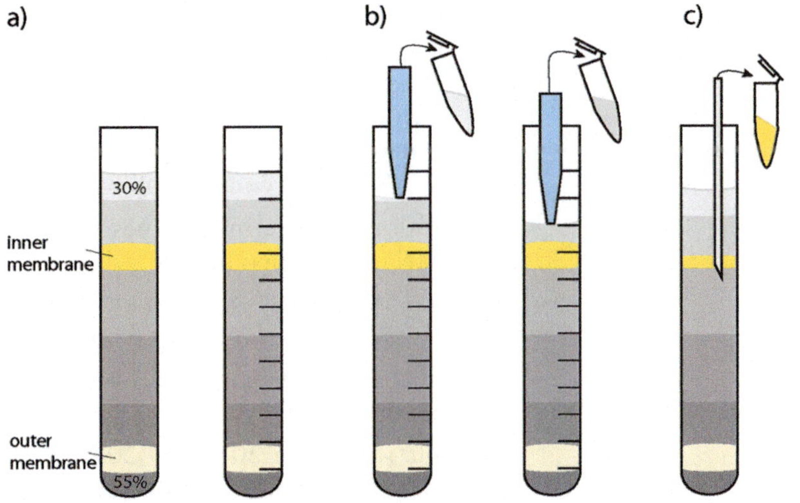

Fig. 2 Collection of fractions of sucrose six-step gradient by hand. (**a**) Labeling of fraction size. (**b**) Taking fractions from top by pipetting. (**c**) Taking individual fractions using syringe. For details *see* Subheading 3.1.5

9. Mark fractions on sides of tubes: use a fine marker, mark steps starting from meniscus of highest filled sample. Mark other tubes identically to first one, even if meniscus is slightly lower. Note which fractions contain membrane bands (Fig. 2a).
10. Collect fractions from top by carefully aspirating 1 mL from just underneath meniscus using a 1 mL pipette. Transfer fractions to collection tube. Use fresh pipette tip for every fraction (Fig. 2b).
11. Alternative: Discrete bands can be fractionated using syringe and a long needle. Place tip of needle just underneath the desired band and aspirate until the entire band is taken up. Transfer fractions to collection tube. Use fresh syringe and needle for every fraction (Fig. 2c).
12. Keep collected fractions on ice until further use.
13. To prepare membrane fractions for downstream experiments, proceed to Subheading 3.1.6

3.1.6 Preparing Membrane Fractions from Sucrose Gradient Fractionations for Downstream Experiments

1. Dilute sucrose at least 1:3 with buffer 1× M. Mix well.
2. Pellet membranes by centrifugation for 45 min at 230,000 × *g* and 4 °C (e.g., in Beckman Type 70.1 Ti).
3. Aspirate supernatant.
4. Resuspend each membrane pellet in 200 μL buffer L.
5. Measure protein concentration by BCA assay according to manufacturer's instructions (*see* **Note 38**).

6. Adjust protein concentration of desired fractions. Extraction of membrane proteins from *Salmonella* or *E. coli* inner membranes by DDM is optimal at a protein equivalent of 3 mg/mL bovine serum albumin (BSA); however, the optimal value must be determined empirically for every combination of detergent and protein or complex of interest (*see* **Note 39**).
7. Optional: Store membranes at −80 °C until further use and thaw on ice when needed.
8. Add 10% DDM (w/v) to desired amount of sample at 3 mg/mL protein to achieve final concentration of 1% DDM.
9. Incubate rotating in cold room for 1 h to extract proteins from membranes.
10. Spin sample for 20 min at 20,000 × *g* and 4 °C to pellet unsolubilized material (*see* **Note 17**).
11. Transfer supernatant to fresh tube.
12. For subsequent BN PAGE, add BN loading buffer to final concentration of 0.5% coomassie dye.
13. Load 5–10 μL of the suspension per well of a BN mini gel (*see* **Note 24**). Larger amounts of up to 50 μL must be used for the analysis of membrane protein complexes by two-dimensional BN/SDS-PAGE (*see* dedicated protocol in Subheading 3.2.2).

3.1.7 Immunoprecipitation of Membrane Protein Complexes Using 3× FLAG Epitope Tags

This protocol describes the purification of DDM-solubilized membrane protein complexes by immunoprecipitation of a 3× FLAG epitope-tagged fusion protein (*see* **Note 40**). Purified complexes can be analyzed by one-dimensional BN PAGE or two-dimensional BN/SDS-PAGE, respectively, as described previously [13–15, 17].

1. All steps are to be performed at 4 °C or on ice unless otherwise stated.
2. Continue with membrane samples from **step 8** in Subheading 3.1.6.
3. Incubate rotating in cold room for 30 min to extract proteins from membranes.
4. Prepare 5 μL packed beads of anti-FLAG M2 affinity gel per 1 mg of membrane sample. Wash beads according to the manufacturer's instructions, twice with PBS and once with PBS/0.1% DDM.
5. Add washed beads to sample and incubate for 4 h at 4 °C (*see* **Note 41**).
6. Pellet beads by centrifugation for 1 min at 500 × *g* and 4 °C.
7. Wash beads 3× 15 min with 10-fold bead volume PBS/0.1% DDM.
8. Resuspend pelleted beads in 1 mL PBS/0.1% DDM, transfer to 1.5 mL tube, and pellet again.

9. Aspirate supernatant completely.
10. Resuspend beads in one packed bead volume of PBS/0.04% DDM/150 μg/mL 3× FLAG peptide.
11. Incubate beads rotating 30 min at 4 °C to elute immunoprecipitated protein.
12. Pellet beads by centrifugation for 1 min at 500 × *g* and 4 °C.
13. Transfer supernatant to fresh 1.5 mL tube.
14. Repeat **steps 10–12** and pool supernatant.
15. Store at 4 °C until further use (*see* **Note 42**).
16. For subsequent BN PAGE, add BN loading buffer to a final concentration of 0.5% coomassie dye.
17. Load suspension into a well of a BN gel. For downstream analysis by MS, it is advised to load the maximum amount the well can hold.

3.2 Blue Native PAGE

3.2.1 One-Dimensional BN PAGE Using Precast Mini Gels

One-dimensional BN PAGE using precast mini gels accommodates all previously described sample preparations but is best suited for analytical amounts of membrane protein complexes and downstream immunoblotting. If preparative amounts of membrane complexes are needed, or the investigation of entire complexomes is sought, larger gel format may be better suited. Preparation and run of larger gel formats are described in Subheading 3.2.2. 3–12% gradient gels are suited for the analysis of protein complexes between 100 kDa and 5 MDa, 4–16% gradient gels for the analysis of protein complexes between 30 kDa and 1 MDa.

1. Assemble mini PAGE (10 × 10 cm gels) equipment according to manufacturer's instructions.
2. Completely fill anode buffer tank with cold BN anode buffer.
3. Pour cold BN cathode buffer A into wells of the precast gel. Do not fill up entire cathode buffer tank because the blue stain makes loading of sample very difficult.
4. Load suitable protein standard, e.g., 10 μL NativeMark from Thermo Fisher, into one well of the gel.
5. Load solubilized and coomassie-treated samples from **step 12** from Subheading 3.1.1, **step 26** from Subheading 3.1.2, **step 12** from Subheading 3.1.6, or **step 16** from Subheading 3.1.7 into the desired wells of the gel.
6. Carefully fill up cathode buffer tank with cold BN cathode buffer A. Take care not to wash out samples from wells.
7. Run electrophoresis at 4 °C as follows: 130 V and 300 mA for 50 min, 250 V and 300 mA for 2 h or until thick blue front runs out of gel (*see* **Note 43**).

8. Optional (*see* **Note 44**): For downstream analysis by immunoblotting, pause the run when the blue front has run one-third into the gel (typically after 50 min). Mix one volume of BN cathode buffer A with nine volumes of BN cathode buffer B. Empty cathode buffer tank completely. Refill cathode buffer tank with prepared BN cathode buffer 1A/9B mix.
9. Disassemble gel once the run has completed and continue with desired downstream analysis.

3.2.2 Two-Dimensional BN/SDS-PAGE

Two-dimensional BN/SDS-PAGE allows for the electrophoretic resolution of individual components of protein complexes. The protocol presented here uses a previously described approach with plastic-backed first-dimension BN gels that facilitates the transfer of BN lane strips onto second-dimension gels [6, 22, 23] (please also refer to these publications for illustrations of the methodology) (*see* **Notes 45** and **46**). An exemplary result of a coomassie-stained two-dimensional BN/SDS-PAGE gel of inner membranes of *S. typhimurium* wt and *spaS* (T3SS export apparatus component) mutant is shown in Fig. 3.

1. Cut polyacrylamide-coated plastic film to size of glass plates. Take care that the length of the plastic film perfectly fits the length of the glass plate; the width of the film is allowed to be somewhat smaller than the glass plate.

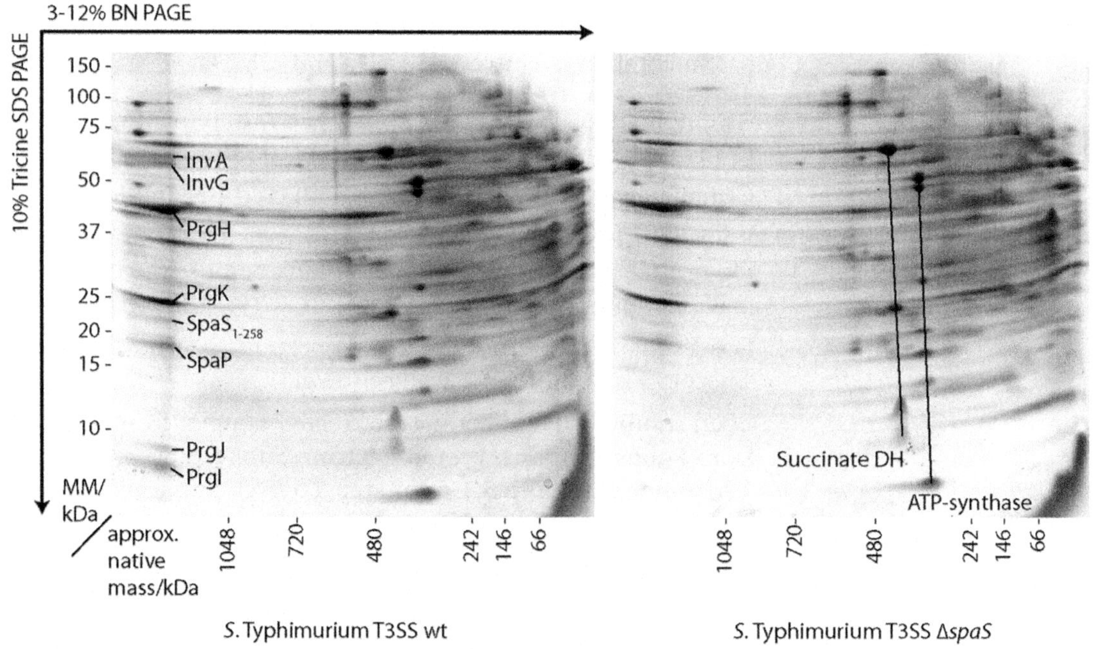

Fig. 3 Two-dimensional BN/SDS-PAGE showing coomassie-stained inner membrane proteins of *S. typhimurium*. *Left*: *S. typhimurium* wild type; T3SS components are indicated. *Right*: *S. typhimurium* Δ*spaS* mutant; other abundant and clearly observed complexes are indicated

Table 2
Solutions for casting of BN gradient gel

	Resolving gel				Stacking gel
	3%	12%	4%	16%	3%
ddH_2O	6.0 mL	1.05 mL	5.6 mL	–	5.1 mL
3× BN gel buffer	3.5 mL	3.5 mL	3.5 mL	3.5 mL	3.0 mL
Acrylamide (30% T, 3% C)	1.05 mL	4.2 mL	1.4 mL	5.6 mL	0.9 mL
Glycerol (80%)	–	1.75 mL	–	1.4 mL	–
TEMED	5.25 μL	5.25 μL	5.25 μL	5.25 μL	4 μL
APS (10%)	52.5 μL	52.5 μL	52.5 μL	52.5 μL	40 μL

2. Fix the plastic film on the lower edge of one of the glass plates using double-sided Scotch tape. Make sure that the hydrophilic side of the plastic film is facing up.
3. Place a few drops of ddH_2O between the glass plate and plastic film. Wipe tightly along film with tissue to remove excess water and trapped air.
4. Slightly grease 1.0 mm spacers on one side with gel seal. Place spacers with greased side on plastic film.
5. Finish assembly of gel casting assembly according to manufacturer's instructions.
6. Prepare solutions for casting of BN gradient gel according to recipes in Table 2. Do not add APS at this point (*see* **Note 47**).
7. Assemble gradient maker including magnet stirrer and peristaltic pump. Squeeze end of outlet tubing between glass plates of gel casting assembly.
8. Add APS to gel solution of lower percentage, mix well, and pour solution into distal chamber of gradient maker. Add APS to gel solution of higher percentage, mix well, and pour solution into proximal chamber of gradient maker (*see* **Note 48**).
9. Switch on peristaltic pump. Open outlet valve and then chamber-connecting valve. Pour gradient gel (*see* **Note 49**).
10. When all the gel solution has entered the gel, add APS to stacking gel solution, mix, and pour solution into proximal chamber of gradient maker (*see* **Note 50**). Continue pouring gel.
11. Carefully place comb (10 or 15 wells) into stacking gel (*see* **Note 51**). Take care not to destroy gradient.
12. Allow gel to polymerize for at least 4 h at room temperature. Gels can be stored after polymerization up to 1 week (4 °C and moist).

13. Remove comb from gel, rinse gel and wells with ddH_2O, and assemble gel running equipment in cold room (*see* **Note 52**). Fill anode buffer tank with 2 L cold BN anode buffer.
14. Pour cold BN cathode buffer A into wells of gel. Do not fill up entire cathode buffer tank because the blue stain will make loading of sample very difficult.
15. Load suitable protein standard, e.g., 10 μL NativeMark from Thermo Fisher, into one well of gel (*see* **Note 43**).
16. Load solubilized and coomassie-treated samples from **step 12** from Subheading 3.1.1, **step 26** from Subheading 3.1.2, **step 12** from Subheading 3.1.6, or **step 16** from Subheading 3.1.7 into desired wells of gel.
17. Carefully fill up cathode buffer tank with cold BN cathode buffer A. Take care not to wash out samples from wells.
18. Run electrophoresis at 4 °C as follows: 130 V and 30 mA for 1 h, 150 V and 30 mA for 14 h, or until thick blue front runs out of gel (*see* **Note 43**).
19. Proceed to second dimension—SDS-PAGE (*see* **Note 53**). Pour as many polyacrylamide gels (resolving and stacking gel) as needed to analyze each lane of BN PAGE. Use a polyacrylamide concentration for resolving gel that allows resolution of all proteins expected to compose complex of interest (e.g., 12% polyacrylamide).
20. After pouring stacking gel, place a preparative comb (one long well matching length and width of one lane of first-dimension BN gel) in gel and allow polymerization at room temperature (*see* **Note 54**).
21. Disassemble first-dimension BN PAGE assembly (obtained from **step 15**). Remove glass plate facing gel (not plastic film) and spacers.
22. Using razor blades, scrape off partition walls of wells and thickly stained coomassie front if not everything has run out of gel.
23. Lift up plastic-backed gel and remove double-sided Scotch tape.
24. Using scissors, cut out lanes of first-dimension run.
25. Place lane strips into SDS equilibration buffer to denature protein complexes and reduce disulfide bonds. Incubate for 20 min (*see* **Note 55**).
26. Remove comb of second-dimension gel and overlay top of gel with hot low-melting agarose solution.
27. Place lane strips of first-dimension gel onto second-dimension gel (*see* **Notes 56** and **57**).
28. Drop 5 μL of protein size ladder onto a small piece of filter paper (3 × 3 mm). Place marker-soaked filter paper on top of second-dimension gel next to BN gel lane strip.

29. Run electrophoresis at 4 °C as follows: 150 V and 300 mA for 1 h, 300 V and 300 mA, until coomassie front runs out of gel (*see* **Note 43**).
30. Disassemble gel assembly, take out gels, and proceed with downstream analysis as needed (*see* Subheading 3.3).

3.3 Protein Detection and Analysis

The described protocol according to Neuhoff et al. [24] allows for the highly sensitive detection of proteins by coomassie G.

3.3.1 Colloidal Coomassie Staining

1. After electrophoresis, place gel in a clean plastic container and rinse briefly in ddH_2O. Reduce gel manipulation to minimum and always wear powder-free gloves.
2. Decant ddH_2O. Add enough fixing solution to completely immerse gel and incubate o/n at room temperature with gentle agitation (*see* **Note 58**).
3. Decant fixing solution and wash gel three times in ddH_2O, 30 min for each wash.
4. Decant ddH_2O. Equilibrate gel in Neuhoff's solution for 1 h. The exact volume of Neuhoff's solution must be known in order to add the right amount of Serva Blue G in the next step.
5. Add Serva Blue G powder to a final concentration of 0.1% (w/v) (e.g., 0.1 g/100 mL Neuhoff's solution) and stain for up to 48 h (*see* **Note 59**).
6. Transfer gel into a clean container and wash several times with ddH_2O. Gently remove coomassie particles from gel surface.
7. Decant ddH_2O and store gel in 5% acetic acid solution at 4 °C until further usage (*see* **Note 60**).

3.3.2 Silver Staining (MS Compatible)

The described protocol according to Shevchenko et al. [25] allows for a highly sensitive detection of proteins.

1. After electrophoresis, place gel in tray and rinse briefly with ddH_2O. Always manipulate gel wearing powder-free gloves that have been rinsed with deionized water to avoid fingerprint contamination and pressure marks.
2. Decant ddH_2O. Immerse gel in fixing solution for 20–30 min; gently agitate (*see* **Note 61**).
3. To remove acid, rinse gel with ddH_2O for at least 20 min (*see* **Note 62**).
4. Soak gel in sensitizing solution for 1–2 min (*see* **Note 63**).
5. Decant sensitizing solution. Rinse gel twice with ddH_2O, 1 min for each wash.
6. Pour out ddH_2O and impregnate gel in chilled silver nitrate solution for 20–40 min, preferably at 4 °C (*see* **Note 64**).

7. Discard solution and rinse gel twice in water, 1 min for each wash (see **Note 65**).
8. Develop by adding a sufficient amount of developing solution to completely cover gel. Agitate until intensity of staining is as desired (*see* **Note 66**).
9. Once the adequate degree of staining is achieved, decant developing solution. Add stop solution and incubate for 30 min. Agitate slowly.
10. Store silver-stained gel in stop solution at 4 °C until scanning or preparation of bands for MS.

3.3.3 Immunoblotting Using Dual-Color Detection

BN PAGE facilitates the analysis of protein complexes that often contain more than one kind of protein component. Quantitative analysis of the composition of protein complexes by BN PAGE is very powerful when a simultaneous detection of two different proteins within the same complex band is achieved by dual-color immunoblotting. The protocol described here uses infrared-fluorescent secondary antibodies and a LiCor Odyssey scanner. Although the protocol is described for immunoblotting of BN gels, it can be easily adapted for two-dimensional BN/SDS-PAGE or regular SDS-PAGE. Figure 4 shows an example of a dual-color immunoblotting of a BN PAGE-separated T3SS needle complex.

1. Equilibrate BN gels in SDS-PAGE running buffer for 15–20 min (*see* **Note 67**).
2. Assemble wet blot sandwich according to manufacturer's instructions. Use only PVDF membranes because coomassie G binds irreversibly to nitrocellulose.
3. Transfer proteins according to the instructions of the wet blot equipment manufacturer.
4. When the transfer is finished, take out the PVDF membrane, place it in a dark container (*see* **Note 68**), and rinse it three or four times in 100% MetOH to remove all bound coomassie G (*see* **Note 69**).
5. Decant MetOH and rinse PVDF membrane with TBS until wetted.
6. Stain PVDF membrane with Ponceau S solution to visualize unstained native protein standard.
7. When sufficient staining is achieved, rinse PVDF membrane with ddH_2O until bands become visible and mark bands with a pencil.
8. Destain PVDF membrane completely using ddH_2O.
9. Block unoccupied binding sites of PVDF membrane by incubating in 5% skim milk/TBS or 3% BSA/TBS for 1 h (*see* **Note 70**).
10. Incubate PVDF membrane with primary antibody diluted to desired concentration TBST ($1\times$TBS plus 0.05% Tween-20) for 1 h (*see* **Note 71**).

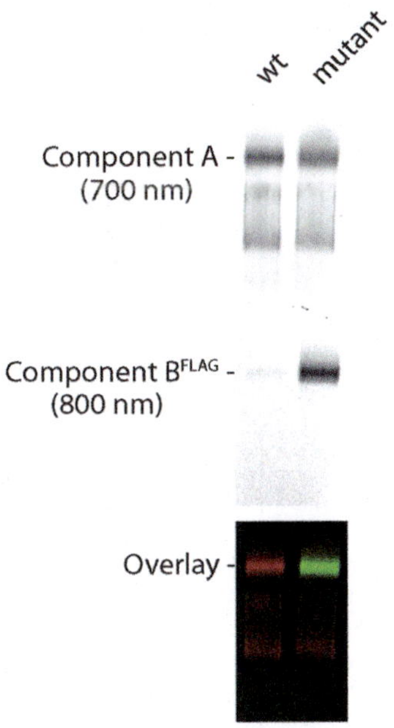

Fig. 4 Dual-color immunoblotting of a BN PAGE-separated T3SS needle complex. Crude membranes of wild-type and mutant *S. typhimurium* were extracted with DDM and analyzed by BN PAGE and immunoblotting. Component A was probed with a primary rabbit antibody and a secondary DyLight goat anti-rabbit 680 nm (*red*). The FLAG-tagged component B was probed with a primary anti-FLAG M2 mouse monoclonal antibody and a secondary DyLight goat anti-mouse 800 nm (*green*)

11. Wash the PVDF membrane three times 15 min with TBST.
12. Incubate PVDF membrane with secondary infrared-fluorescent antibody diluted to desired concentration in TBST for 1 h (*see* **Note 72**).
13. Wash PVDF membrane three times for 15 min with TBST.
14. Scan PVDF membrane in LiCor Odyssey scanner according to manufacturer's instructions.
15. Analyze protein bands using western blot analysis tool of Image Studio software (LiCor). A typical result of dual-color immunoblotting of a type III secretion system is shown in Fig. 4.

3.3.4 Preparation of BN PAGE-Separated Complexes for Analysis by Mass Spectrometry

MS is well suited to detect the composition of protein complexes separated by BN PAGE; however, it can be difficult to distinguish a true complex component from an unrelated comigrating protein. Distinction can be achieved using suitable controls and bioinformatic analysis, as described by Fischer et al. [15], but also by the mass spectrometrical profiling of complete lanes of BN gels [26], as described in what follows. Figure 5 shows the MS-based analysis of lane profiles of immunoprecipitated T3SS needle complexes.

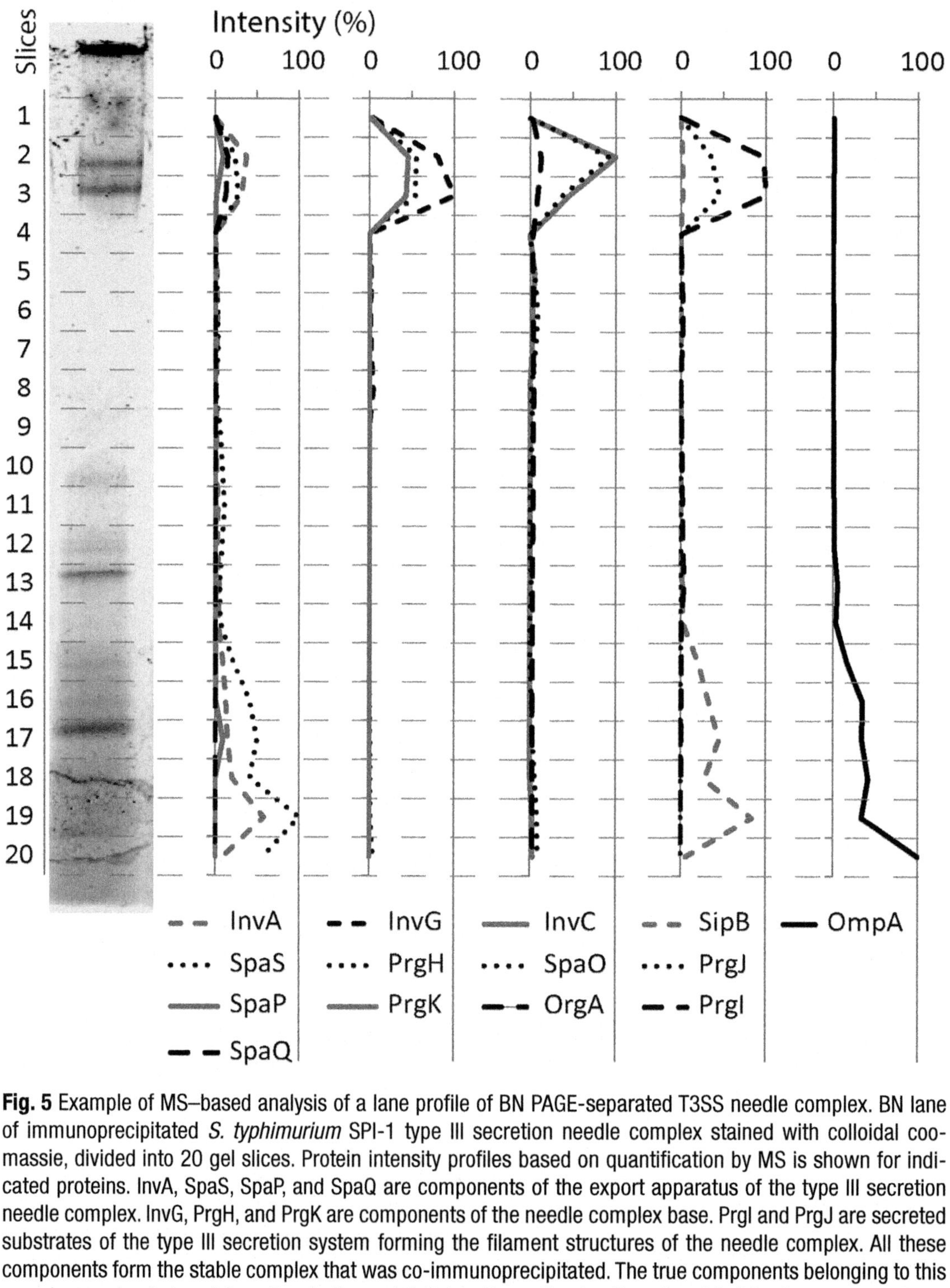

Fig. 5 Example of MS–based analysis of a lane profile of BN PAGE-separated T3SS needle complex. BN lane of immunoprecipitated *S. typhimurium* SPI-1 type III secretion needle complex stained with colloidal coomassie, divided into 20 gel slices. Protein intensity profiles based on quantification by MS is shown for indicated proteins. InvA, SpaS, SpaP, and SpaQ are components of the export apparatus of the type III secretion needle complex. InvG, PrgH, and PrgK are components of the needle complex base. PrgI and PrgJ are secreted substrates of the type III secretion system forming the filament structures of the needle complex. All these components form the stable complex that was co-immunoprecipitated. The true components belonging to this complex can be deduced from the similar protein intensity profiles in slices 2 and 3. It is also apparent that a large fraction of non-needle-complex-associated SpaS and InvA are present in the sample (slices 15–20). InvC, SpaO, and OrgA are cytosolic components of the type III secretion system that are not stably associated with the needle complex and just coprecipitate during purification. This is reflected by the sharp peaking of the intensity of these proteins in slice 2 but not slice 3. The profile of the unrelated outer membrane protein OmpA is shown as a control

1. Make a picture of your stained BN gel before cutting any bands.
2. To avoid contamination, clean surface used to cut protein bands out of gel with fresh 100% ethanol. Work with fresh, clean, powder-free gloves. Avoid touching outside of gloves while putting them on.
3. Measure total length of BN lane and decide on a number of bands.
4. Prepare a 1.5 mL tube for storage of each band (*see* **Note 73**). Fill tubes with 0.5 mL 5% acetic acid (v/v).
5. Remove most of the 5% acetic acid from the translucent tray the gel is stored in. Leave just enough to keep gel covered. Place tray on a light table (*see* **Note 74**).
6. Clean a razor blade and tweezers with ethanol.
7. Use razor blade to make horizontal cuts (*see* Fig. 6a). Cut all replicates at once by pressing the razor blade down over full length of replicate lanes. Clean razor blade with ethanol in between cuts (*see* **Note 75**).
8. Use a new, cleaned razor blade for cutting vertical lines, starting at top (*see* **Note 76**).
9. After each cut, collect bands with tweezers and place them in prepared tubes containing 5% acetic acid (Fig. 6b, c) (*see* **Note 77**). Always clean razor blades and tweezers in between cuts using 100 EtOH.
10. Store cut bands at 4 °C until further processing for MS.

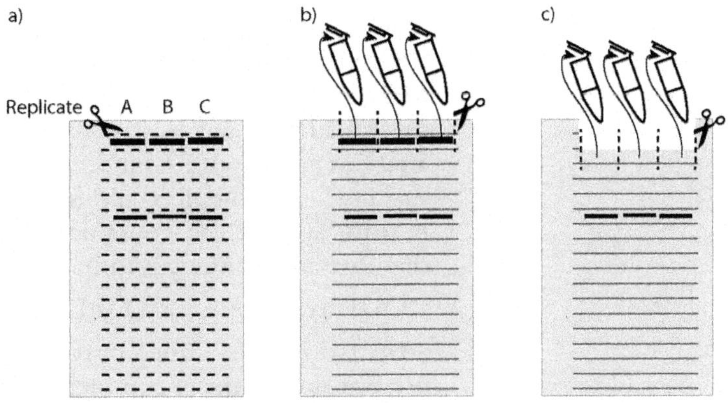

Fig. 6 Preparation of samples for MS-based analysis of lane profiles of BN PAGE-separated samples. (**a**) Cut horizontal lines while leaving border on both sides intact. (**b**, **c**) Cut vertical lines from top to bottom and collect gel bands subsequently in 1.5 mL tubes. For details *see* Subheading 3.3.4

4 Notes

1. Also include 1 mM dithiothreitol if reducing conditions are desired.
2. Serva Blue G guarantees the highest quality of the coomassie dye. If replaced by a dye of another vendor, make sure it is of high quality.
3. It is best to add the dye 30 min before use. Vortex vigorously to dissolve dye. Just before use, spin briefly to pellet unsolubilized dye.
4. The solutions can be prepared in advance and stored at 4 °C but should be at room temperature when making the gradient.
5. Depending on the organism, the percentage of the sucrose gradient may have to be adjusted for an optimal separation of inner and outer membranes.
6. Biocomp Instruments offers certified Seton tubes to guarantee fine sealing when using the Biocomp gradient station.
7. Before making gradients, check each tube for prominent edges at seams, differences in length or wall thickness (happens rarely), and fissures.
8. For smaller gradients 13 × 51 mm tubes (Seton) for, e.g., SW 55 Ti rotor (Beckman) can be used.
9. Or similar depending on protein tag.
10. Choosing the right double-sided Scotch tape is crucial. We use double-sided Scotch tape from Tesa. Other double-sided Scotch tapes lose their adhesive properties during electrophoresis.
11. When used for the first time, clean containers exhaustively with acetone and afterward with alcohol to remove traces of plasticizers that may interfere with mass spectrometric analysis. Make sure the containers are used exclusively for coomassie or silver staining, and wash them with soap and ultrapure water after every staining round. To ensure free movement while shaking and complete immersion in staining solutions, the bottom area of the container should be at least 20% bigger than the area of the gel to be stained.
12. 1 ODU corresponds to 1 mL of bacterial culture at $A_{600} = 1.0$.
13. Snap freezing in liquid nitrogen or dry-ice ethanol helps to prevent damaging of complexes by ice crystal formation.
14. The basic buffer recipe used for cell wall digestion may require empirical evaluation depending on the bacterium or protein complex of interest. We have also successfully used PBS, pH 7.4, supplemented with protease inhibitor, lysozyme, and DNase. Alternative cell-wall-degrading enzymes may be needed for other bacteria, e.g., lysostaphin for *Staphylococcus aureus.*

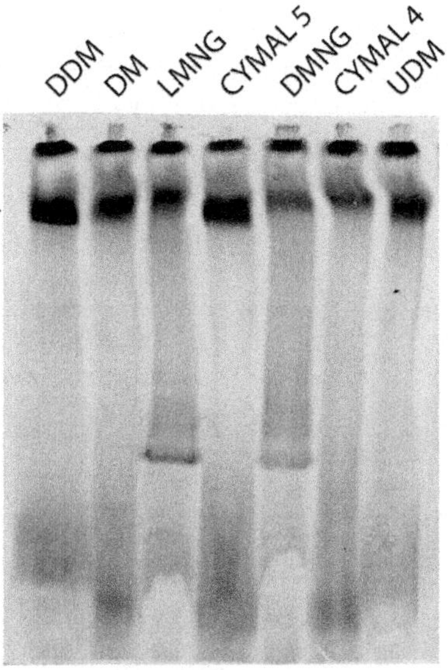

Fig. 7 BN PAGE analysis of T3SS needle complex using a set of different detergents. Crude membranes of *S. typhimurium* were extracted with indicated detergents and subsequently analyzed by BN PAGE and immunoblotting. The FLAG-tagged T3SS export apparatus component SpaS was probed with a primary anti-FLAG M2 mouse monoclonal antibody and a secondary DyLight 800 nm goat anti-mouse antibody. The band of SpaS in the type III secretion needle complex is indicated

15. Aminocapronic acid was reported to facilitate extraction of proteins. Its benefit at this step may be empirically evaluated. Use PBS or your buffer of choice.
16. The detergent used for extraction may require empirical evaluation depending on the protein complex of interest. In addition to DDM, we have had good experience with Cymal 4, Cymal 5, DM, UDM, DMNG, LMNG, LDAO, and Triton X-100 (*see* Fig. 7).
17. The speed of this spin depends on the molecular weight of the complex of interest. T3SS needle complexes are already pelleted at speeds of 100,000 × *g*; hence, it is not advisable to remove unsolubilized material at this speed. If the focus is on smaller complexes, spins up to 30 min at 100,000 × *g* are recommended.
18. Bacterial lysis is based on cell wall digestion and glass bead milling, an approach suitable for the preparation of several samples in parallel.
19. The expected yield of crude membranes is 10 µg protein per ODU of bacterial cells (*E. coli*, *Salmonella*). Of this, about 50% can be extracted by DDM. Loading 10 µg of crude membranes per well is sufficient for analysis by BN mini gels; however, a

larger preparation ensures a clearly visible pellet after ultracentrifugation.

20. The volumes for glass bead milling do not depend on the amount of bacterial cells when working in a range of 2–10 ODU.
21. It is important to use screw-cap tubes because snap-cap tubes may pop open during bead milling.
22. During bead milling, the label may be abraded. Check which labeling position is most suitable.
23. Do not transfer a higher volume since excess liquid may leak out of the tube during ultracentrifugation.
24. Suitable amount needs to be empirically evaluated.
25. Bacterial lysis in this protocol is based on French pressing but can be tailored to specific needs and laboratory equipment.
26. The maximum loading capacity of one gradient is crude membrane extracted from about 2000 ODU of culture. If membranes from more material need to be separated, it is better to divide the sample on several gradients. Samples can be pooled after **step 4** from Subheading 3.1.6.
27. This spin can be done in 50 mL Falcon tubes if using the Fiberlite F15-8x50C rotor. Otherwise, transfer lysate to suitable centrifugation bottles.
28. This speed will also remove a considerable amount of outer membranes. If these are needed, spin at 8000 × *g* instead.
29. Use of a Biocomp gradient station results in six identical continuous gradients, which can be fractionated in defined steps with very limited cross contamination of fractions. Both gradient formation and fractionation are highly reproducible across different experiments. Buffers and gradients described in this protocol were optimized for separating membranes of *E. coli* and *Salmonella*, but the protocol can easily be adapted for other organisms. The protocols are described for use with a Beckman SW 41 Ti swing-out rotor but can be adapted to other rotors accordingly.
30. After filling the syringe, clean the outside of the needle with a lint-free tissue to prevent mixing of 57 and 32% sucrose solutions.
31. Overfilling the tubes (less than 3 mm left at the top) can cause spilling during the centrifugation step, which makes breaking of the tubes more likely.
32. The swinging buckets for the centrifuge should be cleaned before use.
33. If a tube breaks during the overnight step, you can collect the membrane–sucrose mixture, dilute it with the buffer used dur-

ing the membrane extraction (at least 1:2), and spin down the crude membrane pellet again to repeat the gradient.

34. If the membrane bands are not at an equal height in the different samples, the most common problem is the pipetting of the gradient. Either the half-full mark for the wrong cap type (long/short) was used or the sucrose solutions were not properly dissolved before layering the gradient. Accidental stirring of the solutions while underlaying can also lead to aberrations in gradient formation.
35. For the following samples: Move down position of piston until position value is exactly 0.00. If you move too far, make note of the position and reset to 0 to continue; however, fractions will not be identical to the first sample. **Do NOT** press "UP" at this point. Pressing "UP" will move the piston all the way to the top again, mixing at least the upper part of your gradient.
36. Fractionation of a continuous gradient does not require rinsing of sample tubing after each fraction. If a discrete fraction is targeted, rinsing and blowing off tubing before start of fractionation is recommended.
37. The phase boundaries should be visible from an angle. If you cannot see any boundaries at all, you have mixed the layers and should make new gradients.
38. The BCA assay is largely tolerant to lipids in the sample and therefore preferred over Bradford or Lowry assays. For better comparability, always use the same standard protein to make a standard curve, e.g., BSA.
39. The extraction power of a given detergent toward a protein of interest can be tested by extracting crude membranes with a protein concentration equivalent of 1–10 mg/mL BSA in 1 mg increments with 1% of a given detergent. The highest concentration that is still correlated to a linear increase in extracted protein (testable, for example, by immunoblotting) is the optimal concentration to use for extraction.
40. Immunoprecipitation can be performed using suitable antibodies or any other epitope tags that tolerate the detergents used. In an optimal case, the 3× FLAG epitope-tagged fusion protein is encoded in its natural context, e.g., on the chromosome or on its virulence plasmid. However, also expression of plasmid-based artificial operons encoding all or some complex partners of interest (of which one is 3× FLAG-tagged) is possible.
41. During immunoprecipitation with beads, choose a tube according to your sample volume. It should be one-third to two-thirds full and rotated over its own head to guarantee a good distribution of beads in the sample over the whole incubation time.

42. Freezing of solubilized complexes is not recommended as this may cause damage.
43. Voltage clamp limits the run.
44. The coomassie G contained in the BN gel efficiently binds to PVDF membranes upon transfer and blocks protein binding sites. To enable downstream immunoblotting, a BN cathode buffer is used after one-third of the run that contains only one-tenth of the initial coomassie concentration.
45. The first dimension described in what follows utilizes a polyester backing to facilitate the first- to second-dimension transfer of BN lane strips but can in principle also be performed as a single one-dimensional gel without the polyester backing.
46. Here we describe this approach using 16 cm long gels run in a Hoefer SE600 system while the original references describe BN gels of 24 cm length followed by a second dimension in GE Ettan Dalt tanks.
47. The acrylamide concentration of the BN gradient gel should be chosen to suit the molecular mass of the analyzed complex: For complexes beyond 1000 kDa, 3–12% gels should be used; smaller complexes are separated best in 4–16% gels.
48. Be aware that polyacrylamide polymerization can be very quick and clog the gradient maker and tubing when casting gels at warm room temperatures (>25 °C) in the summer. Consider casting gradient gels in the cold room if this is a problem.
49. Make sure the levels of the two gel solutions are at an equal height during the pouring process. If necessary, tilt the gradient maker to put more hydrostatic pressure on the solution of lower percentage.
50. Do not wait for the resolving gel to polymerize before casting the stacking gel. The function of the stacking gel is only to hold the comb, not to stack the sample proteins.
51. Best results are obtained when the comb is released from its support and pushed into the gel solution such that the gel's entire surface is covered by the comb and does not contact any air. Air inhibits polyacrylamide polymerization, in particular when very low polyacrylamide percentages are used.
52. Putting some gel seal on top of the spacers helps to prevent leakage of cathode buffer during the run.
53. For the second dimension the choice of electrophoresis equipment is not critical as long as its width allows the placement of one BN lane on the gel. The thickness of the second-dimension gel needs to be 1.5 mm (0.5 mm thicker than the first-dimension BN gel). For gel recipes, please refer to your standard SDS-PAGE protocol and adjust the volumes accordingly. Alternatively, Tricine SDS-PAGE can be used as a second

dimension, which has been described in detail [23] and was the basis for the gel shown in Fig. 3.

54. If no suitable comb is at hand, one may pour the stacking gel up to one BN lane width below the edge of the glass plate and overlay the acrylamide solution with isobutanol.
55. Equilibration for longer periods may lead to a loss of small proteins due to diffusion out of the gel.
56. Take care that no air bubbles are trapped between the first-dimension lane strip and the second-dimension gel.
57. Take care not to damage the BN gel during this procedure and push only on the plastic film, not on the gel itself.
58. Fixation can last from 4 h to 4 days.
59. Usually bands are visible after 24 h, but optimal staining lasts for 3–4 days.
60. The gel can be stored in 5% acetic acid for several weeks. However, for analysis of protein bands by MS, it is recommended to immediately excise and process the bands of interest. Alternatively, bands can be stored at −20 °C until further processing.
61. Short fixation improves sequence coverage in subsequent MS but results in poor detection of small proteins. Longer fixation may increase staining sensitivity but lower sequence coverage.
62. Extensive washing increases sensitivity and reduced background staining.
63. Avoid the use of glutaraldehyde as sensitizing agent. Although it increases sensitivity and uniformity of staining, glutaraldehyde crosslinks lysine residues and prevents complete trypsinization.
64. Impregnation with silver nitrate can last from 20 min to 2 h without affecting the quality of the results.
65. Collect used silver solution in a dedicated waste container containing sodium chloride to precipitate the silver.
66. Replace developing solution as soon as it turns yellow.
67. It is critical that only BN gels be used for immunoblotting that were finally run with BN cathode buffer containing 0.02% coomassie G to prevent blocking of the PVDF membrane by excess coomassie G.
68. The infrared-fluorescent secondary antibodies are sensitive to light. To avoid bleaching, immunodetection is done in dark containers.
69. Work in fume hood. Collect MetOH in special waste container.
70. Do not use Tween 20 in the blocking buffer. When bound to the membrane, it may give rise to an increased background fluorescence in the 700 nm channel.

71. Two different proteins can be detected in the same complex if mouse antibodies are available against one protein (e.g., against an epitope tag) and rabbit antibodies are available against a second protein.
72. Dual-color detection is achieved by using a secondary infrared-fluorescent antibody with an emission at 680 nm against one species, e.g., mouse, and another secondary infrared-fluorescent antibody with an emission at 800 nm against another species, e.g., rabbit. It is recommended to use the antibody of higher quality in the 700 nm channel and the antibody of lower quality in the 800 nm channel because the background is typically lower at 800 nm than at 700 nm.
73. Ask your MS facility if they have special requirements/preferences for sample tubes.
74. Depending on your light table, the samples may be heated up. Especially in the lower percentage part of the gel, the gel will be very sticky; the warmer it is, the harder it will be to cut. To avoid this, the gel can be kept at 4 °C prior to cutting.
75. While cutting, make sure that your hair is tied back if necessary and do not inhale too much ethanol.
76. Starting at the top ensures that you cut the stickiest part of the gel while it is still cool.
77. It is easier and cleaner if a second person, also working with clean gloves, opens, handles, and closes the tubes and double checks the lane and band numbers.

Acknowledgments

Work performed in the laboratory of SW was supported by the Alexander von Humboldt Foundation in the framework of the Sofja Kovalevskaja Award endowed by the Federal Ministry of Education and Research (BMBF) and in the framework of the Georg Forster Research Fellowships (to J.V.M.F.), and by the Deutsche Forschungsgemeinschaft (DFG) as part of the Collaborative Research Center (SFB) 766 Bacterial cell envelope, project B14.

References

1. Schägger H, von Jagow G (1991) Blue native electrophoresis for isolation of membrane protein complexes in enzymatically active form. Anal Biochem 199:223–231
2. Bessonneau P, Besson V, Collinson I, Duong F (2002) The SecYEG preprotein translocation channel is a conformationally dynamic and dimeric structure. EMBO J 21:995–1003
3. Oates J, Barrett CML, Barnett JP, Byrne KG, Bolhuis A, Robinson C (2005) The *Escherichia coli* twin-arginine translocation apparatus incorporates a distinct form of TatABC complex, spectrum of modular TatA complexes and minor TatAB complex. J Mol Biol 346:295–305
4. Wiedemann N, Kozjak V, Chacinska A, Schönfisch B, Rospert S, Ryan MT, Pfanner N, Meisinger C (2003) Machinery for protein sorting and assembly in the mitochondrial outer membrane. Nature 424:565–571

5. Kikuchi S, Hirohashi T, Nakai M (2006) Characterization of the preprotein translocon at the outer envelope membrane of chloroplasts by blue native PAGE. Plant Cell Physiol 47:363–371
6. Wagner S, Baars L, Ytterberg AJ, Klussmeier A, Wagner CS, Nord O, Nygren PA, van Wijk KJ, de Gier JW (2007) Consequences of membrane protein overexpression in *Escherichia coli*. Mol Cell Proteomics 6:1527–1550
7. Baars L, Wagner S, Wickström D, Klepsch M, Ytterberg AJ, van Wijk KJ, de Gier JW (2008) Effects of SecE depletion on the inner and outer membrane proteomes of *Escherichia coli*. J Bacteriol 190:3505–3525
8. Wagner S, Klepsch MM, Schlegel S, Appel A, Draheim R, Tarry M, Högbom M, van Wijk KJ, Slotboom DJ, Persson JO, de Gier JW (2008) Tuning *Escherichia coli* for membrane protein overexpression. Proc Natl Acad Sci U S A 105:14371–14376
9. Wickström D, Wagner S, Baars L, Ytterberg AJ, Klepsch M, van Wijk KJ, Luirink J, de Gier JW (2011) Consequences of depletion of the signal recognition particle in *Escherichia coli*. J Biol Chem 286:4598–4609
10. Wickström D, Wagner S, Simonsson P, Pop O, Baars L, Ytterberg AJ, van Wijk KJ, Luirink J, de Gier JWL (2011) Characterization of the consequences of YidC depletion on the inner membrane proteome of *Escherichia coli* using 2D blue native/SDS-PAGE. J Mol Biol 409:124–135
11. Krall L, Wiedemann U, Unsin G, Weiss S, Domke N, Baron C (2002) Detergent extraction identifies different VirB protein subassemblies of the type IV secretion machinery in the membranes of *Agrobacterium tumefaciens*. Proc Natl Acad Sci U S A 99:11405–11410
12. Kuroda T, Kubori T, Thanh Bui X, Hyakutake A, Uchida Y, Imada K, Nagai H (2015) Molecular and structural analysis of *Legionella* DotI gives insights into an inner membrane complex essential for type IV secretion. Sci Rep 5:10912
13. Wagner S, Königsmaier L, Lara-Tejero M, Lefebre M, Marlovits TC, Galán JE (2010) Organization and coordinated assembly of the type III secretion export apparatus. Proc Natl Acad Sci U S A 107:17745–17750
14. Lara-Tejero M, Kato J, Wagner S, Liu X, Galán JE (2011) A sorting platform determines the order of protein secretion in bacterial type III systems. Science 331:1188–1191
15. Fischer M, Zilkenat S, Gerlach RG, Wagner S, Renard BY (2014) Pre- and post-processing workflow for affinity purification mass spectrometry data. J Proteome Res 13:2239–2249
16. Monjarás Feria JV, Lefebre MD, Stierhof YD, Galán JE, Wagner S (2015) Role of autocleavage in the function of a type III secretion specificity switch protein in *Salmonella enterica* serovar Typhimurium. MBio 6:e01459–e01415
17. Zilkenat S, Franz-Wachtel M, Stierhof YD, Galán JE, Macek B, Wagner S (2016) Determination of the stoichiometry of the complete bacterial type III secretion needle complex using a combined quantitative proteomic approach. Mol Cell Proteomics
18. Houben ENG, Bestebroer J, Ummels R, Wilson L, Piersma SR, Jiménez CR, Ottenhoff THM, Luirink J, Bitter W (2012) Composition of the type VII secretion system membrane complex. Mol Microbiol 86:472–484
19. Wittig I, Braun HP, Schägger H (2006) Blue native PAGE. Nat Protoc 1:418–428
20. Schraidt O, Marlovits TC (2011) Three-dimensional model of *Salmonella's* needle complex at subnanometer resolution. Science 331:1192–1195
21. Loquet A, Sgourakis NG, Gupta R, Giller K, Riedel D, Goosmann C, Griesinger C, Kolbe M, Baker D, Becker S, Lange A (2012) Atomic model of the type III secretion system needle. Nature 486:276–279
22. Klepsch M, Schlegel S, Wickström D, Friso G, van Wijk KJ, Persson JO, de Gier JW, Wagner S (2008) Immobilization of the first dimension in 2D blue native/SDS-PAGE allows the relative quantification of membrane proteomes. Methods 46:48–53
23. Schlegel S, Klepsch M, Wickström D, Wagner S, de Gier JW (2010) Comparative analysis of cytoplasmic membrane proteomes of *Escherichia coli* using 2D blue native/SDS-PAGE. Methods Mol Biol 619:257–269
24. Neuhoff V, Arold N, Taube D, Ehrhardt W (1988) Improved staining of proteins in polyacrylamide gels including isoelectric focusing gels with clear background at nanogram sensitivity using Coomassie Brilliant Blue G-250 and R-250. Electrophoresis 9:255–262
25. Shevchenko A, Wilm M, Vorm O, Mann M (1996) Mass spectrometric sequencing of proteins silver-stained polyacrylamide gels. Anal Chem 68:850–858
26. Sessler N, Krug K, Nordheim A, Mordmüller B, Macek B (2012) Analysis of the *Plasmodium falciparum* proteasome using blue native PAGE and label-free quantitative mass spectrometry. Amino Acids 43:1119–1129

Chapter 27

In Situ Imaging of Bacterial Secretion Systems by Electron Cryotomography

Gregor L. Weiss, João M. Medeiros, and Martin Pilhofer

Abstract

The unique property of electron cryotomography (ECT) is its capability to resolve the structure of macromolecular machines in their cellular context. The integration of ECT data with high-resolution structures of purified subcomplexes and live-cell fluorescence light microscopy can generate pseudo-atomic models that lead to a mechanistic understanding across size and time scales. Recent advances in electron detection, sample thinning, data acquisition, and data processing have significantly enhanced the applicability and performance of ECT. Here we describe a detailed workflow for an ECT experiment, including cell culture, vitrification, data acquisition, data reconstruction, tomogram analysis, and subtomogram averaging. This protocol provides an entry point to the technique for students and researchers and indicates the many possible variations arising from specific target properties and the available instrumentation.

Key words Tilt series, Plunge freezing, Electron cryomicroscopy, Reconstruction, Cryogen, Segmentation

1 Introduction

Bacterial cell–cell interactions are often mediated by the secretion of effector proteins that act on target cells. Macromolecular machines in the bacterial cell envelope are crucial for the translocation of effectors from the bacterial cytoplasm into the extracellular space or directly into the target cell [1]. Much insight into the mechanisms of these secretion systems are the result of high-resolution structure determination of purified subcomplexes and live-cell fluorescence light microscopy. Structural studies of secretion systems, however, are often challenged by the complexity of the systems, the involvement of membrane proteins, and the dependence on the removal of the system from its cellular context. On the other hand, fluorescence light microscopy is limited to the visualization of labeled components, by the dependence on functional fluorescent tags, and by the achievable resolution. Electron cryotomography (ECT) offers solutions to these problems. ECT

Laure Journet and Eric Cascales (eds.), *Bacterial Protein Secretion Systems: Methods and Protocols*, Methods in Molecular Biology, vol. 1615, DOI 10.1007/978-1-4939-7033-9_27, © Springer Science+Business Media LLC 2017

resolves unique structures in situ, in a lifelike state, in three dimensions, and at a resolution of a few nanometers [2–6]. Flash freezing a sample in noncrystalline ice avoids the introduction of artifacts arising from conventional electron microscopy (EM) sample preparation [7].

In a typical ECT experiment, a few microliters of a bacterial culture are applied to an EM grid and excess liquid is removed. The grid is plunged into a cryogen and transferred to a transmission electron microscope operating at cryogenic temperature. A series of 2D projection images of a bacterial cell from different angles is recorded and the data are reconstructed into a 3D image—the tomogram. The resolution of an individual tomogram is around 2–5 nm. For instance, high-quality tomograms typically resolve the double leaflet of bacterial cellular membranes or cytoskeletal structures in the crowded cytoplasm. Extracting, aligning, and averaging of subvolumes (subtomogram averaging) allows for the improvement of contrast and enables atomic resolution for suitable targets [8]. The generated density maps can be integrated with high-resolution structures of subcomplexes, resulting in pseudo-atomic models of the machines in their cellular context.

ECT has facilitated groundbreaking progress regarding the understanding of diverse bacterial secretion systems [9–20], but it faces a number of technical challenges. The incident electron beam damages the sample, which limits the applicable total dose, resulting in data with low contrast. With increasing sample thickness, noise is introduced by higher fractions of inelastic and multiple scattering events. The physics of image formation requires data acquisition at a defocus in order to be able to observe low-resolution features; however, this reduces high-resolution information. Subcellular targets, including secretion systems, can only be detected after data collection and reconstruction. Tilt series acquisition is time consuming (20–75 min per series), and the limited tilt range causes anisotropic resolution. Fortunately, many of these issues have been addressed by recent methodological developments. One development is the introduction of direct electron detectors, by which the field of electron cryomicroscopy entered a new era [21]. These detectors offer much improved detective quantum efficiency [22], and the fast readout allows for the correction of sample motion occurring during exposure [23]. Phase plates were introduced to modulate the contrast transfer function (CTF) in such a way that tilt series can be acquired in focus, significantly enhancing contrast [24]. Cryo-light microscopy allows for the detection of rare events prior to ECT data acquisition and for the identification of subcellular structures [25, 26]. Cryo-focused ion beam (FIB) milling is a potent approach to thin samples prior to ECT [27]. It is likely that the aforementioned developments, together with novel data collection schemes [28] and the possibility of subtomogram classification [29], will significantly enhance the power and applicability of ECT. The integration with data from cell and structural biology will resolve the mechanisms of secretion machines across scales. This is particularly significant for secretion systems that are dependent on cell–cell contact since their function is likely only to be fully understood by including the secreting as well as the receiving cell in the analysis.

It would require going far beyond the scope of this chapter to detail all possible variations of the ECT workflow. The technology is under active development, and the exact protocol is highly dependent on the target and the available instrumentation. Here we present one possible step-by-step workflow that can help students and researchers to obtain a first insight into the technique.

2 Materials

2.1 Plunge Freezing

1. Sample support: EM grids with a holey carbon support such as Quantifoils (Quantifoil Micro Tools, Germany) or C-Flats (Protochips, USA). Here we use Quantifoil R2/2 copper grids (200 mesh). For grid specifications *see* **Note 1**.
2. Cleaning and hydrophilization of grids: Glow discharge system such as Emitech K100X (Quorum Technologies, UK) and a glass slide as grid support.
3. Preparation of gold fiducial markers: Bovine serum albumin (BSA), double-distilled (dd) water, 10 nm gold nanoparticles, and a tabletop centrifuge.
4. Vitrification of cells on grids: Plunge-freezing device such as EM GP (Leica Microsystems, Germany), Cryoplunge 3 (Gatan, USA), or a custom-built model. Here we use a Vitrobot MK II (FEI, USA).
5. Blotting of excess liquid: Whatman filter paper (diameter: 47 mm, grade #1) and a device for punching holes into the filter paper.
6. Cryogen for plunge freezing: Ethane/propane (37%/63% vol/vol; ±2%) gas mixture.
7. Cooling agent: Liquid nitrogen (LN_2) in a 4 L cryo-dewar.
8. Grid storage: Cryo-grid storage boxes and transfer release tool (TGS Technologies) or custom-made systems. Long-term LN_2 storage dewars and 50 mL Falcon tubes attached to strings.
9. Cryo-transfer dewars (350 mL) for grid box transfer.

2.2 Electron Cryotomography

1. 300 kV transmission electron cryomicroscope, equipped with a field emission gun (FEG), imaging filter, and direct electron detector. Here we use an FEI Polara G2 (FEI) with a postcolumn Gatan Imaging Filter (GIF) 2002 (Gatan) and a K2 Summit detector (Gatan). For alternative instrumentation *see* **Note 2**.

2. Data collection software for automated acquisition of tilt series such as Latitude (Gatan), Leginon [30], SerialEM [31], or Tomography 4.0 (FEI). Here we use UCSF Tomography [32].
3. Cooling agent: 20–40 L LN_2 for a 2 day session.

2.3 Tomogram Reconstruction, Analysis, and Subtomogram Averaging

1. Computation: Workstation equipped with a modern Intel or AMD processor and at least 8 GB memory. Current tomography software suites can take advantage of CUDA-capable graphics processing units for reconstruction, so an NVIDIA card from the GeForce 700 family or newer is recommended. Here we use an iMac 4 GHz Intel Core i7 with 16 GB memory size and an NVIDIA GeForce GTX 780 M (Apple, USA). For the reconstruction of unbinned data or large subtomogram averaging jobs, we use a Linux workstation with 2 × 10 core Intel Xeon 2.2 GHz processor, 256 GB memory, and NVIDIA GTX 1080 8 GB graphics controller.
2. Drift correction software such as DigitalMicrograph (Gatan) or MotionCorr [33]. Here we use Alignframes, which is part of the IMOD software package [34, 35].
3. Reconstruction software: Several reconstruction algorithms are available, such as supersampling SART [36], NUFFT [37], Tomo3D [38], or Tomography 4.0 (FEI). Here we use the IMOD software package. The package requires a Java runtime environment and for Windows additionally the Unix toolkit Cygwin.
4. Generating movies from image sequences: QuickTime 7 Pro (Apple) or Fiji [39].
5. Subtomogram averaging software, such as Dynamo [40] or Relion [29]. Here we use PEET [41], which is part of the IMOD package and requires the MATLAB Compiler Runtime.

3 Methods

As a complementary source of information on the theoretical background, we recommend the online course "Getting Started in Cryo-EM" (http://cryo-em-course.caltech.edu/).

3.1 Cultivation of Bacteria

Culture the bacteria. Optimize the growth conditions (e.g., medium, temperature, gas atmosphere, shaking/static incubation, liquid/solid medium) to obtain a high expression level for the secretion system of interest (*see* **Note 3**) while at the same time minimizing the cell diameter, resulting in higher-quality data (*see* **Note 4**). Ideally, cells are grown in a few milliliters of liquid medium. For a protocol on how to proceed with cells grown on solid medium, *see* **Note 5**. For a description of how to grow bacteria directly on EM grids, *see* **Note 6**. The expression level of the secretion system of interest is ideally monitored just before freezing, e.g., by fluorescence light microscopy.

3.2 Plunge Freezing of Bacterial Cells

To preserve bacterial cells in a frozen-hydrated near-native state, the sample is applied to an EM grid and plunge frozen in liquid ethane/propane [42] using a Vitrobot (FEI) [43]. A video showing the workflow has been published [44].

1. Prepare gold fiducial markers. Fiducials are added to the sample prior to freezing to allow for the alignment of individual tilt images. Mix 400 μL 10 nm gold nanoparticles with 100 μL 5% (w/v) BSA and vortex for 5 s. Centrifuge the mixture at 14,000 × *g* for 15 min, discard the supernatant, and wash the pellet with 200 μL ddH_2O with an additional centrifugation step at 14,000 × *g* for 10 min. Pool the pellets of eight parallel preparations and resuspend them in ddH_2O (final volume 100 μL). Fiducial markers can be stored at 4 °C for up to 2 months.
2. Prepare sample support. Glow discharging is used for cleaning the EM grids and rendering their surface hydrophilic. Measure the distance between the specimen table and electrode inside the glow-discharge device (Emitech K100X) and adjust it to 2 cm. Place the grids with the carbon side facing up onto a glass slide and place the slide onto the specimen table inside the glow discharger. Close the lid and set the following parameters: Negative discharge 15 mA for 60 s, initial pumping to 1×10^{-1} bar, bleed to 2×10^{-1} bar. For different samples and grids, it might be necessary to adjust the table-electrode distance (typically 2–4 cm), current (typically 15–25 mA), and discharge time (typically 15 s to 3 min). Treated grids should be used for freezing within a few hours (otherwise the procedure can be repeated).
3. Prepare the Vitrobot. Fill the Vitrobot humidifier with ddH_2O with a syringe. Mount the prepunched Whatman filter paper carefully on the blotting pads in the Vitrobot chamber. For bacterial cells, set the temperature to 22 °C and the humidity to 100%. Set the following parameters: wait time 1 s (time between applying your sample on the grid and blotting), blotting time 1–10 s (time the filter paper is pressed onto the grid), drain time 1 s (time from blotting to plunging), and blot offset −2 to −3 mm. For more details on parameters *see* **Note 7**.
4. Prepare the cryogen. Cool the Vitrobot LN_2 reservoir and the central cryogen container with LN_2. Keep filling the LN_2 compartment and wait until the cryogen container is free of LN_2. Start filling the cryogen container with ethane/propane gas mixture until it is completely filled with liquid cryogen. *See* **Note 8** for a comment on the choice of cryogen.

5. Mix the bacterial culture with gold fiducial markers (4:1 vol/vol). Prepare this shortly before plunge freezing to avoid alterations of the sample.
6. Pick up a glow-discharged grid with dry Vitrobot tweezers. Mount the tweezers on the Vitrobot and lift the grid up into the blotting chamber by pressing the foot pedal. Apply 3–4 μL of the cell/gold suspension onto the carbon side of the grid and continue by pressing the foot pedal. The grid is then automatically blotted from both sides, removing excess liquid. After the drain time, the grid is plunged into the liquid ethane/propane mixture. Due to the high thermal conductivity of the cryogen, freezing happens very rapidly without the formation of ice crystals, resulting in vitreous ice [42, 45]. *See* **Note 9** for ways to increase the amount of cells on the grid.
7. From this point, the grids must be kept at liquid nitrogen temperature to prevent thawing, ice contamination, and devitrification. Carefully remove the grid from the cryogen and place it into a labeled and precooled grid storage box.
8. In rare cases, the expression level or assembly of a secretion system might be affected by the short incubation time on the grid or blotting of excess liquid. To verify the assembly state of a fluorescently tagged secretion system after plunge freezing, the frozen grid can be analyzed in a cryogenic fluorescence light microscope [25].
9. For storage, grid boxes are dropped into 50 mL Falcon tubes. The tubes are closed with lids attached to strings and kept in long-term liquid nitrogen storage dewars.

3.3 Loading Grids into Electron Microscope

1. Cool down the loading station chamber and cryopump. Place the grid box and Polara cartridges into the chamber. Load a C-clip ring into a C-clip tool and cool the tip using LN_2. Using cooled forceps, transfer a grid into a cartridge and clamp it with the C-clip ring. Repeat for up to six grids/cartridges total.
2. Detach the cooled multispecimen holder (MSH) from the Polara and attach it to the loading station. Vent the holder and insert the cartridge rod into the chamber. Load the cartridges into the parking positions of the MSH with the threads facing inward. Retract the rod and pump the vacuum in the MSH using the rough pump and cryopump. Start the turbomolecular pump of the Polara. Detach the MSH from the loading station and attach it to the Polara. Click "Pump airlock" in the "Setup" tab of the microscope software and open valves A and B when prompted by the software.
3. Push the MSH to the intended cartridge position. The first position refers to the lowest parking position in the cryo loading station. Use the cooled insertion rod to pick up a cartridge from the MSH. Retract the insertion rod and the MSH.

4. Press "Cartridge" in the "Setup" tab of the microscope software and insert the cartridge into the beam path. Screw the cartridge onto the stage by clockwise turns. Stop with the first sign of resistance and retract the insertion rod.

3.4 Basic Microscope Alignments/ Preparations

Microscope alignments change over time and must be adjusted before data collection. Here we describe the procedure for a cooled FEI Polara instrument equipped with a Gatan imaging filter and K2 Summit direct electron detector.

1. Choose a magnification for data collection. The pixel size on the specimen level should be at least two times smaller than the intended resolution. Keep in mind that it might be necessary to bin the data during processing, resulting in a larger final pixel size. The following microscope preparations should be made at the chosen magnification.
2. To center the condenser (C2) aperture, move the stage in X/Y to an area of broken sample support (hole) and save this position in the "Stage" tab of the microscope software. Drop down the fluorescent screen, condense, and spread the beam repeatedly. Adjust the condenser aperture screws so that the beam condenses and spreads centered without lateral movement.
3. To correct for an astigmatic beam, activate the condenser stigmators in the "Direct alignments" tab and adjust them using the X/Y multifunction knobs.
4. Pivot points are aligned to ensure that the beam stays centered on the camera when the focus is being changed. In "Direct alignments," select "Beam tilt ppx." Condense the beam and minimize movements by adjusting the X/Y multifunction knobs. Repeat for "Beam tilt ppy," click "done," and spread the beam. Lift the viewing screen.
5. To adjust the position of the grid so that the target does not move laterally during tilting (eucentric height), move the stage in X/Y to center a feature, e.g., an ice particle on carbon. Start the "Wobbler" in the "Stage" tab and adjust the height of the grid by pressing the "+/−" Z-axis buttons to minimize movements of the centered feature (use the K2 detector to view with an exposure time of 0.1 s at data collection magnification).
6. Focus the feature using the focus knob.
7. Press the "Eucentric focus" button on the microscope control panel
8. To center the beam along the optical axis of the objective lens, select "Tomo rotation center" in "Direct alignments." Minimize movements of the image (viewed on K2 detector in linear mode) using the X/Y multifunction buttons. Using the focus step size knob, the amplitude of the objective lens current oscillation can be adjusted, which results in different intensities of image movement.

9. To center the objective aperture, drop the viewing screen, switch to diffraction mode, and adjust the objective aperture screws so that the bright spot is centered inside the illuminated disk. Turn off diffraction mode.
10. To correct for objective lens astigmatism, lift the viewing screen and start live-FFT in the DigitalMicrograph software (DM). Activate the objective stigmators in "Direct alignments" and adjust them using the X/Y multifunction knobs, so that the Thon rings on the live-FFT are circular.
11. Move the stage back to the saved "hole"-position (broken area in carbon support) using the "Stage" tab.
12. To align the crossover, switch to an intermediate magnification (e.g., 3000×). Select "Cross-over" in the "Tune" tab. Turn multifunction knob X until the edge of the pumping aperture is visible. Move multifunction knob X in the reverse direction and count the turns until the opposite edge is visible. Center by going back half the number of turns. Repeat procedure for multifunction knob Y and press "Done."
13. The Gatan imaging filter (GIF) is used to remove inelastically scattered electrons from the beam by zero-loss filtering. GIF calibration is performed on the Orius charge-coupled device camera to avoid damage to the K2 chip (for newer models such as the Quantum LS, calibration is performed using the K2). Insert the Orius camera in DM, switch to data collection magnification, set spot size to 3, start "Tune GIF" in DM, and select "Full tune."
14. To acquire gain and dark reference images, start "Prepare Gain Reference" in DM. Follow the program's instructions for recording reference images for linear mode (at spot size 3) and counting/superresolution mode (at spot size 8). Note that this procedure should be repeated daily or if any fixed noise pattern appears in the acquired images.

3.5 Automated Sequential Tilt Series Acquisition

The next step is the acquisition of tilt series (2D projection images from different angles). The sequential recording of tilt series for a number of targets is performed automatically using a suitable software package (*see* Subheading 2.2, **step 2**). Here we describe the use of UCSF Tomography [32]. The program operates by recording a low-magnification atlas of a grid region for the localization of targets, intermediate-magnification images for the inspection of potential targets, and final data collection on a selection of targets. For this purpose, the program specifies five different modes that correspond to different settings of the microscope and the camera: Atlas, Search, Track, Focus, and Collect. Table 1 lists the main function for each mode and a set of parameters that can serve as a good starting point for data collection.

Table 1
UCSF tomography modes and example parameters

UCSF tomogr. mode	Main function	Magnification	Defocus (μm)	Slit width (eV)	Exposure time (s)	Binning	Spot size	Pixel size (nm)	Camera mode
Atlas	Recording of a montage image of a defined grid area	4500×	−50	80	2	1	8	67.5	Counted
Search	Recording of images of potential targets for closer inspection	13,500×	−50	40	2	1	8	15.0	Counted
Track	Tuning eucentricity, targeting, and tracking specimen shift after first leg	13,500×	−20	40	1	1	8	15.0	Counted
Focus	Automated focusing	42,000×	0	20	1	1	8	0.50	Counted
Collect	Final data acquisition	42,000×	−4 to −10	20	1–4	1	8	0.50	Counted

1. Start the "TecnaiServer" on the microscope PC and UCSF Tomography on the K2 computer.
2. Set the parameters from Table 1 for all modes. Select the "Image" tab and the desired mode, click "Configure," and change parameters.
3. Set the magnification, beam intensity, and beam shift for each mode. Select Atlas mode, set magnification, center the beam on the detector, and spread it. The spot size should be chosen so that the desired intensity is obtained by spreading the beam just beyond the edges of the detector. Press "From Scope" to save the beam settings for the Atlas mode. Repeat this procedure for all other modes.
4. A series of calibrations is required. Move the stage to center an ice particle on the carbon support. Go to the "Calibration" tab in UCSF Tomography. Set the defocus values for all modes to 0. Bring the specimen to eucentric height (*see* Subheading 3.4, **step 5**). Focus the specimen. Select Collect mode, click "Configure," and press "Read true focus." Select the "Stage shift" radio button and run this calibration for Atlas and Search modes. Run "Image Shift" calibration for all modes. Run "Focus" calibration for Track and Focus modes. Run "Eucentric" calibration for Track mode. Run "Optical axis"

calibration for Focus mode. All calibrations should be repeated in case the alignments of the microscope or the magnification for any of the modes change.

5. Set the desired defocus values for all modes in the "Configure" window. *See* **Note 10** for comments on the defocus value for final data acquisition.
6. To align the different modes to each other, center a feature (ice particle or crack) that is visible in all modes. Switch to Collect mode. Select the "Align modes" radio button and press "Start." Follow the instructions in the log window. Repeat the procedure for Search mode.
7. The goal now is to record a map in Atlas mode covering around 4–16 squares of the grid (*see* Fig. 1a for an example). Screen your grid at low magnification to identify a region with suitable ice thickness and target density. Move the stage to the center of the identified grid area. Select the "Montage" tab in UCSF Tomography. Enter the size of the atlas map in microns (250 × 250 covers ~4 grid squares on a 200 mesh grid) and a file name. Click "Build" and monitor the automated recording of the atlas map. Note that the recording can be aborted at any time in case a sufficient grid area has been imaged.
8. To identify targets on the atlas, load the atlas in the "Montage" tab. Uncheck "Go to Target." Double click an interesting tile on the map to zoom in. Within this image, double click a spot on the carbon support, which will be used for focusing and finding eucentricity. With every additional double click, potential targets can be selected for closer inspection (collection of Search images in the next step). Click "Zoom Out." A red circle is added on the atlas map to indicate that this area has been visited. Repeat procedure for other tiles of interest.
9. To acquire intermediate-magnification search images of the potential targets marked on the atlas map, switch to the "Target" tab and select the corresponding atlas map file in the "Pre-Rotation" field. Click "Acquire Targets" and monitor the progress (*see* Fig. 1b for an example).
10. To inspect the potential targets, switch to the "Target Review" tab. Target images are saved in a new file (named like the atlas and appended with "tgt"). Load the file. Inspect the potential targets and double click positions that will be used for final data collection. Selected targets can be removed by right clicking. Note that on some images no targets can be selected since they are dedicated to finding eucentricity and focusing.
11. Set tilt series parameters in the "Tomography" tab. Enter the tilt range (−60° to +60°), tilt increment (+1°), and starting angle (−20°). *See* **Note 11** for comments on tilting schemes.

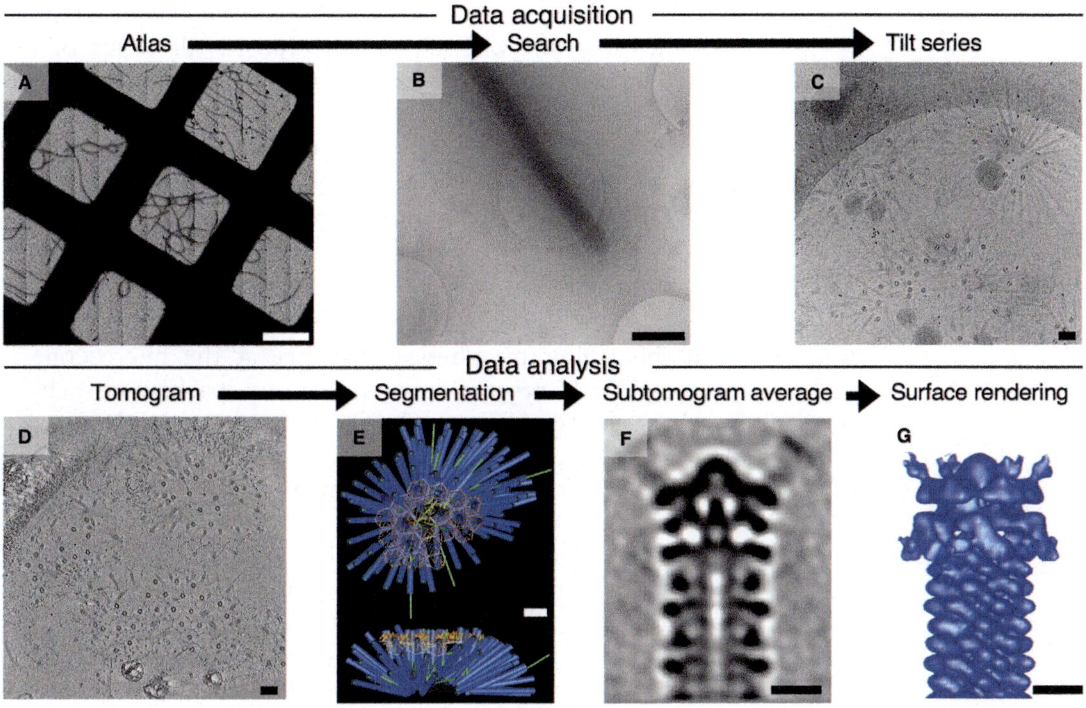

Fig. 1 Steps in ECT data acquisition and analysis. Examples are shown for (**a**) an atlas covering several squares, (**b**) a search image of a bacterial cell lying across the carbon support (**c**) an individual tilt image of arrays of metamorphosis-associated contractile structures, (**d**) a slice through a cryotomogram that was reconstructed from the tilt series shown in **c**, (**e**) two views of a segmentation of the tomogram shown in **d**, (**f**) a slice through an average generated from subvolumes of tomograms similar to **d**, (**g**) an isosurface of the average shown in **f**. **c**–**e** were generated with data from [16]. Bars, 1000 nm in **a** and **b**, 100 nm in **c**–**e**, 10 nm in **f** and **g**

Insert the base name for tilt series into the "File" field and locate the target file in the "Target" field. Press "More" to set additional imaging parameters. Check "Dose Fract" to enable dose fractionation and enter total exposure time and subframe exposure time (*see* **Note 12**). Select "UShort" and "Cryo" for "MRC Data Type." Check "Align" for "ZLP Alignment" (to align zero-loss peak before each tilt series) and "Close at End" (to close column valves after data acquisition).

12. To set the electron dose, move to a hole in the carbon support and switch to Collect mode. Acquire an image with DM and read out the dose for the image (e^-/pixel). By adjusting the beam intensity, spot size, and exposure time, set the dose for individual tilt images. *See* **Note 12** for considerations on total dose, dose rate, spot size, exposure time, and dose-fractionation parameters.

13. Start data collection by clicking "Start" in the "Tomography" tab. The program will sequentially record tilt series of the selected targets (*see* Fig. 1c for an example). Note that data collection can be paused ("Pause") or aborted ("Stop All"). Pressing "Stop" will abort the current tilt series and the program will move on to the next target in the list. Since batch tomography can last many hours or even days, make sure that the LN_2 dewar of the microscope is always filled. Additional grids can be kept at cryogenic temperature inside the MSH (*see* **Note 13** for long-term cooling of the MSH).
14. From this point, no physical presence at the microscope is required. With remote-control software packages, such as TeamViewer (TeamViewer GmbH, Germany) or VNC Viewer (RealVNC Ltd., UK), the progress of tilt series acquisition can be monitored and controlled remotely.

3.6 Data Processing

1. Motion correction. Imperfections in the microscope stage and the incident electron beam result in sample movement during exposure [23]. To correct for this motion, the projection images for each tilt image are read out as subframes (dose fractionation) and saved as image stacks. The subframes of a given tilt image are then computationally aligned to each other, averaged, and saved. Move the uncorrected tilt series and the corresponding subframes to an empty folder. Rename each of the numbered subframe stacks from "(…).mrc001" to "(…)_001.mrc," "(…).mrc002" to "(…)_002.mrc," and so on. Open a terminal window within that folder and enter the command "*alignframes –stack <uncorrected tilt series.mrc> <sub-frames-basename_*> <output file.mrc>*." The measured drift for each frame set can be monitored in the terminal window.
2. The next step is the reconstruction of the tilt series into a 3D image (tomogram). Move the motion-corrected tilt series to a new directory. Open a terminal window, change to the preceding directory, and start the IMOD program eTomo. The documentation for programs of the IMOD package can be accessed via the Help menu. Locate the tilt series (stack) under the "Build Tomogram" tab. *See* **Note 14** for automated data reconstruction.
3. Specify tilt series parameters. Enter the diameter of the gold fiducials (10 nm), enter tilt series axis type (single axis), and choose "cryosample.adoc" as the system template. Click "Scan header" to read pixel size and image rotation from the tilt series file. Click "View raw image stack" to inspect the selected tilt series (*see* Subheading 3.7, **step 1** on using 3dmod for data visualization). Specify tilt images to be excluded from reconstruction (e.g., images in which the grid bar blocks the beam at high tilts) to the "Exclude views" field. Click "Create Com Scripts."

4. The procedure now follows the ETOMO panels from top ("Pre-processing") to bottom ("Clean Up"). Note that intermediate files are saved and it is possible to go back and rerun certain steps with changed parameters. Also note that moving the mouse over an input field in IMOD will show more information on the required input parameters.
5. "Pre-processing." To remove pixels with aberrantly high intensity. Set 12 for "Peak criterion" and 9 for "Difference criterion." Click "Create Fixed Stack" and inspect the processed tilt series by clicking "View Fixed Stack." Click "Use Fixed Stack" and proceed by clicking "Done."
6. "Coarse Alignment." To align successive tilt images, click "Calculate Cross-Correlation." Bin the coarsely aligned image stack by 2 and check "Reduce size with antialiasing filter" (note that this binning will not affect the size of the final reconstruction). Click "Generate Coarse Aligned Stack" and inspect the aligned stack by clicking "View Aligned Stack in 3dmod." Use "Midas" to fix any misaligned image pairs. Proceed with "Done."
7. "Fiducial Model Gen." The selection of gold markers and the tracking of the selected markers throughout the tilt series can be performed automatically using the program RAPTOR, or manually. Here, we manually select the fiducial markers after selecting "Make seed and track" and "Make seed model manually." Open the aligned tilt series by clicking "Seed Fiducial Model." Select 10–20 gold fiducial markers distributed over the entire 0° tilt image, and save the model before closing 3DMOD. Select the "Track Beads" tab and track the fiducials over the whole tilt series with "Track Seed Model." Open the fiducial model by clicking "Fix Fiducial Model" and modify the model to minimize the number of gaps and to fix mis-tracked fiducials. Save the model, click "Done," and proceed.
8. "Fine Alignment." Calculate the fine alignment with "Compute Alignment." Inspect by clicking "View/Edit Fiducial Model." Inspect fiducials by clicking "Go to Next Big Residual" and if appropriate fix the model by clicking "Move Point by Residual." Repeat until there are no more residuals. Save model and repeat "Compute alignment." Proceed with "Done."
9. "Tomogram Positioning." Enter "1500" as the preliminary tomogram Z-height in the field "Positioning tomogram thickness" and click "Create Whole Tomogram." Click "Create Boundary Model" and flip the tomogram 90° using "Edit > Image > Flip/Rotate." Indicate the boundaries of the cell on a central slice ("View axis position") by drawing two horizontal lines. Repeat this for one higher and one lower slice ("View axis position"). Save the model by typing "s," close 3DMOD, and press "Compute Z Shift & Pitch Angles." Click "Create Final Alignment" and proceed with "Done."

10. “Final Aligned Stack.” Enter the binning factor in the “Aligned image stack binning” field. We usually use a factor of 2, resulting in data with 1920 pixels in X and 1854 pixels in Y. Check “Use linear interpolation” and “Reduce size with antialiasing filter.” Click “Create Full Aligned Stack.” Other, optional processing steps in this tab allow for the erasing of gold beads, CTF correction, and 2D filtering.
11. “Tomogram Generation.” Choose between two different tomogram reconstruction algorithms: weighted backprojection in Fourier space [46] or the Simultaneous Iterative Reconstruction Technique (SIRT) [47]. Click “Generate Tomogram” and proceed with “Done.”
12. “Post-processing.” Open the tomogram with “3dmod Full Volume.” Select a slice with structures of interest. Use the contrast sliders to determine the maximum and minimum values for “black” and “white,” needed to visualize the structure of interest. Enter the numbers into the “Scale to match contrast” fields. For “Reorientation,” choose either “Rotate around X” or “Swap Y and Z dimensions.” Note that this option will affect the handedness of the data. The correct option must be determined as described previously [48]. Click “Trim Volume” and “Done.”
13. “Clean Up.” Select all listed intermediate files and click “Delete Selected” to save disk space. Finish by clicking “Done.” Note that the “Com Scripts Interface” for a specific reconstruction can be reopened using the “.edf” file.

3.7 Data Visualization

1. View the tomogram. Open the “.rec” file in 3DMOD of the IMOD package using the command *“3dmod <input file.rec>”* (*see* Fig. 1d for an example). 3DMOD offers different viewing tools, including “ZAP,” “XYZ,” and “Slicer.” “Slicer” is particularly useful to detect macromolecular complexes (such as secretion systems), based on the possibilities of rotating the tomogram around all three axes (use sliders “X-, Y-, Z-rotation”), pan up and down slice by slice (use slider “View axis position”), and average multiple slices to enhance contrast (enter value in “Img”). To measure distances, select “model” mode in the 3DMOD information window, click the left mouse button on point A, move the mouse to point B, and type “q.” The A-B distance (in 3D) will be reported in the log of the 3DMOD information window.
2. Generate 3D models by segmentation and visualize them. Segmentation (visualization of selected pixels) can be performed automatically, e.g., by applying a density threshold (use 3DMOD’s “Isosurface” in the “Image” drop-down menu) or by algorithms that trace specific features such as membranes or filaments (e.g., SHAPE, at bio3d.colorado.edu, or other automated segmentation approaches [49–51]). The poor con-

trast in low-dose, low-defocus cryotomograms, however, often challenges available programs to deliver meaningful models. Manual segmentation can be performed in such cases or for structures for which no programs are available (*see* Fig. 1e for an example). AMIRA (FEI) and 3DMOD are frequently used for manual segmentation.

Open the tomogram in 3DMOD, select the "ZAP" viewer, and open the "Drawing Tools" under the "Special" menu. To segment membranes, use "sculpt" to outline the membrane in a given slice (select "Edit > Object > Type... > Closed"). Repeat segmentation for the same membranous structure in every approximately tenth slice. Interpolate the contour for the intermediate slices ("Special > Interpolator"). Inspect and refine the model (can be viewed without tomogram slices in "Image > Model View"). Use meshing (select "Model View" window followed by "Edit > Objects... > Meshing") to generate a surface based on the modeled contours. For filamentous structures, change the object type to "open" and draw lines with the middle mouse button. Apply "meshing" to represent structures as rods or tubes. Save the model file as a ".mod" file.

3. Make a movie. 3D data (tomograms and models) can be effectively presented as movies. Start 3DMOD in a new folder (a sequence of images will be saved inside this folder). Open a tomogram. Remove red/yellow navigation markers by typing "shift-T." Select the "Movie/Montage" option under "File" in the menu bar. Choose start and end frame and other options. Make a test run by pressing the command key and middle mouse button at the same time. Save an image sequence by selecting an output format under "Snapshot" and start sequence as described earlier. The "Movie/Montage..." option for the Model viewer functions similarly; however, it offers the possibility of introducing rotations and switching off tomogram slices. Use QUICKTIME 7 PRO or FIJI to generate a movie from the image sequence.

3.8 Subtomogram Averaging

Contrast and resolution of cryotomographic data can be improved by aligning and averaging repeating subvolumes. This is particularly effective for secretion systems. Here we describe a protocol using PEET [41, 52]. The workflow takes advantage of the cell-envelope-spanning localization of the targets, which allows for a reduced angular search space, resulting in reduced computational effort. Note that PEET tutorials can be found at https://goo.gl/nsXEtn.

1. Mark subvolumes as model points and indicate their orientation. Open the tomogram in 3DMOD and choose "Edit > Object > Type" from the menu bar. Select object type "Open" and add a "Symbol" to visually indicate modeled points. Select the "Model" radio button in the small 3DMOD information win-

dow. Using the "Slicer," rotate the tomogram to the best possible view of a particular secretion system. Click with the middle mouse button on the feature of interest to define the center of a subvolume. Click a second point to indicate the orientation of the subvolume (e.g., perpendicular to the cell envelope or along a tube). The program draws a line between both model points (green line and circles in Fig. 2.). Type "n" (for new Contour) and repeat for all subvolumes in this tomogram. Record the number of points within the object and save the model as a ".mod" file. Repeat the procedure for additional tomograms and save the corresponding model files. Note that all tomograms should have the same pixel size.

2. Compute an initial motive list. This list contains information about the orientation of the modeled subvolumes. Run the IMOD command "*stalkInit <input file.mod>*" in the folder where the model is located.
3. Open PEET via the eTomo graphical user interface and select a base name and a folder. Parameters from previous projects can be imported.
4. Load the tomograms and corresponding models ("head.mod" file, generated by the *stalkInit* command) into the respective fields of the "Volume Table."
5. As "Reference" for the first round of alignment, choose a random subvolume.

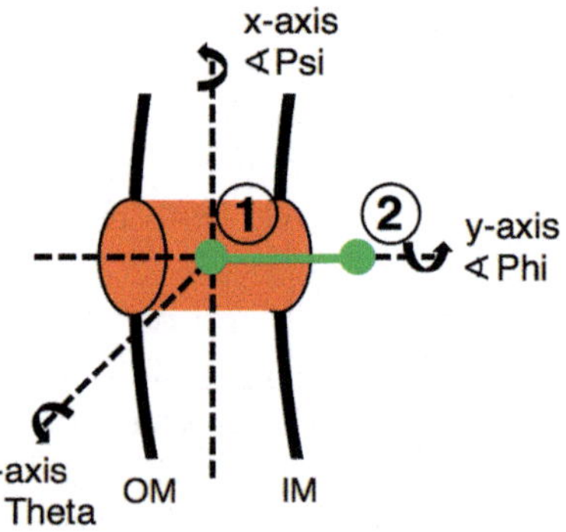

Fig. 2 Modeling of subvolumes for subsequent subtomogram averaging. Shown is a schematic of a cell envelope-spanning secretion system (*OM* outer membrane, *IM* inner membrane). Points 1 and 2 and the connecting vector (*green*) represent the model that is manually generated for each subvolume. *Dashed lines*: axes assigned to each subvolume based on orientation of vector; *arrows*: assignment of search angles in PEET

6. Enter the "Volume size" in voxels (volumetric pixels). Use the distance measurement tool (*see* Subheading 3.7, **step 1**) in 3DMOD to estimate a suitable box size.
7. In the Particle Y axis box, select "user supplied csv files." This option defines the Y-axis for all subvolumes as the vector that was modeled for each subvolume, which allows for a smaller angular search, saving computational resources. Make sure that the "..._RotAxes.csv" file for each tomogram, generated by the *stalkInit* command, is in the same folder as the corresponding tomogram and that both share the same name besides the "_RotAxes.csv" extension.
8. Click "User supplied csv files" in the "Initial Motive List" box to load the "..._In it MOTL.csv" file generated by *stalkInit* into the appropriate field in the "Volume Table." This file specifies rotations or translations required for an approximate alignment of each particle to the reference.
9. Generate an "Iteration Table" in the "Run" tab. PEET aligns individual subvolumes to a reference volume by rotational and translational movements over several iterations (reducing search space and step size for each iteration). A new reference is generated from a subset of the aligned and averaged subvolumes at the end of each run for use in the next iteration. The "Iteration Table" specifies the search and alignment parameters (*see* Table 2 for values that are suitable for the approach

Table 2
PEET subtomogram averaging iteration table

	Angular search range							High-frequency filter		
	Phi		Theta		Psi					
Run #	Max	Step	Max	Step	Max	Step	Search distance	Cutoff	Sigma	Reference threshold
1	60	20	7.5	2.5	7.5	2.5	15	0.2	0.01	2/3 of all particles
2	30	10	7.5	2.5	7.5	2.5	15	0.2	0.01	2/3 of all particles
3	15	5	7.5	2.5	7.5	2.5	15	0.2	0.01	2/3 of all particles
4	7.5	2.5	7.5	2.5	7.5	2.5	10	0.5	0.005	2/3 of all particles
5	3.75	1.25	3.75	1.25	3.75	1.25	10	0.5	0.005	2/3 of all particles
6	1.875	0.625	1.875	0.625	1.875	0.625	10	0.5	0.005	2/3 of all particles
7	1.0	0.3	1.0	0.3	1.0	0.3	5	0.5	0.005	2/3 of all particles

described here). Specify the maximum range of rotation ("Max") as well the "Step" for "Phi" (angle around particle Y-axis), "Theta" (angle around particle Z-axis), and "Psi" (angle around particle X-axis). Angles and axes are indicated in Fig. 2. Enter the "Search Distance," high frequency filter "Cutoff" and "Sigma," as well as "Reference Threshold." Move the mouse over an input field to get more information about the parameters. Press "Insert" to enter another iteration.

10. Calculate the total number of subvolumes and set the numbers of particles to be averaged ("Start," "Incr," "End," and "Additional numbers"). For example, with 53 subvolumes you can define 10 as the start, 15 as increment, 40 as end, and 53 as additional. This will result in averages with 10, 25, 40, and 53 subvolumes, respectively. The remaining parameters ("Optional/Advanced Features") can be left at their default settings. Activate parallel processing under "Options > Settings" as subtomogram averaging is computationally intensive. Click "Run" to start the computation.
11. View averages by clicking "Open averages in 3dmod" and select the "Slicer" window (*see* Fig. 1f for an example). Switch between averages with different numbers of particles using the arrowheads at the top of the window. For an intensity threshold rendering view select "Image > Isosurface" (*see* Fig. 1g for an example).

4 Notes

1. The choice of the hole size and the distance between holes in the carbon film depends on the sample and the data collection magnification. For Quantifoils, both parameters are specified in the grid type, e.g., R2/1 designates a grid with 2 μm hole diameter and 1 μm hole distance. Larger hole diameters and smaller hole distances provide more area for imaging without a carbon background; however, they result in higher fragility (resulting in more breakage during transfers and more beam-induced sample movement during data collection). The mesh number specifies the number of squares on the grid. A higher mesh number results in smaller area per square, which in turn provides higher stability.
2. Alternative 300 kV instruments are the Titan Krios (FEI), Titan Halo (FEI), and JEOL3200 (JEOL, Japan). A 200 kV instrument, such as a Tecnai F20 (FEI) with cryosample holder (Gatan) or Talos Arctica (FEI), can be used for thin specimens or for screening freezing conditions. The imaging filter is critical for improving the signal-to-noise ratio by removing inelastically scattered electrons, in particular for thicker samples.

3. It is usually impossible to discern whether a given cell will express a certain secretion system prior to collecting a tilt series and reconstructing the tomogram. It is therefore crucial to maximize both the percentage of cells expressing the secretion system and the number of secretion systems per cell. Some secretion systems are inducible by specific growth conditions or genetic manipulations of regulatory genes [53–55].
4. Thicker samples produce noisier data based on inelastic and multiple scattering events. Certain growth media (e.g., starvation media) or genetic manipulations [56] can be used to reduce the cell diameter. Gentle lysozyme treatment will also result in a better signal-to-noise ratio as the cells lose some of their cytoplasmic content [57].
5. If cells are cultured on a solid medium, collect some colonies with an inoculation loop, resuspend cells in 200 μL liquid medium, and immediately continue with plunge freezing.
6. If cells are directly grown on grids, gold mesh material is preferred over copper to avoid cytotoxic effects. Grids are sterilized under ultraviolet light in a sterile workbench for 15 min and subsequently glow discharged (*see* Subheading 3.2, **step 2**). Using sterile forceps, the grids are then placed on the bottom of a bacterial liquid culture in a 12 well plate (Thermo Fisher Scientific, USA). The plate is incubated, and the cell density on the grid can be checked using a light microscope.
7. The critical parameters for the freezing process are blot time and blot offset. Blot time implies the time (in seconds) the blotting paper is pressed against the grid. Longer blotting results in thinner ice, although excessive blotting can be detrimental. The blot time for bacterial cells usually ranges from 1 to 10 s. Blot offset is the vertical location of a grid before the blot is applied. This will change the grid's position on the wedge, affecting the gradient in ice thickness across the grid. A good starting point for bacterial cells is a blot time of 2 s and an offset of −3 mm.
8. Pure ethane is also frequently used as a cryogen. The advantage of ethane/propane, however, is that the mixture does not solidify even in close thermal contact with LN_2. This ensures that the cryogen can be kept at low temperatures to achieve optimal vitrification without the need for thawing solidified cryogen [42].
9. The number of cells on the grid can be increased by repeated cycles of applying a sample on the grid and blotting the grid (change the number of blottings in the option panel of the Vitrobot software). Alternatively, blotting from only one side (opposite side of sample) can be very effective. This can be achieved by blotting manually with a forceps-held filter paper

or by replacing one Whatman paper inside the Vitrobot chamber with a teflon sheet (Miroslava Schaffer, personal communication).

10. The defocus should be chosen as a compromise based on the following considerations: For low defocus values, the CTF oscillates slowly, resulting in good information transfer for higher spatial frequencies and poor information transfer for lower spatial frequencies. The goal is therefore to choose a defocus value as close as possible to focus while still being able to detect the secretion systems of interest in the individual tomograms. In a typical ECT experiment with bacterial cells, a defocus value between −4 and −10 μm is usually chosen.

11. The tilt increment needed to obtain a certain resolution depends on the diameter of the sample and can be approximated by the Crowther criterion [58]. In practice, tilt increments of <1° are not applicable owing to dose limitation and the resulting low contrast in individual tilt images. Bacterial cells are typically imaged with an increment of 1°. For subtomogram averaging approaches, increments can be increased to 2° or even 4°. The tilt range is limited by the grid holder and the extremes are typically chosen in the ranges of −60° to −70° and +60° to +70°.

    Low-tilt images provide higher-quality data compared to high-tilt images owing to specimen thickness. It therefore became popular to start the tilt series at −30° (instead of 0°), tilting toward 60°, followed by a second leg from −30° to −60°. This scheme results in reduced beam damage for the most informative low-tilt projection images and allows for the computational removal of high-tilt information from the final reconstruction [28, 59, 60].

12. Choose data acquisition parameters based on the following considerations: Choose the total electron dose between 60 and 180 $e^-/Å^2$. Higher doses result in more beam damage and can be detrimental for resolving high-resolution features (which is particularly important for subtomogram averaging approaches). Choose the tilt increment and tilt range (*see* **Note 11**) and calculate the dose per tilt image. Choose a dose rate <15 e^-/pix/s (<10 for high-resolution approaches) to avoid coincidence loss during electron counting [34]. Choose a high spot size to obtain a coherent beam (typically 8–11) that makes it possible to illuminate the entire detector. Choose an exposure time that results in the intended total dose (typically 1–5 s). The fast readout of the K2 detector allows for dose fractionation (reading out subframes) and correction for sample motion during image acquisition (which particularly occurs at high tilts). To ensure enough contrast to allow for the proper alignment of the subframes, choose a sufficient subframe exposure time (typically 0.2–0.5 s).

13. Because the standard multispecimen holder LN_2 dewar lasts for only ~4 h, we usually replace it with a 1.5 L coffee thermos placed on a lifting platform, which will last for at least 14 h.
14. For screening experiments or during data collection, it can be useful to run automated tomogram reconstruction using programs such as BATCHRUNTOMO (part of IMOD), RAPTOR [61], or TOMOAUTO [62].

Acknowledgments

We thank D. Böck, R. Kooger, and P. Szwedziak for comments on the manuscript. G. L. Weiss was supported by a Boehringer Ingelheim Fonds PhD Fellowship. The Pilhofer Lab is supported by grants from ETH Zürich, the European Research Council, the Swiss National Science Foundation, and the Helmut Horten Foundation.

References

1. Costa TD, Felisberto-Rodrigues C, Meir A, Prevost MS, Redzej A, Trocker M, Waksman G (2015) Secretion systems in Gram-negative bacteria: structural and mechanistic insights. Nat Rev Microbiol 13:343–359
2. Gan L, Jensen GJ (2012) Electron tomography of cells. Q Rev Biophys 45:27–56
3. Harapin J, Eibauer M, Medalia O (2013) Structural analysis of supramolecular assemblies by cryo-electron tomography. Structure 21:1522–1530
4. Briggs JAG (2013) Structural biology in situ—the potential of subtomogram averaging. Curr Opin Struct Biol 23:261–267
5. Lučič V, Rigort A, Baumeister W (2013) Cryo-electron tomography: the challenge of doing structural biology in situ. J Cell Biol 202:407–419
6. Asano S, Engel BD, Baumeister W (2016) In situ cryo-electron tomography: a post-reductionist approach to structural biology. J Mol Biol 428:332–343
7. Pilhofer M, Ladinsky MS, McDowall AW, Jensen GJ (2010) Bacterial TEM. Methods Cell Biol 96:21–45
8. Schur FKM, Obr M, Hagen WJH, Wan W, Jakobi AJ, Kirkpatrick JM, Sachse C, Kräuslich HG, Briggs JAG (2016) An atomic model of HIV-1 capsid-SP1 reveals structures regulating assembly and maturation. Science 353:506–508
9. Basler M, Pilhofer M, Henderson GP, Jensen GJ, Mekelanos JJ (2012) Type VI secretion requires a dynamic contractile phage tail-like structure. Nature 483:182–186
10. Abrusci P, Vergara-Irigaray M, Johnson S, Beeby MD, Hendrixson DR, Roversi P, Friede ME, Deane JE, Jensen GJ, Tang CM, Lea SM (2013) Architecture of the major component of the type III secretion system export apparatus. Nat Struct Mol Biol 20:99–104
11. Kawamoto A, Morimoto YV, Miyata T, Minamino T, Hughes KT, Kato T, Namba K (2013) Common and distinct structural features of *Salmonella* injectisome and flagellar basal body. Sci Rep 3:3396
12. Kudryashev M, Stenta M, Schmelz S, Amstutz M, Wiesand U, Castaño-Diez D, Degiacomi MT, Münnich S, Bleck CKE, Kowal J, Diepold A, Heinz DW, Dal Peraro M, Cornelis GR, Stahlberg H (2013) In situ structural analysis of the Yersinia enterocolitica injectisome. elife 2:e00792
13. Nans A, Saibil HR, Hayward RD (2014) Pathogen-host reorganization during Chlamydia invasion revealed by cryo-electron tomography. Cell Microbiol 16:1457–1472
14. Pilhofer M, Aistleitner K, Ladinsky MS, König L, Horn M, Jensen GJ (2014) Architecture and host interface of environmental chlamydiae revealed by electron cryotomography. Environ Microbiol 16:417–429
15. Radics J, Königsmaier L, Marlovits TC (2014) Structure of a pathogenic type 3 secretion system in action. Nat Struct Mol Biol 21:82–87

16. Shikuma NJ, Pilhofer M, Weiss GL, Hadfield MG, Jensen GJ, Newman DK (2014) Marine tubeworm metamorphosis induced by arrays of bacterial phage tail-like structures. Science 343:529–533
17. Hu B, Morado DR, Margolin W, Rohda JR, Arizmendi O, Picking WL, Picking WD, Liu J (2015) Visualization of the type III secretion sorting platform of Shigella flexneri. Proc Natl Acad Sci U S A 112:1047–1052
18. Kudryashev M, Diepold A, Amstutz M, Armitage JP, Stahlberg H, Cornelis GR (2015) Yersinia enterocolitica type III secretion injectisomes form regularly spaced clusters, which incorporate new machines upon activation. Mol Microbiol 95:875–884
19. Nans A, Kudryashev M, Saibil HR, Hayward RD (2015) Structure of a bacterial type III secretion system in contact with a host membrane in situ. Nat Commun 6:10114
20. Chang YW, Rettberg LA, Treuner-Lange A, Iwasa J, Segaard-Anderson L, Jensen GJ (2016) Architecture of the type IVa pilus machine. Science 351:1165–1172
21. Kühlbrandt W (2014) The resolution revolution. Science 343:1443–1444
22. McMullan G, Faruqi AR, Clare D, Henderson R (2014) Comparison of optimal performance at 300 keV of three direct electron detectors for use in low dose electron microscopy. Ultramicroscopy 147:156–163
23. Campbell MG, Cheng A, Brilot AF, Moeller A, Lyumkis D, Veesler D, Pan J, Harrison SC, Potter CS, Carragher B, Grigorieff N (2012) Movies of ice-embedded particles enhance resolution in electron cryo-microscopy. Structure 20:1823–1828
24. Danev R, Buijsse B, Khoshouei M, Plitzko JM, Baumeister W (2014) Volta potential phase plate for in-focus phase contrast transmission electron microscopy. Proc Natl Acad Sci U S A 111:15635–15640
25. Briegel A, Chen S, Koster AJ, Plitzko JM, Schwartz CL, Jensen GJ (2010) Correlated light and electron cryo-microscopy. Methods Enzymol 481:317–341
26. Chang YW, Chen S, Tocheva EI, Treuner-Lange A, Löbach S, Søgaard-Anderson L, Jensen GJ (2014) Correlated cryogenic photoactivated localization microscopy and cryo-electron tomography. Nat Methods 11:737–739
27. Rigort A, Bäuerlein FJB, Villa E, Eibauer M, Laugks T, Baumeister W, Plitzko JM (2012a) Focused ion beam micromachining of eukaryotic cells for cryoelectron tomography. Proc Natl Acad Sci U S A 109:4449–4454
28. Hagen WJH, Wan W, Briggs JAG (2016) Implementation of a cryo-electron tomography tilt-scheme optimized for high resolution subtomogram averaging. J Struct Biol
29. Bharat TAM, Russo CJ, Löwe J, Passmore LA, Scheres SHW (2015) Advances in single-particle electron cryomicroscopy structure determination applied to sub-tomogram averaging. Structure 23:1743–1753
30. Suloway C, Pulokas J, Fellmann D, Cheng A, Guerra F, Quispe J, Stagg S, Potter CS, Carragher B (2005) Automated molecular microscopy: the new Leginon system. J Struct Biol 151:41–60
31. Mastronarde DN (2005) Automated electron microscope tomography using robust prediction of specimen movements. J Struct Biol 152:36–51
32. Zheng SQ, Keszthelyi B, Branlund E, Lyle JM, Braunfeld MB, Sedat JW, Agard DA (2007) UCSF tomography: an integrated software suite for real-time electron microscopic tomographic data collection, alignment, and reconstruction. J Struct Biol 157:138–147
33. Li X, Mooney P, Zheng S, Booth CR, Braunfeld MB, Gubbens S, Agard DA, Cheng Y (2013) Electron counting and beam-induced motion correction enable near-atomic-resolution single-particle cryo-EM. Nat Methods 10:584–590
34. Mastronarde DN (2008) Correction for non-perpendicularity of beam and tilt axis in tomographic reconstructions with the IMOD package. J Microsc 230:212–217
35. Kremer JR, Mastronarde DN, McIntosh JR (1996) Computer visualization of three-dimensional image data using IMOD. J Struct Biol 116:71–76
36. Kunz M, Frangakis AS (2014) Super-sampling SART with ordered subsets. J Struct Biol 188:107–115
37. Chen Y, Förster F (2014) Iterative reconstruction of cryo-electron tomograms using non-uniform fast Fourier transforms. J Struct Biol 185:309–316
38. Agulleiro JI, Fernandez JJ (2015) Tomo3D 2.0—exploitation of advanced vector eXtensions (AVX) for 3D reconstruction. J Struct Biol 189:147–152
39. Schindelin J, Arganda-Carreras I, Frise E, Kaynig V, Longair M, Pietzsch T, Preibisch S, Rueden C, Saalfeld S, Schmid B, Tinevez JY, White DJ, Hartenstein V, Eliceiri K, Tomancak P, Cardona A (2012) Fiji: an open-source platform for biological-image analysis. Nat Methods 9:676–682

40. Castaño-Díez D, Kudryashev M, Arheit M, Stahlberg H (2012) Dynamo: a flexible, user-friendly development tool for subtomogram averaging of cryo-EM data in high-performance computing environments. J Struct Biol 178:139–151
41. Nicastro D, Schwartz C, Pierson J, Gaudette R, Porter ME, McIntosh JR (2006) The molecular architecture of axonemes revealed by cryo-electron tomography. Science 313:944–948
42. Tivol WF, Briegel A, Jensen GJ (2008) An improved cryogen for plunge freezing. Microsc Microanal 14:375–379
43. Iancu CV, Tivol WF, Schooler JB, Dias PD, Henderson GP, Murphy GE, Wright ER, Li Z, Yu Z, Briegel A, Gan L, He Y, Jensen GJ (2006) Electron cryotomography sample preparation using the Vitrobot. Nat Protoc 1:2813–2819
44. Chen S, McDowall A, Dobro MJ, Briegel A, Ladinsky M, Shi J, Tocheva EI, Beeby M, Pilhofer M, Ding HJ, Li Z, Gan L, Morris DM, Jensen GJ (2010) Electron cryotomography of bacterial cells. J Vis Exp 39:e1943
45. Brüggeller P, Mayer E (1980) Complete vitrification in pure liquid water and dilute aqueous solutions. Nature 288:569–571
46. Radermacher M (2007) Weighted back-projection methods. In: Electron tomography. Springer, New York
47. Wolf D, Lubk A, Lichte H (2014) Weighted simultaneous iterative reconstruction technique for single-axis tomography. Ultramicroscopy 136:15–25
48. Briegel A, Pilhofer M, Mastronarde DN, Jensen GJ (2013) The challenge of determining handedness in electron tomography and the use of DNA origami gold nanoparticle helices as molecular standards. J Struct Biol 183:95–98
49. Rigort A, Günther D, Hegerl R, Baum D, Weber B, Prohaska S, Medalia O, Baumeister W, Hege HC (2012b) Automated segmentation of electron tomograms for a quantitative description of actin filament networks. J Struct Biol 177:135–144
50. Volkmann N (2002) A novel three-dimensional variant of the watershed transform for segmentation of electron density maps. J Struct Biol 138:123–129
51. Baker ML, Yu Z, Chiu W, Bajaj C (2006) Automated segmentation of molecular subunits in electron cryomicroscopy density maps. J Struct Biol 156:432–441
52. Heumann JM, Hoenger A, Mastronarde DN (2011) Clustering and variance maps for cryo-electron tomography using wedge-masked differences. J Struct Biol 175:288–299
53. Galán JE, Curtiss R (1990) Expression of Salmonella typhimurium genes required for invasion is regulated by changes in DNA supercoiling. Infect Immun 58:1879–1885
54. Basler M, Ho BT, Mekalanos JJ (2013) Tit-for-Tat: Type VI secretion system counterattack during bacterial cell-cell interactions. Cell 152:884–894
55. Eichelberg K, Galán JE (1999) Differential regulation of Salmonella typhimurium type III secreted proteins by pathogenicity island 1 (SPI-1)-encoded transcriptional activators InvF and hilA. Infect Immun 67:4099–4105
56. Farley MM, Hu B, Margolin W, Liu J (2016) Minicells, back in fashion. J Bacteriol 198: 1186–1195
57. Briegel A, Wong ML, Hodges HL, Oikonomou CM, Piasta KN, Harris MJ, Fowler DJ, Thompson LK, Falke JJ, Kiessling LL, Jensen GJ (2014) New insights into bacterial chemoreceptor Array structure and assembly from electron cryotomography. Biochemistry 53:1575–1585
58. Crowther RA, DeRosier DJ, Klug A (1970) The reconstruction of a three-dimensional structure from projections and its application to electron microscopy. Proc R Soc A 317: 319–340
59. Wan W, Briggs JAG (2016) Cryo-electron tomography and subtomogram averaging. Methods Enzymol 579:329–367
60. Pfeffer S, Burbaum L, Unverdorben P, Pech M, Chen Y, Zimmermann R, Beckman R, Förster F (2015) Structure of the native Sec61 protein-conducting channel. Nat Commun 6:8403
61. Amat F, Moussavi F, Comolli LR, Elidan G, Downing KH, Horowitz M (2008) Markov random field based automatic image alignment for electron tomography. J Struct Biol 161:260–275
62. Morado DR, Hu B, Liu J (2016) Using Tomoauto: a protocol for high-throughput automated cryo-electron tomography. J Vis Exp 107:e53608

Chapter 28

Structural Analysis of Protein Complexes by Cryo Electron Microscopy

Tiago R.D. Costa, Athanasios Ignatiou, and Elena V. Orlova

Abstract

Structural studies of biocomplexes using single-particle cryo-electron microscopy (cryo-EM) is now a well-established technique in structural biology and has become competitive with X-ray crystallography. The latest advances in EM enable us to determine structures of protein complexes at 3–5 Å resolution for an extremely broad range of sizes from ~200 kDa up to hundreds of megadaltons (Bartesaghi et al., Science 348(6239):1147–1151, 2051; Bai et al., Nature 525(7568):212–217, 2015; Vinothkumar et al., Nature 515(7525):80–84, 2014; Grigorieff and Harrison, Curr Opin Struct Biol 21(2):265–273, 2011). The majority of biocomplexes comprise a number of different components and are not amenable to crystallisation. Secretion systems are typical examples of such multi-protein complexes, and structural studies of them are extremely challenging. The only feasible approach to revealing their spatial organisation and functional modification is cryo-EM. The development of systems for digital registration of images and algorithms for the fast and efficient processing of recorded images and subsequent analysis facilitated the determination of structures at near-atomic resolution. In this review we will describe sample preparation for cryo-EM, how data are collected by new detectors, and the logistics of image analysis through the basic steps required for reconstructions of both small and large biological complexes and their refinement to nearly atomic resolution. The processing workflow is illustrated using examples of EM analysis of a Type IV Secretion System.

Key words Cryo-electron microscopy, Sample preparation, Single particle analysis, Image processing, Type IV secretion system

1 EM Advances in Studies of Macro-Complexes (Type IV Secretion Systems)

The complexity of experimental and computational procedures used for studies of the structure–function relationships of biological complexes is growing significantly. One has to use different approaches to uncover conformational changes linked to the functional activity of the complexes. X-ray, nuclear magnetic resonance (NMR), and electron microscopy (EM), combined with biochemical and biophysical methods, allow for a deeper understanding of the mechanisms which underlie such macromolecular complex functions. This was clearly demonstrated by studies of a

Laure Journet and Eric Cascales (eds.), *Bacterial Protein Secretion Systems: Methods and Protocols*, Methods in Molecular Biology, vol. 1615, DOI 10.1007/978-1-4939-7033-9_28, © Springer Science+Business Media LLC 2017

ribosome [1]. Recent advances in EM, such as the invention of direct electron detection cameras, systems for automated data collection, and the development of new powerful image processing algorithms, have dramatically expanded the range of biological macromolecules suitable for study by this technique. The main advantages of EM are that it does not require crystallisation of samples and it is able to work with biocomplexes within a large range of sizes: from ~150 kDa to several hundreds of megadaltons [2–5]. Additionally, image processing software packages have been significantly improved to make it possible to analyse the quality of images and distortions caused by the microscope that prevent obtaining high-resolution structures. New approaches were developed to reveal more consistently the sample quality: to assess its homogeneity, to separate different conformations within the sample, and evaluate distributions of particles between different conformations [6–8]. A number of different packages with sophisticated image processing algorithms are used for the analysis of macromolecular complexes with different symmetries or asymmetry. Another positive aspect of modern achievements is that computing power is steadily increasing, making it possible to analyse many hundreds of thousands of particle images from heterogeneous samples. Nonetheless, the basic workflow of sample imaging and image processing remains the same (Fig. 1) [9].

This tremendous success in improving the resolution of EM structures during the last decade would not have been possible without cryo sample preparation and what we now call cryo-EM. This approach for sample preparation and the combination of methods of X-ray crystallography with EM have made it possible to achieve near-atomic-resolution details for some of the Gram-negative bacterial secretion systems. These structures have given us unprecedented understanding of the mechanistic details of how bacteria assemble these highly specialised nano-machines to secrete proteins and DNA to the bacterial extracellular space and to the eukaryotic or bacterial target cell. Among the Gram-negative bacterial secretion systems, Type IV Secretion Systems (T4SS) possess the unique ability to secrete proteins, DNA, or protein–DNA complexes in an adenosine triphosphate–dependent process. Given the competency for T4SSs to secrete such a variety of substrates involved in the pathogenesis and spreading of conjugative plasmids encoding antibiotic resistance genes, this secretion system become an important target for structural biology studies [10].

Among all Gram-negative T4SSs, those encoded by the pTi plasmid of *Agrobacterium tumefaciens*, together with the conjugative pKM101 and pR388 plasmids from *E. coli*, are the best characterized. This macromolecular structure is composed of 12 proteins: VirB1–VirB11 and VirD4 [11]. The first main advance in the understanding of the general architecture of a T4SS took place when the cryo-EM structure of the so-called core outer-membrane

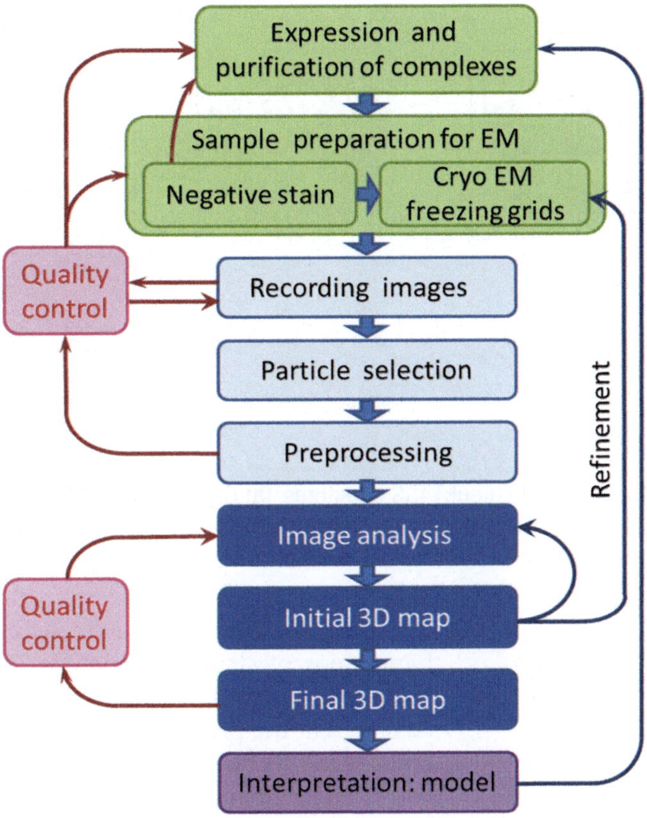

Fig. 1 Workflow of EM structural analysis. In *green* is the experimental part of structural analysis. The computational part is shown in *light* and *dark blue*; the initial steps of processing are shown in *light blue*. They include image frame alignment, CTF correction, normalisation, and filtering. The subsequent steps—alignment, statistical analysis, determination of particle orientations, and initial three-dimensional reconstruction (3D)—are shown in *dark blue*. The final step (*light purple*) is the interpretation of the maps obtained

complex (OMC, encoded by the conjugative pKM101 plasmid) was solved with a resolution of 15 Å. This 1.1 MDa structure, which spans both the outer and inner membrane, is made of 14 copies of VirB7, VirB9, and VirB10 proteins (Fig. 2a) [12]. Further, the resolution of the same core OMC was improved to 12.4 Å, which provided further details on the structural organisation of the proteins that form this complex (Fig. 2b) [13]. Recently, the almost complete full structure of the T4SS (VirB3–VirB10) encoded by the conjugative R388 plasmid was solved by negative staining (NS). This remarkable structure provided the first view of both the outer and the bipartite inner-membrane complex (IMC) and how these are linked by a structure called stalk (Fig. 2c) [14].

This review might not be complete from the point of view of a specialist and it might not provide sufficient mathematical background for the reader. However, we will try to give a general

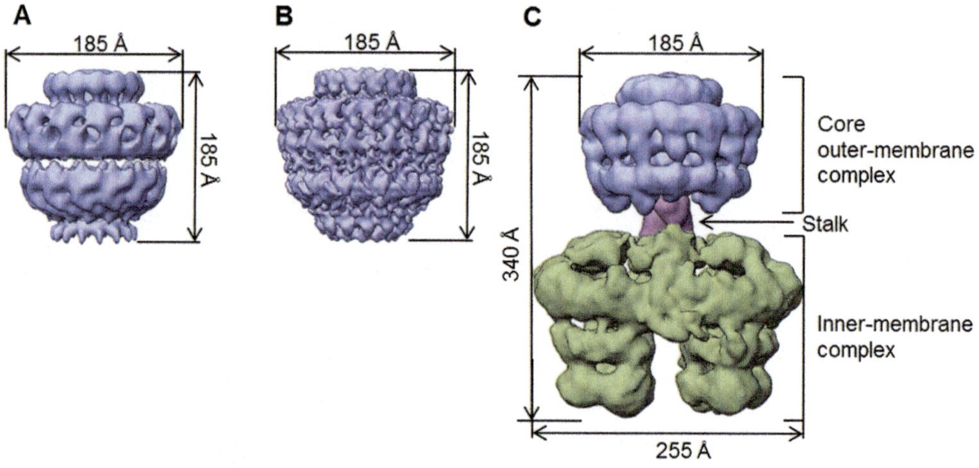

Fig. 2 EM maps of different T4SS structures. (**a**) Cryo-EM structure of core outer-membrane complex at 15 Å resolution. (**b**) Cryo-EM structure of the core outer-membrane complex at 12.4 Å resolution. (**c**) Structure of almost complete the T4SS complex (in negative stain) at overall resolution of ~20 Å

overview of imaging using an electron microscope and the current basic steps of structural analysis. This will include the outline of sample cryo preparation, the effects of radiation damage, and advances in the procedure of data collection. We will describe steps considered to be pre-processing determination of orientations of the particle images, and methods used to obtain structures and how they can be evaluated. Since this is a rather short review, we will not describe here how an image is obtained in an electron microscope. This information can be found in other reviews and books [8, 15]; readers interested in more details on the topics described here may consult the references provided at the end of the chapter.

2 Sample Preparation in Cryo-EM

Although EM provides much better resolution than light microscopy, it has the disadvantage that samples must be imaged in a vacuum. This is due to the fact that images are created by a beam of electrons in a transmission electron microscope. Without a vacuum, electrons become very quickly absorbed by air since they collide with air molecules and lose their energy and direction of scattering, therefore to obtain a high-quality image of a sample, it is necessary to keep the electron path free of air molecules (under a vacuum) to allow electrons to move directly to the sample. In their native conditions biological objects (macromolecules and cells) are immersed in water solutions. To be visualised in an electron microscope under vacuum the bio complexes have to be made rigid and stable to avoid drying out or undergo structural changes during the exposure time at data collection.

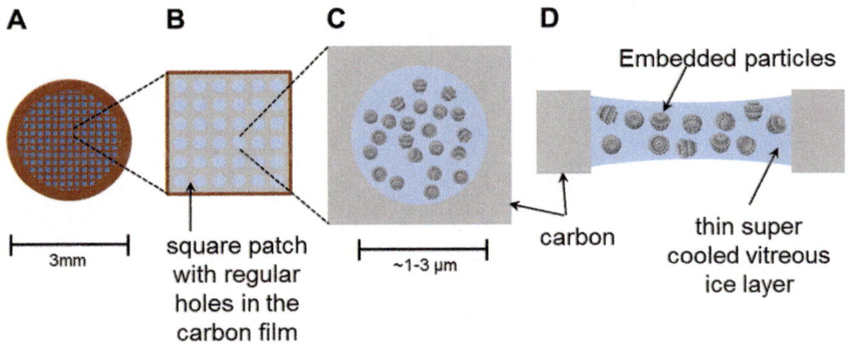

Fig. 3 Cryo-EM sample preparation. (**a**) 3 mm copper mesh grid covered with a film of holey carbon. (**b**) Magnified image of square patch showing microscopic holes in carbon. (**c**) Enlarged image of a single hole containing a layer of vitrified ice with protein molecules. (**d**) Cross section of a hole with particles embedded in ice

Cryo-EM methods of sample preparation allow to preserve the structural integrity of biocomplexes, keeping them in a nearly native hydrated state in the vacuum system of the microscope. The method proposed by Dubochet et al. [16, 17] is now a well-established, standard technique for freezing aqueous solutions of samples on cryo-EM grids. The EM grid is a metallic round plate (~3 mm in diameter and usually made of copper) with a fine mesh. The size of the mesh is typically chosen depending on the experiment, but the most commonly used type has 400 squares per inch (Fig. 3). Depending on the sample, a continuous thin layer of carbon film or one perforated with irregular (lacey grids) or regular holes should be put on the top of the metal grid. One can use manufactured grids with regular holes in carbon films. The grids should be chosen according to the size and shapes of holes and distance between holes, which are most suitable for the given sample (e.g. Quantifoil grids, Quantifoil Micro Tools GmbH; C-flat grids, Protochips, Inc.) with regularly arranged holes for automated and manual data collection (www.protochips.com; http://www.agarscientific.com).

A drop of a sample (~3 μL) is applied to a glow-discharged (to make the surface more hydrophilic) grid; the sample is kept for a short time on the grid (0.5–2 min, depending on the sample) and then the gird is maintained on a plunger. Excess sample is blotted to make a thin layer of sample solution and then the grid immediately plunged into liquid ethane (or propane) that has a temperature of −182 °C. The ethane must be cooled prior sample freezing by liquid nitrogen (Fig. 4). Plunge freezing in liquid ethane takes place in ~10^{-5} s, trapping the biological molecules in their native, hydrated state embedded in amorphous ice that is like solid water. Cooling by plunging into liquid ethane is much faster than plunging directly into liquid nitrogen because the liquid ethane is used

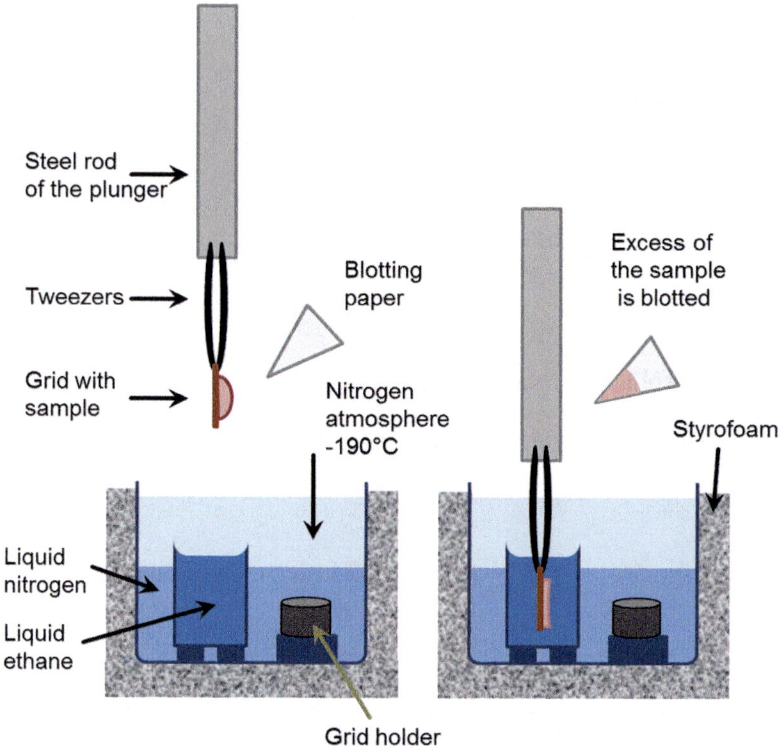

Fig. 4 Sample vitrification. *Left panel*: grid with sample applied being held in tweezers; *right panel*: after blotting excess of a sample, grid is plunged into container filled with liquid ethane. The top level of the container should be immersed in the nitrogen atmosphere that has a temperature slightly above the temperature of liquid nitrogen. The grid is then transferred into the grid holder. The transfer must be done without taking the grid out of the nitrogen atmosphere

near its freezing point so it does not evaporate and produce an insulating gas layer. This fast freezing prevents the formation of ice crystals and keeps samples in a nearly native hydrated state [18, 19]. More detailed information on the vitrification of samples can be found in papers by M. Samsó and R.A. Grassucci [20, 21].

The grids must be kept all the time at temperatures not higher as –170 °C (in a storage, during transfer to the microscope using a cryo-transfer holder, and at imaging in the electron microscope); otherwise, ice will change its conformation and start to make crystals that will destroy the sample and contaminate the grids. Another important advantage of cryo-EM is that liquid nitrogen temperatures reduce the radiation damage induced by electron beam when it passes through samples [22, 23]. Nowadays automated and controlled devices (Vitrobots) have been developed, thereby allowing higher reproducibility in grid preparation [24, 25]. However, it is recommended that the first estimation of the sample quality should be done using the NS technique, which is fast, robust, and reliable [26]. It allows a quick assessment of sample quality, concentration, and suitability for the subsequent cryo preparations.

3 Image Acquisition by Digital Detectors

3.1 CCD Cameras

EM, like all modern photographic systems, is related to the acquisition of images and now uses digital cameras. All digital cameras convert analogue optical signals into digital format so the steps in developing and scanning film have become redundant. The first digital cameras used charge-coupled device (CCD) sensors, which had been invented in 1969 by W.S. Boyle and G.E. Smith at Bell Telephone Laboratories (Nobel Prize, 2009) [27]. The concept behind this device was based on the transformation of an analogue signal, such as photon energy, into an electrical charge using specially designed photo sensors. The magnitude of the charge registered by the sensor is proportional to the energy of photons absorbed by the sensor. CCD chips consist of an array of photosensitive elements. In the readout mechanism, charges are successively transferred to a reading register, amplified, and converted into a digital signal. The number of reading registers determines the speed of the CCD image recording. Since there are only a few of them, the readout of these cameras is not very high.

However, in electron microscopes, the process of recording electrons is more complicated than with photosensors, which are not able to register electrons. Therefore, the sensors were modified in such a way that a mono- or polycrystalline scintillator that converts electron energy into photons was placed on top of the photosensor, and only then were photons converted into an electrical signal (Fig. 5a) [28, 29]. Unfortunately, a CCD camera's sensitivity decreases with increased voltage of an electron microscope, so thicker scintillator layers are needed to improve the electron detection efficiency. Thick layers of scintillators affect the image quality, which is degraded because the higher-energy electrons are scattered through several adjacent sensors, reducing the image resolution. Nonetheless, these cameras provided experience and understanding for the development of automated data collection in EM.

3.2 Direct Electron Detectors

In the last decade, new digital detectors have been designed enabling the detection of electrons without the intermediate step of transforming electrons into photons and then into an electrical signal (Fig. 5b). Direct detector devices (DDDs) use an array of radiation-hardened active pixel sensors (a pixel circuit) which are integrated into a silicon complementary metal-oxide semiconductor (CMOS) chip [30, 31]. In this case the electron energy is transformed directly into an electrical signal. Another advance in this

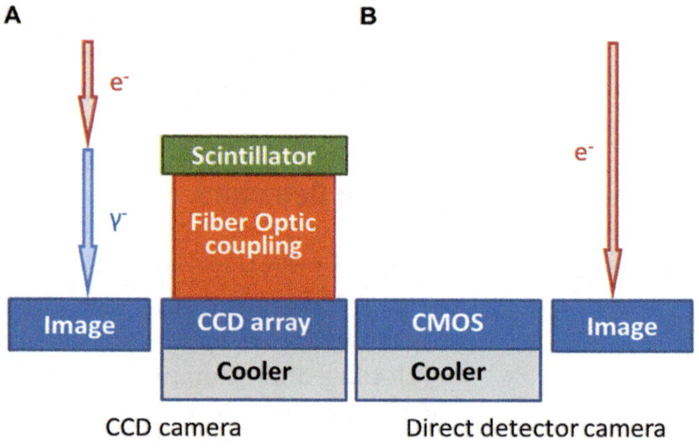

Fig. 5 Digital cameras. (**a**) In CCD-based cameras, the electrons hit a scintillator, generating light, which is partially captured by fibre optics, and directed onto the cooled CCD chip. (**b**) In direct detectors (two *right panels*), active pixel sensors, mostly based on CMOS technology, are capable of capturing and directly detecting incident electrons

technology is that an amplifier is built into each pixel and allows fast signal readout from each individual sensor (or pixel) nearly simultaneously. This makes it possible to separate a single exposure into a set of smaller subexposures. In cryo-EM, this provides a valuable option for electron dose fractionation that is important in studies of radiation-sensitive biological samples. The resultant image subframes can then be used for specimen drift correction, which is not possible with conventional CCD cameras.

The photon-electron conversion step is removed by the CMOS semiconductor technology, where the fibre optics became unnecessary allowing to improve the signal-to-noise ratio (SNR) in images registered by DDD compared to those from CCD. The quality of DD detectors is best described in terms of the detective quantum efficiency (DQE) [32, 33]. The DQE is a measure of the efficiency of signal transfer by the camera and defined as a ratio of the SNR in the output image registered by the camera sensors to the SNR at the input image:

$$DQE = (SNR_{out})^2 / (SNR_{in})^2$$

The ratio depends on the spatial frequency (sizes of the details) of the image. A perfect detector would not distort the input signal, so in an ideal system the output should be the same as the input. Therefore, the DQE of an ideal system would be equal to one for all frequencies. In reality, cameras distort the fine details in images, and this is reflected by a significant decline in the DQE at high frequencies [34, 35].

Direct detectors make it possible to register electrons within a high range of energy and are now used in 300 keV microscopes.

The high sensitivity of such a system has made it possible to reduce the size of sensors, and additional improvements in software have offered a new mode of image exposure, such as counting mode, where the system records single electrons, as implemented in Gatan's K2 cameras [33, 34, 36, 37].

3.3 Micrograph Subframe Alignment

Cryo-EM images using DDD cameras can record from 7 subframes on Falcon (FEI) to 40–50 subframes using direct electron (DE) or K2 (Gatan) cameras. Thus, the data recorded with DDD cameras represent sets of image frames (movies) that can be motion-corrected. Such high rates of image recording can reveal distortions of images induced by drift of the grid (sample) within the EM. Typically, frame alignment starts from frame N−1 that is aligned to the last frame. Then these two images are summed, and frame N−2 will be aligned to this sum. Then frames N−1 and N−2 will be summed, and frame N−3 is aligned to this new sum. The process is repeated in the same way towards the first frame. There are variations in algorithms when summation is done not for two frames for the following alignment but for four or five frames or all frames. Alignment is refined iteratively when on the next round of alignment the total sum obtained during the previous round is used as a reference. The entire procedure improves, firstly, SNR of the reference and then the quality of the alignment. Nowadays a number of software packages can be used for the frame alignment [36, 38–41]. The image shown in Fig. 6a represents the sum of the original frames without any correction. The trajectory of the image shift during these several exposures indicates that the movement of initial shifts of the sample is large initially but then slows down (Fig. 6b). A power spectrum from the sum of the frames without motion correction demonstrates that the Thon rings are not very sharp: they fade fast owing to the small shifts in different directions shown in Fig. 6b (red dots). When the movie frames are aligned (the motion correction), the Thon rings become symmetrical, going up to 3 Å (Fig. 6c). This indicates the presence of high-resolution details in the images. Summation of the motion-corrected subframes generates the final sharper image (Fig. 6d).

3.4 Radiation Damage

Images in an electron microscope are generated by the electron beam that illuminates the sample and then the image is formed by electromagnetic lenses in the plane of the camera. While a short wavelength of the electron beam improves dramatically the resolution of images of biological molecules, it was proved that biosamples are very sensitive to the high-energy electron irradiation that takes place during imaging. Changes in biological complexes depend on the time of the overall exposure (cumulative) dose and were estimated using spot fading diffraction experiments on two-dimensional (2D) crystals [42–44]. Therefore, a 3D structure derived from experiments when a sample was overexposed can

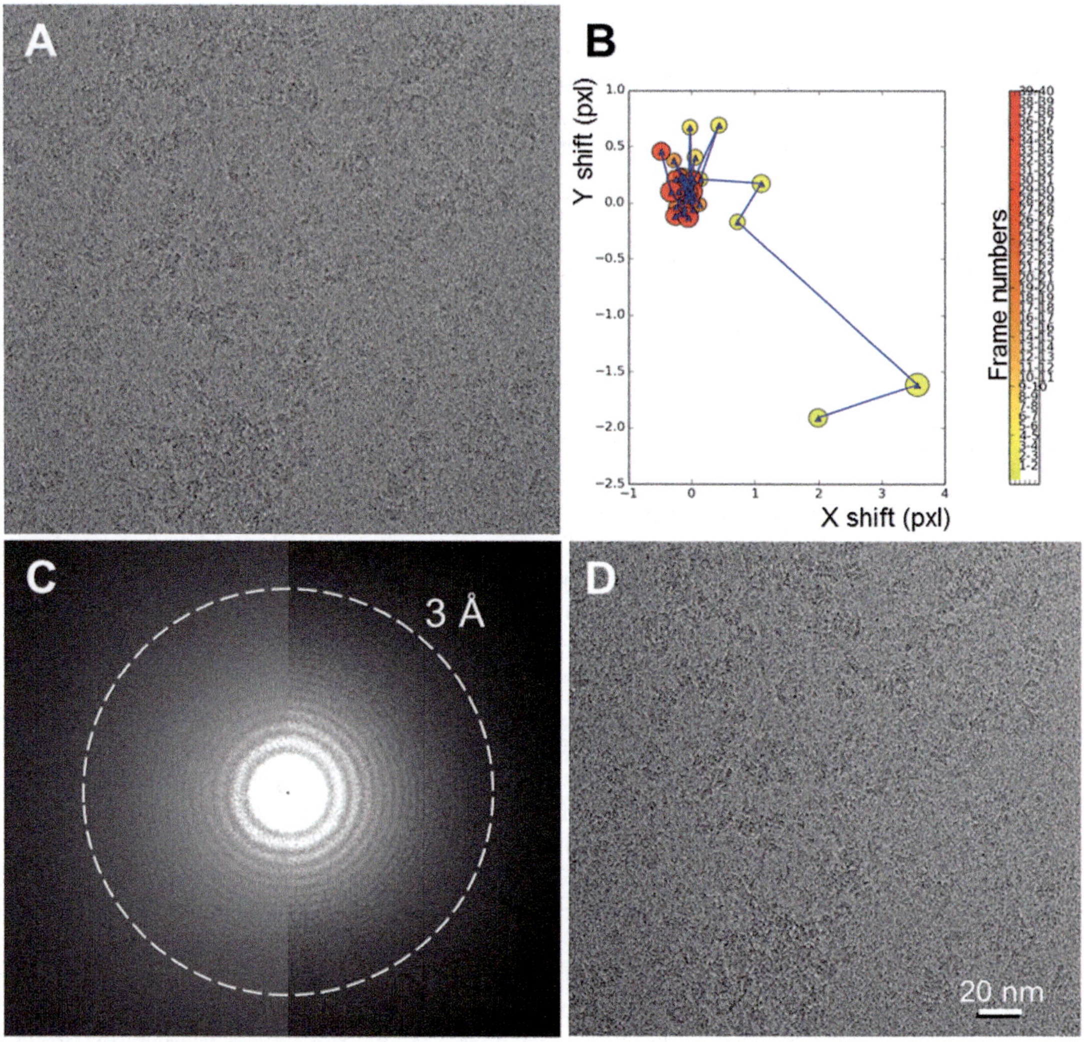

Fig. 6 Cryo-EM images with motion correction. (**a**) Representative cryo-EM image of vitrified T4SS particles. (**b**) Trace of motion in *X* and *Y* directions of frames. (**c**) *Left*: power spectrum from the sum of raw movie frames without motion correction. *Right*: power spectrum from the sum of movie frames after motion correction. (**d**) Sum of movie frames that were shifted according to determined shifts shown in (**b**). Protein is *black* in these images

differ remarkably from the structure of the native molecule. High-energy electrons of the electron beam in EM may induce displacements, bond breakage, and mass loss of low-atomic-number elements such as carbon, nitrogen, and oxygen [45]. It has been shown by crystallography that exposure of crystals to X-rays induces decarboxylation of glutamate and aspartate residues, the breakage of disulphide bonds, and the loss of hydroxyl groups from tyrosine and the methylthio group of methionine [46].

Cryo-EM imaging carries the major benefit of reducing radiation damage as samples are kept at cryogenic temperatures during imaging. Vitrified samples preserve their native structure and are imaged well at liquid nitrogen temperatures [22, 47]. Such low temperatures increase tolerance to ionising radiation damage [44, 48] since the free radicals generated from inelastic scattering events are unable to diffuse through the sample and cause secondary damage [49]. In addition, the freezing also constrains the movement and degrees of freedom of the atoms of a molecule after a bond is broken, thereby limiting the structural rearrangement produced during irradiation [44]. As a result, keeping and imaging samples at liquid nitrogen temperature improves radiation tolerance two- to sixfold over room temperature imaging [44, 48].

Another important approach has been used for a number of years in EM which is the usage of the low-dose mode during data collection. Low-dose imaging is based on reducing the amount of time a sample is exposed to electrons by focusing on an adjacent area that is sufficiently close to the area of interest but does not overlap it. All modern electron microscopes that are used for biological studies come with pre-installed low-dose software allowing for efficient exchange between imaging modes. The search mode is a low-magnification overview image used to identify areas of interest while the imaging (or photo) mode is used for actual data collection at high magnification. The focus mode is typically set at a higher magnification than imaging mode, but the beam is shifted to an adjacent area. Such an interchange between these modes is implemented in systems for automated data collection and allows significant reduction of radiation damage.

The next and now very fast evolving method is the usage of dose fractionation, which is provided by current direct detector technology. DDDs have a very high speed of frame readouts. Depending on the detector type (FEI, Gatan, or DE) and the available software, it is possible to record from 7 to 60 subframes per exposure. Typically, the images in the first two or three frames demonstrate large sample shifts, while later on the movement slows down. However, the last frames indicate often that the sample has been damaged by the beam (Fig. 6b). Bartesaghi and co-authors compared the density maps reconstructed from different fractions of the total exposure (10, 20, or 30 $e^-/Å^2$). Analysis of these high-resolution cryo-EM structures show that densities for residues with positively charged and neutral side chains are well resolved, while the residues with negatively charged side chains, having weaker densities, were less resolved [50]. The negatively charged glutamate and aspartate show on average 30% less density than the similarly sized neutral glutamine and asparagines [50], which is consistent with observations in X-ray analysis [46]. Therefore, the user can use all frames for the quality assessment of images and samples and then only the first half or first two thirds of subframes (depending on the type of the DDD used in experiments) for the reconstruction of the native complex [42, 43, 48, 50, 51].

4 Image Analysis of Micrographs

4.1 Contrast Transfer Function

The aim of single-particle reconstruction is to obtain an accurate representation of the 3D structure of a molecule using a set of 2D projection image data. The macromolecular complexes are considered as thin objects so their images can be described as linear projections of the Coulomb potential of the molecular complex [44]. This is a primary condition necessary for the subsequent reconstruction procedure. However, images produced by electron microscopes do not directly represent projections of the molecules under study. Deviations from the real densities of projections are induced by aberrations in the optical system of the microscope [52].

The function that describes the actual representation of every single point in the registered image of a theoretically correct projection is called the contrast transfer function (CTF) of the microscope [44, 52, 53]. The CTF is defined by the acceleration voltage (electron wavelength), the type of electron source (beam coherence), and aberrations of the objective lens (C_s, C_c, and astigmatism). The major factors affecting the CTF are the degree of spherical aberration (C_s) of the objective lens and level of defocus (Δf). As a result, the CTF modulates the amplitudes and phases of the electron diffraction pattern formed in the back focal plane of the objective lens. For any given defocus setting the features of a specimen are modulated through positive and negative contrast. The CTF limits the amount of information which can be obtained from electron images. At zero crossings of the CTF no information is transmitted, and specimen features corresponding to such spatial frequencies will not be visible in the final image.

A transmission electron microscopy (TEM) image can be represented as a power spectrum (Fourier space), which demonstrates the magnitude of the various frequency components contained within the image (Figs. 7 and 8). The effect of the CTF on an image is that the power spectrum looks like it is oscillating and appears as concentric rings or Thon rings [54], which indicate the location of the minima and maxima in frequency space. Dark regions show the positions of all zero crossings of the CTF, and bright regions correspond to areas where the CTF has either positive or negative contrast (Figs. 7 and 8).

The major effect on images of biological samples from the spherical aberration of the objective lens is to cause phase changes, and therefore the representation of densities in image is altered significantly. On the other hand, biological samples viewed in ice under close to focus conditions demonstrate very little amplitude contrast since the difference in their densities and water density is very small [8, 44]. Thus, the images are typically taken far from focus (in underfocus mode) to increase the weight of low frequencies and thereby improve the visibility of particles [8, 44]. Here it

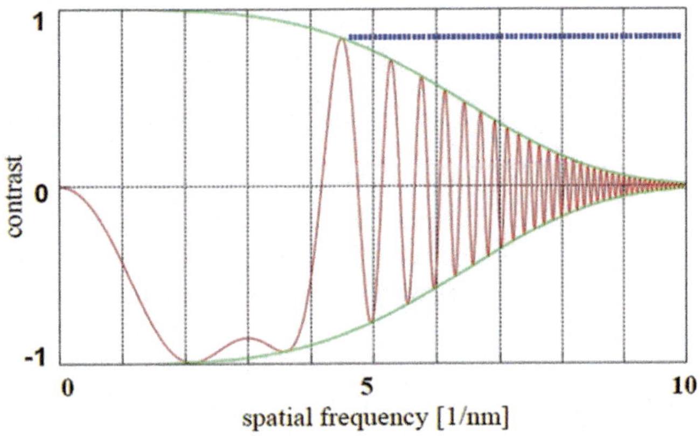

Fig. 7 CTF with envelope function. *Dotted blue line* : amplitude of all frequencies in perfect microscope; *green line*: effect of envelope function on CTF (*red*) resulting in suppression of high spatial frequencies

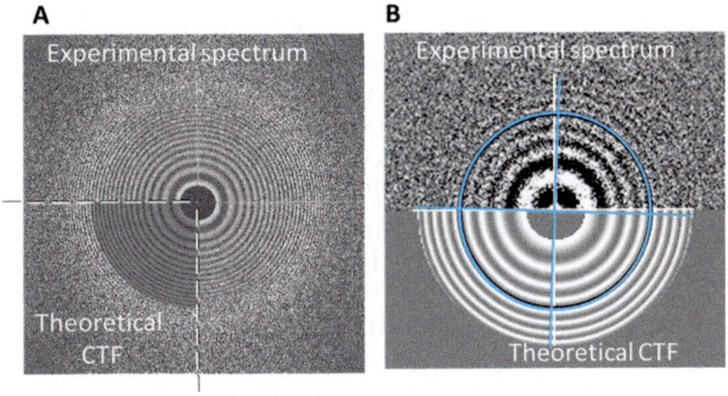

Fig. 8 Assessment of CTF parameters. (**a**) Comparison of theoretically calculated CTF (*left bottom quadrant*) with CTF seen in experimental spectrum. For an accurate CTF determination the Thon rings from both image parts should match accurately. (**b**) Identification of axes of astigmatism which are superimposed over Thon rings of an actual observed power spectrum and compared with the theoretical spectrum. The spectrum of a micrograph shown here indicates that there is a small astigmatism, ~2%, and the axes of ellipse are slightly tilted, shown in *light blue*

should be mentioned that low frequencies are responsible for the overall shape and appearance of particles in images. However, high defocusing induces changes in the distribution of density information related to fine details that could be lost owing to the attenuation of amplitudes at high frequencies. The level of defocus used for imaging depends on the size of the biocomplex. The images of small particles (~100–300 kDa) are taken with a large defocus, sometimes up to 6–7 μm, while viruses with diameters of at least 50 nm can be imaged at 0.5–1.0 μm.

The CTF for biological samples can be described by the formula

$$Phase\,CTF = -2\sin\left[\pi\left(\Delta f \lambda q^2 - C_s \lambda^3 q^4 / 2\right)\right],$$

Phase CTF C_s = spherical aberration constant; Δf = defocus; q = spatial frequency; λ = electron wavelength. The spherical aberration coefficient and the electron wavelength are the only constants, and these values remain fixed for each electron microscope [52].

4.2 Defocus Determination and Correction of CTF

To correct an image for CTF effects and obtain an image that corresponds to the projection, it is necessary to determine its defocus and to check it for astigmatism and drift. The nominal value of defocus set on the microscope does not usually represent the actual defocus obtained in the final digital image or micrograph. This occurs because, although the acceleration voltage and spherical aberration remain constant, the common deviations in sample thickness and the position of the supporting film will cause local variations in defocus. As a result, the CTF related to the defocus should be determined for each image frame. Finding the exact level of defocus and astigmatism in cryo-EM images is of absolutely crucial importance when working on the production of a high-resolution structure.

CTF determination is performed by calculating the sum of power spectra (or amplitudes) of small patches (256 × 256 or slightly larger) from the sum of all subframes. This spectrum is correlated with a number of CTFs theoretically calculated in a range of possible defocus values. A maximum correlation between the observed and a theoretical CTF would indicate the actual defocus of the values and will define frequencies where the phases must be flipped (Fig. 8). Different options for (semi-) automated defocus determination are available in a number software packages, such as EMAN2, CTFIND, and IMAGIC5 [55–57].

Astigmatic images have power spectra which are not rotationally symmetric, and this can complicate and reduce the accuracy of CTF determination. Generally cryo-EM images which have greater than 5% astigmatism are not used for further processing except in special cases where strong astigmatism can be used to recover information in areas where the CTF crosses zeroes. The level of astigmatism can be calculated as follows:

$$Astigmatism = \left(Defocus_{\max} - Defocus_{\min} / Defocus_{avg}\right)$$

An EM projection image is only considered a faithful representation of the observed object of interest if it has been corrected for the CTF modulation effects of the microscope. This can only be done after CTF determination. Phase correction is carried out in reciprocal space by multiplying the alternating rings of the CTF by

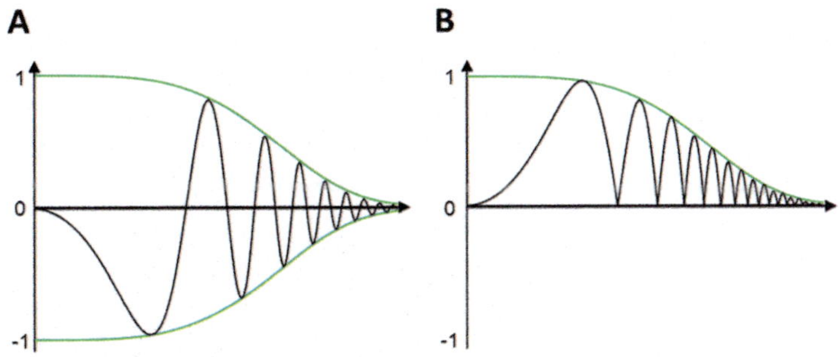

Fig. 9 CTF correction. (**a**) CTF oscillates changing contrast from negative to positive depending on frequencies. Information is lost only where CTF crosses zero line. (**b**) Negative lobes of uncorrected CTF are flipped over to positive (correction of CTF by phase flipping). The missing information can be recovered by collecting images at different defocus levels which fill these zero regions with information

−1 at positions where the contrast transfer is negative and +1 where the contrast transfer is positive. This has the effect of reversing or "flipping" the negative lobes of the CTF into positive contrast, thereby restoring the correct image phases (Fig. 9).

The missing information where the CTF crosses zeroes are restored by combining images at different defocuses, so that where some images lack spatial information at a particular frequency others will provide the complementary missing information. High spatial frequencies are suppressed by envelope decay, so amplitude correction is also important for maximising high-resolution details. This operation usually involves applying a Wiener filter [58] to remove noise from the CTF prior to amplitude amplification.

4.3 Particle Selection

The structural analysis process in EM begins with selecting images of individual particles from micrographs. This involves recording their unique locations (*x*,*y*) within the image field and saving these coordinates in a data file that will be used in the next steps of processing. This can be done interactively using packages like Xmipp [59], EMAN2 [55], Ximdisp [60], RELION (semi-automated selection of cryo-EM particles in RELION-2 [61], and others. The simplest way to do this is to select a single particle image by clicking on the image with a mouse. The coordinates of these points will be stored and then used to extract individual particles within a square box of designated dimension, for example, 500 × 500 pixels. The cut-out area must be large enough to retain all the image data around the object with as little background as possible. Particles can also be selected automatically with particle identification/selection programs, for example, Autopicker [62], BShow [63], and FindEM [64]. These programs use the assessment of local correlation to measure the degree of similarity between reference images and then a small area of the raw micrograph. Areas which show maximum correlation to the references are boxed out.

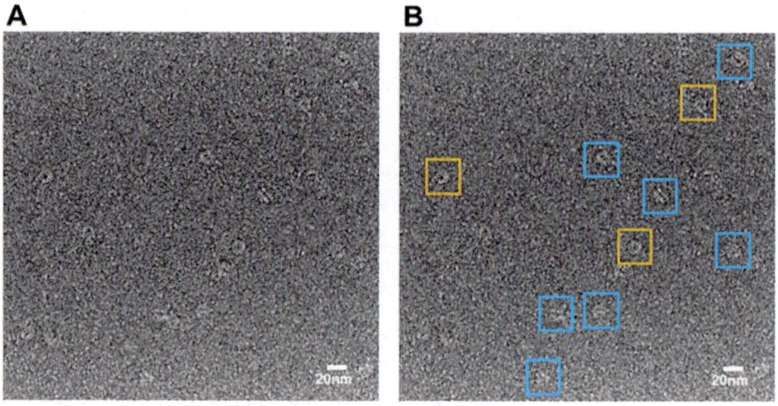

Fig. 10 Particle picking. (**a**) Cryo-EM micrograph of T4SS core outer-membrane complex. (**b**) Particles outlined by *yellow squares* represent end views; side views are outlined by *squares* in *cyan*

The poor contrast in ice images and the presence of artefacts which could resemble target particles can generally increase the inaccuracy of automatic selection.

The images of selected particles must not overlap with other particles and particles in images should not be distorted. In Fig. 10 we show an example of a micrograph of a vitrified sample of the T4SS core OMC. Images were recorded on a 4096 × 4096 Gatan CCD camera with a low electron dose on a Tecnai F20 FEG microscope operating at a voltage of 200 kV, a magnification of 68,100, and a defocus range of 1250–3500 nm.

4.4 Normalisation of Data

The normalisation of all images is an essential pre-processing step. The contrast and intensity can vary from image to image during data acquisition, even when all the EM settings are the same. This effect arises because of a number of factors, including differences in the thickness of the carbon support film or ice, particle orientations, uneven staining, or the merging of images from different data collection sessions. Normalisation standardises the densities of images by setting the mean pixel grey value of each image to the same level, commonly zero, and also rescaling the standard deviation to an equal value for every particle image. Without normalisation the density variations, such as very bright or very dark regions within an image, could bias the cross-correlation procedures which are later used for alignment and calculation of particle classes. Typically, normalisation is based on the following formula:

$$\rho_{i,j}^{norm} = \left(\left(\rho_{i,j} - \rho_{avg}\right) / \sigma_{old}\right) \sigma_{new},$$

σ_{old} and σ_{new} are the standard deviations of the original and target images respectively, and $\rho_{i,j}$ is the density of a pixel in the image array coordinates.

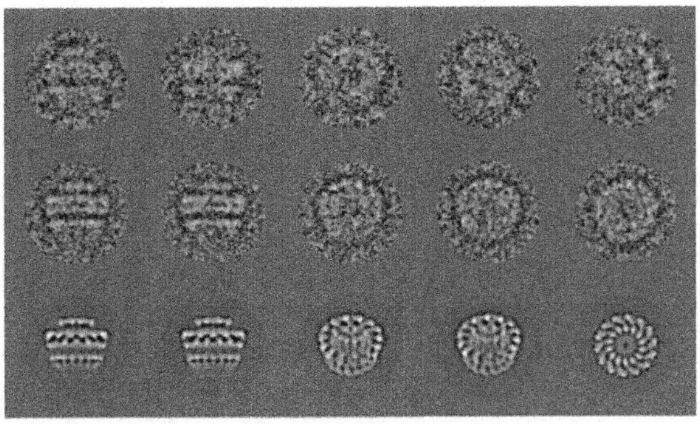

Fig. 11 Alignment and classification of T4SS core outer-membrane complex images. *Upper panel*: representative images of core outer-membrane complex; *middle panel*: class averages of images at nearly the same orientations; *bottom panel*: corresponding reprojections from final 3D model

4.5 Alignment of Particle Images

One of the most important steps in image processing is the reduction of noise and enhancement of the signal. Different factors, such as insufficient coherence of the electron beam, quality of amorphous ice (due to uneven distribution of salts and some other effects in buffers), supporting films, and the noise of registering cameras, contribute to decreasing the SNR in particle images. Since these types of noise are not related to a signal from the sample, the averaging of particle images improves the SNR. However, to retrieve reliable information using averaging, images must represent the same particles in the same orientations [44]. Therefore, images of particles in the same orientations should be identified, and before averaging, the images should be aligned rotationally and translationally with respect to each other. Alignment compares all images to a reference image and shifts them so that they are in the same position as the reference. Usually, normalised particle images should be centred or aligned to the reference that represents a typical view of the complex.

One of the possible options for starting analysis is to align all images of a data set to the rotationally averaged total sum of all images (with no alignment). The centred images are then subjected to multivariate statistical analysis (MSA) (*see* following discussion) to obtain a number of averages of images which are aligned only translationally and grouped according to common features. The best characteristic views (averages of groups of images with the lowest variations among them) are used for multi-reference alignment. The most reliable classes are centred and used as new references for the next round of alignment (Fig. 11). The procedure can be repeated several times alternating with MSA [65]. In other cases one can use alignment in Fourier space [66] or so-called reference-free alignment implemented in EMAN2 and SPIDER [55, 67].

4.6 Statistical Analysis and Classification

To improve the SNR, aligned images which have high similarity between each other should be grouped together, and for this purpose, statistical analysis and classification are used. Different approaches have been developed to reduce a large number of variables to a limited number of important parameters [68].

4.6.1 Principal Component Analysis

Principal component analysis (PCA) reduces the number of variables to find the most significant variations in the measurements [44, 65]. The essence of the procedure is a transformation of a set of observations of possibly correlated variables (in our case images) into a set of values of uncorrelated variables called principal components. In the complete representation, a number of the principal components are equal to the number of original variables. However, since images contain a high level of noise, the number of the meaningful components becomes much smaller. The principal components are described by the eigenvectors of the data matrix. PCA is the simplest of the true eigenvector-based multivariate analyses [65, 68].

4.6.2 Factor Analysis

Factor analysis is designed to identify variations in a number of original variables using predefined "significant" factors that are often defined by a researcher [44]. That requires specific assumptions about the underlying features of the object under study, such as average density or the perimeter and size of specific domains.

4.6.3 Maximum Likelihood Estimation

Maximum likelihood estimation (ML) is a method that assesses parameters that would correspond to a statistical model. When applied to a data set (such as our image data set) and given a statistical model (the initial 3D model), ML provides estimates of how our new reconstruction would correspond to the proposed model and what sort of deviations could be observed. This concept can be stated in different words: once a model is specified with parameters (to a certain extent) and data have been collected (our EM images), one is able to evaluate how well the model fits the observed data. The quality of this correlation is assessed by finding parameter values of a model that best fit the data—a procedure called *parameter estimation*. In the EM case, that would be a 3D model that corresponds in the best way to the data set; otherwise, the model must be modified. ML has many properties which should be taken into account during estimation: sufficiency (complete information about parameters reflecting features of interests), consistency (numbers of images related to this or some other 3D model), efficiency (lowest possible variance of parameter estimates), and perhaps other practical parameters [69, 70]. This method is successfully used in the analysis of 3D reconstructions and implemented in RELION [71].

4.6.4 Classification

Classification is done once the principal components or important factors of the data are defined. Cluster analysis is a tool to identify groups of similar objects. This kind of analysis is used for grouping similar images (particles in the same orientations), and in EM, two implementations are used. One is K-means (used in SPIDER, EMAN, and XMIPP [55, 59, 67]), where the user defines a number of classes (K, typically not bigger than 10) that should be obtained and the algorithm randomly assigns each image to one of the classes [72]. These starting (random) points are called centroids or seeds. The centroids should be placed as far away from each other as possible. The next step is to take each point belonging to a given data set and associate it to the nearest centroid. Averages are calculated for each class, and the distance between each image and the obtained averages are calculated (Fig. 12a). A new class will be formed by the images that were closest to one of the averages. Then the class averages are recalculated. This process is done iteratively until the images stop moving between classes. The K-means method is reasonably fast and works better in low-dimensional space since dimensionality increases time and a local minima problem may occur.

Another method of classification is hierarchical ascendant classification (HAC) (implemented in IMAGIC, SPIDER, and EMAN2). There are two main streams: agglomerative, which is a "bottom-up" approach where each observation starts in its own cluster and pairs of clusters are merged as one moves up the hierarchy, and divisive, a "top-down" approach where all observations start in one cluster and splits are performed recursively as one moves down the hierarchy. HAC is based on the Ward criterion [73], which minimises the intra-class variance while maximising the inter-class variance. In IMAGIC a version of the agglomerative approach is used. This criterion is used in the pair-wise merging of the classes that are supposed to be obtained to form a HAC tree (Fig. 12b). The user chooses a number of classes that would be needed for subsequent processing, and the HAC tree is cut at that level. Sometimes it is hard to achieve the lowest intra-class variance since this algorithm does not allow one to move images from one class to another. However, this can be achieved by weighting parameters during reclassification.

4.7 Determination of Particle Orientation

To obtain the 3D structure of a biocomplex from EM images, the orientation of each of the individual particle images must be determined. The location of an individual molecule can be identified by the X, Y, and Z coordinates, and the shifts of different particles with respect to each other can be described in the same way as shifts in X, Y, and Z. The particles can also be rotated by α, β, and γ angles, called *Euler angles.* This means that a molecule has six degrees of freedom in space. In the microscope images correspond to the projections along the Z-axis of the translational system of coordinates, so the shift in the Z direction is not significant (we assume that the electron beam is parallel), but the shifts in the X and Y directions should be determined. During translational

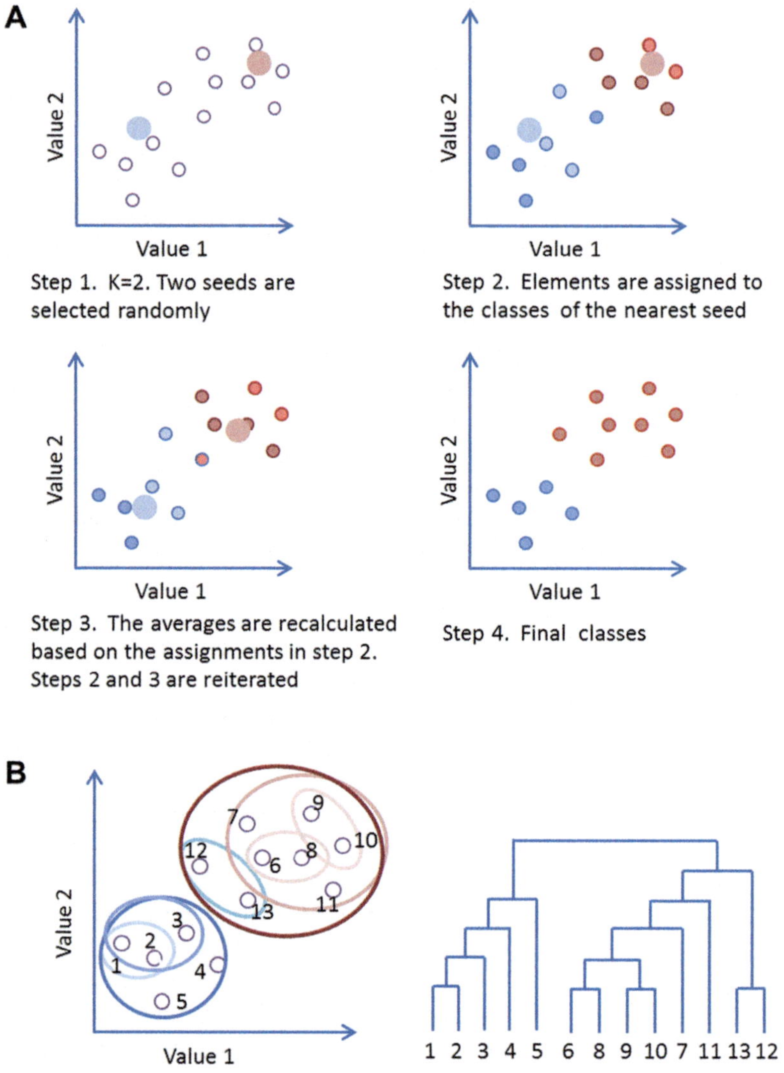

Fig. 12 Principles of image classification. (**a**) K-means classification. Experimental data represented by *empty circles* with their characteristic parameters v1 and v2. The initial randomly selected seeds are shown in *light blue* and *pink*. Results of classification steps are shown in *circles* coloured correspondingly. (**b**) Hierarchical classification. The *left panel* shows position of points in plane and their successive forming of classes; the *right panel* shows the classification tree

alignment, the centre of the molecule is set to $X = 0$ and $Y = 0$. To calculate the 3D map from individual images or class sums, it is necessary to determine the orientations of the characteristic views (classes) relative to each other.

4.7.1 Random Conical Tilt

The random conical tilt (RCT) technique is used to obtain an initial model of a new complex about which little is known. The method is based on a typical property of samples having a preferential orientation on a grid when negatively stained. The RCT approach is a reliable method for generating an unbiased initial 3D model obtained by experimental measurements. Two electron micrographs of the same part of the grid (at magnification ~30–40K) are required; the first one is typically at a high tilt (45–60°, where a goniometer is used), and the second image of the same area is taken without tilt [74]. The tilt respectively to the *Z* axis is known (the same as that used for the tilt of the first image), and rotation around the *Z* axis is obtained from images of the same particles but in the untilted micrograph. As soon the tilted images are centred, angles are assigned to them (Euler angles). With the relative orientation in space determined, a 3D reconstruction of the object can be produced [44]. Although such a model could be far from prefect, it will be a good start for subsequent refinement.

4.7.2 Projection Matching

Projection matching requires an initial model and is based on a simple principle of comparison of images with projections of the model [44]. As a 3D template (an initial map), one can use the low-resolution negative-stain EM 3D reconstruction, the low-pass filtered X-ray model, or the EM map of a homolog. This template is projected in all possible directions covering an entire Euler sphere with a certain angular increment. Then the images or class averages of the data set are compared with these references, and the angles corresponding to the reference with the best cross-correlation will be assigned to the image [67]. Projection matching helps to determine out-of-plane rotations of the object. During the angular determination refinement, the angular increment between projections is reduced, or additional projections with a small increment can be calculated around the initial angle. This method is easy to use; however, it is extremely time consuming due to a long computation during which it is necessary to try all possible in-plane alignments and to compare each image to a set of references. Nonetheless, multi-processor computers can be used to speed up the process. Once the Euler angles are assigned to all images or class averages, a new 3D reconstruction will be calculated and a new set of refined higher resolution model projections computed for the next round of projection matching. Several different software programs, such as IMAGIC [57], EMAN2 [55], and SPIDER [67], offer the option of projection matching.

4.7.3 Angular Reconstitution

EM images represent projections of embedded molecules in random orientations. If there is no initial model, the angular reconstitution technique is a method of choice to determine the orientation of images relative to the other images. The common line projection theorem postulates that every pair of 2D projections of the same 3D object has at least one mutual 1D line projection called a *common line projection* [8, 75]. Thus, by matching 1D line projections for different images, we can identify the relationship between the 2D projections and determine the angles between the common lines and, therefore, determine the relative Euler angles of the images.

A set of 1D projections of an image is obtained when the image is rotated 360° at 1° intervals. The set of 1D projections forms a sinogram (because a trajectory of projections of one point at successive rotations from 1° to 360° corresponds to a sine function). During angular reconstitution [75] each 1D projection of the first image is compared with each 1D projection calculated from the second 2D image (Fig. 13). The line that has the highest correlation (in theory it should be equal to 1) is considered as a common line of these two projections. Sinograms of different images are compared pair-wise line by line, checking the correlation to find common 1D projections (the most similar 1D projections) between two selected 2D images. At least three images are required to determine an initial orientation with respect to the object. Sinograms are generated for all classes, and the search for common lines is performed for all images. The angles between common 1D projections are used to assign an orientation to the class average. Then other classes are added to the initial set of three [75, 76]. It is also possible to determine orientations using common lines in Fourier space [77, 78]. The central section theorem states that the 2D Fourier transform of a 2D projection represents a 2D central section through the 3D Fourier transform of the 3D density. In Fourier space, a common line corresponds to the cross section of Fourier transforms of images. This means that two Fourier transforms from two different 2D projections of the same 3D object have one common central line [77, 78]. Here a comparison of the radial lines of the Fourier transform of one image with all possible radial lines of the Fourier transform of the other image is performed. Again, as in real space, the angle between common lines of the two images with respect to the third one gives the angle between these two views. A combination of EM and angular reconstitution has become an important method for analysing 3D structures of non-crystallised molecules.

4.8 3D Reconstruction

EM images are considered as 2D projections of a 3D object [44]. This is due to the large depth of focus in TEM images. The depth of defocus is related to the acceleration voltage: the higher the voltage, the greater the depth of focus, which can be up to 200 nm. Therefore, the image should represent a projection (the total sum of electron densities along the beam rays) in the image plane produced. However, for the images to be considered as real projections, they must be corrected for CTF effects (*see* earlier discussion). Once the CTF correction is done and the Euler angles have been assigned to each projection image or class average, the 3D electron density for the particle can be determined.

Several approaches are used to calculate the 3D densities of the molecules from their projections. Since the current trend is towards complete automation of the image processing, two approaches are commonly used in EM owing to the efficiency of their

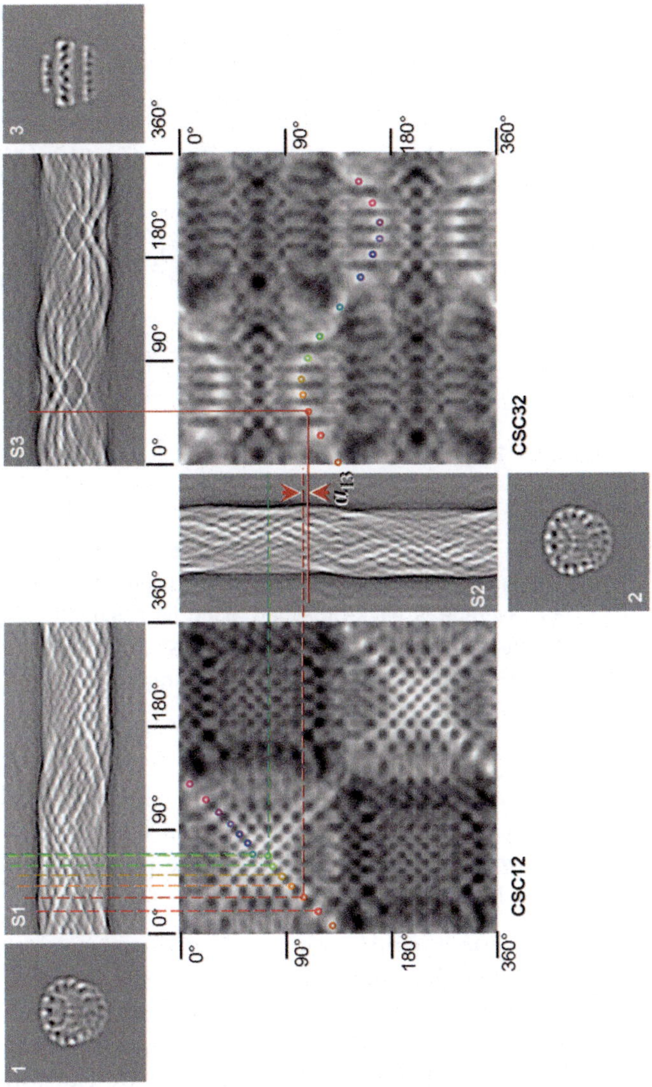

Fig. 13 Sinograms and sinocorrelation functions for three projections. A sample object is a model of the T4SS core outer-membrane complex, which has 14-fold symmetry. Images corresponding to projections are numbered 1–3. S1, S2, and S3 (sinograms) are sets of 1D projections of corresponding projections 1–3. CSC12 and CSC32 correspond to cross-sinogram correlation (CSC) functions between projections 1 and 2 and projections 3 and 2, respectively. Each point of the CSC function represents the correlation coefficient of a pair of lines from the two sinograms. There are 14 common lines between each pair of images since the object has 14-fold symmetry. The highest correlations (*rainbow circles*) point to the position of common lines. Some of them are shown by *dashed lines* between projections 2 and 1, and the *solid red line* shows one common line between projections 2 and 3 (the other are not shown). Each CSC function has all peaks doubled because projections from 180° to 360° mirror those from 0° to 180°. The angular distance between the common lines (*red solid* and *dashed*) shows the angle α_{13} between projections 1 and 3

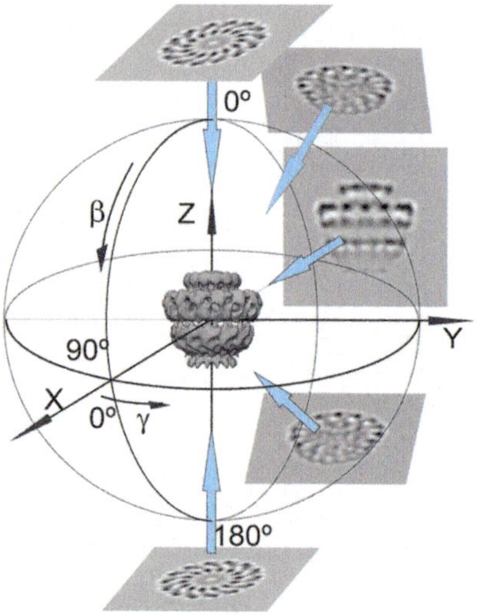

Fig. 14 3D reconstruction in real space. Different class averages of the core outer-membrane complex are shown and situated around the Euler sphere. Each class average is back-projected along its assigned Euler angle in real space. Densities of each 2D class average are stretched as rays through 3D space and the crossover points of the intersecting rays will sum together defining the 3D electron density of the entire object

implementations. In the first one, the reconstruction is calculated in real space and based on filtered back-projection algorithms; in the other one, reconstructions are performed using Fourier space [79–81].

Three-dimensional reconstructionin real space. These methods calculate the 3D distribution of densities in the space of objects. In EM, the exact filtered back-projection method [82] is used more often. In this method, each image of the data set (or classes) is stretched along a direction defined by the found orientations of images. Electron densities in three dimensions are obtained by the summation of rays from stretched projections. The electron density generated by these summations produces densities for each voxel. As more projections are included in the 3D reconstruction, the voxels become better defined (Fig. 14). The angular distribution of different images should evenly cover the Euler sphere or the asymmetric triangle (for particles with symmetry). This is essential to achieve an even representation of all details in the structure, or at least the set of images chosen for the reconstruction should have a distribution of angles that covers a great circle of the Euler sphere [83]. A patchy distribution of angles leads to the appearance of stripes in the 3D electron density map. To avoid an additional low-frequency background induced by the projection-stretching procedure, images are filtered in advance (although in some packages the filter is applied on the resulting 3D reconstruction). A high-pass filtering of the input 2D projections corrects the overweighting of the low-frequency components, thereby restoring amplitude balance and, thus, minimisation of blurring. The exact filter algorithm used in IMAGIC computes a specific filter unique to each 2D projection [82, 84].

Three-dimensional reconstructionin Fourier space is based on a theorem which states that the Fourier transform of a 2D projection of a 3D object constitutes a central section of the 3D Fourier transform of the object [85]. This means that the Fourier transform of projections from different angular views can be merged to fill up the Fourier space with different 2D sections calculated from the images (or classes). Recovery of the 3D structure of an object in real space is done by reverse transformation of its 3D Fourier transform (Fig. 15) [85–87].

The size of trustworthy resolved details can be assessed using a formula derived by A. Cowther [86] assuming that projections are evenly distributed:

$$R = D / N$$

where N is the number of views, D the particle diameter, and R the target resolution. N can be significantly decreased for the same resolution if the complex has a high order of point group symmetry.

In Fourier space the large number of central sections used causes overlapping of the central sections near and at the point of origin. This leads to overweighting of the low-frequency components in the Fourier transform and therefore to an effect similar to a simple back-projection in real space, such as blurring of the reconstruction. Thus, currently used algorithms based on Fourier methods employ down-weighting of low frequencies or a high-pass map filter.

4.9 Structure Refinement

All single-particle EM packages use nearly the same procedure for the refinement of structures after the first 3D model is obtained (Fig. 1). It is done by a realignment of single-particle images with new references obtained from the new model. It is often combined with the determination of angles: projection matching with a smaller angular increment or a local refinement of the angles [88]. Reprojections of the new model can be used as a new anchor set in angular reconstitution to refine the orientation of the classes. The anchor set is a set of projections calculated from the first 3D model which are used to determine orientations for new classes or realigned images. The angular increment between projections used as the anchor set is usually chosen so that 100–150 projections (which is much less compared to projection matching) will be calculated and used for the refinement of angles. New Euler angles are found and assigned to the new class averages. Sorting classes

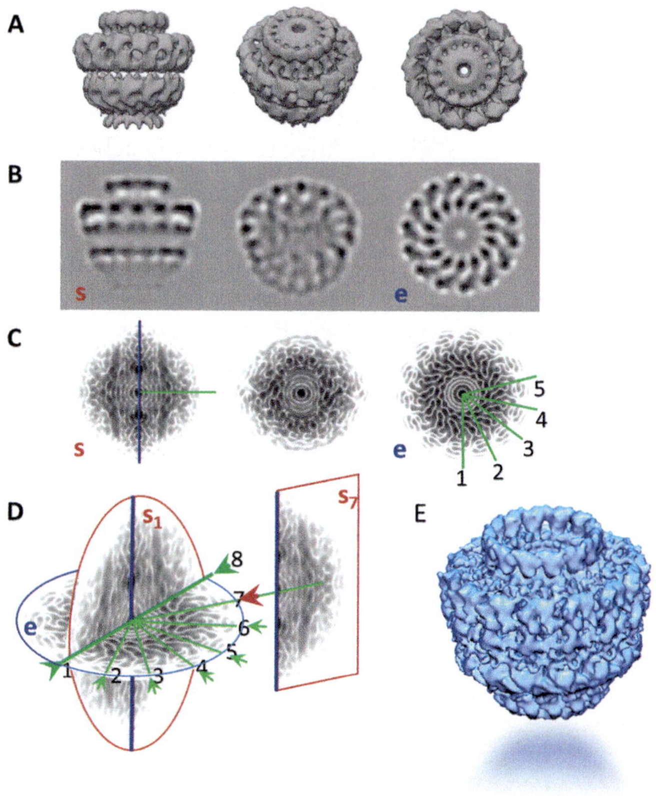

Fig. 15 Reconstruction using Fourier transform (FT) of classes (images). (**a**) T4SS core outer-membrane complex (14-fold symmetry) is observed in different orientations on the supporting film. (**b**) Projections of these particles that are equivalent to EM images in the direction of the electron beam. (**c**) FT from projections shown in **b**. *s* : FT of side view; *e* : FT of end view. The corresponding images are shown in **b**. (**d**) A pair of 2D transforms that share at least one common line in reciprocal space. Common lines between the side view (*s*) projection and the end view (*e*) projection are indicated by *green lines* ; *purple line*: common line between side views (shown in s_1 and s_7). The angles between pairs of common lines determine the relative Euler angle orientations. (**e**) An inverse Fourier transformation of the combined 2D transforms generates an improved real-space structure seen as electron density

according to the errors between classes and reprojections helps to facilitate the refinement procedure. While all implementations share the same principles of refinement, the details of the algorithms and the degree to which the user can control the process vary significantly [9].

4.10 Evaluation of Quality Structure

When the map of a new structure is obtained, the molecular mass and its oligomeric state should be estimated. The map should be inspected at the 1σ threshold of the density level, which typically corresponds to the molecular weight of the complex in the study. If the complex consists of several interacting proteins, the map should not have disconnected fragments of densities; they should be continuous at densities well above the background noise. The concept of resolution is based on an assessment of the minimal distance between two points in an image at which they can still be distinguished from one another. That criterion was formulated as the Rayleigh criterion: when the centre of a peak of one point image falls exactly on the first zero of the image of the second point. In electron crystallography, the signal-related Fourier components of the image are linked to the reflections on a regular lattice, the *reciprocal lattice*, and a resolution is defined by frequencies of the reflections that are above the background noise and therefore available for Fourier synthesis [89]. This crystallographic resolution, R_c, and Raleigh's point-to-point resolution distance, d, are related by $d = 0.61/R_c$. How can this be done in an objective way in single-particle analysis? For several decades researchers have used a concept of Fourier shell correlation (FSC) (*see* following discussion). However, in recent years, owing to the tremendous achievements of single-particle cryo-EM, several modified approaches have been proposed and methods of resolution assessment have moved closer to criteria used in X-ray crystallography.

4.10.1 Fourier Shell Correlation and Gold Standard Approach

In single-particle analysis there is no well-defined periodical pattern in Fourier spectra. Common practice today is to look for data consistency by splitting the data set randomly in half and compare the two resulting averages (3D reconstructions). The resolution of a 3D map (the size of reliable details) could be assessed by FSC. Two Fourier transforms are calculated, and the corresponding spherical shells are compared using normalised cross-correlation as a function of spatial frequency (R) (radius in the Fourier space). The value of the cross-correlation is used to assess the frequencies (or size of details) at which these two maps start to differ. If the correlation falls below the 0.5 threshold, then the details are considered to be different. Currently, several criteria are used to determine the threshold used in FSC, and a threshold of 0.1432 has become rather popular [90, 91]. It should be noted that FSC resolution depends on how the data were split and which threshold of FSC is used for the assessment. Scheres and Chen proposed a "gold standard" method for the evaluation of structural quality [92]. According to this approach, the initial data set is divided into two halves from the very beginning, and two models are refined independently. As soon as the structures are obtained, the FSC curves can be determined as usual. The FSC between two independent reconstructions shows that when the gold-standard procedure is used, the resolution of the final results depends on how well the structures are aligned and what sort of mask was used to remove surrounding noise (Fig. 16). This separation of the data set into two equal subsets and the independent refinement helps to avoid bias towards the same model used on initial steps for the alignment and determination of angles.

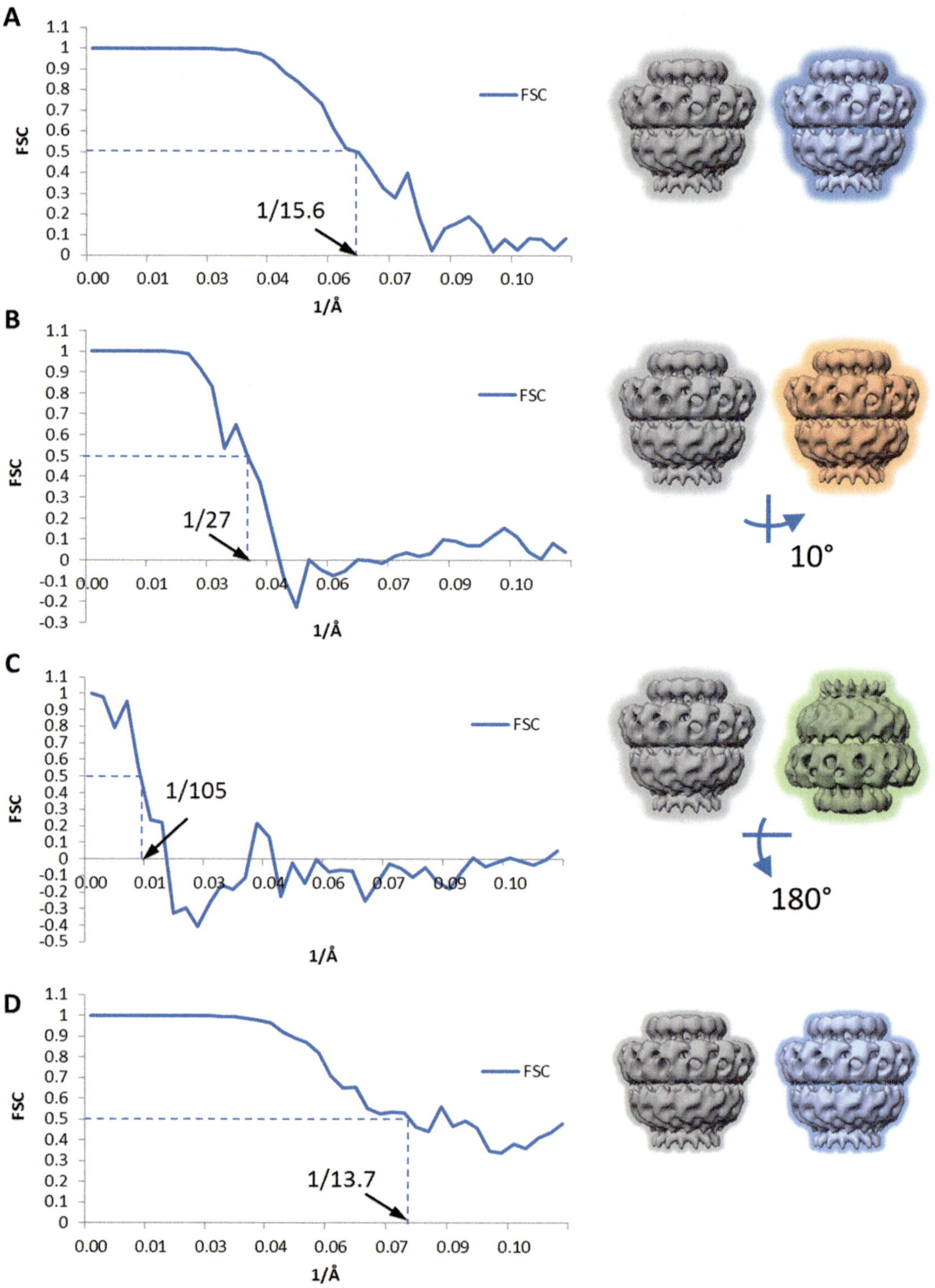

Fig. 16 Examples of Fourier shell correlation (FSC) and resolution assessments. (**a**) FSC curvature of two independent 3D structures that have been well aligned and masked with a loose soft mask shown as a wide halo around the particle images. A resolution assessed at the 0.5 level is 15.6 Å. (**b**) Here the second structure is

The disadvantage of the FSC is that for the resolution assessment one needs to split the data into two halves, but this reduces the resolution of the final 3D model since the entire image data set was not used for the same 3D reconstruction. The other approach to the evaluation of detail reliability reconstructed in 3D is the randomisation of phases (or one can use both randomisation of phases and amplitudes) above the frequencies critical for detail assessment. The main concept behind the approach is to modify an original data set of particle images in such a way that the amplitudes and phases beyond a certain chosen frequency are substituted by random values. This randomisation of phases (and sometimes amplitudes) at high frequencies is equivalent to replacing high-frequency structural details by noise. The modified data set is subsequently subjected to the same image processing procedure used for the original experimental data. The FSC between these two structures usually demonstrates a sharp drop at the same frequency where substitution of phases and amplitudes was carried out [93]. Any non-zero FSC values beyond the resolution where the noise was introduced reflects a level of bias during image processing. It is important to note that a data set with high-frequency noise contains little information about the real structure compared to the original data set, so the particle orientations may be less accurately defined and may affect the value of FSC at frequencies close to the threshold selected for the phase substitution. The behaviour of the FSC curves at high frequencies may also be affected by 3D masking. The FSC can be improved if the featureless regions are masked out. However, a tight mask with a very sharp boundary can produce strange artefacts in FSC such as rising up to the Nyquist frequencies, indicating that there are non-reliable details in the structure (Fig. 16).

4.10.2 Spectral Signal-to-Noise Ratio

The concept of resolution estimation using an assessment of the SNR in spectra of the reconstruction (SSNR) where an entire data set was used has been suggested by Unser and collaborators [94–97] and is similar to the *Q*-factor [98, 99]. The basis of the method is to measure the consistency between the input data for the reconstructed 3D map and a corresponding set of reprojections computed. The method estimates the relative energy

Fig. 16 (continued) rotated by 10° around the rotational axis. While the overall shape still coincides well between both structures small details are not in the register. It is reflected in the FSC curve which declines much faster and the resolution at the 0.5 threshold indicates only 27 Å, corresponding to a size of major domains in the structure. (**c**) The structures are not aligned. FSC falls down even earlier at lower frequencies indicating only consistency in overall sizes. (**d**) FSC between two 3D structures when the same tight mask was applied. The increase at high frequencies indicates the correlation between masks imposed on the structures, and here the resolution is overestimated contribution of the reconstructed signal and noise components by calculating two independent reconstructions [96]. SSNR characterises the quality of the reconstruction as a function of the radial frequency. The bottom line is that one will only trust those signal frequency components whose energy is above what would have been obtained had the algorithm been applied to noise only.

4.10.3 Local Estimation of Resolution

The third approach whose popularity has risen in recent years analyses the 3D density maps obtained from a whole data set by computing the cross-correlation between neighbouring voxels in the Fourier domain, the Fourier Neighbor Correlation (FNC). A 3D mask is applied on a 3D structure in such a manner that values outside the mask are altered for zeroes and pixels remain unchanged only in part of the cube. Any density inside the structure mask counts as signal plus noise, whereas any density outside this mask counts as noise. This operation can be represented in Fourier space as a convolution of the Fourier transform of the mask with the Fourier transform of the noise. The convolution provides an assessment of correlations between the Fourier terms. The method has been implemented in a computer program called RMEASURE [100]. It is used for 3D reconstructions only.

The ResMap algorithm [101] is based on initialising a local sinusoid model at $r = 2d$, where d is the voxel spacing in angstroms (Å). Likelihood ratio tests are conducted at all voxels in the volume. At a fixed wavelength equal to d, the standard likelihood ratio test can detect whether a local sinusoid is a meaningful part of the model approximation. The test requires an estimate of the noise variance, which can be evaluated from the region surrounding the structure. The smallest r at which the likelihood ratio test passes at a given p-value defines the resolution. The p-value is the measure of whether the outcome of the attempt is due to an actual effect or a mere random chance. Voxels that pass the test are assigned a resolution r, while those that fail the test will be examined at a larger r. The algorithm produces a local resolution map with a number assigned to every voxel in the density map.

5 Interpretation and Fitting of Atomic Models

The final verification of the quality of the obtained EM map is done by an analysis of fitting or "docking" of known atomic structures, homologous atomic models, or results obtained by de novo tracing of the polypeptide chain into subunits or protein component domains (Fig. 17). In recent decades the resolution (the smallest reliable details in a structure) of most EM maps produced by single-particle EM analysis was between 20 and 30 Å. At such a low resolution large domains can be recognised according to their overall shape. But this level of detail does not provide sufficient information on the interaction between proteins and possible

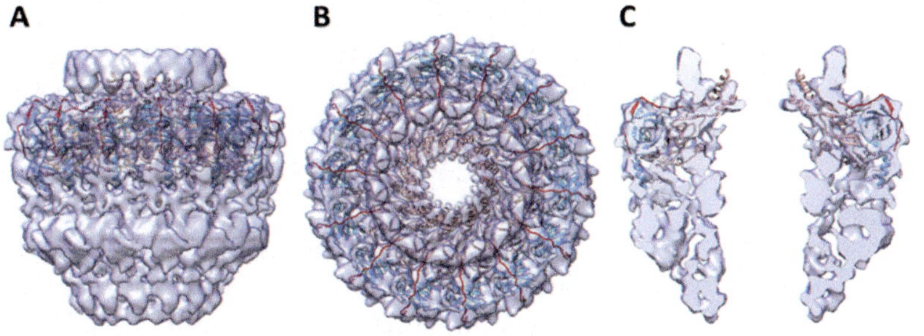

Fig. 17 Fitting of atomic model (PDB: 3JQO) to cryo-EM map (EMD-2232) of core outer-membrane complex. (**a**) Side view. (**b**) Top view. (**c**) Central cross section

active sites. The use of antibodies and different methods of labelling specific domains allow to localise domain positions and construct reasonable pseudo-atomic models based on EM maps; nonetheless, these results require substantial additional biochemical research to verify the interpretation. In subnanometer-resolution maps (6–9 Å range), densities corresponding to α-helices reveal characteristic cylindrical densities with a twist that makes more accurate fittings and determination of structural handiness. At the 4.5 Å level one can see a separation of strands in the β-layers. A resolution of around 4 Å reveals densities corresponding to large amino acid side chains [50, 102, 103], and at resolutions of around 3.7 Å one can do de novo tracing of a polypeptide chain using methods developed in X-ray crystallography [104].

An efficient approach is when pseudo-atomic models obtained by homology modelling are fitted into cryo-EM density maps to build atomic models of individual proteins. The basic principle behind the fitting procedure is the assessment of the correlation between a density map and the model. The maximum cross-correlation value between the EM map and modelled densities of the atomic structure indicates the best fit of the model. Such fitting is performed for maps at a resolution greater than 3.7 Å. Depending on the software used, a search can be carried out in reciprocal or real space. The initial stage of the fitting procedure is carried out either manually or automatically as a so-called rigid fit if a homologous atomic model exists. If there is no such a model, it can be built up with online homology servers like Phyre2 [105] or I-Tasser [106]. These models can be initially fitted into the EM density map using Chimera [107]; and then locations of flexible domains can be identified in Coot [108] or in a more automated way with FlexEM [109] or IMODFIT [110]. The flexible fitted structure can be optimised and checked for clashes using PHENIX [111]. This last step makes it possible to fix the positions of secondary elements simultaneously while conforming to geometrical restraints.

Recently, a new method, electron microscopy–iterative modular optimisation (EM–IMO) was published [112]. It endeavours to build, modify, and refine the local structures of protein models using cryo-EM maps as constraints. A multi-parameter refinement strategy which combines EM–IMO and molecular dynamics with fine-tuning parameters allows one to build backbone models for the different conformations of proteins at near-atomic resolution in cryo-EM maps. The use of EM–IMO demonstrates that homology modelling and a multi-parametric refinement protocol offer a practical strategy for building atomic models based on medium- to high-resolution cryo-EM density maps [113]. Recent developments in single-particle cryo-EM now allow structures to be resolved at resolutions close to 3 Å. To facilitate the interpretation of EM reconstructions, X-ray packages such as REFMAC and PHENIX were modified for optimal fitting of atomic models to EM maps because external structural information can enhance the reliability of the derived atomic models, stabilise refinement, and reduce overfitting [104].

An atomic model obtained as a result of flexible fitting should be evaluated for its correctness and consistency with requirements of interactions between atoms. A Ramachandran plot [114] is routinely used to visualise the distribution of dihedral torsion angles. These angles of a polypeptide backbone are the most dominant local structural factors which dictate protein folding. Geometrical constraints and steric clashes between atoms of the main chain and side chains of each residue generate sometimes incorrect orientations of neighbouring amino acids. The angles are grouped into favoured and disallowed regions within the plot, which indicates the overall structural quality. Also, for each type of secondary structure, i.e. α-helix or β-sheet, there will be a characteristic range of allowed torsion angles, and these are shown on the plot.

6 Conclusions

Although impressive improvements have been made recently in the field of EM, many challenges remain in structural studies like large multi-protein complexes with low or no symmetry. Large complexes are typically flexible or can be unstable, so better approaches to dealing with sample heterogeneity are needed, which means more computer power will be required. Flexible, multi-domain protein structures could be uncovered by an iterative approach where their spatial organisation is determined by solving structures of domains step by step. Here, the largest domain could be tackled first, then "subtracted" computationally from the experimental image to get at the next largest domain; this process could then be repeated sequentially until the complete structure and overall architecture have been determined. Cryo-EM and methods of image analysis have become important and powerful tools in the analysis of biocomplexes. Recent advances in EM have greatly contributed to unravelling the so far elusive structural and mechanistic details of Gram-negative bacterial secretion systems. Such a level of detailed information has not only led to a better understanding of the mechanistic traits underlying the process of substrate secretion through the T4SS apparatus but also given us the unique opportunity to visualise how this bacterial nano-machine is structurally organised within both outer and inner membranes. This valuable structural and mechanistic knowledge can now be used to design and develop new antibacterial compounds which target critical bacterial processes and could help us to curb the spread of pathogenicity and antibiotic resistance.

Acknowledgments

The authors thank Dr. H. White for reading the manuscript and useful discussions that led to improvements in the manuscript. This work was funded by MRC Grant MR/K012401/1 to E.V.O. The authors apologise for not covering all methods fully owing to space constraints.

References

1. Steitz TA (2008) A structural understanding of the dynamic ribosome machine. Nat Rev Mol Cell Biol 9(3):242–253
2. Bartesaghi A, Merk A, Banerjee S, Matthies D, Wu X, Milne JL, Subramaniam S (2015) 2.2 A resolution cryo-EM structure of beta-galactosidase in complex with a cell-permeant inhibitor. Science 348(6239):1147–1151
3. Bai XC, Yan C, Yang G, Lu P, Ma D, Sun L, Zhou R, Scheres SH, Shi Y (2015) An atomic structure of human gamma-secretase. Nature 525(7568):212–217
4. Vinothkumar KR, Zhu J, Hirst J (2014) Architecture of mammalian respiratory complex I. Nature 515(7525):80–84
5. Grigorieff N, Harrison SC (2011) Near-atomic resolution reconstructions of icosahedral viruses from electron cryo-microscopy. Curr Opin Struct Biol 21(2):265–273
6. Grant T, Grigorieff N (2015) Automatic estimation and correction of anisotropic magnification distortion in electron microscopes. J Struct Biol 192(2):204–208
7. Scheres SH (2010) Classification of structural heterogeneity by maximum-likelihood methods. Methods Enzymol 482:295–320
8. Orlova EV, Saibil HR (2010) Methods for three-dimensional reconstruction of heterogeneous assemblies. Methods Enzymol 482:321–341
9. Cheng Y, Grigorieff N, Penczek PA, Walz T (2015) A primer to single-particle cryo-electron microscopy. Cell 161(3):438–449
10. Costa TR, Felisberto-Rodrigues C, Meir A, Prevost MS, Redzej A, Trokter M, Waksman G (2015) Secretion systems in Gram-negative bacteria: structural and mechanistic insights. Nat Rev Microbiol 13(6):343–359
11. Ilangovan A, Connery S, Waksman G (2015) Structural biology of the Gram-negative bacterial conjugation systems. Trends Microbiol 23(5):301–310
12. Fronzes R, Schafer E, Wang L, Saibil HR, Orlova EV, Waksman G (2009) Structure of a type IV secretion system core complex. Science 323(5911):266–268
13. Rivera-Calzada A, Fronzes R, Savva CG, Chandran V, Lian PW, Laeremans T, Pardon E, Steyaert J, Remaut H, Waksman G, Orlova EV (2013) Structure of a bacterial type IV secretion core complex at subnanometre resolution. EMBO J 32(8):1195–1204
14. Low HH, Gubellini F, Rivera-Calzada A, Braun N, Connery S, Dujeancourt A, Lu F, Redzej A, Fronzes R, Orlova EV, Waksman G (2014) Structure of a type IV secretion system. Nature 508(7497):550–553
15. Spence JCH (2003) High resolution microscopy, 3rd edn. Oxford University Press, New York

16. Dubochet J, Adrian M, Chang JJ, Homo JC, Lepault J, McDowall AW, Schultz P (1988) Cryo-electron microscopy of vitrified specimens. Q Rev Biophys 21(2):129–228
17. Jaffe JS, Glaeser RM (1987) Difference Fourier analysis of "surface features" of bacteriorhodopsin using glucose-embedded and frozen-hydrated purple membrane. Ultramicroscopy 23(1):17–28
18. Dubochet J, Lepault J, Freeman R, Berriman A, Homo JC (1982) Electron microscopy of frozen water and aqueous solutions. J Microsc 124(3):219–237
19. Lepault J, Dubochet J (1986) Electron microscopy of frozen hydrated specimens: preparation and characteristics. Methods Enzymol 127:719–730
20. Cabra V, Samso M (2015) Do's and don'ts of cryo-electron microscopy: a primer on sample preparation and high quality data collection for macromolecular 3D reconstruction. J Vis Exp 95:52311
21. Grassucci RA, Taylor DJ, Frank J (2007) Preparation of macromolecular complexes for cryo-electron microscopy. Nat Protoc 2(12):3239–3246
22. Adrian M, Dubochet J, Lepault J, McDowall AW (1984) Cryo-electron microscopy of viruses. Nature 308(5954):32–36
23. Baker LA, Rubinstein JL (2010) Radiation damage in electron cryomicroscopy. Methods Enzymol 481:371–388
24. Tivol WF, Briegel A, Jensen GJ (2008) An improved cryogen for plunge freezing. Microsc Microanal 14(5):375–379
25. Vos MR, Bomans PH, Frederik PM, Sommerdijk NA (2008) The development of a glove-box/Vitrobot combination: air-water interface events visualized by cryo-TEM. Ultramicroscopy 108(11):1478–1483
26. Jensen GJ (2010) Cryo-EM. Part A: sample preparation and data collection. Preface. Methods Enzymol 481:xv–xvi
27. Boyle WS, Smith GE (1970) Charge coupled semiconductor devices. J Bell Syst Tech 49(4):587–593
28. McMullan G, Cattermole DM, Chen S, Henderson R, Llopart X, Summerfield C, Tlustos L, Faruqi AR (2007) Electron imaging with Medipix2 hybrid pixel detector. Ultramicroscopy 107(4–5):401–413
29. Faruqi AR, Henderson R (2007) Electronic detectors for electron microscopy. Curr Opin Struct Biol 17(5):549–555
30. Ramachandra R, Bouwer JC, Mackey MR, Bushong E, Peltier ST, Xuong NH, Ellisman MH (2014) Improving signal to noise in labeled biological specimens using energy-filtered TEM of sections with a drift correction strategy and a direct detection device. Microsc Microanal 20(3):706–714
31. Veesler D, Campbell MG, Cheng A, Fu CY, Murez Z, Johnson JE, Potter CS, Carragher B (2013) Maximizing the potential of electron cryomicroscopy data collected using direct detectors. J Struct Biol 184(2):193–202
32. Cunningham IA (1999) Practical digital imaging and PACS. Advanced Medical Publishing for American Association of Physicists in Medicine, USA
33. McMullan G, Chen S, Henderson R, Faruqi AR (2009) Detective quantum efficiency of electron area detectors in electron microscopy. Ultramicroscopy 109(9):1126–1143
34. McMullan G, Faruqi AR, Clare D, Henderson R (2014) Comparison of optimal performance at 300keV of three direct electron detectors for use in low dose electron microscopy. Ultramicroscopy 147:156–163
35. Ruskin RS, Yu Z, Grigorieff N (2013) Quantitative characterization of electron detectors for transmission electron microscopy. J Struct Biol 184(3):385–393
36. Bammes BE, Rochat RH, Jakana J, Chen DH, Chiu W (2012) Direct electron detection yields cryo-EM reconstructions at resolutions beyond 3/4 Nyquist frequency. J Struct Biol 177(3):589–601
37. Milazzo AC, Moldovan G, Lanman J, Jin L, Bouwer JC, Klienfelder S, Peltier ST, Ellisman MH, Kirkland AI, Xuong NH (2010) Characterization of a direct detection device imaging camera for transmission electron microscopy. Ultramicroscopy 110(7):744–747
38. Campbell MG, Cheng A, Brilot AF, Moeller A, Lyumkis D, Veesler D, Pan J, Harrison SC, Potter CS, Carragher B, Grigorieff N (2012) Movies of ice-embedded particles enhance resolution in electron cryo-microscopy. Structure 20(11):1823–1828
39. Li X, Mooney P, Zheng S, Booth CR, Braunfeld MB, Gubbens S, Agard DA, Cheng Y (2013) Electron counting and beam-induced motion correction enable near-atomic-resolution single-particle cryo-EM. Nat Methods 10(6):584–590
40. Abrishami V, Vargas J, Li X, Cheng Y, Marabini R, Sorzano CO, Carazo JM (2015) Alignment of direct detection device micrographs using a robust optical flow approach. J Struct Biol 189(3):163–176
41. Afanasyev P, Ravelli RB, Matadeen R, De Carlo S, van Duinen G, Alewijnse B, Peters PJ, Abrahams JP, Portugal RV, Schatz M, van Heel M (2015) A posteriori correction of camera characteristics from large image data sets. Sci Rep 5:10317
42. Glaeser RM (1971) Limitations to significant information in biological electron microscopy as a result of radiation damage. J Ultrastruct Res 36(3):466–482

43. Chiu W, Jeng TW (1982) Electron radiation sensitivity of protein crystals. Ultramicroscopy 10(1–2):63–69
44. Frank J (2006) Three dimensional electron microscopy of macromolecular assemblies: visualization of biological molecules in their native state, 2nd edn. Oxford University Press, New York
45. Egerton RF, Li P, Malac M (2004) Radiation damage in the TEM and SEM. Micron 35(6):399–409
46. Burmeister WP (2000) Structural changes in a cryo-cooled protein crystal owing to radiation damage. Acta Crystallogr D Biol Crystallogr 56(Pt 3):328–341
47. Taylor KA, Glaeser RM (1976) Electron microscopy of frozen hydrated biological specimens. J Ultrastruct Res 55(3):448–456
48. Chiu W (1986) Electron microscopy of frozen, hydrated biological specimens. Annu Rev Biophys Biophys Chem 15:237–257
49. Knapek E, Dubochet J (1980) Beam damage to organic material is considerably reduced in cryo-electron microscopy. J Mol Biol 141(2): 147–161
50. Bartesaghi A, Matthies D, Banerjee S, Merk A, Subramaniam S (2014) Structure of beta-galactosidase at 3.2-a resolution obtained by cryo-electron microscopy. Proc Natl Acad Sci U S A 111(32):11709–11714
51. Carlson DB, Evans JE (2012) Low-dose imaging techniques for transmission electron microscopy. The transmission electron microscope. InTech, China
52. Erickson HP, Klug A (1971) Measurement and compensation of defocusing and aberrations by Fourier processing of electron micrographs. Philos Trans R Soc B 261(837):105–118
53. Wade RH (1992) A brief look at imaging and contrast transfer. Ultramicroscopy 46:145–156
54. Thon F (1966) Zur Defokussierungsabh ä ngigkeit des Phasenkontrastes bei der elektronenmikroskopischen Abbildung. Naturforschg 21a:476–478
55. Tang G, Peng L, Baldwin PR, Mann DS, Jiang W, Rees I, Ludtke SJ (2007) EMAN2: an extensible image processing suite for electron microscopy. J Struct Biol 157(1):38–46
56. Rohou A, Grigorieff N (2015) CTFFIND4: fast and accurate defocus estimation from electron micrographs. J Struct Biol 192(2): 216–221
57. van Heel M, Harauz G, Orlova EV, Schmidt R, Schatz M (1996) A new generation of the IMAGIC image processing system. J Struct Biol 116(1):17–24
58. Wiener N (1964) Extrapolation, interpolation, and smoothing of stationary time series. Wiley, New York
59. de la Rosa-Trevin JM, Oton J, Marabini R, Zaldivar A, Vargas J, Carazo JM, Sorzano CO (2013) Xmipp 3.0: an improved software suite for image processing in electron microscopy. J Struct Biol 184(2):321–328
60. Smith JM (1999) Ximdisp–a visualization tool to aid structure determination from electron microscope images. J Struct Biol 125(2–3):223–228
61. Scheres SH (2015) Semi-automated selection of cryo-EM particles in RELION-1.3. J Struct Biol 189(2):114–122
62. Langlois R, Pallesen J, Ash JT, Nam Ho D, Rubinstein JL, Frank J (2014) Automated particle picking for low-contrast macromolecules in cryo-electron microscopy. J Struct Biol 186(1):1–7
63. Heymann JB, Belnap DM (2007) Bsoft: image processing and molecular modeling for electron microscopy. J Struct Biol 157(1):3–18
64. Roseman AM (2004) FindEM--a fast, efficient program for automatic selection of particles from electron micrographs. J Struct Biol 145(1–2):91–99
65. Van Heel M, Portugal RV, Schatz M (2009) Multivariate statistical analysis in single particle (Cryo) electron microscopy. In: Verkley A, Orlova E (eds) An electronic textbook: electron microscopy in life science. 3D-EM Network of Excellence, London
66. Grigorieff N (2007) FREALIGN: high-resolution refinement of single particle structures. J Struct Biol 157(1):117–125
67. Frank J, Radermacher M, Penczek P, Zhu J, Li Y, Ladjadj M, Leith A (1996) SPIDER and WEB: processing and visualization of images in 3D electron microscopy and related fields. J Struct Biol 116(1):190–199
68. Bartholomew DJ, Steele F, Galbraith J, Moustaki I (2008) Analysis of multivariate social science data. Statistics in the social and behavioral sciences series, 2nd edn. Taylor & Francis, USA
69. Myung IJ (2003) Tutorial on maximum likelihood estimation. J Math Psyc 47(1):90–100
70. Sigworth FJ (1998) A maximum-likelihood approach to single-particle image refinement. J Struct Biol 122(3):328–339
71. Scheres SH (2012) A Bayesian view on cryo-EM structure determination. J Mol Biol 415(2):406–418
72. Macqueen J (1967) Some methods for classification and analysis of multivariate observations. Proc Fifth Berkeley Symp Math Stat Prob 1:281–297
73. Ward JHJ (1963) Hierarchical grouping to optimize an objective function. J Am Stat Assoc 58(301):236–244

74. Guan W, Lockwood A, Inkson BJ, Mobus G (2011) A piezoelectric goniometer inside a transmission electron microscope goniometer. Microsc Microanal 17(5):827–833
75. Van Heel M (1987) Angular reconstitution: a posteriori assignment of projection directions for 3D reconstruction. Ultramicroscopy 21(2):111–123
76. van Heel M, Orlova EV, Harauz G, Stark H, Dube P, Zemlin F, Schatz M (1997) Angular reconstitution in three-dimentional electron microscopy: historical and theoretical aspects. Scanning Microsc 11:195–210
77. Crowther RA (1971) Procedures for three-dimensional reconstruction of spherical viruses by Fourier synthesis from electron micrographs. Philos Trans R Soc Lond Ser B Biol Sci 261(837):221–230
78. Fuller SD (1987) The T=4 envelope of Sindbis virus is organized by interactions with a complementary T=3 capsid. Cell 48(6): 923–934
79. Herman GT (1980) Image reconstruction from projections: the fundamentals of computerized tomography. Academic, New York
80. Penczek PA (2010) Fundamentals of three-dimensional reconstruction from projections, vol 482. Methods in enzymology: Cryo-EM, part B, 3-D reconstruction. Academic, Elsevier, San Diego, CA
81. Orlova EV, Saibil HR (2011) Structural analysis of macromolecular assemblies by electron microscopy. Chem Rev 111(12):7710–7748
82. Harauz G, van Heel M (1986a) Exact filters for general geometry three-dimensional reconstruction. Optik 73:146–156
83. Orlov SS (1976) Theory of three dimensional reconstruction—conditions of a complete set of projections. Sov Phys Crystallogr 20:312–314
84. Radermacher M (1988) Three-dimensional reconstruction of single particles from random and nonrandom tilt series. J Electron Microsc Tech 9(4):359–394
85. De Rosier DJ, Klug A (1968) Reconstruction of three dimensional structures from electron micrographs. Nature 217(5124):130–134
86. Crowther RA, DeRosier DJ, Klug A (1970) The reconstruction of a three-dimensional structure from projections and its application to electron microscopy. Proc R Soc A 317 (1530)
87. DeRosier DJ, Moore PB (1970) Reconstruction of three-dimensional images from electron micrographs of structures with helical symmetry. J Mol Biol 52(2):355–369
88. Penczek PA (2008) Single particle reconstruction. In: Shmueli U (ed) International tables for crystallography. Springer, New York, pp 375–388
89. Glaeser RM, Downing KH, DeRosier DJ, Chiu W, Frank J (2007) Electron crystallography of biological macromolecules. Oxford University Press, New York
90. van Heel M, Schatz M (2005) Fourier shell correlation threshold criteria. J Struct Biol 151(3):250–262
91. Rosenthal PB, Henderson R (2003) Optimal determination of particle orientation, absolute hand, and contrast loss in single-particle electron cryomicroscopy. J Mol Biol 333(4): 721–745
92. Scheres SH, Chen S (2012) Prevention of overfitting in cryo-EM structure determination. Nat Methods 9(9):853–854
93. Chen S, McMullan G, Faruqi AR, Murshudov GN, Short JM, Scheres SH, Henderson R (2013) High-resolution noise substitution to measure overfitting and validate resolution in 3D structure determination by single particle electron cryomicroscopy. Ultramicroscopy 135:24–35
94. Unser M, Trus BL, Steven AC (1987) A new resolution criterion based on spectral signal-to-noise ratios. Ultramicroscopy 23(1):39–51
95. Unser M, Trus BL, Frank J, Steven AC (1989) The spectral signal-to-noise ratio resolution criterion: computational efficiency and statistical precision. Ultramicroscopy 30(3):429–433
96. Unser M, Sorzano CO, Thevenaz P, Jonic S, El-Bez C, De Carlo S, Conway JF, Trus BL (2005) Spectral signal-to-noise ratio and resolution assessment of 3D reconstructions. J Struct Biol 149(3):243–255
97. Penczek PA (2002) Three-dimensional spectral signal-to-noise ratio for a class of reconstruction algorithms. J Struct Biol 138(1–2): 34–46
98. Kessel M, Radermacher M, Frank J (1985) The structure of the stalk surface layer of a brine pond microorganism: correlation averaging applied to a double layered lattice structure. J Microsc 139(Pt 1):63–74
99. van Heel M, Hollenberg J (1980) The stretching of distorted images of two-dimensional crystals. Electron microscopy at molecular dimensions. Springer, Berlin
100. Sousa D, Grigorieff N (2007) Ab initio resolution measurement for single particle structures. J Struct Biol 157(1):201–210
101. Kucukelbir A, Sigworth FJ, Tagare HD (2014) Quantifying the local resolution of cryo-EM density maps. Nat Methods 11(1):63–65
102. Zhang R, Alushin GM, Brown A, Nogales E (2015) Mechanistic origin of microtubule dynamic instability and its modulation by EB proteins. Cell 162(4):849–859

103. Clare DK, Orlova EV (2010) 4.6A cryo-EM reconstruction of tobacco mosaic virus from images recorded at 300 keV on a 4k x 4k CCD camera. J Struct Biol 171(3): 303–308
104. Brown A, Long F, Nicholls RA, Toots J, Emsley P, Murshudov G (2015) Tools for macromolecular model building and refinement into electron cryo-microscopy reconstructions. Acta Crystallogr D Biol Crystallogr 71(Pt 1):136–153
105. Kelley LA, Mezulis S, Yates CM, Wass MN, Sternberg MJ (2015) The Phyre2 web portal for protein modeling, prediction and analysis. Nat Protoc 10(6):845–858
106. Yang J, Yan R, Roy A, Xu D, Poisson J, Zhang Y (2015) The I-TASSER suite: protein structure and function prediction. Nat Methods 12(1):7–8
107. Pettersen EF, Goddard TD, Huang CC, Couch GS, Greenblatt DM, Meng EC, Ferrin TE (2004) UCSF chimera—a visualization system for exploratory research and analysis. J Comput Chem 25(13):1605–1612
108. Emsley P, Cowtan K (2004) Coot: model-building tools for molecular graphics. Acta Crystallogr D Biol Crystallogr 60(Pt 12 Pt 1):2126–2132
109. Topf M, Lasker K, Webb B, Wolfson H, Chiu W, Sali A (2008) Protein structure fitting and refinement guided by cryo-EM density. Structure 16(2):295–307
110. Lopez-Blanco JR, Chacon P (2013) iMODFIT: efficient and robust flexible fitting based on vibrational analysis in internal coordinates. J Struct Biol 184(2):261–270
111. Adams PD, Afonine PV, Bunkoczi G, Chen VB, Davis IW, Echols N, Headd JJ, Hung LW, Kapral GJ, Grosse-Kunstleve RW, McCoy AJ, Moriarty NW, Oeffner R, Read RJ, Richardson DC, Richardson JS, Terwilliger TC, Zwart PH (2010) PHENIX: a comprehensive python-based system for macromolecular structure solution. Acta Crystallogr D Biol Crystallogr 66(Pt 2):213–221
112. Zhu J, Cheng L, Fang Q, Zhou ZH, Honig B (2010) Building and refining protein models within cryo-electron microscopy density maps based on homology modeling and multiscale structure refinement. J Mol Biol 397(3):835–851
113. Lindert S, Alexander N, Wotzel N, Karakas M, Stewart PL, Meiler J (2012) EM-fold: de novo atomic-detail protein structure determination from medium-resolution density maps. Structure 20(3):464–478
114. Ramachandran GN, Ramakrishnan C, Sasisekharan V (1963) Stereochemistry of polypeptide chain configurations. J Mol Biol 7:95–99

Chapter 29

Bacterial Filamentous Appendages Investigated by Solid-State NMR Spectroscopy

Birgit Habenstein and Antoine Loquet

Abstract

The assembly of filamentous appendages at the surface of bacteria is essential in many infection mechanisms. The extent of mechanical, dynamical, and functional properties of such appendages is very diverse, ranging from a structural scaffold of the pathogen–host cell interaction to cell motility, surface adhesion, or the export of virulence effectors. In particular, the architectures of several bacterial secretion systems have revealed the presence of filamentous architectures, known as pili, fimbriae, andneedles. At the macroscopic level, filamentous bacterial appendages appear as thin extracellular filaments of several nanometers in diameter and up to several microns in length. The structural characterization of these appendages at atomic-scale resolution represents an extremely challenging task because of their inherent noncrystallinity and very poor solubility. Here, we describe protocols based on recent advances in solid-state NMR spectroscopy to investigate the secondary structure, subunit–subunit protein interactions, symmetry parameters, and atomic architecture of bacterial filaments.

Key words Solid-state nuclear magnetic resonance, Structure determination, Pilus, Needle, Protein assembly, Protein complex, Helical symmetry

1 Introduction

Filamentous appendages are present at the surface of numerous bacteria to execute crucial functions ranging from DNA uptake and toxin secretion to cell adhesion. Gram-negative bacteria and their associated secretion systems display various extracellular appendages, termed *pili*, *fimbriae*, and *needles*, which play essential roles during infection. For instance, the type III secretion system contains an extracellular filament called a needle, made by the helical assembly of multiple copies of a single protein subunit [1, 2] and acting as a conduit to export virulence factors from the periplasm to the extracellular space [3]. The chaperone–usher pathway forms an extracellular pilus rod, the type 1 pilus [4–6], that mediates surface attachment [7]. Type IV secretion systems can assemble at their surface filamentous assemblies called pili

Laure Journet and Eric Cascales (eds.), *Bacterial Protein Secretion Systems: Methods and Protocols*, Methods in Molecular Biology, vol. 1615, DOI 10.1007/978-1-4939-7033-9_29, © Springer Science+Business Media LLC 2017

that promote conjugation, DNA uptake, or the export of effectors [8]. Furthermore, type IV pili are used to adhere to surfaces or to generate twitching motility [9]. Filamentous appendages are often involved in cell or surface attachment, and other proteins, for instance adhesin proteins, are commonly observed to be present along the surface of filaments [10] or at the tips of filaments as with type III secretion needles [11, 12] and chaperone–usher pili [13].

These filamentous appendages are formed by the noncovalent assembly of several dozens to hundreds of copies of protein subunits, self-organized in a hierarchical and highly symmetrical arrangement. The resulting supramolecular complex consists of a thin and unbranched filament, usually observed with a diameter in a range of 5–50 nm and up to several microns in length. Mechanically resistant, they endure the molecular tumbling and steric collisions that make up the extracellular environment. From a structural point of view, bacterial filaments present various important challenges to structural biologists with respect to solving their architectures at an atomic-scale resolution. Indeed, their elongated shape does not exhibit the long-range order required to crystallize, limiting the use of X-ray crystallography in studying the native assembled conformation. Moreover, the size of intact filaments restricts their molecular tumbling in solution, hampering solution nuclear magnetic resonance (NMR) spectroscopy. Several strategies have been developed to circumvent these limitations, and integrative approaches combining local structural information on the monomeric subunits (from X-ray diffraction data or solution NMR) and the molecular envelope determined by electron microscopy (EM) or high-order symmetry parameters obtained from diffraction techniques have proven suitable for obtaining near-atomic-resolution 3D models of filaments, as recently demonstrated for the P pilus [6], the type II secretion pseudopilus [14], or the type III secretion needle [2]. The fitting of monomeric subunits into cryo-EM data or the combination with X-ray diffraction data nevertheless has several drawbacks: (1) the monomeric subunit structure might undergo conformational changes between the soluble or crystalline state and the assembled state; (2) some subunit proteins need to be truncated or mutated to obtain a soluble or crystalline monomer and prevent its aggregation into filaments, leading to a partial loss of structural information; and (3) atomic-resolution reconstruction requires high-resolution EM density maps, which are often not available for filamentous assemblies exhibiting structural heterogeneity at the macroscopic scale.

In the past two decades, magic-angle spinning (MAS) solid-state NMR spectroscopy (SSNMR) has emerged as a powerful technique in structural biology to solve the 3D structure of complex biomolecular systems [15–23]. The design of new SSNMR probes and pulse sequences involves the setup of high field magnets (>800 MHz up to 1 GHz), and, by adapting solution NMR-based strategic isotope-labeling schemes and developing SSNMR-based hybrid approaches including complementary techniques (e.g., X-ray, EM, solution NMR, electron paramagnetic resonance spectroscopy (EPR), computational modeling), SSNMR has established a cutting-edge approach to investigating the structures of biological samples. One main attractive feature of SSNMR in structural biology lies in its power to perform atomic-resolution characterization of insoluble samples (e.g., precipitates, aggregates, nanoparticles, fibers, filaments, capsids) lacking crystalline order or homogeneity at the macroscopic level. Nevertheless, the technique is subject to several prerequisites: samples should have a descent structural order at the local level, i.e., poor static disorder; structural studies are limited to medium-size protein subunits (<400 residues); and ^{13}C- or ^{15}N-labeled proteins need to be produced and the assemblies reconstituted in vitro. Bacterial filamentous appendages are very often built by the noncovalent assembly of a major protein subunit through high-order symmetries, in particular helical symmetries [24, 25], and usually present an impressive structural homogeneity at the local level. They therefore represent promising targets for high-resolution structure determination by SSNMR techniques. Several atomic 3D models of bacterial appendages have been solved by SSNMR-based approaches

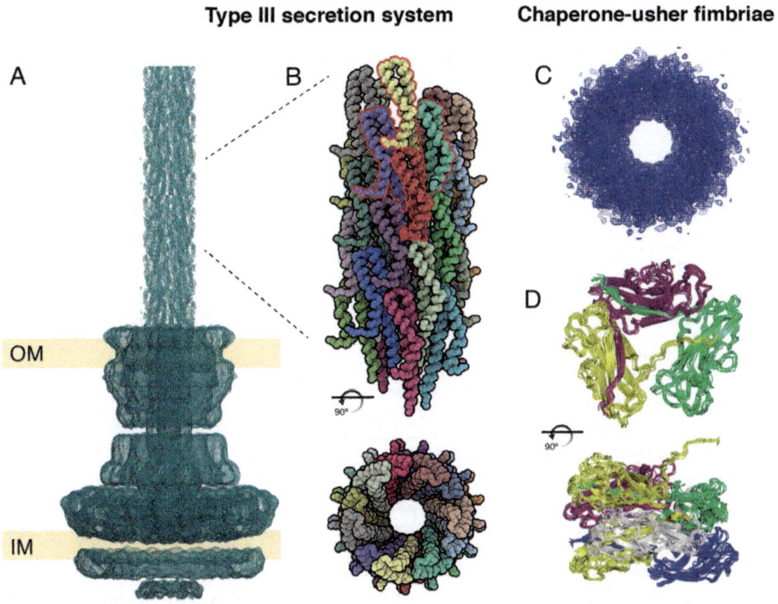

Fig. 1 Overview of structural investigations of type III secretion needle and type 1 pilus by SSNMR-based approaches. (**a**) Schematic representation of type III injectisome based on cryo-EM maps of *Shigella flexneri* needle (accession number EMD-5352 [2]) and *Salmonella typhimurium* needle complex (accession number EMD-1875 [70]). (**b**) SSNMR-based atomic model of *S. typhimurium* needle filaments [1], side perspective and top view, PDB accession number 2LPZ. (**c**) Top view of chaperone–usher pilus structure obtained by cryo-EM (accession number EMD-3222 [6]). (**d**) SSNMR-based atomic resolution model (PDB accession number 2N7H [5])

(Fig. 1), including the type III secretion needles of *S. typhimurium* [1] (Fig. 1a, b) and *S. flexneri* [26] and the type 1 pilus of *E. coli* [5] (Fig. 1c, d). In this chapter, we describe a protocol based on SSNMR spectroscopy to investigate the structure and assembly architecture of bacterial filamentous appendages. The protocol relies on the production and purification of isotopically $^{13}C/^{15}N$ labeled proteins, their in vitro assembly into filamentous samples, and their analysis by SSNMR. The protocol described here can be adapted to numerous filamentous assemblies.

SSNMR studies are commonly based on ^{13}C and ^{15}N detection because of the sensitivity and high spectral dispersion of these nuclei. The monomeric protein subunit is produced in isotopically labeled media to obtain ^{13}C and ^{15}N enrichment necessary for high NMR sensitivity and then self-assembled into filaments. For each filamentous sample, the protein subunits are expressed, purified, and assembled differently to obtain a maximal quantity of pure protein filaments, and the assembly conditions are optimized to avoid structural polymorphism or heterogeneity. For a structural characterization of bacterial filamentous appendages, it is important to produce pure protein and to optimize the assembly conditions to avoid structural polymorphism or heterogeneity. We here globally describe the steps that are needed to obtain a protein filament sample for SSNMR analysis, even though the purification techniques and assembly conditions are different for each protein assembly. Figure 2 represents a general scheme for in vitro filament reconstitution for SSNMR analysis.

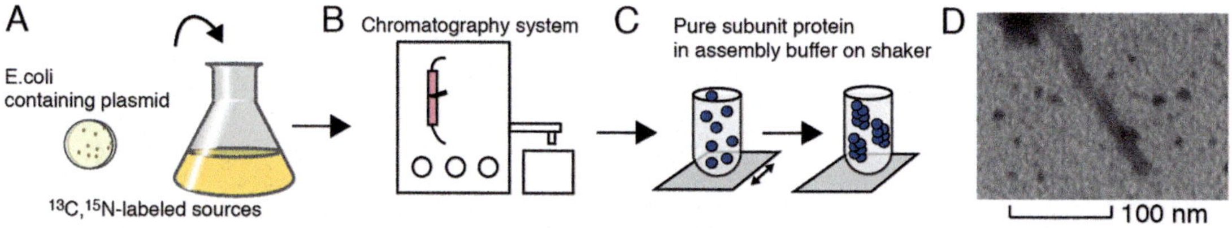

Fig. 2 In vitro bacterial filament reconstitution for SSNMR analysis. (**a**) Expression of labeled protein subunits in minimal medium containing labeled ^{13}C, ^{15}N sources. One colony of the transformed *E. coli* bacteria must be selected from an agar plate to inoculate a preculture, which is further used for inoculating the main culture. (**b**) The cells are destroyed and the protein subunit is harvested and purified on an adequate column system. (**c**) The pure protein subunits are concentrated (or diluted) to a reasonable assembly concentration in the assembly buffer, which should be meticulously tested and optimized at the optimal pH and salt concentration. The assembly conditions are usually optimal under slow shaking. The subunits then adopt their native conformation by assembling (represented here). (**d**) Filament formation should be visualized by transmission electron microscopy, as illustrated here for a protein filament (subunit size of 14 kDa)

2 Materials

2.1 In Vitro Protein Expression, Purification, and Self-Assembly of Isotopically ^{13}C-/^{15}N-Labeled Protein Subunits

SSNMR structural investigations require the use of uniformly (or selectively) $^{13}C/^{15}N$ samples. In this protocol, the sample production is based on the in vitro polymerization of protein subunits into filaments. Here, the purpose is to give an overall description of the requirements and their particularities throughout the SSNMR sample preparation process using *Escherichia coli* as a standard expression system (*see* **Notes 1–3**).

1. Expression vector encoding the protein construct with, for example, a $(His)_7$-tag. Expression vector, typically pET21, containing the DNA fragment encoding the protein sequence of interest; a $(His)_7$-tag is often used for protein purification.
2. *E.coli* expression strain. A commonly used protein expression strain for NMR studies is *E. coli* BL21 (DE3).
3. Isopropyl-thio-β-D-galactopyranoside (IPTG), stock solution (1 M).
4. A culture medium. For producing unlabeled protein standard Lysogeny Broth LB is used. For isotope-labeled proteins and expression tests, M9 minimal medium is used, following the composition (*see* Table 1).
5. In case an isotopically labeled sample is produced, different labeled ^{13}C sources will be needed (*see* Subheading 3.1.2). To obtain a uniformly $^{13}C/^{15}N$ ([U-$^{13}C/^{15}N$])-labeled sample, the minimal medium is supplemented with uniformly ^{13}C-labeled glucose (glc) and ^{15}N-labeled NH_4Cl to correspondingly labeled ^{13}C and ^{15}N atoms. As part of the structural studies it is necessary to selectively label the proteins by supplementing the M9 medium instead of uniformly ^{13}C-labeled glc with selectively ^{13}C-labeled glc or glycerol (gly) [27–30]. The selectively ^{13}C-labeled precursors are [1-^{13}C]-glc and [2-^{13}C]-glc or [1,3-^{13}C]-glyc and [2-^{13}C]-glyc. The two distinct selective labeling schemes, two for glc and two for glyc labeling, are complementary in their resulting amino acid labeling patterns, which arise from the metabolic pathways [27–29].
6. Incubator/shaker.
7. Purification elements: usually Äkta (GE Healthcare) or Bio-rad chromatography system, purification chromatography column(s), desalting column, centrifuges, centrifugal filter units.
8. pH meter.
9. Purification buffers.

Table 1
M9 Minimal medium composition

Component	M9 medium
NaCl	0.5 g/L
KH_2PO_4	3 g/L
Na_2HPO_4	6.7 g/L
$MgSO_4$	1 mM
$ZnCl_2$	10 μM
$FeCl_3$	1 μM
$CaCl_2$	100 μM

10. Buffer for protein assembly (*see* **Note 4**).
11. Ultracentrifuge.
12. 4,4-dimethyl-4-silapentane-1-sulfonic acid (DSS).

2.2 Solid-State NMR Spectroscopy

1. Table centrifuge or ultracentrifuge and ultracentrifugal device [31, 32] for rotor filling procedure.
2. SSNMR rotors, with diameter ranging from 3.2 to 4 mm.
3. Spectrometer(s): a detailed structural analysis of macromolecular protein assembly such as bacterial filaments should be carried out with an NMR spectrometer at a magnetic field ≥500 MHz proton frequency (11.75 T) for sensitivity and spectral resolution purposes. The spectrometer will be equipped with MAS probes with double ($^1H/X$) and triple channels ($^1H/^{13}C/^{15}N$) allowing the detection of ^1H, ^{13}C, and ^{15}N nuclei.

2.3 Solid-State NMR Analysis

1. NMR data processing software: NMRpipe [33].
2. Graphical and analysis programs to visualize multidimensional SSNMR spectra and perform resonance assignment.
3. CCPNMR [34], SPARKY (*see* Table 2).
4. Databases of ^{13}C, ^{15}N, and ^1H chemical shifts: Biological Magnetic Resonance Data Bank (BMRB) [35] (*see* Table 2) or published data [36].
5. Computational routines to derive dihedral angles from SSNMR chemical shifts:

   TALOS+ [37], PREDITOR [38] (*see* Table 2).
6. Prediction of chemical shifts if monomer structure is available:

   SPARTA+ [39], SHIFTX2 [40], CamShift [41] (*see* Table 2).

Table 2
Websites of computational programs and databases

Name	Website address
NMRpipe	http://spin.niddk.nih.gov/NMRPipe/
CcpNmr analysis	http://www.ccpn.ac.uk/software/analysis
BMRB	http://bmrb.wisc.edu/
SPARKY	https://www.cgl.ucsf.edu/home/sparky/
TALOS+	http://spin.niddk.nih.gov/bax/software/TALOS/
PREDITOR	http://wishart.biology.ualberta.ca/shiftor/cgi-bin/preditor_current.py/
SPARTA+	http://spin.niddk.nih.gov/bax/software/SPARTA+/
SHIFTX2	http://www.shiftx2.ca/
CamShift	http://www-vendruscolo.ch.cam.ac.uk/camshift/camshift.php
XPLOR-NIH	http://nmr.cit.nih.gov/xplor-nih/
CNS	http://cns-online.org/v1.3/
ARIA	http://aria.pasteur.fr/
CS-ROSETTA	http://spin.niddk.nih.gov/bax/software/CSROSETTA/
ROSETTA	https://www.rosettacommons.org/
HADDOCK	http://haddock.science.uu.nl/services/HADDOCK2.2/
CING	https://code.google.com/archive/p/cing/
PSVS	http://psvs-1_4-dev.nesg.org/
PROCHECK-NMR	http://www.ebi.ac.uk/thornton-srv/software/PROCHECK/
Pymol	https://www.pymol.org/
Swiss PDB Viewer	http://spdbv.vital-it.ch/

2.4 Structure Modeling

1. Several computational programs can be employed in the modeling process depending on the extent of available structural information, in particular if SSNMR restraints are combined with structural data obtained from other biophysical techniques:

   XPLOR-NIH [42], CNS [43], ARIA [44], CS-ROSETTA [45], HADDOCK [46] (*see* Table 2).

2. Subsequent to structure modeling, structure validation programs are used:

   CING, PSVS [47], PROCHECK-NMR [48] (*see* Table 2).

3. Protein visualization software:

   Pymol, Swiss PDB Viewer (*see* Table 2).

3 Methods

3.1 Isotope-Labeled In Vitro Protein Expression, Purification, and Self-Assembly

3.1.1 Expression of Unlabeled Subunit Proteins

1. Transform *E.coli* strain (e.g., BL21 DE3) with vector, including protein-coding DNA, and plate it on LB agar plates containing appropriate antibiotic.
2. Inoculate a preculture of ≤10 mL LB medium with one isolated culture picked from the agar plate (*see* **Note 5**). Incubate at 37 °C while shaking (rotation values around 200 rpm) until exponential growth phase is reached ($0.6 < OD_{600} < 1$).
3. Inoculate main culture (LB) with preculture. To optimize protein expression, purification, and assembly conditions, as well as to obtain a preliminary NMR signature (*see* Subheading 3.3), it is appropriate to use unlabeled LB medium. With this low-cost filament production, the conditions can be optimized and the produced quantity estimated. Based on a 1D ^{13}C spectrum recorded at natural abundance (unlabeled material), a preliminary analysis of the structural complexity in the filament can be obtained.
4. Induce protein expression with 0.75–1 mM IPTG at an optical density measured at a wavelength of 600 nm (OD_{600}) of 0.8–1. Test the optimal expression time by sodium dodecyl sulfate (SDS) gels, usually between 3 and 6 h at 37 °C, but sometimes protein expression can be optimal at low temperatures overnight. Harvest cells by centrifugation on a table centrifuge (6000 rpm, 30 min, 4 °C) and store at −80 °C until purification.

3.1.2 Expression of Isotopically Labeled Subunit Proteins

1. Transform *E. coli* strain (e.g., BL21 DE3) with vector, including protein-coding DNA, and plate it on LB agar plates containing appropriate antibiotic.
2. A preculture of ≤10 mL LB medium should be inoculated with one isolated culture picked from the agar plate (*see* **Note 5**), incubated at 37 °C while shaking (rotation values around 200 rpm) until the exponential growth phase is reached ($0.6 < OD_{600} < 1$). Centrifuge down the preculture and remove supernatant.
3. Inoculate main culture—M9 medium with unlabeled (**step 4**) or labeled carbon and nitrogen sources (**step 5**)—with the harvested cells (preculture of ≤10 mL LB per liter main culture).
4. First perform a protein production in M9 medium supplemented with unlabeled glucose and NH_4Cl to quantify the production yield in M9 medium (*see* **Note 6**).
5. For the subsequent production of uniformly $^{13}C/^{15}N$-labeled ([U-$^{13}C/^{15}N$]-labeled) protein in M9 medium (*see* Fig. 3a),

supplement the M9 medium with uniformly labeled ^{13}C glucose and $^{15}NH_4Cl$. As mentioned in Subheadings 2.1 and 3.1.4, selective labeling schemes can be achieved via different ^{13}C-labeled precursors: selectively [1-^{13}C]-glc and [2-^{13}C]-glc or [1,3-^{13}C]-glyc and [2-^{13}C]-glyc (Fig. 3b–d). For the ^{15}N labeling, $^{15}NH_4Cl$ is always supplemented. The two distinct selective labeling schemes for glc or glyc are complementary in their resulting amino acid labeling pattern. After evaluation of SSNMR data on uniformly labeled protein filament samples, a selective labeling scheme can be chosen (*see* **Note 7**) [49].

6. Induce protein expression with 0.75–1 mM IPTG at OD_{600} 0.5–0.8. Express protein for period determined in Subheading 3.1.1. Harvest cells by centrifugation on table centrifuge (6000 rpm, 30 min, 4 °C) and store at −80 °C until purification.

3.1.3 Purification and Polymerization of Subunit Proteins

The purification protocol should be set up using unlabeled proteins to avoid unnecessary expenses.

1. Thaw cells on ice and resuspend in appropriate lysis buffer.
2. Lyse *E. coli* cells by appropriate lysis method (e.g., sonication, mechanical disruption or chemical lysis).
3. Set up a purification protocol. If the protein is produced in inclusion bodies, the purification generally needs to be carried out under denaturing conditions (e.g., 6–8 M urea or guanidine). In this case, a refolding step needs to be scheduled, which might be concurrent with the desalting step. If the protein stays folded in the cytoplasm, the desalting step might not be required and a simple concentration after purification might be sufficient.
4. Concentrate the soluble protein fraction to a final concentration favorable to correct assembly. Typical values for protein concentration range between 0.1 and 1 mM.
5. Keep protein solution under smooth shaking conditions for at least 1 week. The solution becomes turbid upon filament formation. After 1 week, filament formation should be checked by EM. If filaments have formed, it is recommended to keep assembled fibrils at 4 °C and add an antibacterial agent such as sodium azide (0.02% (w/v)) to avoid sample contamination.

3.1.4 Homogeneous Versus Asymmetric Labeling Strategies for Intermolecular Restraint Detection

1. To distinguish between intra- and intermolecular restraints during the assignment process (*see* Subheading 3.5.3), several strategies have been developed based on asymmetric labeling schemes. These approaches differ from homogeneous labeling strategies where all protein subunits are produced with the same isotope labeled precursor.
2. (1:1) mixed labeled filaments correspond to the equimolar mixture of two protein batches with different isotope-labeling

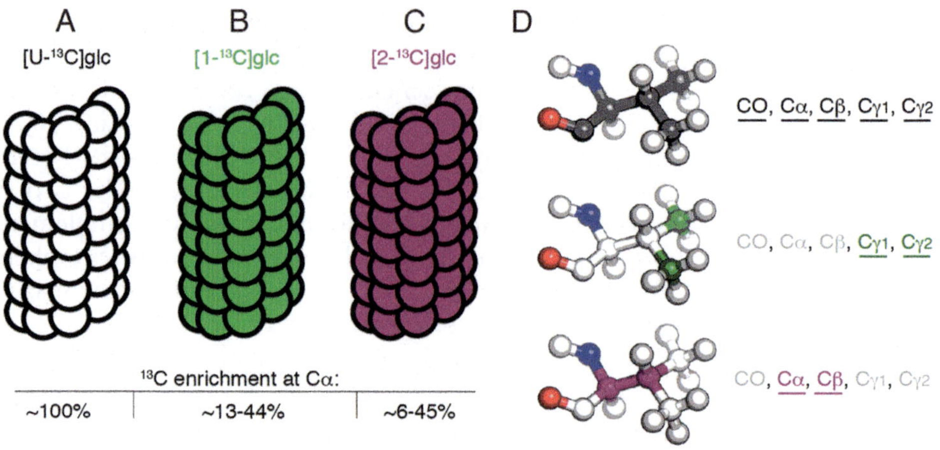

Fig. 3 Strategies for isotopic labeling of protein subunits. Homogeneous labeling schemes, based on (**a**) uniformly ^{13}C-labeled subunits (*white*), (**b**) selectively [1-^{13}C]-glucose-labeled subunits (*green*), and (**c**) selectively [2-^{13}C]-glucose-labeled subunits (*pink*). [1-^{13}C]- and [2-^{13}C]-glucose labeling schemes are homogeneous in the sense that all protein subunits are labeled with the same scheme, although the isotopically labeled carbon positions are selective. (**d**) Labeling patterns for the amino acid valine for each of the labeling schemes. Isotopically labeled atomic positions are underlined and colored according to the cases **a**–**c**

schemes prior to the assembly process. Two types of modus operandi are conceivable, based on the detection of ^{13}C-^{13}C or ^{15}N-^{13}C intermolecular subunit–subunit interactions. The discrimination between intra- and intermolecular ^{13}C-^{13}C proximities is based on the use of a mixture of (1:1) [1-^{13}C]-glc/[2-^{13}C]-glc [30] or (1:1) [1,3-^{13}C]-glyc/[2-^{13}C]-glyc-labeled filaments. For both labeling precursors, two complementary schemes in terms of the resulting labeling pattern are available ([1-^{13}C]-glc [27] and [2-^{13}C]-glc [29] or [1,3-^{13}C]-glyc and [2-^{13}C]-glyc [28, 50]).

3. To detect intermolecular ^{15}N-^{13}C proximities, a mixed labeled filament sample containing (1:1) [U-^{15}N]- and [U-^{13}C]-labeled subunits should be prepared [51]. The mixed labeled samples are prepared by mixing purified differently labeled subunits before filament assembly by gently vortexing.
4. A possibility arises aimed at highlighting intramolecular long-range contacts. To do so, a mixed labeled filament sample containing 1:X (X = 3–5) [U-^{13}C/^{15}N]-labeled and unlabeled subunits should be prepared, also termed a ***diluted sample***. The use of the mixed labeled and diluted samples for identifying long-range distances is described in Subheading 3.5.3.

3.2 Solid-State NMR Setup

3.2.1 Rotor Packing

1. Centrifuge aggregated filament sample and remove and keep the supernatant at 4 °C until SSNMR experiments are successful. Wash sample one to five times with ultrapure water if a high concentration of salts was used in the assembly buffer (*see* **Note 8**). Centrifugation is needed between the washing steps, and

resuspension of the filaments should be done by pipetting without vortexing to avoid filament damage.

2. Centrifuge filament sample using an ultracentrifuge. The duration of this step depends on the consistency of the sample, with samples having a high water content requiring longer centrifugation (up to 24 h). Remove supernatant using a pipette.
3. Introduce pellet inside SSNMR rotor. Different sizes of MAS rotors are illustrated in Fig. 4a. Depending on sample consistency, rotor filling can be done using different devices. In case of a solid sample (crystals, powder, lyophilized sample), a regular spatula is used. For gel-like samples, as is the case for bacterial filamentous samples, a pipet or a capillary pipet is suitable.
4. Add DSS for temperature and chemical shift calibration during SSNMR experiments (*see* Subheadings 3.2.3).
5. Close rotor with cap. If rotor is not fully filled, a top (or a top and a bottom) spacer should be used to equilibrate the content and ensure stable spinning.

3.2.2 Magic-Angle and Pulse Calibration

Several preparatory and optimization steps are required to carry out MAS SSNMR spectroscopy on biological samples. SSNMR experiments are designed using a so-called NMR pulse sequence, which contains a single or a series of radiofrequency pulses and one or more acquisition times to record the NMR signal. The procedures and parameters described in what follows are typical in the operation of a Bruker spectrometer in MAS configuration. A probehead for 3.2 mm MAS rotors of an 800 MHz spectrometer is shown in Fig. 4b.

1. Adjust angle of rotor relative to magnetic field (called the magic angle) at a value of ~54.7°. A sample of KBr in powder form is used to record a ^{79}Br detected 1D spectrum under a spinning frequency of 5 kHz; a single scan is enough to observe the signal on most spectrometers. The spinning frequency gives rise to so-called spinning sidebands at the distance of the spinning frequency (in this case 5 kHz) from the ^{79}Br signal. Adjust the angle by optimizing the linewidth of the spinning sidebands, i.e., the sideband signal to the maximum intensity. The beneficial effect of MAS is illustrated in Fig. 4c.
2. Calibrate the hard NMR pulses on standard reference samples, usually Adamantane for calibrating the ^{1}H pulse and a ^{13}C/^{15}N-labeled amino acid (e.g., ^{13}C/^{15}N-labeled histidine hydrochloride monohydrate in polycrystalline powder form) for ^{13}C and ^{15}N pulses. 90° pulses are optimized by adjusting the pulse length or the pulse power. For example, to obtain the 90° pulse values, the 180° pulse can be found by optimizing the signal decrease until it disappears.

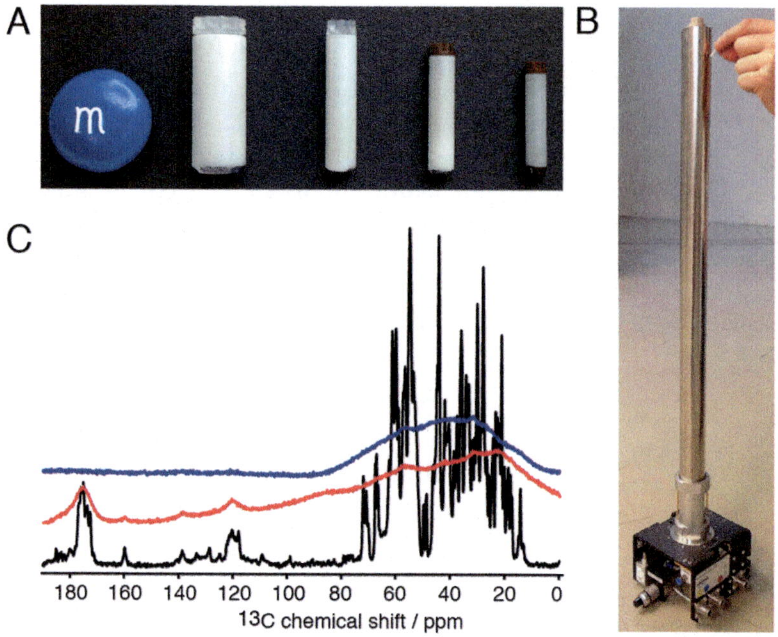

Fig. 4 Solid-state experimental setup. (**a**) SSNMR MAS rotors with 7, 4, 3.2, and 2.5 mm diameter. (**b**) Bruker triple resonance $^1H/^{13}C/^{15}N$ MAS 3.2 mm probehead. (**c**) Carbon-detected ^1H-^{13}C CP spectrum of a protein filament recorded under MAS condition with ^1H decoupling during acquisition (SPINAL64) [54] (*black*), under MAS condition without ^1H decoupling (*red*), and without MAS (*blue*)

3. The optimization of the different polarization transfer steps involving through-space transfer, such as ^1H-X Hartmann–Hahn cross polarization (CP) [52] and ^{15}N-^{13}C-specific CP (DCP) [53], ^{15}N to $^{13}C\alpha$ and ^{15}N to ^{13}CO, are realized on a $^{13}C/^{15}N$-labeled histidine (or a different amino acid) sample. Proton decoupling during evolution and acquisition times is performed with SPINAL64 [54] and optimized setting the pulse frequency to 90 kHz (applied on ^1H). The beneficial effect on the spectral resolution and sensitivity of proton decoupling is illustrated in Fig. 4c.

3.2.3 Sample Temperature and Chemical Shift Calibration

Once the NMR setup on reference samples has been done, the sample of interest is introduced into the spectrometer to adjust the sample temperature and calibrate the chemical shift.

1. To obtain a precise estimation of the sample temperature during an SSNMR experiment, it is recommended to use an internal calibration to avoid inaccurate measurements of the probe thermocouple. On hydrated samples such as the filament samples described in this chapter, the sample temperature can be calibrated using the difference between the DSS ^1H resonance (observed at 0 ppm) and the supernatant water resonance

[31]. Following the relationship $\delta(H_2O) = 7.83 - T/96.9$ ppm (with the temperature T in Kelvin), the temperature can be measured with a precision of ±1–2 °C. The experimental sample temperature is commonly set to 1–10 °C for all SSNMR experiments.

2. Because DSS is an inert compound in most biological samples (except at very low pH), its ^{1}H frequency (representative of a chemical shift value of 0 ppm) can be used directly to calibrate the ^{1}H frequency and indirectly to calibrate the ^{13}C and ^{15}N chemical shifts following International Union of Pure and Applied Chemistry (IUPAC) recommendations [55].

3.3 Preliminary Characterization of an Unlabeled Sample

A first global structural characterization of a sample of interest should be carried out on an unlabeled filament sample to avoid unnecessary expenses arising from isotope-labeled precursors. Several standard SSNMR experiments can be carried out on an unlabeled sample. The aim is to obtain rapidly (1–2 days of NMR spectrometer time) a rough evaluation of the structural order and to determine the presence of mobile segments within the protein assembly. Only for a more detailed analysis of mobile segments (described in Subheading 3.3.3) is a [U-^{13}C]- or [U-^{13}C/^{15}N]-labeled sample necessary.

3.3.1 CP-Based One-Dimensional Experiment

1. Introduce the unlabeled filament sample into the spectrometer.
2. Set the spinning frequency to 11 kHz and simultaneously cool down the probehead to maintain a sample temperature in the probehead of around 5–10 °C (*see* Subheading 3.2.3).
3. Setup a 1D ^{1}H-^{13}C CP experiment following the hard pulse optimization on a reference sample (*see* Subheading 3.2.2). For most natural abundance filament samples, the ^{13}C signal is visible only after several hours of acquisition, making the optimization of pulse lengths, decoupling, and CP transfers extremely time consuming and practically almost impossible. For this experiment, the values should therefore be taken from the values obtained in the reference sample optimization. Set the number of scans to 20k (20,480), the CP contact time to 1 ms, and the decoupling strength to 90 kHz. Different causes might be at play if the spectrum shows no signal (*see* **Note 9**).

3.3.2 Interpretation of a CP-Based One-Dimensional Experiment: Structural Elements and Homogeneity

The carbon-detected 1D ^{1}H-^{13}C CP spectrum is very informative since the linewidths are directly correlated to the degree of molecular order and the resonance peak distribution indicates the secondary structure content. Figure 5 illustrates the information content of a 1D ^{1}H-^{13}C CP versus a 1D INEPT-based experiment recorded on a bacterial filament.

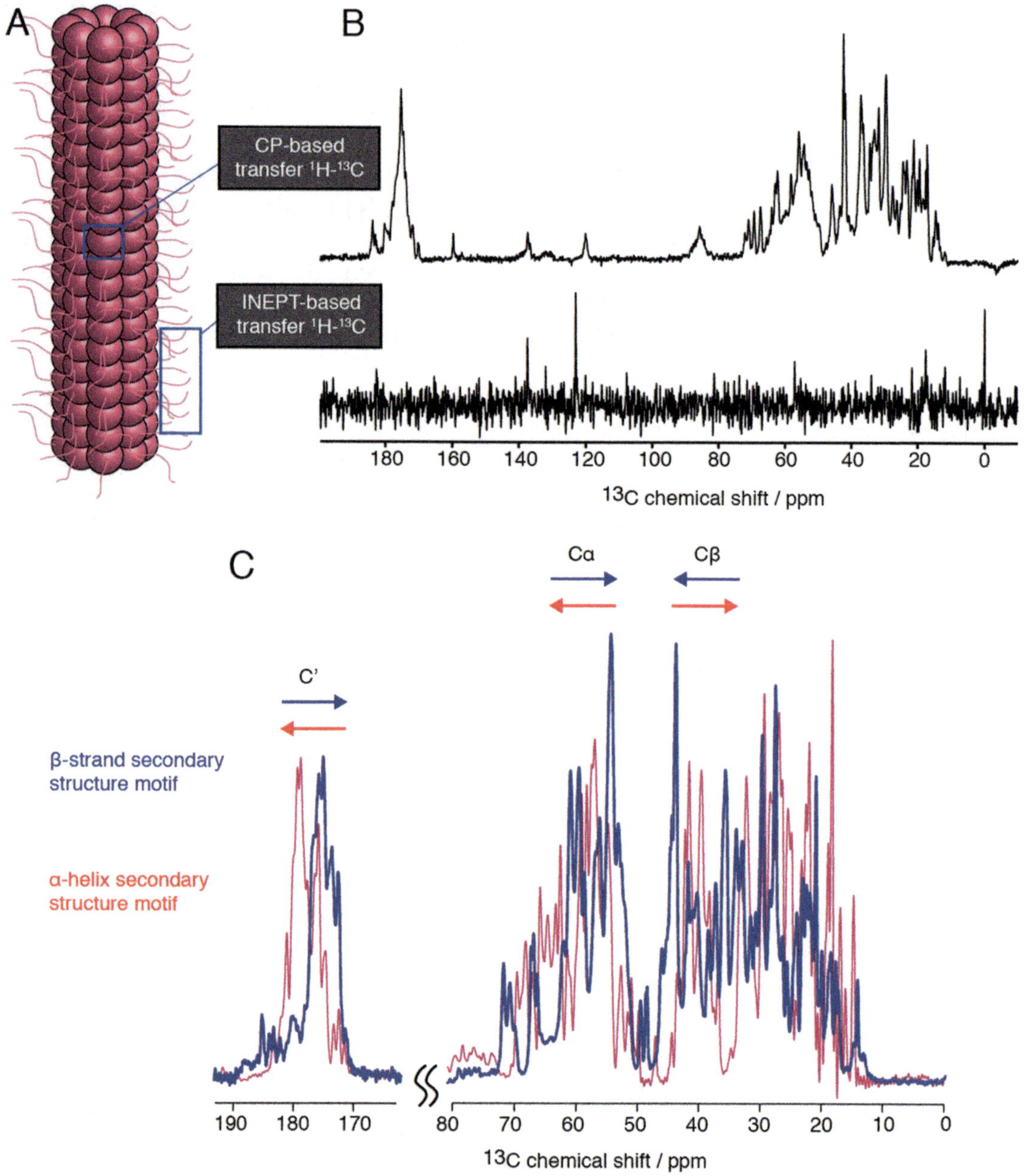

Fig. 5 SSNMR characterization of rigid and mobile protein segments in a protein filamentous assembly. (**a**) Schematic representation of a bacterial filament, including rigid and mobile protein segments. (**b**) Cross polarization (CP)-based experiments reveal residues contributing to the rigid core of the assembly. INEPT-based experiments probe the presence of mobile segments, illustrated on a protein filament with a subunit size of 14 kDa. (**c**) The chemical shift value of Cα, Cβ, and carbonyl C′ atoms is sensitive to the secondary structure, and the chemical shift distribution is indicative of the secondary structure composition; illustrated on a 1D CP spectrum of two protein filaments (both subunit sizes of 9 kDa) containing β-strand (*blue*) and α-helical (*red*) conformation

1. ^{13}C linewidths experimentally observed in biological samples under MAS and high field are typically between 0.2 and 5 ppm, depending on the structural homogeneity of the protein subunit conformation. Figure 5a shows a typical ^{13}C-detected 1D CP recorded for a filamentous protein assembly with linewidths <70–80 Hz. Protein assemblies exhibiting a high structural heterogeneity (i.e., the protein subunits in the assembly adopt several slightly different atomic structures or are not in the same local environment) would lead to a ^{13}C linewidth >~2 ppm.
2. If the examined sample exhibits linewidths >2 ppm in a spectral region where nonoverlapped peaks can be examined, readjusting/optimizing the purification or assembly conditions to obtain a more ordered protein structure in the filaments could be helpful. Nonetheless, the observation of large linewidths for some distinguishable peaks might reveal a less homogeneously ordered protein segment in the assembly and would not necessarily imply that all protein segments contributing to the assembly have adopted a less-well-ordered structure. It is therefore important to carefully judge all visible signals.
3. In insoluble and noncrystalline protein assemblies, the structural homogeneity can be intrinsically poor if the protein subunit molecules have slightly different conformations within the assembly. In this case, an atomic 3D structure determination is extremely challenging, and the identification of the residues participating in the rigid core, as well as the determination of the secondary structure, should become the main objectives of the SSNMR study.
4. The signal distribution and dispersion are representative of the amino acid composition in the rigid structural segments and of the secondary structure elements.

   Each amino acid has its ^{13}C SSNMR spectral fingerprint, which arises from all carbon atoms present in the specific amino acid. The signal distribution in a 1D CP ^{1}H-^{13}C spectrum already provides some indications about residues in a rigid conformation; for example, resonance peaks at a chemical shift position of 75 ppm only arise from threonine Cβ resonances (*see* Fig. 5b). Similar observations can also be carried out for other carbon resonance signals.
5. As presented in Fig. 5c, the signal dispersion gives a first indication of the secondary structure elements contributing to the rigid protein structure. Figure 5c compares two protein assemblies, in one case built of near-complete α-helical proteins (red) and in the second case of near-complete β-strand subunits (blue). The position of C′, Cα, and Cβ carbon signals depends on the secondary structure elements and are therefore shifted toward specific spectral regions.

3.3.3 INEPT-Based One-Dimensional Experiment: Mobility

Although a CP-based experiment can reveal rigid residues, SSNMR also has the ability to probe residues with higher mobility. INEPT-based experiments are used to perform a through-bond transfer and probe protein segments with increased molecular mobility (from picoseconds to nanoseconds). Such INEPT-based experiments are useful for detecting the presence of mobile protein segments in comparison with the rigid core probed by CP-based experiments; both approaches are complementary in the delineation of different protein segments or domains with different mobility (*see* Fig. 5). The following procedure is used to carry out the INEPT-based approach.

1. Optimize a ^{1}H-^{13}C INEPT transfer using GARP [56] proton decoupling with a radio frequency of 5 kHz. Set up a 1D spectrum with 4000 scans and a recycle delay of 1 s.
2. Analysis: The interpretation of a 1D ^{13}C-detected INEPT spectrum (e.g., Fig. 5b) depends mostly on the amount of visible signals. If a low amount of signals is present, the signal assignment to the amino acid carbons is possible (residue-type assignment). The shifts in an INEPT experiment are usually very close to chemical shifts typical for random coil conformation and can be assigned using the published ^{13}C random coil chemical shifts (e.g., [36]). Always carefully check the resonances of your buffer components that might remain after washing. If signals do not correspond to protein signals and cannot be identified, it is recommended to record a solution NMR spectrum of the buffer only to clearly assign buffer signals.
3. Analysis: Based on the ^{1}H/^{13}C/^{15}N chemical shift database (e.g., the useful paper by Jardetzky et al. [36]), residue spin systems can mostly be assigned unambiguously. As described earlier, the residues observed in INEPT-based experiments often exhibit random coil chemical shifts. Therefore, a comparison of the chemical shifts assigned in the INEPT spectrum to the random coil standard values allows for estimating the degree of mobility of the residue present in this dynamic regime. The amino acids observed in the INEPT spectrum compared to the primary sequence of a protein might make it possible to identify the mobile protein segment, sometimes with high confidence if a single copy of one residue type is present in the primary sequence.
4. For all other described analyses of the mobile elements a [U-^{13}C]- or [U-^{13}C/^{15}N]-labeled sample is necessary.
5. If multiple residues of the same residue type are present showing slightly different chemical shifts or difficulties in disambiguating signals, 2D ^{1}H-^{13}C INEPT and ^{1}H-(^{13}C)-^{13}C or (^{1}H-)^{13}C-^{13}C INEPT-TOBSY [57, 58] can be recorded.

2D ^1H-^{13}C INEPT can connect the ^{13}C signal visible in the 1D ^{13}C INEPT to its bonded ^1H atom (^1H random coil chemical shift values can be used). An additional ^{13}C-^{13}C polarization transfer with the TOBSY sequence then makes it possible to connect the ^1H to the neighboring carbon positions, establishing the typical $^1H\alpha$-$^{13}C\alpha$-$^{13}C\beta$ spin system aimed at the residue-type assignment.

6. As reported by Baldus et al. [59], the peak intensity of each residue type Cα-Cβ correlation in a through-bond (e.g., INEPT-TOBSY) or through-space (e.g., CP-PDSD) spectrum can be used to compare the amino acid composition within the mobile and the rigid protein segments, respectively.

3.4 Structural Characterization of Rigid Core of Filamentous Assembly

A major task consists in assigning experimentally observed resonances to the ^{13}C atoms where they originate. The optimal procedure is to first familiarize oneself with the amino acid composition, the signal dispersion, and linewidth by identifying a maximum of amino acid spin systems (*see* Subheadings 3.4.1 and 3.4.2 and as an example Fig. 6). Once a maximum of spin systems has been identified, integrating information from selective labeling schemes if available (Fig. 7), and the necessary spectra have been recorded to perform the sequential assignment, the subsequent step consists in assigning the amino acid spin systems to the primary sequence and to identify missing spin systems using the sequential connections of the N, C′, Cα, and Cβ resonances (*see* Subheading 3.4.3 and Fig. 8). At this stage, create a project in your assignment program of choice

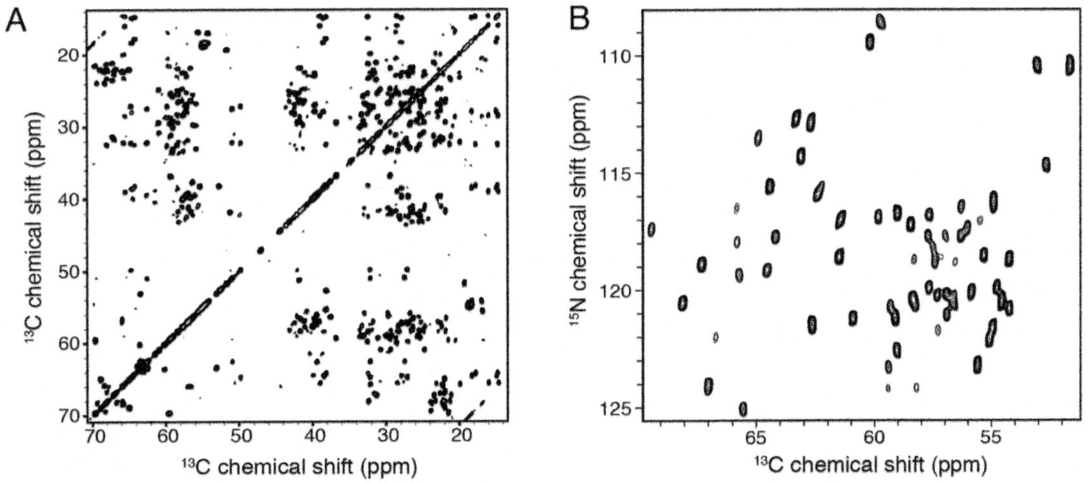

Fig. 6 Fingerprint SSNMR experiments for structural investigation of filamentous appendages. (**a**) 2D ^{13}C-^{13}C PDSD spectrum recorded with a short mixing time to observe intraresidue correlations. (**b**) 2D ^{15}N-$^{13}C\alpha$ spectrum to probe intraresidue backbone correlations. These experiments are illustrated on the *S. typhimurium* type III secretion needle [1]

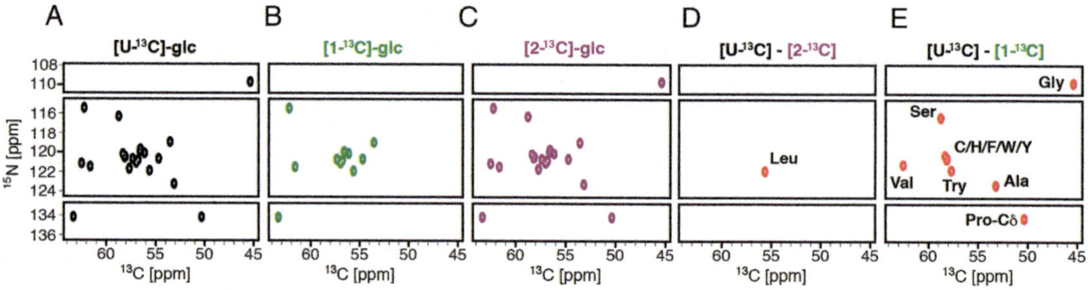

Fig. 7 Comparison of theoretical 2D NCA spectra for each amino acid, for three different labeling schemes: [U-^{13}C]-glc (**a**), [1-^{13}C]-glc (**b**), and [2-^{13}C]-glc (**c**). (**d**) Spectral comparison of [U-^{13}C] and [2-^{13}C]-glc leads to the identification of leucine correlations. (**e**) Spectral comparison of [U-^{13}C] and [1-^{13}C]-glc leads to the identification of several amino acid correlations

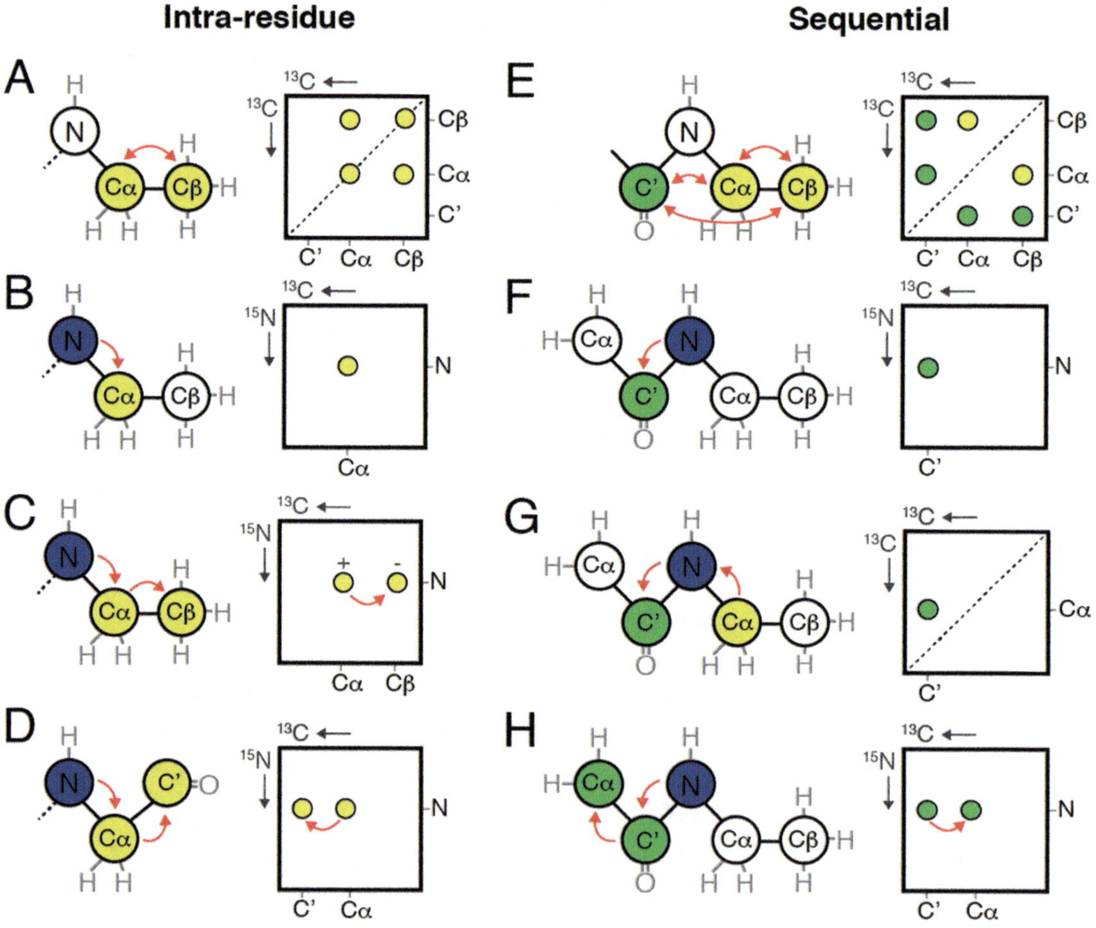

Fig. 8 Strategy for SSNMR resonance assignment. Intraresidue (**a–d**) and interresidue (**e–h**) correlations are established through the connectivity between the carbon–carbon and the nitrogen–carbon atoms. (**a**) C-C PDSD experiment for all intraresidue ^{13}C-^{13}C correlations (short mixing time). (**b**) N-C specific-CP for intraresidual ^{15}N-^{13}Cα correlations. (**c**) N-(Cα)-Cβ (^{15}N-^{13}Cα specific-CP, DREAM) for intraresidual ^{15}N-^{13}Cα-^{13}Cβ correlations. (**d**) N-(Cα)-C′ (^{15}N-^{13}Cα specific-CP, MIRROR) for intraresidual ^{15}N-^{13}Cα-^{13}C′ correlations. (**e**) C-C PDSD experiment for predominantly intraresidue and sequential ^{13}C-^{13}C correlations (intermediate mixing time). (**f**) N-C′ (^{15}N-^{13}C′ specific-CP) for sequential ^{15}N-^{13}C′ correlations. (**g**) Cα-(N)-C′ (^{13}Cα-^{15} N specific-CP, ^{15}N-^{13}C′ specific-CP) for sequential ^{13}Cα-^{13}C′ correlations. (**h**) N-(C′)-Cα (^{15}N-^{13}C′ specific-CP, MIRROR) for sequential ^{15}N-^{13}C′-^{13}Cα correlations

(*see* **item 1** in Subheading 2.3) and carefully set up all molecular parameters of your system in the project.

Several system parameters obtained during the characterization based on a [U-$^{13}C/^{15}N$]-labeled sample are important in terms of choosing an adequate precursor for the selective labeling scheme. As described in **Step 3** in Subheading 3.2.1, glycerol-based selective labeling should be chosen if the signal-to-noise ratio in the spectra is a limiting factor. If the spectral overlap is the most limiting factor, glucose-based selective labeling should be the correct choice [49].

3.4.1 Identification of Amino Acid Spin Systems

1. Set up a 2D ^{13}C-^{13}C proton-driven spin diffusion (PDSD) [60] experiment using a mixing time of 50 ms on a [U-$^{13}C/^{15}N$]-labeled sample. This experiment is illustrated with a bacterial secretion needle in Fig. 6a. The relatively short mixing time allows for the observation of intraresidue correlations (Fig. 8a). The 2D spectrum is processed using the program NMRpipe (*see* **item 3** in Subheading 2.1).
2. Compare the resulting ^{13}C-^{13}C spectrum with standard chemical shift values for amino acids based on the BMRB average chemical shift statistics. Several amino acid spin systems, such as isoleucine, threonine, serine, alanine, or proline, show a typical resonance pattern that can be identified in a straightforward manner if the resolution is sufficiently high (narrow signal linewidths and adequate molecular weight). Spectral analysis should be carried out using one of the visualization programs listed in **item 2** of Subheading 2.3.
3. Set up a 2D ^{13}C-^{13}C DREAM experiment with a short mixing time (typically less than 5 ms) [61]. The DREAM experiment provides intraresidue correlations, such as the PDSD; however, positive and negative signals are observed owing to the adiabatic double quantum polarization transfer. This experiment has the advantage of high polarization transfer efficiency and is therefore extremely useful in sensitivity problems. The inversion of the signal intensity leads to the observation of typical $C\alpha$-$C\beta$-$C\gamma$ correlations for which the $C\gamma$ signal is again positive (as the diagonal is) and helps in identifying spin systems.
4. If available, set up a 2D ^{13}C-^{13}C PDSD experiment using a mixing time of 75 ms on a selectively ^{13}C-labeled sample. The relatively short mixing time allows for the observation of intraresidue correlations, and the labeling scheme decreases the number of signals, because several carbons are unlabeled in the selectively ^{13}C-labeled sample and no signal will be observed, and therefore disambiguates signal assignments. To assign the signals, refer to the metabolic labeling pattern of the corresponding scheme ([1-^{13}C]-glc [27] and [2-^{13}C]-glc [29] or [1,3-^{13}C]-glyc and [2-^{13}C]-glyc [50]).

5. The complementarity of the selective labeling schemes can be used to rapidly identify specific amino acid types, as illustrated in Fig. 7 in an N-Cα spectrum. For instance, leucine is the only amino acid that is not mainly ^{13}C labeled on its Cα position in the [2-^{13}C]glc labeling scheme. Comparison of N-Cα spectra recorded for [U-^{13}C]glc and [2-^{13}C]glc allows for a rapid identification of leucine residues (Fig. 7d). A considerable number of amino acid ambiguities can be lifted in this manner as their N-Cα correlation will arise in either or both [1-^{13}C]glc and [1,3-^{13}C]glyc or in the [2-^{13}C]glc, [2-^{13}C]glyc-labeled sample. When performed on a ^{13}C-^{13}C correlation spectrum, this type of spectrum comparison between the different labeling schemes is likewise useful for lifting assignment ambiguities.

3.4.2 ^{15}N-^{13}C Intraresidue Linking

1. Set up a ^{15}N-^{13}C-specific CP with a mixing time ranging from 2 to 5 ms (to be optimized on a 1D spectrum). This experiment is illustrated with a bacterial secretion needle in Fig. 6b. Although the experiment is based on through-space magnetization transfer, short mixing times of 2–5 ms and relatively low-power pulses (e.g., 5–20 kHz on ^{13}C and ^{15}N) allow for the observation of specific intraresidue ^{15}N to $^{13}C\alpha$ correlations (Fig. 8b).
2. For amino acids with isolated Cα resonances, the Cα resonance can be unambiguously linked to their nitrogen resonance frequency.
3. The NCC-type experiment (e.g., N-(Cα)-Cβ, N-(Cα)-Cx, N-(Cα)-C′) adds an additional dimension to identify the amino acid spin systems (already described on a ^{13}C-^{13}C basis in Subheading 3.4.1) (*see* **Note 10**).
4. For identifying N-Cα-Cβ correlations, the optimal transfer efficiency is obtained by adding a DREAM $^{13}C\alpha$-$^{13}C\beta$ transfer after the ^{15}N-^{13}C-specific CP, the resulting spectrum containing mostly positive $^{13}C\alpha$ and negative $^{13}C\beta$ signals [62]. The ^{13}C resonances can now be linked to their corresponding intraresidual ^{15}N signal, which will further be used for the sequential assignment (Fig. 8c).
5. To identify the carbonyl resonances, an N-(Cα)-C′ experiment (MIRROR transfer [63]) (Fig. 8d) is used to link Cα to their attached carbonyl.

3.4.3 Sequential Assignment

The assignment procedure described in what follows requires uniformly or uniformly and selectively $^{13}C/^{15}N$-labeled samples for **steps 1** and **2** and a [U-$^{13}C/^{15}N$]-labeled sample for **steps 3** and **4**.

1. Set up a 2D ^{13}C-^{13}C PDSD experiment using a mixing time of 200 ms. The intermediate mixing time allows for the interresidue

correlations to build up. The comparison of an intermediate-mixing with a short-mixing PDSD makes it possible to identify spatial ^{13}C-^{13}C correlations that arise mostly from contacts with neighboring residues, i.e., the signals visible only in the long-mixing PDSD arise from interresidue contacts (Fig. 8e). Nevertheless, this information must be treated with caution and needs to be used only in combination with exclusively sequential spectra (*see* following discussion) since the signals in a ^{13}C-^{13}C PDSD spectrum can arise from all spatially close carbons. For example, in an α-helix, the carbons of residue *i* + 3 are very close to those of residue *i* and the contact signals between the two residues can build up rapidly in a PDSD spectrum if the helix is part of a very rigid segment in the protein.

2. If available, set up a 2D ^{13}C-^{13}C PDSD experiment using an intermediate mixing time of 300 ms on a selectively $^{13}C/^{15}N$-labeled sample. The intermediate mixing time allows for the interresidue correlations to build up. As described in paragraph 1 of this subheading, the comparison of an intermediate-mixing with a short-mixing PDSD makes it possible to identify spatial ^{13}C-^{13}C correlations that arise from contacts with carbons that are close in space, as are the neighboring residues, i.e., the signals visible only in the intermediate-mixing PDSD arise from interresidue contacts. Nevertheless, the information needs to be treated with caution, as mentioned earlier. To assign the signals, refer to the metabolic labeling pattern of the corresponding scheme.
3. Set up the spectra linking the resonances between the residues (*see* **Note 10**). Depending on the system complexity (residue number, spectral dispersion, amino acid composition, linewidth, available spectrometer field strength), the spectra need to be recorded in a 2D or 3D setup, and the number of necessary spectra to perform the complete sequential assignment will vary. The minimal amount of indispensable 2D spectra is N-C′ (Fig. 8f), N-(Cα)-Cβ, N-(Cα)-C′, Cα-(N)-C′ (Fig. 8g), N-(C′)-Cα (Fig. 8h) using the following magnetization transfer schemes: ^{15}N-^{13}C and ^{13}C-^{15}N-specific CP; $^{13}C\alpha$-$^{13}C\beta$ DREAM; $^{13}C'$-$^{13}C\alpha$ DARR/MIRROR or PDSD (less specific toward $^{13}C\alpha$). These spectra can be set up with three acquisition times (N-Cα-Cββ, N-Cα-C′, Cα-N-C′, N-C′-Cα) if required by the system. If necessary, supplemental spectra can be recorded in a 2D or 3D setup. The most useful are the following: N-C′-(Cα)-Cβ or N-(C′)-Cα-Cβ; Cα-(N)-(C′)-Cα or Cα-N-(C′)-Cα; N-(Cα)-Cx (DARR or PDSD $^{13}C\alpha$-^{13}Cx transfer); 3D CCC (with a short ^{1}H-$^{13}C\alpha$ CP toward the Cα, followed by a DREAM and a DARR or PDSD.
4. Choose an amino acid spin system with identified ^{15}N, $^{13}C\alpha$, and $^{13}C'$ resonances (residue *i*) and proceed to link those

resonances to the neighboring residue. At each linking step two resonance frequencies should be constant (i.e., identified). ^{15}N, ^{13}Cα of residue *i* can then be found in the Cα-N-C′ linked to the C′ of residue *i* – 1. In the N-C′-Cα, the N(*i*) and C′(*i* – 1) are linked to the Cα of residue *i* – 1. The N-Cα-C′ then provides the N(*i* – 1) and the N-Cα-Cβ the Cβ (*i* – 1), so that Cα and Cβ resonances of the spin system *i* – 1 are identified and the side-chain resonances can be obtained from the 2D ^{13}C-^{13}C or the 3D ^{13}C-^{13}C-^{13}C. The supplemental spectra mentioned earlier can help remove ambiguities and solve difficult assignment stretches. For example, if the N, C′, Cα resonances of two spin systems are identical, then using the sequential connection that includes the Cβ resonance (present in N-Cα-Cβ and N-(C′)-Cα-Cβ) could resolve the ambiguity.

3.4.4 Determination of Secondary Structure and Topology of Protein Subunit

1. The assigned ^{13}Cα, ^{13}Cβ resonances make it possible to identify the secondary chemical shift [64] ΔδCα-ΔδCβ, indicative of the secondary structure.
2. Calculate ^{13}Cα(assigned)-^{13}Cα(random coil) and ^{13}Cβ(assigned)-^{13}Cβ(random coil). Chemical shift values of amino acids in random coil conformation can be obtained from [36].
3. Plot ΔδCα-ΔδCβ, negative values for >3 residues in a row are indicative of β-strands and positive for α-helical conformation. It appears that amino acids such as glycine and proline show unusual chemical shift values, especially when they occur as *secondary structure breakers*. Caution must be taken when defining the limits of secondary structure elements from the secondary chemical shifts.
4. Use TALOS+ [39] or PREDITOR [38] to predict the protein dihedral angles from the assigned chemical shifts. The phi/psi dihedral angle restraints will be used later as structural input during the modeling process.

3.5 SSNMR Structural Restraints

Several types of structural restraints are used in the structural modeling of macromolecular protein filaments. The following list (Subheadings 3.5.1–3.5.3) follows chronological order in a basic SSNMR structure determination process of a macromolecular assembly.

3.5.1 Dihedral Angle Restraints

The dihedral angles can be derived from the SSNMR chemical shifts based on the TALOS+ routine (**step 2** in Subheading 3.4.4) and are introduced as structural restraints in the protocol of the selected modeling software to perform the structure determination.

3.5.2 Collection of Distance Restraints

Following the nomenclature used in solution NMR, SSNMR distance restraints are classified in different categories depending on the number of residues separating the residues involved in the contact. With the spectral assignment, the sequential (residue *i* – residue *i* ± 1) and medium-range (residue *i* – residue *i* ± 2, 3, or 4) contacts will be identified first because they can be derived directly from the sequential assignment process. Long-range (residue *i* – residue *i* > 4) contacts establish the 3D fold of the protein subunit by defining the carbon network and thereby connecting the different secondary structure elements (intramolecular long-range), as well as by defining the subunit–subunit interfaces (intermolecular long-range). In this chapter, we focus on the collection of ^{13}C-^{13}C and ^{15}N-^{13}C long-range distance restraints. While identifying a long-range contact, several considerations must be taken into account (*see* **Note 11**).

1. Set up a 2D ^{13}C-^{13}C PDSD experiment using a mixing time of 400 ms on a [U-^{13}C]- or [U-$^{13}C/^{15}N$]-labeled sample. The long mixing time allows for long-range distance correlations to build up. The comparison of this long-mixing (400 ms) PDSD with the other intermediate-mixing (200 ms) PDSD (*see* Subheading 3.4.3) makes it possible to identify additional spatial ^{13}C-^{13}C correlations that might arise from contacts with the neighboring residues or from residues that are far away in the primary sequence. If the protein filament allows for it, i.e., if the spectral resolution is high enough that the signal can be unambiguously assigned, long-range contacts might be encoded in the signals and be identified in a straightforward manner.
2. Set up a 2D ^{13}C-^{13}C PDSD experiment using a mixing time of 800 ms on a selectively ^{13}C-labeled sample. The labeling scheme will result from the selectively labeled precursors used for the protein production (*see* Subheading 3.1.2). In selectively labeled samples, the buildup of long-range contact signals is considerably enhanced when compared to uniformly labeled protein because an SSNMR magnetization transfer phenomenon called *dipolar truncation* is reduced. In most cases, these spectra are the most promising in terms of identifying long-range ^{13}C-^{13}C contacts. The long mixing time allows for long-range distance correlations to build up. The comparison of this long-mixing (800 ms) PDSD with the intermediate-mixing (200 ms) PDSD makes it possible to identify additional spatial ^{13}C-^{13}C correlations that might arise from contacts with neighboring residues or from residues that are far away in the primary sequence. If the protein filament allows for it, i.e., if the spectral resolution is high enough that the signal can be unambiguously assigned, long-range contacts might be encoded in the signals and be identified in a straightforward manner.

3.5.3 Intrasubunit and Subunit–Subunit Protein Interactions

1. For the detection of intermolecular ^{15}N-^{13}C long-range contacts, set up a 2D ^{15}N-^{13}C PAIN-CP [65] on a filament sample prepared with a mixture of [U-^{13}C]- and [U-^{15} N]-labeled subunits (*see* Subheading 3.1.4 for sample preparation). The signals arise unambiguously from the intermolecular ^{15}N-^{13}C contacts experiment (Fig. 9a).
2. To discriminate between intra- and intermolecular ^{13}C-^{13}C long-range contacts, set up a 2D ^{13}C-^{13}C PDSD experiment using a mixing time of 800 ms on a filament prepared with a 1:1 mixture of selectively ^{13}C-labeled subunits (*see* Subheading 3.1.4 for sample preparation). The correlation peaks detected between two carbons that are labeled in one labeling scheme and unlabeled in the other must arise from an intermolecular contact [30]. Therefore, a comparison of this spectrum with the two spectra recorded on each labeling scheme ([1-^{13}C]-glc

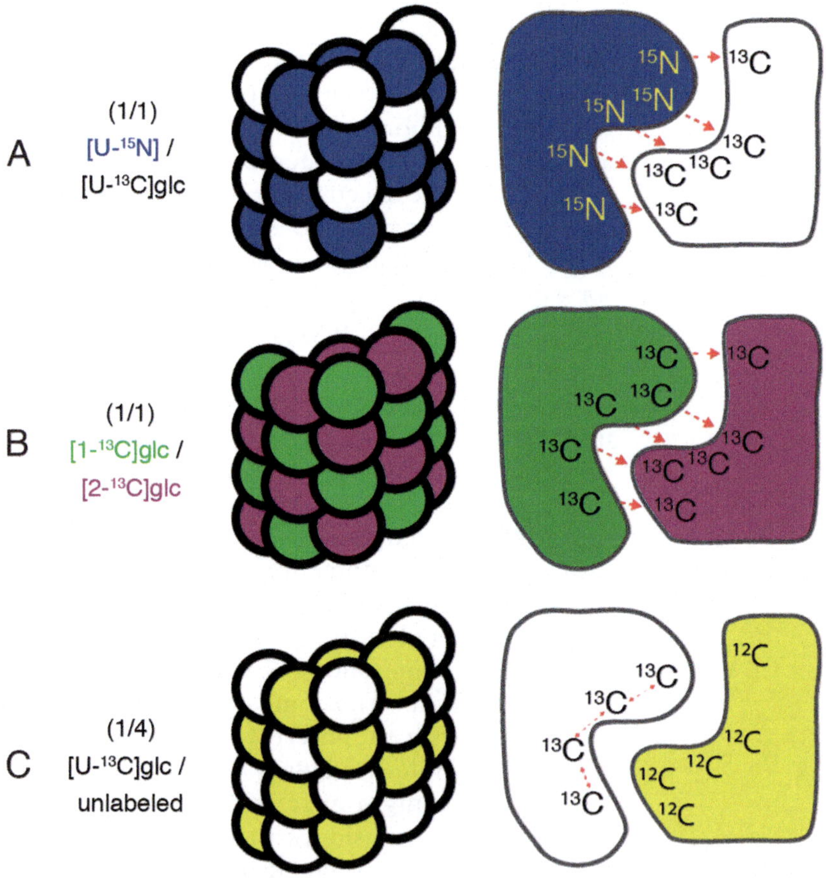

Fig. 9 Strategies for heterogeneous isotopically labeled filament samples. Heterogeneous labeling schemes based on (**a**) an equimolar mixture of uniformly ^{13}C and uniformly ^{15}N labeled subunits (*blue* and *white*, respectively), (**b**) an equimolar mixture of [1-^{13}C]-glucose- and [2-^{13}C]-glucose-labeled subunits (green and *magenta*, respectively), and (**c**) a diluted mixture (1/4) of ^{13}C-labeled subunits in an unlabeled subunit background (*white* and *yellow*, respectively)

and [2-^{13}C]-glc or [1,3-^{13}C]-glyc and [2-^{13}C]-glyc) makes it possible to identify the intermolecular contacts (Fig. 9b).

3. Highlighting intramolecular long-range contacts. A mixed labeled filament sample containing 1:*X* (with *X* = 3 to 5) [U-^{13}C/^{15}N]-labeled and unlabeled subunits should be prepared, also termed ***diluted sample***. If long-range contact signals are detected in a ^{13}C-^{13}C spectrum of a diluted sample, they arise from intramolecular contacts. Signals present in the ^{13}C-^{13}C long-range correlation spectrum on a nondiluted [U-^{13}C/^{15}N]-labeled and absent in the equivalent spectrum recorded on the diluted sample can potentially arise from intermolecular long-range contacts (Fig. 9c). This information should be treated with caution since the overall decrease in signal intensity might also cause the disappearance of intramolecular contact peaks.

3.6 SSNMR Structure Calculation

Modeling the 3D structure aims at the generation of a PDB file containing all atom coordinates for each protein subunit within the filament assembly (Fig. 10). However, depending on the quality of the SSNMR spectra, the amount of SSNMR structural restraints or the access to complementary structural data from other techniques, the modeling process can lead to 3D models with different degrees of precision, ranging from pseudo-atomic resolution structures to cartoonlike 3D models based on the molecular topology.

1. Prepare the SSNMR distance restraint lists (unambiguous and ambiguous lists) encoding for intramolecular (i.e., intrasubunit) distances and TALOS-based dihedral restraints.

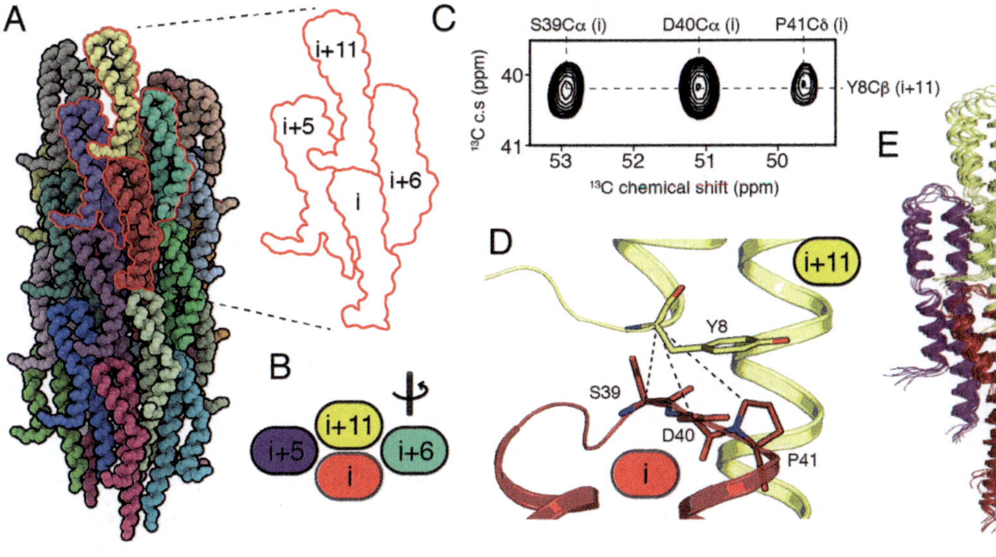

Fig. 10 (**a**, **b**) T3SS needle filament and its basic building block, consisting of the smallest asymmetric unit containing all subunit–subunit interfaces. (**c**) Based on the detection of intermolecular SSNMR restraints, (**d**) distance restraints at subunit–subunit interfaces can be derived. (**e**) SSNMR atomic structure of T3SS building block

2. Generate subunit monomeric structures using standard CNS or XPLOR-NIH routines based on simulated annealing protocols with torsion angles as internal degrees of freedom. At this stage, only unambiguous SSNMR restraints are used in the calculation.
3. Carefully observe whether the structure converges to one monomer fold. Select the ten lowest-energy monomers.
4. Assign the ambiguous restraints from the ambiguous restraint list based on the preliminary experimentally determined 3D fold. Rerun the structure calculation as described in **step 1** by adding the disambiguated restraints. This procedure can be carried out manually, but it can also be automatized in programs such as ARIA [66], UNIO [67], or HADDOCK [46].
5. Prepare the SSNMR distance restraint lists (unambiguous and ambiguous lists) encoding for intermolecular (i.e. intersubunit) distances.
6. Choose the minimal number of subunit monomers that will be calculated for the building block modeling. Typically, the building block should be considered as the smallest multimeric protein unit encoding each existing subunit–subunit interface of the assembly [49]. For example, for the T3SS needle filament, composed of three ($n = 3$) different subunit–subunit interfaces, the building block is an ($N = n + 1 = 4$) tetramer.
7. According to the molecular topology encoded in the secondary chemical shifts and the monomeric 3D fold, structurally ambiguous intermolecular distance restraints can already be disambiguated.
8. Generate a homomultimeric subunit made of N subunits using CNS or XPLOR-NIH routines. At this stage of the modeling, the TALOS and intramolecular restraints are used as input in addition to the intermolecular restraints. To guarantee the symmetry between the different subunits, noncrystallographic symmetries (NCS) [43] are used to reinforce the superimposability of the monomeric subunits. Please note that the quasi-symmetry of the local subunit structure is detected by the presence of a single SSNMR resonance set.
9. Select the ten lowest-energy multimeric structures to perform the quality assessment using PROCHECK-NMR [48] .

3.7 Integrative Structural Analysis

The structure determination of the filamentous protein assembly follows the previously described procedure. In this section, we globally describe how and which data from other sources can be integrated and used throughout the study. The integration of the data can be performed at different stages of the structural study (*see* Fig. 11).

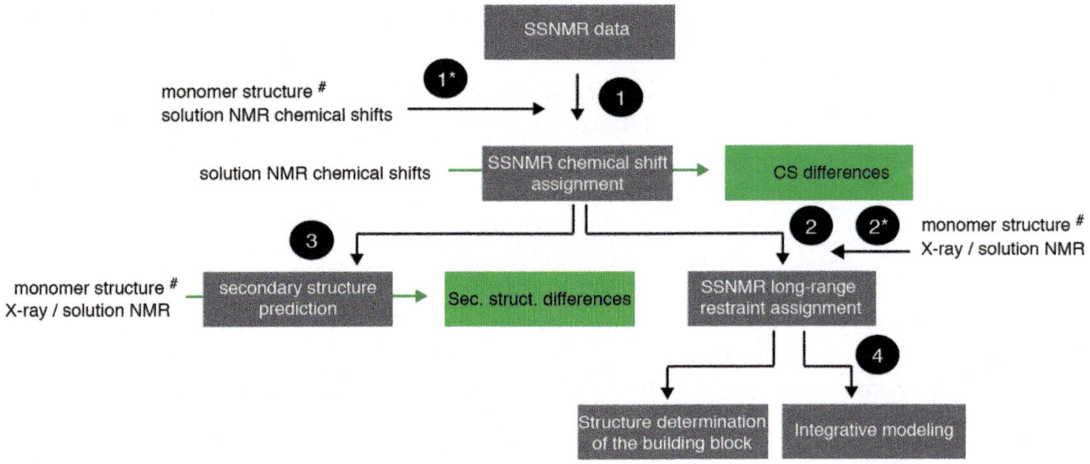

Fig. 11 Integrative structure determination process based on SSNMR. The different tasks are *numbered* and *asterisks* indicate the data that can be integrated. *Green arrows* and *underlining* highlight intermediate results, if the necessary data are available. Step 1 assignment process; 2 assignment process of long-range contacts in SSNMR data; 3 secondary structure determination; 4 structure modeling with integration of different structural data obtained from biophysical techniques. # The monomer structure can be obtained from x-ray crystallography or solution NMR

3.7.1 SSNMR Analysis Guided by Solution NMR Chemical Shifts

This subheading corresponds to steps 1 and 3 illustrated in Fig. 11.

1. If the solution NMR ^{13}C chemical shifts of a monomer or a truncated version of the monomer are available, predict a 2D ^{13}C-^{13}C spectrum and superimpose it to the experimental SSNMR ^{13}C-^{13}C spectrum recorded on the assembly, using, for example, the CCPNMR analysis prediction tool. If a monomer crystal structure is available, predict the ^{13}C chemical shifts (*see*prediction programs listed in Subheading 2.3) to further predict a 2D SSNMR ^{13}C-^{13}C.
2. Plot the predicted 2D ^{13}C-^{13}C on the experimental SSNMR spectrum and compare the two spectra. If an experimental peak and a predicted peak correspond in an unoverlapped region of the spectrum, you can use with caution the predicted peak as assignment to start a sequential assignment stretch. Be aware that this assignment is only predicted and very liable to be incorrect. Nevertheless, it can facilitate/accelerate the sequential assignment process (described in Subheading 3.4.3).
3. If the solution NMR chemical shifts of a monomeric subunit are available, plot the chemical shift differences between solution NMR and SSNMR data obtained on the protein sequence. Where differences larger than the error bar are observed, structural differences between the monomeric state and the monomer in the assembly might occur or an interaction

site between the monomers in the assembly might be located in this segment.

4. The secondary structure of the subunit in the assembly can be obtained following Subheading 3.4.4. If the monomeric protein structure is available from X-ray crystallography or solution NMR, the secondary structures of the monomeric subunit and the subunit in the assembly can be compared and possible structural rearrangements delineated.

3.7.2 SSNMR Restraint Assignment Guided by a Solution NMR/Crystal Monomeric Structure or EM Data

This subheading corresponds to step 2 illustrated in Fig. 11. The assignment of long-range SSNMR contacts in the ^{13}C-^{13}C spectra can be facilitated by different biophysical data, which can be used alone or in combination.

1. If a monomeric subunit structure is available and the secondary structure is revealed to be mostly conserved, the carbon–carbon distances can be extracted using protein visualization software (e.g., Pymol or SwissPDB viewer). Two different carbon–carbon distance lists should be extracted: all distances below 4 Å or below 8 Å.
2. The distances should then be used to simulate a ^{13}C-^{13}C peak list based on the assigned chemical shifts, which can be displayed on the ^{13}C-^{13}C spectra recorded for the detection of long-range signals.
3. Intramolecular carbon–carbon distances below 4 Å should be detected and below 8 Å could be visible in the spectra; ^{13}C-^{13}C peaks can therefore be assigned based on the simulated peak list. Nevertheless, the peaks should be assigned with caution and the potential assignments discarded in regions where the secondary structure is not conserved between the monomeric and the assembled subunits. The remaining unassigned peaks should arise from intermolecular carbon–carbon contacts.
4. EM provides data on the assembly dimensions and can thereby exclude certain protein arrangements in the filament. On the basis of scanning transmission electron microscopy (STEM), mass-per-length data can be extracted; they give access to a crucial filament parameter: the number of monomers per length unit. First modeling based on the available data can reduce the long-range intra- and intermolecular SSNMR contact assignment ambiguities. If a cryo-EM map is available, a first model based on the cryo-EM map will reduce significantly the ambiguities arising during SSNMR long-range contact assignment.

3.7.3 Integrative Modeling Procedure

Several integrative modeling approaches have been developed to efficiently deliver 3D models of filamentous assemblies based on SSNMR and complementary data.

1. The protocol described in Subheading 3.6 can be used to generate an atomic resolution structure of the filament building block. This building block structure can serve as a rigid body core to introduce symmetry restraints (e.g., axial rise and helical angle per subunit) for energy minimization and refinement to obtain a 3D model of the filament objet. Such a protocol, already used without SSNMR data but starting from a monomeric crystal structure to determine the type II secretion pilus structure, is currently being developed by Nilges et al. [14].
2. On the other hand, Baker et al. proposed a Rosetta protocol [1, 26] to start from an ensemble of unfolded subunit polypeptide chains that are simultaneously minimized under all available experimental restraints (local SSNMR data such as chemical shifts and distance restraints, EM density map, symmetry restraints). This protocol has been used to determine 3D atomic models of the T3SS needles of *Salmonella* [1], *Shigella* [26], and the M13 bacteriophage [68].
3. Also starting from extended subunit monomers, Habeck, Lange et al. proposed a protocol [5] based on inferential structure determination (ISD) [69] to integrate solution NMR, SSNMR, and MPL data to simultaneously assign SSNMR restraints from ambiguous data and model the *E. coli* type 1 pilus structure in an iterative way. By disambiguating the SSNMR signals with the structural model, which becomes more precise at each step, the integration of ambiguous SSNMR data is facilitated.

4 Notes

1. All glassware, tubes, pipet tips, stock solutions, media, and buffer should be sterile (e.g., autoclave passage or filter sterilization).
2. Throughout the entire expression and purification hand gloves should be used.
3. If possible, expression and purification should be carried out in a sterile environment to avoid contamination
4. To obtain optimal results during protein assembly, different assembly conditions should be tested on unlabeled proteins (salt concentrations, pH, protein concentration, seeding effects). The resulting filament samples should be compared, if possible recording a 1D SSNMR fingerprint spectrum (^{1}H-^{13}C CP spectrum) and otherwise by visual comparison of the filament pellet after centrifugation and EM.
5. Prewarmed medium (37 °C) can help the bacteria to take up growth again.

6. Between a protein production in LB and M9 medium the quantity decreases usually by around 10–60%.
7. In the case of selectively labeled glc, only one-sixth of all carbon atoms in the filament subunit will be ^{13}C-labeled, i.e., visible in SSNMR spectra. This gives rise to a maximum resolution but simultaneously decreases the spectral sensitivity, one of the most limiting factors in SSNMR analysis, by a factor of 6 and an even higher factor with increasing dimensionality. In the case of [1,3-^{13}C]-glyc and [2-^{13}C]-glyc, two-thirds and one-third of all carbon atoms in the filament subunit will be ^{13}C-labeled, leading to a diminished sensitivity compared to [U-^{13}C/^{15}N]-labeled protein assemblies but a significantly higher sensitivity than for selectively glc-labeled proteins. For a detailed comparison see [49].
8. High salt concentrations are a problem in NMR as they make probe matching and tuning difficult, and they also increase pulse durations.
9. Check any hardware or software issues such as cable connections. The protein could still be in solution and salts could have crystallized (improbable if washing has been performed). Check the supernatant fraction of the centrifugation step after filament recovery for protein content (UV spectroscopy or SDS gel) or subsequently check a minimal amount of the filament sample for the protein of interest by SDS-PAGE (polyacrylamide gel electrophoresis). The pulse strengths could be different from those required for experiments on this filament sample. This situation can arise if the optimization was done in a powder sample since NMR hard pulses in powder samples require less power (or shorter duration) compared to those in hydrated samples, this effect being reinforced for salty samples. Slightly higher values may be tried for the biological samples to compensate for these effects.
10. In a 2D setup of experiments containing a ^{13}C-^{13}C polarization transfer, for example N-(Cα)-Cβ, the ^{13}C frequency signals in brackets (i.e., without acquisition before next transfer) are also observed owing to the incomplete polarization transfer.
11. If ambiguity arises between a sequential/medium-range contact and a long-range (residue i – residue $i > 4$) contact, the signal should be assigned to the sequential or medium-range contact as this usually corresponds to a shorter internuclear distance. We recommend extensively assigning all possible sequential and medium-range contacts prior to the long-range assignment in the spectra designated for long-range distance detection. The identification of a cross peak as an "unambiguous long-range contact" requires that no other intraresidue, sequential, or medium-range contact can explain the chemical

shift values with respect to the chosen chemical shift tolerance window. Likewise, only a single long-range contact should explain the observed cross peak. In case of several assignment possibilities, the assignment should be considered spectrally ambiguous. The NMR user defines the value of the chemical shift tolerance window. It corresponds to the range in parts per million for which assignment possibilities will be considered while assigning a cross peak. Typical values can be ±0.1–0.25 ppm, depending on the experimental linewidth. The detection of a long-range contact in homogeneously labeled samples ([U-^{13}C]glc; [1-^{13}C]glc; [2-^{13}C]glc; [1,3-^{13}C]glyc; [2-^{13}C]-glyc) is always associated with an ambiguity arising from the distinction between intramolecular and intermolecular interactions. From an NMR point of view, both possibilities are not distinguishable and require the use of asymmetric labeling strategies (*see* Subheading 3.1.4). In the present chapter, distance restraints will be considered to encode for a single distance range of 2–8.5 Å. This relatively large upper distance (8.5 Å) is used to compensate multiple relayed polarization transfers that can occur during a long recoupling time, especially for ^{13}C-^{13}C mixing.

Acknowledgments

The authors thank their past and present colleagues, in particular Prof. Adam Lange at the Leibniz-Institut für Molekulare Pharmakologie for his guidance during the author postdoctoral periods and his main intellectual contribution for the T3SS needle and type I pilus projects. This work was further supported by the Fondation pour la Recherche Médicale (FRM-AJE20140630090 to A.L.), the ANR (13-PDOC-0017-01 to B.H. and ANR-14-CE09-0020-01 to A.L.), the FP7 program (FP7-PEOPLE-2013-CIG to A.L.), the IdEx Bordeaux University (Chaire d'Installation to B.H.)and the European Research Council (ERC) under the European Union's Horizon 2020 research and innovation program (ERC Starting Grant to A.L., agreement 105945). Erick Dufourc is acknowledged for his continuous support.

References

1. Loquet A, Sgourakis NG, Gupta R, Giller K, Riedel D, Goosmann C et al (2012) Atomic model of the type III secretion system needle. Nature 486:276–279
2. Fujii T, Cheung M, Blanco A, Kato T, Blocker AJ, Namba K (2012) Structure of a type III secretion needle at 7-A resolution provides insights into its assembly and signaling mechanisms. Proc Natl Acad Sci U S A 109: 4461–4466
3. Blocker AJ, Deane JE, Veenendaal AK, Roversi P, Hodgkinson JL, Johnson S et al (2008) What's the point of the type III secretion system needle? Proc Natl Acad Sci U S A 105:6507–6513

4. Sauer FG, Futterer K, Pinkner JS, Dodson KW, Hultgren SJ, Waksman G (1999) Structural basis of chaperone function and pilus biogenesis. Science 285:1058–1061
5. Habenstein B, Loquet A, Hwang S, Giller K, Vasa SK, Becker S et al (2015) Hybrid structure of the type 1 pilus of Uropathogenic *Escherichia coli*. Angew Chem Int Ed Engl 54:11691–11695
6. Hospenthal MK, Redzej A, Dodson K, Ukleja M, Frenz B, Rodrigues C et al (2016) Structure of a chaperone-usher pilus reveals the molecular basis of rod uncoiling. Cell 164:269–278
7. Geibel S, Waksman G (2014) The molecular dissection of the chaperone-usher pathway. Biochim Biophys Acta 1843:1559–1567
8. Chandran Darbari V, Waksman G (2015) Structural biology of bacterial type IV secretion systems. Annu Rev Biochem 84:603–629
9. Melville S, Craig L (2013) Type IV pili in Gram-positive bacteria. Microbiol Mol Biol Rev 77:323–341
10. Stones DH, Krachler AM (2015) Fatal attraction: how bacterial adhesins affect host signaling and what we can learn from them. Int J Mol Sci 16:2626–2640
11. Cheung M, Shen DK, Makino F, Kato T, Roehrich AD, Martinez-Argudo I et al (2015) Three-dimensional electron microscopy reconstruction and cysteine-mediated crosslinking provide a model of the type III secretion system needle tip complex. Mol Microbiol 95:31–50
12. Rathinavelan T, Lara-Tejero M, Lefebre M, Chatterjee S, McShan AC, Guo DC et al (2014) NMR model of PrgI-SipD interaction and its implications in the needle-tip assembly of the Salmonella type III secretion system. J Mol Biol 426:2958–2969
13. Jones CH, Pinkner JS, Roth R, Heuser J, Nicholes AV, Abraham SN et al (1995) FimH adhesin of type 1 pili is assembled into a fibrillar tip structure in the Enterobacteriaceae. Proc Natl Acad Sci U S A 92:2081–2085
14. Campos M, Nilges M, Cisneros DA, Francetic O (2010) Detailed structural and assembly model of the type II secretion pilus from sparse data. Proc Natl Acad Sci U S A 107:13081–13086
15. Habenstein B, Loquet A. (2015). Solid-state NMR: an emerging technique in structural biology of self-assemblies. Biophys Chem.
16. Meier BH, Bockmann A (2015) The structure of fibrils from 'misfolded' proteins. Curr Opin Struct Biol 30:43–49
17. Miao Y, Cross TA (2013) Solid state NMR and protein-protein interactions in membranes. Curr Opin Struct Biol 23:919–928
18. Tang M, Comellas G, Rienstra CM (2013) Advanced solid-state NMR approaches for structure determination of membrane proteins and amyloid fibrils. Acc Chem Res 46: 2080–2088
19. Weingarth M, Baldus M (2013) Solid-state NMR-based approaches for supramolecular structure elucidation. Acc Chem Res 46: 2037–2046
20. Loquet A, Habenstein B, Lange A (2013) Structural investigations of molecular machines by solid-state NMR. Acc Chem Res 46: 2070–2079
21. Yan S, Suiter CL, Hou G, Zhang H, Polenova T (2013) Probing structure and dynamics of protein assemblies by magic angle spinning NMR spectroscopy. Acc Chem Res 46:2047–2058
22. Tycko R, Wickner RB (2013) Molecular structures of amyloid and prion fibrils: consensus versus controversy. Acc Chem Res 46:1487–1496
23. Hong M, Zhang Y, Hu F (2012) Membrane protein structure and dynamics from NMR spectroscopy. Annu Rev Phys Chem 63:1–24
24. Egelman EH (2015) Three-dimensional reconstruction of helical polymers. Arch Biochem Biophys 581:54–58
25. Egelman EH (2010) Reducing irreducible complexity: divergence of quaternary structure and function in macromolecular assemblies. Curr Opin Cell Biol 22:68–74
26. Demers JP, Habenstein B, Loquet A, Kumar Vasa S, Giller K, Becker S et al (2014) High-resolution structure of the Shigella type-III secretion needle by solid-state NMR and cryo-electron microscopy. Nat Commun 5:4976
27. Hong M (1999) Determination of multiple ***φ***-torsion angles in proteins by selective and extensive (13)C labeling and two-dimensional solid-state NMR. J Magn Reson 139:389–401
28. Castellani F, van Rossum B, Diehl A, Schubert M, Rehbein K, Oschkinat H (2002) Structure of a protein determined by solid-state magic-angle-spinning NMR spectroscopy. Nature 420:98–102
29. Lundstrom P, Teilum K, Carstensen T, Bezsonova I, Wiesner S, Hansen DF et al (2007) Fractional 13C enrichment of isolated carbons using [1-13C]- or [2-13C]-glucose facilitates the accurate measurement of dynamics at backbone Calpha and side-chain methyl positions in proteins. J Biomol NMR 38:199–212
30. Loquet A, Giller K, Becker S, Lange A (2010) Supramolecular interactions probed by 13C-13C solid-state NMR spectroscopy. J Am Chem Soc 132:15164–15166

31. Bockmann A, Gardiennet C, Verel R, Hunkeler A, Loquet A, Pintacuda G et al (2009) Characterization of different water pools in solid-state NMR protein samples. J Biomol NMR 45:319–327
32. Bertini I, Engelke F, Luchinat C, Parigi G, Ravera E, Rosa C et al (2012) NMR properties of sedimented solutes. Phys Chem Chem Phys 14:439–447
33. Delaglio F, Grzesiek S, Vuister GW, Zhu G, Pfeifer J, Bax A (1995) NMRPipe: a multidimensional spectral processing system based on UNIX pipes. J Biomol NMR 6:277–293
34. Vranken WF, Boucher W, Stevens TJ, Fogh RH, Pajon A, Llinas M et al (2005) The CCPN data model for NMR spectroscopy: development of a software pipeline. Proteins 59:687–696
35. Ulrich EL, Akutsu H, Doreleijers JF, Harano Y, Ioannidis YE, Lin J et al (2008) BioMagResBank. Nucleic Acids Res 36:D402–D408
36. Wang Y, Jardetzky O (2002) Probability-based protein secondary structure identification using combined NMR chemical-shift data. Protein Sci 11:852–861
37. Shen Y, Delaglio F, Cornilescu G, Bax A (2009) TALOS+: a hybrid method for predicting protein backbone torsion angles from NMR chemical shifts. J Biomol NMR 44:213–223
38. Berjanskii MV, Neal S, Wishart DS (2006) PREDITOR: a web server for predicting protein torsion angle restraints. Nucleic Acids Res 34:W63–W69
39. Shen Y, Bax A (2010) SPARTA+: a modest improvement in empirical NMR chemical shift prediction by means of an artificial neural network. J Biomol NMR 48:13–22
40. Han B, Liu Y, Ginzinger SW, Wishart DS (2011) SHIFTX2: significantly improved protein chemical shift prediction. J Biomol NMR 50:43–57
41. Kohlhoff KJ, Robustelli P, Cavalli A, Salvatella X, Vendruscolo M (2009) Fast and accurate predictions of protein NMR chemical shifts from interatomic distances. J Am Chem Soc 131:13894–13895
42. Schwieters CD, Kuszewski JJ, Tjandra N, Clore GM (2003) The Xplor-NIH NMR molecular structure determination package. J Magn Reson 160:65–73
43. Brunger AT (2007) Version 1.2 of the crystallography and NMR system. Nat Protoc 2:2728–2733
44. Rieping W, Habeck M, Bardiaux B, Bernard A, Malliavin TE, Nilges M (2007) ARIA2: automated NOE assignment and data integration in NMR structure calculation. Bioinformatics 23:381–382
45. Shen Y, Vernon R, Baker D, Bax A (2009) De novo protein structure generation from incomplete chemical shift assignments. J Biomol NMR 43:63–78
46. Dominguez C, Boelens R, Bonvin AM (2003) HADDOCK: a protein-protein docking approach based on biochemical or biophysical information. J Am Chem Soc 125:1731–1737
47. Bhattacharya A, Tejero R, Montelione GT (2007) Evaluating protein structures determined by structural genomics consortia. Proteins 66:778–795
48. Laskowski RA, Rullmannn JA, MacArthur MW, Kaptein R, Thornton JM (1996) AQUA and PROCHECK-NMR: programs for checking the quality of protein structures solved by NMR. J Biomol NMR 8:477–486
49. Loquet A, Habenstein B, Chevelkov V, Vasa SK, Giller K, Becker S et al (2013) Atomic structure and handedness of the building block of a biological assembly. J Am Chem Soc 135:19135–19138
50. Higman VA, Flinders J, Hiller M, Jehle S, Markovic S, Fiedler S et al (2009) Assigning large proteins in the solid state: a MAS NMR resonance assignment strategy using selectively and extensively 13C-labelled proteins. J Biomol NMR 44:245–260
51. Etzkorn M, Bockmann A, Lange A, Baldus M (2004) Probing molecular interfaces using 2D magic-angle-spinning NMR on protein mixtures with different uniform labeling. J Am Chem Soc 126:14746–14751
52. Hartmann SR, Hahn EL (1962) Nuclear double resonance in rotating frame. Phys Rev 128:2042
53. Baldus M, Petkova AT, Herzfeld J, Griffin RG (1998) Cross polarization in the tilted frame: assignment and spectral simplification in heteronuclear spin systems. Mol Phys 95:1197–1207
54. Fung BM, Khitrin AK, Ermolaev K (2000) An improved broadband decoupling sequence for liquid crystals and solids. J Magn Reson 142:97–101
55. Harris RK, Becker ED, Cabral De Menezes SM, Granger P, Hoffman RE, Zilm KW et al (2008) Further conventions for NMR shielding and chemical shifts IUPAC recommendations 2008. Solid State Nucl Magn Reson 33:41–56
56. Shaka AF, Frenkeil T, Freeman R (1983) NMR broadband decoupling with low radiofrequency power. J Magn Reson. 52:159–163
57. Baldus M, Geurts DG, Hediger S, Meier BH (1996) Efficient 15N-13C polarization transfer

by adiabatic passage Hartmann-Hahn cross-polarization. J Magn Reson Ser A 118: 140–144

58. Andronesi OC, Becker S, Seidel K, Heise H, Young HS, Baldus M (2005) Determination of membrane protein structure and dynamics by magic-angle-spinning solid-state NMR spectroscopy. J Am Chem Soc 127:12965–12974
59. Ader C, Frey S, Maas W, Schmidt HB, Gorlich D, Baldus M (2010) Amyloid-like interactions within nucleoporin FG hydrogels. Proc Natl Acad Sci U S A 107:6281–6285
60. Szeverenyi NM, Sullivan MJ, Maciel GE (1982) Observation of spin exchange by two-dimensional Fourier-transform C-13 cross polarization-magic-angle spinning. J Magn Reson 47:462–475
61. Verel R, Ernst M, Meier BH (2001) Adiabatic dipolar recoupling in solid-state NMR: the DREAM scheme. J Magn Reson 150:81–99
62. Westfeld T, Verel R, Ernst M, Bockmann A, Meier BH (2012) Properties of the DREAM scheme and its optimization for application to proteins. J Biomol NMR 53:103–112
63. Scholz I, Hodgkinson P, Meier BH, Ernst M (2009) Understanding two-pulse phase-modulated decoupling in solid-state NMR. J Chem Phys 130:114510
64. Luca S, Filippov DV, van Boom JH, Oschkinat H, de Groot HJ, Baldus M (2001) Secondary chemical shifts in immobilized peptides and proteins: a qualitative basis for structure refinement under magic angle spinning. J Biomol NMR 20:325–331
65. Lewandowski JR, De Paepe G, Griffin RG (2007) Proton assisted insensitive nuclei cross polarization. J Am Chem Soc 129:728–729
66. Bardiaux B, Malliavin T, Nilges M (2012) ARIA for solution and solid-state NMR. Methods Mol Biol 831:453–483
67. Guerry P, Herrmann T (2012) Comprehensive automation for NMR structure determination of proteins. Methods Mol Biol 831: 429–451
68. Morag O, Sgourakis NG, Baker D, Goldbourt A (2015) The NMR-Rosetta capsid model of M13 bacteriophage reveals a quadrupled hydrophobic packing epitope. Proc Natl Acad Sci U S A 112:971–976
69. Rieping W, Habeck M, Nilges M (2005) Inferential structure determination. Science 309:303–306
70. Schraidt O, Marlovits TC (2011) Three-dimensional model of Salmonella's needle complex at subnanometer resolution. Science 331:1192–1195

Chapter 30

Energy Requirements for Protein Secretion via the Flagellar Type III Secretion System

Marc Erhardt

Abstract

Protein transport across the cytoplasmic membrane is coupled to energy derived from adenosine triphosphate hydrolysis or the protein motive force (pmf). A sophisticated, multi-component type III secretion system exports substrate proteins of both the bacterial flagellum and virulence-associated injectisome system of many Gram-negative pathogens. The type-III secretion system is primarily a pmf-driven protein exporter. Here, I describe methods to investigate the export of substrate proteins into the culture supernatant under conditions that manipulate the pmf.

Key words Type III secretion system, Bacterial flagellum, Protein export, Proton motive force, ΔpH gradient, ΔΨ gradient, Ionophore, Carbonyl cyanide *m*-chlorophenylhydrazone (CCCP), Valinomycin

1 Introduction

Bacterial protein transportation systems utilize energy derived from the protein motive force (pmf) or adenosine triphosphate (ATP) hydrolysis for the translocation of substrate proteins across biological membranes [1].

Protein substrates of the bacterial flagellum and the evolutionarily related virulence-associated injectisome or needle complex are secreted by a homologous protein transportation system, termed the type III secretion system (T3SS). Secretion of substrate proteins via the flagellar-specific type III secretion system (f-T3SS) or virulence-associated type III secretion system (v-T3SS) of the injectisome complex is essential for the assembly of the corresponding nanomachine, as well as the secretion of effector proteins in the case of the injectisome system (reviewed in detail in [2–4]). The T3SS is intrinsically a pmf-driven protein exporter that exploits the activity of an associated ATPase to facilitate substrate secretion [5–8].

The core T3SS consists of eight or nine proteins that directly participate in the process of substrate protein translocation across the inner membrane [2–4]. Five integral membrane proteins

Laure Journet and Eric Cascales (eds.), *Bacterial Protein Secretion Systems: Methods and Protocols*, Methods in Molecular Biology, vol. 1615, DOI 10.1007/978-1-4939-7033-9_30, © Springer Science+Business Media LLC 2017

(f-T3SS: FlhA FlhB FliO FliP FliQ FliR; v-T3SS: SctV SctU SctR SctS SctT) form the export gate and are involved in the primary substrate recognition, substrate unfolding, energy transduction and protein transport across the cytoplasmic membrane. An associated ATPase complex (f-T3SS: FliH FliI FliJ; injectisome: SctL SctN SctO) has a facilitating role in substrate recognition, substrate unfolding and energy transduction, but is not strictly required for protein export [9].

A scaffold structure formed by an integral membrane ring (MS-ring; flagellum: FliF; injectisome: SctD SctJ) is essential for the assembly of a functional core export apparatus [10]. In addition, accessory proteins form a cytoplasmic ring that facilitates substrate recognition and binding of ATPase complex components (flagellum: FliG FliM FliN; injectisome: SctQ) [11–13].

The contribution of the pmf and ATP hydrolysis to the protein export process via T3SS has been examined using T3SS-dependent substrate protein secretion of mutant strains and after treatment with compounds that modulate the pmf. Here, I describe a methodology to analyse the protein export of an f-T3SS-specific substrate into culture supernatant in the presence or absence of compounds interfering with the pmf.

2 Materials

Standard chemicals are purchased in analytical quality from established commercial suppliers. Prepare all solutions using ultrapure water unless indicated otherwise.

2.1 FlgM-Secretion Assay

1. *Salmonella enterica* serovar Typhimurium strains: TH3730 P*flhDC*5451::*Tn*10*d*Tc[del-25], TH10874 Δ*flgM*5628::FRT Δ*araBAD*923::*flgM*-FKF P*araBAD*924 (*see* **Note 1**).
2. Lysogeny broth (LB): 10 g tryptone, 5 g yeast extract, 5 g NaCl. Add 12 g agar for LB agar plates. Add water to a volume of 1 L and autoclave.
3. Shaking incubator (*see* **Note 2**).
4. Spectrophotometer for OD_{600} determination.
5. Anhydrotetracycline: 0.2 mg/mL stock solution in 50% H_2O 50% ethanol (*see* **Note 3**).
6. L-arabinose: 20% stock in H_2O (*see* **Note 4**).
7. Tabletop centrifuge, refrigerated.

2.2 Protein Fractionation

1. Spectrophotometer for OD_{600} determination.
2. Table-top centrifuge, refrigerated.

2.2.1 Protein Extraction by Filtration Over a Nitrocellulose Filter

1. Nitrocellulose filter, 0.45 μm pore size.
2. 2× sodium dodecyl sulfate (SDS) sample buffer: 100 mM Tris–HCl, pH 6.8, 4% SDS, 20% glycerol, 1% β-mercaptoethanol, 25 mM EDTA, 0.04% bromophenol blue (*see* **Note 5**).
3. Heating block for 1.6 and 2.0 mL centrifugation tubes.

2.2.2 Protein Precipitation Using Trichloroacetic Acid

1. Trichloroacetic acid (TCA). Store at 4 °C.
2. Acetone. Store at 4 °C.
3. Vortexer.
4. 2× SDS sample buffer: 100 mM Tris–HCl, pH 6.8, 4% SDS, 20% glycerol, 1% β-mercaptoethanol, 25 mM EDTA, 0.04% bromophenol blue (*see* **Note 5**).
5. Heating block for 1.6 and 2.0 mL centrifugation tubes.

2.3 Immunoblotting

1. 4–20% precast gels. Store at 4 °C.
2. Mini-gel caster system and SDS-polyacrylamide gel electrophoresis (PAGE) apparatus.
3. SDS running buffer: 25 mM Tris–HCl, 192 mM glycine, 0.1% SDS, pH 8.3.
4. Western blot transfer membranes: 0.2 μm pore size Hybond-P polyvinylidene fluoride (PVDF) or 0.45 μm pore size nitrocellulose (*see* **Note 6**).
5. Western blot transfer buffer: 25 mM Tris–HCl, 192 mM glycine, 20% methanol, pH 8.3.
6. Trans-Blot apparatus for western blot transfer.
7. Serum-purified anti-FlgM rabbit polyclonal antibodies [14], dilution 1:10,000.
8. Horseradish-peroxidase-conjugated anti-rabbit polyclonal antibodies, dilution 1:10,000–1:20,000.

2.4 Assays to Inhibit Proton Motive Force

1. Ionophore carbonyl cyanide *m*-chlorophenylhydrazone (CCCP): 20 mM stock solution: Dissolve 4.1 mg CCCP in 1 mL dimethyl sulfoxide (DMSO) (*see* **Note 7**).
2. Valinomycin: 20 mM stock solution: Dissolve 22.2 mg valinomycin in 1 mL distilled H_2O.
3. Potassium chloride: 1 M stock solution: Dissolve 7.45 g KCl in 100 mL distilled H_2O.
4. Potassium acetate: 1 M stock solution: Dissolve 9.81 g CH_3COOK in 100 mL distilled H_2O (*see* **Note 8**).

3 Methods

Perform experiments at room temperature unless indicated otherwise.

3.1 FlgM-Secretion Assay

1. Streak *S. Typhimurium* strains TH3730 (*see* **Note 9**) or TH10874 (*see* **Note 10**) for single colonies on fresh LB plates (*see* **Note 1**). Incubate overnight at 37 °C.
2. Inoculate a single colony of *S. Typhimurium* strains TH3730 or TH10874 into 1 mL LB and incubate overnight at 37 °C in a water bath incubator, shaking at 200 rpm.
3. Dilute overnight culture of *S. Typhimurium* strains TH3730 or TH10874 1:100 in 3 mL LB and incubate at 37 °C in a water bath incubator, shaking at 200 rpm. Grow approximately 2 h until optical density (OD_{600}) of 0.5. Induce flagellar gene expression by addition of 100 ng/mL anhydrotetracycline and resume incubation at 37 °C for 60 min. Grow TH10874 cultures in LB supplemented with 0.2% L-arabinose to induce expression of FlgM for 60 min at 37 °C.
4. Pellet cells by 5 min centrifugation at 10,000 × *g* and perform permeabilization and washing steps as detailed in Subheading 3.4.
5. Resuspend in 3 mL LB medium containing appropriate dilutions of pmf inhibitors and 100 ng/mL anhydrotetracycline or 0.2% L-arabinose for TH3730 or TH10874, respectively, and resume incubation at 37 °C for 30 min in a water bath incubator, shaking at 200 rpm as outlined in Subheading 3.4.
6. Store bacterial culture on ice until further treatment.

3.2 Protein Fractionation

1. Remove 0.5 mL of bacterial culture and record OD_{600} for normalization purposes (*see* **step 4** in Subheading 3.2.1 and **step 7** in Subheading 3.2.2).
2. Centrifuge 2 mL of bacterial culture at 10,000 × *g* for 5 min in a tabletop centrifuge and transfer 1.8 mL of the supernatant to a new centrifugation tube. Discard remaining supernatant and store pellet on ice until further treatment (label 'cellular fraction').
3. Centrifuge supernatant at 10,000 × *g* for 5 min in a tabletop centrifuge and transfer 1.6 mL of the supernatant to a new centrifugation tube (label 'supernatant fraction') (*see* **Note 11**).
4. Two alternative methods can be used for collecting proteins from the culture supernatant: filtration on nitrocellulose filters and TCA precipitation.

*3.2.1 Protein Extraction by Filtration over Nitrocellulose Filter (See **Note 12**)*

1. Filter supernatant from **step 3** in Subheading 3.2 through a prewetted nitrocellulose filter with a pore size of 0.45 μm for protein binding.
2. Elute proteins by addition of 40 μL 2× SDS sample buffer and heat treatment for 30 min at 65 °C.
3. Resuspend bacterial pellet from **step 2**, Subheading 3.2, in 50 μL 2× SDS sample buffer and heat for 10 min at 95 °C.
4. Adjust cellular and supernatant fractions to 20 OD_{600} equivalents per microlitre by addition of 2× SDS sample buffer and keep on ice or store at −20 °C until further use.

3.2.2 Protein Precipitation Using Trichloroacetic Acid

1. Add TCA to a final concentration of 10% to the supernatant from **step 3**, Subheading 3.2, and incubate on ice for 30 min.
2. Centrifuge at 20,000 × *g* for 30 min in a refrigerated tabletop centrifuge at 4 °C and discard supernatant.
3. Resuspend pellet in 1 mL ice-cold acetone by vortexing.
4. Centrifuge at 20,000 × *g* for 30 min in a refrigerated tabletop centrifuge at 4 °C and discard supernatant.
5. Dry pellet overnight or for 30 min in a laminar flow bench.
6. Resuspend supernatant pellet in 40 μL 2× SDS sample buffer and heat for 10 min 95 °C.
7. Resuspend bacterial pellet from **step 2**, Subheading 3.2, in 50 μL 2× SDS sample buffer and heat for 10 min at 95 °C.
8. Adjust cellular and supernatant fractions to 20 OD_{600} equivalents per microlitre by addition of 2× SDS sample buffer and keep on ice or store at −20 °C until further use.

3.3 Immunoblotting

1. Load 200 OD_{600} equivalents of cellular and supernatant fractions onto 4–20% precast gels and perform protein separation by SDS-PAGE in standard Tris-glycine buffer.
2. Following separation, electrotransfer proteins to a 0.2 μm pore size Hybond-P PVDF transfer membrane (*see* **Note 6**) or a 0.45 μm pore size nitrocellulose membrane using a Trans-Blot transfer apparatus.
3. Perform immunodetection using appropriate concentrations of primary and secondary antibodies. In case of TH3730 or TH10874, secreted and cellular FlgM protein is detected using serum-purified anti-FlgM rabbit polyclonal antibodies (dilution 1:10,000, [14]) and horseradish-peroxidase-conjugated anti-rabbit polyclonal antibodies (dilution 1:10,000, BioRad).

3.4 Assays to Inhibit Proton Motive Force

3.4.1 Disruption of Proton Motive Force Using Carbonyl Cyanide m-Chlorophenylhydrazone (See **Note 13**) (See Fig. 1a)

1. Grow bacterial cultures as described in Subheading 3.1, **steps 1–3**.
2. Pellet bacterial culture by 5 min centrifugation at 10,000 × *g* and resuspend in 3 mL LB medium containing 0–20 μM carbonyl cyanide *m*-chlorophenylhydrazone (*see* **Notes 14** and **15**).
3. Wash bacterial cells by centrifugation at 10,000 × *g* for 5 min. Resuspend pellet in 3 mL LB medium containing 0–20 μM carbonyl cyanide *m*-chlorophenylhydrazone and inducer as in **step 3**, Subheading 3.1.
4. Resume incubation at 37 °C for 30 min in a water bath shaker at 200 rpm and continue with **step 6**, Subheading 3.1.

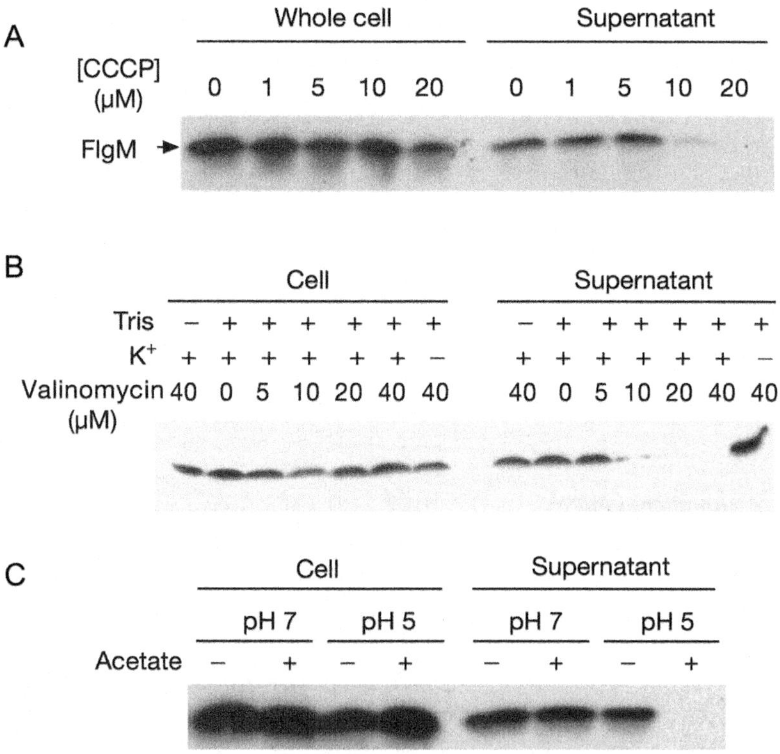

Fig. 1 (**a**) Effect of inhibition of pmf on FlgM export. FlgM secretion was inhibited by addition of 10 μM uncoupling agent carbonyl cyanide *m*-chlorophenyl hydrazone (CCCP) and completely abolished by treatment with 20 μM CCCP in strain TH10874 (arabinose-inducible *flgM*). Cytoplasmic FlgM levels remained constant. (**b**) Effect of inhibition of ΔΨ component of pmf on FlgM export. FlgM secretion was inhibited by addition of valinomycin in the presence of K+. Cells were pretreated with 120 mM Tris–HCl to permeabilize the outer membrane to valinomycin where indicated. (**c**) Effect of inhibition of ΔpH on FlgM export. Secretion of FlgM for cultures grown in pH 5 was inhibited by addition of 34 mM potassium acetate. Adapted with permission from Macmillan Publishers Ltd: Nature [6], copyright 2008

*3.4.2 Disruption of ΔΨ Component of Proton Motive Force by K+/Valinomycin (See **Note 16**) (See Fig. 1b)*

1. Grow bacterial cultures as described in Subheading 3.1, **steps 1–3**.
2. Pellet bacterial culture by 5 min centrifugation at 10,000 × *g* and resuspend in 3 mL LB medium containing 120 mM Tris–HCl, pH 7.3. Incubate for 2 min (*see* **Note 17**).
3. Pellet bacterial culture by 5 min centrifugation at 10,000 × *g* and resuspend in 3 mL LB medium containing 120 mM Tris–HCl, pH 7.3, and 0–40 μM valinomycin in the presence or absence of 150 mM KCl (*see* **Note 15**).
4. Pellet bacterial culture by 5 min centrifugation at 10,000 × *g*, discard supernatant and resuspend in 3 mL LB medium containing 120 mM Tris–HCl, pH 7.3, and 0–40 μM valinomycin in the presence or absence of 150 mM KCl.
5. Resume incubation at 37 °C for 30 min in a water bath shaker at 200 rpm and continue with **step 6**, Subheading 3.1.

*3.4.3 Disruption of ΔpH Component of Proton Motive Force by Potassium Acetate (See **Note 18**) (See Fig. 1c)*

1. Grow bacterial cultures as described in Subheading 3.1, **steps 1–3**.
2. Pellet bacterial culture by 5 min centrifugation at 10,000 × *g* and resuspend in 3 mL LB medium.
3. Wash bacterial cells by centrifugation at 10,000 × *g* for 5 min and discard supernatant.
4. Resuspend pellet in 3 mL LB medium at pH 7 or pH 5 in the presence or absence of 34 mM potassium acetate, respectively. Add inducer of flagellar genes transcription as required for strains TH3730 and TH10874 (*see* **Note 15**).
5. Resume incubation at 37 °C for 30 min in a water bath shaker at 200 rpm and continue with **step 6**, Subheading 3.1.

4 Notes

1. Alternatively, use a mutant strain where ATP synthesis is uncoupled from the pmf: TH11802 Δ*atpA*::*tetRA* Δ*flgM*5628::FRT Δ*araBAD*923::*flgM*-FKF P*araBAD*934, which is deficient in a major subunit of the F_OF_1 ATP synthase. The absence of the F_OF_1 ATP synthase results in a growth defect, which can be partially rescued by growth in media containing 0.2% glucose.
2. For best growth use a shaking water bath incubator.
3. Filter sterilize using 0.2 μm mixed cellulose ester or polyethersulfone filters.
4. Filter sterilize using 0.2 μm mixed cellulose ester or polyethersulfone filters.

5. Add freshly prepared β-mercaptoethanol.
6. Transfer to a 0.2 μm pore size Hybond-P PVDF membrane, as recommended for efficient electrotransfer of FlgM.
7. Dissolve freshly in DMSO.
8. Adjust pH to desired final pH by addition of HCl or NaOH.
9. This strain harbours the flagellar master regulatory operon *flhDC* under control of an anhydrotetracycline-inducible promoter and allows for constant expression of flagellar genes in the presence of inducer.
10. This strain harbours *flgM* as a non-structural reporter substrate of the flagellar T3SS under the control of an arabinose-inducible promoter.
11. This step minimizes potential contamination of the supernatant fraction by residual bacterial cells.
12. Protein binding to nitrocellulose filters is recommended for efficient recovery of secreted FlgM from the culture supernatant.
13. The pmf consists of two components, a proton concentration gradient (ΔpH) and a charge difference between the periplasmic and cytoplasmic faces of the membrane (ΔΨ). The ionophore CCCP disrupts both the proton gradient ΔpH and the membrane potential ΔΨ by causing an influx of H+ into the cytoplasm [6].
14. To control for DMSO-induced effects, add 0.5% DMSO to samples not treated with CCCP.
15. Add inducer 100 ng/mL anhydrotetracycline or 0.2% L-arabinose for continuous expression of flagellar genes (TH3730) or *flgM* (TH10874), respectively.
16. Valinomycin renders membranes permeable to potassium, which dissipates the ΔΨ component of the pmf by balancing the charge difference [6, 15].
17. Appropriate controls include samples not treated with 120 mM Tris–HCl, pH 7.3.
18. Weak acids such as acetate or benzoate cross the cytoplasmic membrane in neutral form and release a proton in the cytoplasm. The resulting decrease in the cytoplasmic pH essentially collapses the proton gradient ΔpH at an external pH of 5 [6, 16].

Acknowledgments

This work was supported by the Helmholtz Association young investigator grant VH-NG-932 and the People Programme (Marie Curie Actions) of the European Union Seventh Framework Programme (grant 334030).

References

1. Wickner W, Schekman R (2005) Protein translocation across biological membranes. Science 310:1452–1456
2. Erhardt M, Namba K, Hughes KT (2010) Bacterial nanomachines: the flagellum and type III injectisome. Cold Spring Harb Perspect Biol 2:a000299
3. Minamino T (2014) Protein export through the bacterial flagellar type III export pathway. Biochim Biophys Acta 1843:1642–1648
4. Diepold A, Wagner S (2014) Assembly of the bacterial type III secretion machinery. FEMS Microbiol Rev 38:802–822
5. Wilharm G, Lehmann V, Krauss K, Lehnert B, Richter S, Ruckdeschel K, Heesemann J, Trulzsch K (2004) *Yersinia enterocolitica* type III secretion depends on the proton motive force but not on the flagellar motor components MotA and MotB. Infect Immun 72:4004–4009
6. Paul K, Erhardt M, Hirano T, Blair DF, Hughes KT (2008) Energy source of flagellar type III secretion. Nature 451:489–492
7. Minamino T, Namba K (2008) Distinct roles of the FliI ATPase and proton motive force in bacterial flagellar protein export. Nature 451:485–488
8. Lee PC, Zmina SE, Stopford CM, Toska J, Rietsch A (2014) Control of type III secretion activity and substrate specificity by the cytoplasmic regulator PcrG. Proc Natl Acad Sci U S A 111:E2027–E2036
9. Erhardt M, Mertens ME, Fabiani FD, Hughes KT (2014) ATPase-independent type-III protein secretion in *Salmonella enterica*. PLoS Genet 10:e1004800
10. Morimoto YV, Ito M, Hiraoka KD, Che YS, Bai F, Kami-Ike N, Namba K, Minamino T (2014) Assembly and stoichiometry of FliF and FlhA in *Salmonella* flagellar basal body. Mol Microbiol 91:1214–1226
11. McMurry JL, Murphy JW, Gonzalez-Pedrajo B (2006) The FliN-FliH interaction mediates localization of flagellar export ATPase FliI to the C ring complex. Biochemistry 45:11790–11798
12. Erhardt M, Hughes KT (2010) C-ring requirement in flagellar type III secretion is bypassed by FlhDC upregulation. Mol Microbiol 75:376–393
13. Diepold A, Kudryashev M, Delalez NJ, Berry RM, Armitage JP (2015) Composition, formation, and regulation of the cytosolic c-ring, a dynamic component of the type III secretion injectisome. PLoS Biol 13:e1002039
14. Hughes KT, Gillen KL, Semon MJ, Karlinsey JE (1993) Sensing structural intermediates in bacterial flagellar assembly by export of a negative regulator. Science 262:1277–1280
15. Minamino T, Morimoto YV, Hara N, Namba K (2011) An energy transduction mechanism used in bacterial flagellar type III protein export. Nat Commun 2:475
16. Minamino T, Imae Y, Oosawa F, Kobayashi Y, Oosawa K (2003) Effect of intracellular pH on rotational speed of bacterial flagellar motors. J Bacteriol 185:1190–1194

Chapter 31

Identification of Effectors: Precipitation of Supernatant Material

Nicolas Flaugnatti and Laure Journet

Abstract

Bacterial secretion systems allow the transport of proteins, called effectors, as well as external machine components in the extracellular medium or directly into target cells. Comparison of the secretome, i.e. the proteins released in the culture medium, of wild-type and mutant cells provides information on the secretion profile. In addition, mass spectrometry analyses of the culture supernatant of bacteria grown in liquid culture under secreting conditions allows the identification of secretion system substrates. Upon identification of the substrates, the secretion profile serves as a tool to test the functionality of secretion systems. Here we present a classical method used to concentrate the culture supernatant, based on trichloroacetic acid precipitation.

Key words Supernatant, TCA precipitation, Secretome

1 Introduction

Bacterial secretion systems are macromolecular machines dedicated to the transport of proteins across the cell envelope. These secretion systems deliver effectors outside the cell, either in the medium (T1SS, T2SS, T5SS, T9SS) or directly into target cells (T3SS, T4SS, T6SS) [1]. Secretion of effector proteins into the milieu can be observed in these systems, and the analysis of secretion supernatant has been widely used either to identify new secreted effectors or to probe the functionality of secretion systems. For contact-dependent systems such as the T3SS, in vitro secretion in the medium can be observed under certain conditions (e.g. Ca^{2+} depletion, acidic pH) [2, 3]. As it is not always possible to predict effectors by bioinformatics approaches (*see* Chapter 2), analysis of the content of the culture media, the so-called secretome, using global proteomic approaches has been widely used to identify secretion system substrates in T2SS [4–7], T6SS [8–10], T3SS [11] and T9SS [12].

Upon the identification of substrates, the secretion profile is used to test the functionality of the secretion system using sodium

Laure Journet and Eric Cascales (eds.), *Bacterial Protein Secretion Systems: Methods and Protocols*, Methods in Molecular Biology, vol. 1615, DOI 10.1007/978-1-4939-7033-9_31, © Springer Science+Business Media LLC 2017

dodecyl sulfate (SDS)-polyacrylamide gel electrophoresis (PAGE) of the supernatant fraction followed by coomassie blue staining or immunostaining by western blot detection of specific effectors or components of the machinery. In some secretion systems, such as T3SS and T6SS, external structural components are released in the milieu upon secretion and can also be used to test the proper assembly of the system. For example, the Hcp release assay is widely used to probe the functionality of T6SS (*see* also Chapter 32).

Such analyses of secretomes require concentrating the dilute solutions that are the culture supernatant or the biological fluids. This can be achieved using trichloroacetic acid (TCA) precipitation and acetone-based protocols [13, 14]. Alternative protocols have been proposed using acetone alone, methanol/chloroform [15], or a combination of pyrogallol red, molybdate, and methanol [16].

Here we detail the most classical assay used to precipitate proteins of bacterial culture supernatant based on TCA precipitation; it is used widely in secretion system studies. First, cells and supernatant are separated by centrifugation. Cell-free culture supernatant fraction samples are then obtained by further centrifugation and filtration and subjected to TCA precipitation before analysis by mass spectrometry or western blot.

2 Materials

1. Lysogeny broth (LB) or recommended medium to grow strain of interest in secreting conditions.
2. TCA (CCl_3COOH, MW: 163.39, TCA): 100% (w/v). Add 227 mL ultrapure water to previously unopened bottle containing 500 g TCA (*see* **Note 1**). Wear personal protective equipment and work under a fume hood.
3. Sodium deoxycholate (DOC): 16 mg/mL (optional, *see* **Note 2**). Store at room temperature.
4. Acetone. Pre-chill before use.
5. Refrigerated centrifuge capable of 21,460 × *g* or tabletop centrifuge (*see* **Note 3**).
6. 0.22-μm-pore-size syringe filters (*see* **Note 4**).
7. 2 mL syringe.
8. 3 M Tris–HCl, pH 8.8
9. SDS-PAGE loading buffer: 60 mM Tris–HCl, pH 6.8, 2% SDS, 10% glycerol, 5% β-mercaptoethanol, 0.01% bromophenol blue.
10. Boiling water bath or thermomixer.
11. Vortexer.
12. 2 mL microtubes (safe-lock) (*see* **Note 3**).

13. Fume hood and personal protective equipment for TCA handling.
14. Spectrophotometer to measure absorbance at $\lambda = 600$ nm.
15. SDS-PAGE and protein transfer apparatus.

3 Methods

1. Grow a 10 mL bacterial strain culture in the appropriate medium and conditions allowing secretion (*see* **Notes 5** and **6**). Measure the optical density at $\lambda = 600$ nm (OD_{600}).
2. Dispose the culture in 2 mL microtubes (*see* **Note 7**) and pellet cells by centrifugation at 6000 × *g* for 5 min.
3. Carefully remove 1.8 mL of supernatant and transfer it to a new microtube and keep it on ice before performing **step 5**.
4. Carefully discard the remaining 200 μL of supernatant from the cell pellet obtained in **step 3**. (Centrifuge again at 6000 × *g* for 5 min if the cells from the cell pellet started to resuspend.) Keep this total cell fraction pellet on ice before resuspending the pellet in an appropriate volume of SDS-PAGE loading buffer (the equivalent of 0.2–0.5 OD_{600} units (ODU)/10 μL). Store on ice (or at −20 °C).
5. Centrifuge the 1.8 mL supernatant fraction obtained in **step 3** at 16,000 × *g* at 4 °C for 5 min. Carefully recover the supernatant and transfer it to a new microtube. Avoid recovering the remaining cells from the pellet, if any.
6. Filter-sterilize the supernatant using a 0.22 μm syringe filter and transfer the filtered supernatant directly to a new microtube. Check the volume (around 1.5 mL). This fraction constitutes the cell-free fraction (*see* **Note 8**).
7. Add TCA to a final concentration of 20% (add 375 μL TCA to 1.5 mL filtered supernatant). Invert four times to mix, vortex and keep on ice for 1 h to overnight.
8. Centrifuge at 21,000 × *g* for 30 min at 4 °C.
9. Discard the as much supernatant as possible (*see* **Note 9**).
10. Resuspend the pellet in 400–500 μL cold acetone. Vortex.
11. Centrifuge at 21,000 × *g* for 15 min at 4 °C. Discard the supernatant with a pipet and further on a paper towel. Leave the tube open at room temperature to dry the pellet (*see* **Note 10**).
12. Resuspend the pellet in appropriate buffer for further analysis (such as mass spectrometry) or go to **step 13** for SDS-PAGE analysis.
13. Resuspend TCA-precipitated pellets of supernatant fractions in an appropriate volume of SDS-PAGE loading buffer

(1 ODU/10 μL). If the TCA-precipitated sample turns yellow, add 1 μL (or more) of Tris–HCl, pH 8.8.

14. Vortex. Heat the samples from **step 4** (whole-cell fraction) and **step 13** (cell-free supernatant precipitated fraction) at 95 °C for 10 min (*see* **Note 11**).
15. Analyse whole-cell samples and cell-free supernatants by SDS-PAGE, followed by coomassie blue staining or immunoblot. If performing western blot, include a control for cell lysis, using antibodies detecting an internal protein. Alternatively, check the coomassie or silver staining profile.

4 Notes

1. For a safe and easy preparation, avoid weighting out the TCA crystalline powder as it becomes easily syrupy upon contact with humidity. The TCA solution must be kept in a dark glass bottle. It is very corrosive and should be handled with care with suitable protection. Do not use plastic containers.
2. DOC may be used as a carrier to assist protein precipitation. If using DOC, add the DOC stock solution at the final concentration of 0.16 mg/mL to the cell-free fraction obtained in **step 6**, vortex and leave on ice for 30 min; then proceed to TCA precipitation as described in **step 7**. DOC should be washed out with further acetone washing steps (repeat **steps 10** and **11** three times). However, this could be a problem with further mass spectrometry analysis.
3. Tabletop centrifuge at maximum speed may be sufficient; however, we generally use a higher speed. TCA-resistant tubes, such as Eppendorf tubes (check with your manufacturer for tube compatibility), should be used.
4. In principle, any 0.22 μm filter may be used. However, we had experience with a secreted protein that was retained on polyvinylidene fluoride filters, so we moved to Polyether sulfone (PES) filters. Be aware that the material of the filter may be of importance.
5. A "non-secreting strain" should be used as a control, such as a mutant in a core component, the ATPase energizing the assembly of the secretion machinery or the substrate transport.
6. You must find conditions where secretion can be detected in vitro. Because effectors can be secreted at low levels, high-sensitivity mass spectrometry methods may be required [10]. If the secretion system is not produced in laboratory conditions, native endogenous promoter(s) may be swapped for an inducible promoter (e.g. P*tac*, P*lac*, P*BAD*) to artificially induce the expression of the secretion system [17].

7. A 5–10 mL of culture is generally sufficient. We usually transfer 2 mL of supernatant in 2 mL microtubes, leading to the recovery of 1.5 mL of cell-free supernatant. An equivalent of 1 OD_{600} unit will be loaded on the gel for supernatant fraction analysis. To scale up experiments, you will have to use tubes with larger volumes compatible with high-speed spin that are resistant to TCA. Appropriate 50 mL tubes (polyether) may be used; check first with your manufacturer for TCA compatibility.
8. At this stage, for bacteria producing high levels of vesicles (e.g. for T9SS in Bacteroidetes), an additional ultracentrifugation (30,000 × *g* for 4 h at 4 °C) will allow separation of vesicles from vesicle-free supernatant [12].
9. Check the orientation of the microtube before the centrifugation step since the pellet is not always visible at this stage.
10. You may use a vacuum concentrator (SpeedVac or equivalent) for 10 min to evaporate the acetone. However, pellets may be more difficult to resuspend if too dry, and this step may decrease recovery of the samples.
11. In some cases, we have observed that an additional freezing at −20 °C in SDS-PAGE loading buffer helps resuspension of TCA precipitates.

Acknowledgments

This work was supported by the Centre National de la Recherche Scientifique, the Aix-Marseille Université and grants from the Agence Nationale de la Recherche (ANR-14-CE14-0006-02 and ANR-15-CE11-0019-01). The doctoral studies of N.F. are supported by the ANR-14-CE14-0006-02 grant.

References

1. Costa TR, Felisberto-Rodrigues C, Meir A, Prevost MS, Redzej A, Trokter M, Waksman G (2015) Secretion systems in Gram-negative bacteria: structural and mechanistic insights. Nat Rev Microbiol 13:343–359
2. Cornelis GR, Biot T, Lambert de Rouvroit C, Michiels T, Mulder B, Sluiters C, Sory MP, Van Bouchaute M, Vanooteghem JC (1989) The Yersinia yop regulon. Mol Microbiol 3: 1455–1459
3. Beuzon CR, Banks G, Deiwick J, Hensel M, Holden DW (1999) pH-dependent secretion of SseB, a product of the SPI-2 type III secretion system of Salmonella typhimurium. Mol Microbiol 33:806–816
4. Coulthurst SJ, Lilley KS, Hedley PE, Liu H, Toth IK, Salmond GP (2008) DsbA plays a critical and multifaceted role in the production of secreted virulence factors by the phytopathogen Erwinia carotovora subsp. atroseptica. J Biol Chem 283:23739–23753
5. Kazemi-Pour N, Condemine G, Hugouvieux-Cotte-Pattat N (2004) The secretome of the plant pathogenic bacterium Erwinia chrysanthemi. Proteomics 4:3177–3186
6. Sikora AE, Zielke RA, Lawrence DA, Andrews PC, Sandkvist M (2011) Proteomic analysis of the Vibrio cholerae type II secretome reveals new proteins, including three related serine proteases. J Biol Chem 286:16555–16566

7. Burtnick MN, Brett PJ, DeShazer D (2014) Proteomic analysis of the Burkholderia pseudomallei type II secretome reveals hydrolytic enzymes, novel proteins, and the deubiquitinase TssM. Infect Immun 82:3214–3226
8. Hood RD, Singh P, Hsu F, Güvener T, Carl MA, Trinidad RR, Silverman JM, Ohlson BB, Hicks KG, Plemel RL, Li M, Schwarz S, Wang WY, Merz AJ, Goodlett DR, Mougous JD (2010) A type VI secretion system of Pseudomonas aeruginosa targets a toxin to bacteria. Cell Host Microbe 7:25–37
9. Russell AB, Singh P, Brittnacher M, Bui NK, Hood RD, Carl MA, Agnello DM, Schwarz S, Goodlett DR, Vollmer W, Mougous JD (2012) A widespread bacterial type VI secretion effector superfamily identified using a heuristic approach. Cell Host Microbe 11:538–549
10. Fritsch MJ, Trunk K, Diniz JA, Guo M, Trost M, Coulthurst SJ (2013) Proteomic identification of novel secreted antibacterial toxins of the Serratia marcescens type VI secretion system. Mol Cell Proteomics 12:2735–2749
11. Deng W, de Hoog CL, Yu HB, Li Y, Croxen MA, Thomas NA, Puente JL, Foster LJ, Finlay BB (2010) A comprehensive proteomic analysis of the type III secretome of Citrobacter rodentium. J Biol Chem 285:6790–6800
12. Veith PD, Chen YY, Gorasia DG, Chen D, Glew MD, O'Brien-Simpson NM, Cecil JD, Holden JA, Reynolds EC (2014) Porphyromonas gingivalis outer membrane vesicles exclusively contain outer membrane and periplasmic proteins and carry a cargo enriched with virulence factors. J Proteome Res 13:2420–2432
13. Hwang BJ, Chu G (1996) Trichloroacetic acid precipitation by ultracentrifugation to concentrate dilute protein in viscous solution. BioTechniques 20:982–984
14. Ozols J (1990) Amino acid analysis. Methods Enzymol 182:587–601
15. Wessel D, Flügge UI (1984) A method for the quantitative recovery of protein in dilute solution in the presence of detergents and lipids. Anal Biochem 138:141–143
16. Caldwell RB, Lattemann CT (2004) Simple and reliable method to precipitate proteins from bacterial culture supernatant. Appl Environ Microbiol 70:610–612
17. Gueguen E, Cascales E (2013) Promoter swapping unveils the role of the Citrobacter rodentium CTS1 type VI secretion system in interbacterial competition. Appl Environ Microbiol 79:32–38

Chapter 32

Screening for Secretion of the Type VI Secretion System Protein Hcp by Enzyme-Linked Immunosorbent Assay and Colony Blot

Brent S. Weber, Pek Man Ly, and Mario F. Feldman

Abstract

The bacterial type VI secretion system (T6SS) is a secretory apparatus encoded by many Gram-negative bacteria. The T6SS facilitates the secretion and injection of toxic effector proteins into host cells, providing a competitive advantage to bacteria encoding this machinery. The activity of the T6SS can be monitored by probing for the conserved tubule component Hcp, which is secreted to the supernatants by the T6SS. Detection of Hcp in culture supernatants is indicative of an active T6SS, but this secretion system is often tightly regulated or inactive under laboratory conditions and different bacterial strains display differing Hcp secretion phenotypes. Herein, we describe an enzyme-linked immunosorbent assay (ELISA) and colony blot methods to facilitate large-scale screening of isolates for Hcp secretion and, thus, T6SS activity.

Key words ELISA, Colony blot, Supernatant, Hcp, Effector

1 Introduction

Secretion systems encoded by Gram-negative bacteria secrete a wide range of proteins and are often essential for virulence [1]. Of the many secretory machines characterized, the type VI secretion system (T6SS) has emerged as a potent mediator of antibacterial and antieukaryotic activity [2, 3]. The T6SS is encoded by approximately 13 conserved proteins, which assemble to form a secretory apparatus capable of secreting effector substrates to adjacent bacterial and eukaryotic cells [4]. The effector repertoire is highly variable, both across different species and in individual isolates of a single species [5]. Furthermore, the T6SS is often tightly regulated, and myriad mechanisms exist among different bacteria to control activation of T6SS [6]. However, activation of T6SS invariably results in the secretion of Hcp, an essential structural component of the secretory apparatus. Thus, Hcp secretion is a molecular marker for T6SS, and detection of Hcp in culture supernatants is indicative of an active T6SS [7]. Typically, Hcp secretion has been

Laure Journet and Eric Cascales (eds.), *Bacterial Protein Secretion Systems: Methods and Protocols*, Methods in Molecular Biology, vol. 1615, DOI 10.1007/978-1-4939-7033-9_32, © Springer Science+Business Media LLC 2017

detected by western blot or mass spectrometry. While some bacterial strains encode a constitutively active T6SS, others maintain tight regulation of this system using a diverse range of mechanisms. Furthermore, different isolates of the same species often show different T6SS activity, for example between clinical and environmental isolates [8–10]. A single bacterial strain can also show variability in T6SS activation [11, 12]. Screening large numbers of species, strains, or colonies by western blot to assess T6SS activity is time and resource consuming. Therefore, we have developed two methods to screen a large number of colonies, either from many different isolates or from many colonies from the same isolate, for Hcp secretion [12, 13]. These assays enable relatively rapid detection of T6SS activity among hundreds or even thousands of isolates or colonies and provide a time-saving alternative to more traditional western blot or mass spectrometry methods. We describe the application of an enzyme-linked immunosorbent assay (ELISA) and a colony blot method to detect Hcp secretion by isolates and transposon mutant libraries of *Acinetobacter baumannii*, which can be adapted for use in other T6SS-encoding organisms. The ELISA protocol is particularly suited to screening many different isolates at once, while the colony blot method has obvious applications for screening transposon mutant libraries. Both assays can be adapted for the detection of any secreted protein in an organism of interest.

2 Materials

2.1 ELISA

1. Lysogeny broth(LB) broth: Dissolve 10 g NaCl, 10 g ryptone, 5 g yeast extract in 1 L distilled water. Autoclave at 121 °C for 15 min to sterilize.
2. LB agar: Dissolve 15 g agar in 1 L LB broth. Autoclave at 121 °C for 15 min to sterilize.
3. Sterile toothpicks or other tool to inoculate colonies into 96-well plates.
4. Binding buffer: 100 mM sodium bicarbonate/carbonate, pH 9.6. Dissolve 3.03 g Na_2CO_3, 6.0 g $NaHCO_3$ in 1 L distilled water.
5. Phosphate buffered saline (PBS) wash buffer: Dissolve 8 g NaCl, 0.2 g KCl, 1.44 g Na_2HPO_4, 0.24 g KH_2PO_4 in 1 L distilled water. Prepare as a 10× solution and dilute prior to use. Autoclave at 121 °C for 15 min to sterilize.
6. PBS-Tween 20 (PBST) wash buffer: Add 1 mL Tween 20 to 1 L PBS.
7. Blocking and antibody diluent solutions: Dissolve 5% skim milk in PBS (initial blocking buffer) or 2.5% skim milk in PBST (diluent for antibodies). Prepare fresh for each experiment.

8. 96-well plates for growth cultures and high-binding ELISA.
9. Antibodies: Primary: Rabbit anti-Hcp (polyclonal, not commercially available, developed against Hcp protein of interest) (*see* **Note 1**).
10. Secondary: Goat anti-rabbit horseradish peroxidase (HRP) conjugate.
11. TMB substrate and stop solution: 3,3′,5,5″-tetramethyl-benzidine.
12. Standard laboratory equipment: Multichannel pipettes, centrifuges, incubators, microplate reader.

2.2 Colony Blot for Hcp Secretion

1. LB agar: Dissolve 10 g NaCl, 10 g tryptone, 5 g yeast extract, 15 g agar in 1 L distilled water. Autoclave at 121 °C for 15 min to sterilize.
2. Circular nitrocellulose blotting membranes: 0.45 μm, diameter: 82 mm. Autoclave in glass Petri dish at 121 °C for 15 min to sterilize.
3. Tris-buffered saline (TBS) wash buffer: Dissolve 8.76 g NaCl, 1.21 g Tris in 1 L distilled water. Adjust to pH 8.0 with HCl.
4. TBST wash buffer: Add 1 mL Tween 20 to 1 L TBS.
5. Odyssey® blocking buffer TBS (LI-COR) or equivalent reagent.
6. Primary antibodies: Rabbit anti-Hcp (polyclonal, not commercially available, developed against Hcp protein of interest) (*see* **Note 1**), mouse anti-*E. coli* RNA polymerase β prime (or any antibody raised against a cytoplasmic protein).
7. Goat anti-mouse and goat anti-rabbit secondary antibodies coupled to infrared fluorescent dyes.
8. Fluorescence imaging system (LI-COR Odyssey or equivalent).
9. Standard laboratory equipment: Metal tweezers, incubator, 50 mL conical tubes, tumbling or rocking platform.

3 Methods

3.1 Hcp ELISA

1. The day prior to beginning the experiment, inoculate LB agar plate(s) with the bacterial strain(s) of interest and incubate overnight at 37 °C. This should be done in a manner that ensures well-defined single colonies are present the next day.
2. Prepare 96-well plates for bacterial growth: Aseptically add 200 μL LB broth to the wells of a 96-well plate. Using sterile toothpicks, inoculate individual colonies from plates grown overnight into each well. If available, inoculate a positive and negative control strain into the last two wells (e.g., strain that secretes Hcp and an *hcp* mutant). Place 96-well plate in a humidified chamber (*see* **Note 2**) at 37 °C with shaking (200 rpm) overnight (*see* **Note 3**).

3. The next day, retrieve 96-well plate and measure the OD_{600} in a plate reader for a measure of bacterial cell growth. Then, centrifuge the plate at 5000 × *g* for 10 min to pellet the bacterial cells.
4. Meanwhile, transfer 25 μL blocking buffer to the wells of a 96-well high-binding ELISA plate.
5. Using a multichannel pipette, carefully transfer 75 μL of the supernatants to the high-binding ELISA plate. Avoid pipetting any cellular material (*see* **Note 4**). Be sure to keep the original plate with the cell pellets as this will serve as the master stock for any positive wells later on. The plate can be kept at 4 °C, or alternatively add glycerol to the pellets and store at −80 °C for the long term.
6. Incubate ELISA plate at room temperature for 1.5 h.
7. After incubation, remove solution from plate and wash three times with PBS. Then completely fill wells with blocking buffer (5% skim milk in PBS). Place on a rocking platform at room temperature for 1 h.
8. Remove blocking solution, wash once with PBS.
9. Add 6 μL anti-Hcp antibody to 12 mL 2.5% skim milk in PBST (1:2000 dilution) (*see* **Note 5**). Pipet 100 μL of this solution into each well of the ELISA plate and place on room-temperature rocking platform for 1 h.
10. Remove solution and wash thoroughly three to five times with PBST.
11. Add 2.4 μL goat anti-rabbit HRP conjugate to 12 mL 2.5% skim milk in PBST (1:5000 dilution) and pipet 100 μL of this solution into each well of the ELISA plate. Incubate for 1 h at room temperature on rocking platform.
12. Remove solution and wash thoroughly three to five times with PBST.
13. Add 100 μL TMB substrate to wells. Gently shake the plate or place on rocking platform. Watch plate for appearance of blue color, which may be very rapid or take several minutes(*see* **Note 6**). At desired time, measure the $A_{650\text{nm}}$ of the plate (*see* **Note 7**). This can be used in conjunction with the $A_{600\text{nm}}$ taken earlier to compare wells relative to growth. Take a photograph of the plate if desired (*see* Fig. 1).

3.2 Colony Blot for Hcp SECRETION

All liquids should be discarded as per biohazard protocols. All steps are performed at room temperature unless otherwise stated.

1. Inoculate LB agar plate(s) with bacteria of interest in a way that allows defined single colonies to grow. For large-scale screening of colonies, LB agar plates are inoculated by spread

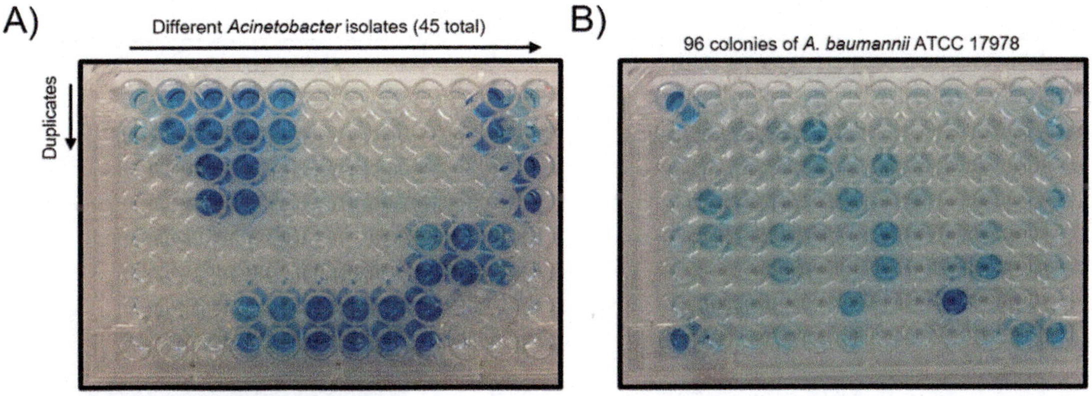

Fig. 1 (**a**) Example of using Hcp ELISA to screen multiple isolates of a given species at once. Here, 45 different isolates of various *Acinetobacter* species were screened for Hcp secretion in duplicate. The results highlight the variability among different species for T6SS activity within a given genus. (**b**) Screening a single strain, *A. baumannii* ATCC 17978, for Hcp secretion. Single colonies of the strain were inoculated and used to perform the Hcp ELISA. For this strain, a plasmid encodes the repressors for T6SS but can be lost upon culture without selection [12]

plating of diluted bacteria cultures in such a way as to allow no more than 200 colony-forming units per plate. Incubate at 37 °C overnight (*see* **Note 8**).

2. Working aseptically, transfer colonies onto a nitrocellulose membrane by gently overlaying one sterile membrane using tweezers on top of the colonies for about 30 s to 1 min, or until entire membrane is moistened (*see* **Note 9**). Gently lift membrane from plate and allow to dry, transfer side up, for 20 min.
3. Place the dried membrane to a 50 mL conical tube, transfer side away from walls, and wash for 5 min (tumbling or rocking) with 25 mL distilled water three times, or until colonies are no longer attached to membrane. Next, wash the membrane twice with 25 mL TBS.
4. In a new 50 mL conical tube, block the membrane with 25 mL Odyssey blocking buffer TBS for 1 h at room temperature or overnight at 4 °C.
5. In a new 50 mL conical tube, add 1.5 mL TBST and 1.5 mL Odyssey blocking buffer TBS (1:1 ratio). Add 3.3 μL anti-Hcp antibody (1:9000 dilution) (*see* **Note 5**) and 1.2 μL anti-RNA polymerase antibody (1:2500 dilution) (*see* **Note 10**). Transfer membrane to this tube, transfer side away from walls, and incubate (tumbling or rocking) for 40 min.
6. Discard solution and wash membrane three times with 10 mL TBST (10 min each).

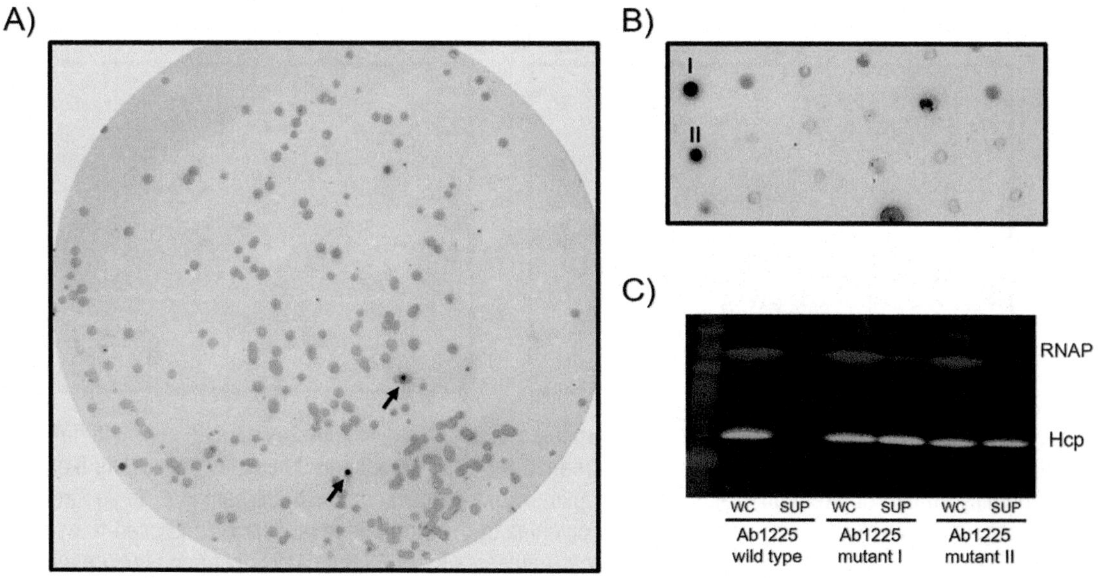

Fig. 2 Use of colony blots to isolate active T6SS mutants from a T6SS-inactive strain. (**a**) Example of an Hcp colony blot used to screen transposon mutants of *A. baumannii* strain 1225, which has an inactive T6SS and does not secrete Hcp under normal laboratory conditions. *Arrows* indicate strong signals from single colonies probed with the anti-Hcp antibody. (**b**) Hcp colony blot used on patch plated *A. baumannii* transposon mutants. Colonies I and II show strong signals for anti-Hcp. (**c**) Western blot for Hcp expression in whole cells (WC) and secretion in supernatant (SUP) of colonies I and II from figure (**b**). RNAP antibody used as lysis and loading control. Wild-type *A. baumannii* 1225 expresses but does not secrete Hcp and is used as a negative control

7. In a new 50 mL conical tube, add 2.5 mL TBST and 2.5 mL Odyssey blocking buffer TBS (1:1 ratio). Add 0.4 μL anti-rabbit and anti-mouse antibodies (1:12,500 dilution). Transfer membrane to this tube, transfer side away from walls, and incubate (tumbling or rocking) in the dark for 40 min (*see* **Note 11**).
8. Discard solution and wash membrane two times with 10 mL TBST (10 min each) in the dark.
9. Wash membrane once with 10 mL TBS (10 min) in the dark.
10. Image membrane using LI-COR Odyssey CLx imaging system (*see* Fig. 2).

4 Notes

1. For ideal results, an anti-Hcp antibody raised against purified Hcp from the bacterial species of interest is preferred, whether polyclonal or monoclonal. Hcp is generally well conserved within a given genus of bacteria, and anti-Hcp antibodies have been shown to cross react with Hcp proteins across different

bacterial genera [14]. We have successfully employed our anti-Hcp antibody raised against *A. baumannii* to detect Hcp in other *Acinetobacter* species, but the Hcp protein is highly conserved. Validation of a given anti-Hcp antibody is an important step before using it for the ELISA.

2. We typically use a large plastic container with a lid and place damp paper towels in the bottom. This prevents excessive evaporation of liquid from plate that would occur in the shaking incubator.
3. Incubation time will vary depending on bacterial species under study. Mid-log phase cultures may be desired rather than overnight incubation, but we have found overnight (~16 h) gives consistent results. Longer incubation times may result in excessive lysis of bacterial cells, which may interfere with downstream analysis.
4. It is important not to transfer any cells into the ELISA plates in order to only measure secreted Hcp. Extra centrifuging may be required. Alternatively, 0.22 μm filters are available for 96-well plates; however, in our experience careful pipetting is usually sufficient to avoid this problem.
5. Antibody titer must be empirically determined.
6. A solution of sulfuric acid can be used to stop the reaction, but we typically do not include this step and simply read the absorbance of the plate at 650 nm. In our experience, the color change that occurs after stopping the reaction can lead to a signal that is too intense to be measured by our plate reader. This is particularly problematic with strains that robustly secrete Hcp.
7. Often bubbles will result during pipetting. These can affect absorbance readings and should be removed (using a flame or pipet tip) before measuring absorbance.
8. Dilution of bacterial cultures for spread plating must be optimized and is species and strain specific. For best results, avoid using inoculated plates over a day old.
9. Be careful not to press down onto agar surface. Avoid disrupting colonies as much as possible to limit cell lysis.
10. We use anti-RNA polymerase as a probe for cell lysis. Colony blots can be performed by only probing for anti-Hcp; however, probing for anti-RNA polymerase eliminates single colonies with strong anti-Hcp signals owing to lysis on the membrane.
11. We use tin foil to wrap around conical tubes.

References

1. Costa TR, Felisberto-Rodrigues C, Meir A, Prevost MS, Redzej A, Trokter M, Waksman G (2015) Secretion systems in Gram-negative bacteria: structural and mechanistic insights. Nat Rev Microbiol 13:343–359
2. Pukatzki S, Ma AT, Sturtevant D, Krastins B, Sarracino D, Nelson WC, Heidelberg JF, Mekalanos JJ (2006) Identification of a conserved bacterial protein secretion system in Vibrio cholerae using the Dictyostelium host model system. Proc Natl Acad Sci U S A 103:1528–1533
3. Mougous JD, Cuff ME, Raunser S, Shen A, Zhou M, Gifford CA, Goodman AL, Joachimiak G, Ordonez CL, Lory S, Walz T, Joachimiak A, Mekalanos JJ (2006) A virulence locus of Pseudomonas aeruginosa encodes a protein secretion apparatus. Science 312:1526–1530
4. Russell AB, Peterson SB, Mougous JD (2014) Type VI secretion system effectors: poisons with a purpose. Nat Rev Microbiol 12:137–148
5. Cianfanelli FR, Monlezun L, Coulthurst SJ (2016) Aim, load, fire: the type VI secretion system, a bacterial nanoweapon. Trends Microbiol 24:51–62
6. Silverman JM, Brunet YR, Cascales E, Mougous JD (2012) Structure and regulation of the type VI secretion system. Annu Rev Microbiol 66:453–472
7. Pukatzki S, McAuley SB, Miyata ST (2009) The type VI secretion system: translocation of effectors and effector-domains. Curr Opin Microbiol 12:11–17
8. Bernardy EE, Turnsek MA, Wilson SK, Tarr CL, Hammer BK (2016) Diversity of clinical and environmental isolates of Vibrio cholerae in natural transformation and contact-dependent bacterial killing indicative of Type VI secretion system activity. Appl Environ Microbiol 82:2833–2842
9. Repizo GD, Gagne S, Foucault-Grunenwald ML, Borges V, Charpentier X, Limansky AS, Gomes JP, Viale AM, Salcedo SP (2015) Differential role of the T6SS in Acinetobacter baumannii virulence. PLoS One 10:e0138265
10. Unterweger D, Kitaoka M, Miyata ST, Bachmann V, Brooks TM, Moloney J, Sosa O, Silva D, Duran-Gonzalez J, Provenzano D, Pukatzki S (2012) Constitutive type VI secretion system expression gives Vibrio cholerae intra- and interspecific competitive advantages. PLoS One 7:e48320
11. Tang L, Liang X, Moore R, Dong TG (2015) The icmF3 locus is involved in multiple adaptation- and virulence-related characteristics in Pseudomonas aeruginosa PAO1. Front Cell Infect Microbiol 5:83
12. Weber BS, Ly PM, Irwin JN, Pukatzki S, Feldman MF (2015) A multidrug resistance plasmid contains the molecular switch for type VI secretion in Acinetobacter baumannii. Proc Natl Acad Sci U S A 112:9442–9447
13. Weber BS, Miyata ST, Iwashkiw JA, Mortensen BL, Skaar EP, Pukatzki S, Feldman MF (2013) Genomic and functional analysis of the type VI secretion system in Acinetobacter. PLoS One 8:e55142
14. Carruthers MD, Nicholson PA, Tracy EN, Munson RS Jr (2013) Acinetobacter baumannii utilizes a type VI secretion system for bacterial competition. PLoS One 8:e59388

Chapter 33

Effector Translocation: Cya Reporter Assay

Suma Chakravarthy, Bethany Huot, and Brian H. Kvitko

Abstract

An accurate and complete roster of the Type III effector (T3E) proteins translocated by the *P. syringae* Type III secretion system (T3SS) into host cells is critical to understanding the pathogen's interactions with plants. The adenylate cyclase (Cya) reporter offers a highly sensitive and robust assay for monitoring the translocation of T3Es. T3Es are fused to the calmodulin-dependent adenylate-cyclase domain of CyaA. The T3E targets Cya for translocation through the T3SS into the host cell at which point it is activated by calmodulin and converts adenosine triphosphate into cyclic adenosine monophosphate (cAMP). The T3SS translocation-dependent increase in cAMP concentration in plant cells is then measured with an enzyme-linked immunosorbent assay kit. The Cya reporter can be used to determine whether a candidate protein is translocated by T3SS or to measure relative levels of T3SS translocation in a semiquantitative manner.

Key words *Pseudomonas syringae*, Type III secretion system, Type III translocation, Translocation reporter, Adenylate cyclase, Calmodulin, cAMP, ELISA

1 Introduction

The plant pathogenic bacterium *Pseudomonas syringae* deploys a Type III secretion system (T3SS) to translocate Type III effector (T3E) proteins directly into plant host cells. The translocation of T3Es is essential for *P. syringae* pathogenicity [1]. Although collectively *P. syringae* strains infect a wide range of plant hosts, creating diverse symptoms, individual strains typically infect a limited range of hosts. The host range of a given *P. syringae* strain is defined largely by its repertoire of translocated T3Es [2]. T3Es act as virulence factors, coordinating their actions to modify host cellular targets and creating a susceptible state amenable to bacterial proliferation. However, individual T3Es may also be detected if cognate plant resistance (R) protein immune receptors are present in the host [3]. Detection of T3Es by R proteins results in a potent effector-triggered immune response that blocks pathogen proliferation. Collectively, the T3E repertoire of a given *P. syringae* strain

Laure Journet and Eric Cascales (eds.), *Bacterial Protein Secretion Systems: Methods and Protocols*, Methods in Molecular Biology, vol. 1615, DOI 10.1007/978-1-4939-7033-9_33, © Springer Science+Business Media LLC 2017

both delineates that strain's capacity to create susceptibility in a given host plant through the T3Es' collective virulence functions, as well as its capacity to induce a virulence through R protein–mediated detection [4]. Therefore, having an accurate and complete roster of a particular *P. syringae* strain's repertoire of translocated T3Es is critical to understanding its interactions with plants.

The use of the Cya translocation reporter has been crucial in confirming the T3E rosters of *P. syringae* strains [5, 6]. T3SS translocation reporter constructs generate a distinct output signal that is only created when the candidate protein fused to the reporter is delivered by T3SS competent bacteria into host cells [7–9]. Constructs used in the Cya translocation reporter assay have two main components to accomplish this: a eukaryotic cell-specific reporter and a T3SS-specific translocation signal. The reporter used is the adenylate cyclase domain ($Cya_{2\text{-}400}$) of the *Bordetella pertussis* CyaA adenylate cyclase toxin. The specificity derives from the enzyme's calmodulin-dependent conversion of adenosine triphosphate to cyclic adenosine monophosphate (cAMP), which cannot occur in bacteria owing to a lack of calmodulin [8]. The $Cya_{2\text{-}400}$ domain alone is incapable of independent exit from bacteria or entry in host cells as it lacks the proper translocation signal, thereby preventing production of a signal not associated with the delivery of T3Es [8]. The $Cya_{2\text{-}400}$ reporter domain is fused to the C-terminus of a candidate translocated effector of interest (EOI) to preserve the N-terminal T3SS translocation signal. *P. syringae* expressing the 3′ $Cya_{2\text{-}400}$–T3E fusion protein is inoculated into leaves of the test host at relatively high concentrations [6, 10]. Within a few hours following infiltration, *P. syringae* will deploy its T3SS, initiating the translocation of T3Es into the plant cells. If the expressed $Cya_{2\text{-}400}$–T3E candidate protein fusion can be translocated via the T3SS, it will also be delivered into the host cells along with the other effectors. Upon exposure to calmodulin within the plant cytoplasm, $Cya_{2\text{-}400}$ adenylate cyclase will be activated resulting in the accumulation of cAMP. The concentration of cAMP in the leaf tissue can then be measured by an enzyme-linked immunosorbent assay (ELISA) assay, which is standardized against soluble protein. A mutant *P. syringae* strain with a defective T3SS is used as a control to confirm that the cAMP accumulation is T3SS-dependent. In addition to its use to verify candidate T3Es as being translocated by *P. syringae*, the Cya reporter assay has also been used to measure relative levels of T3SS translocation in a semiquantitative manner since increased concentrations of cAMP correlates with increased protein translocation [11, 12].

2 Materials

2.1 Construction of Plasmids and Strains

1. 1.5 mL microcentrifuge tubes.
2. Tabletop microcentrifuge.
3. Primers to amplify the EOI, P1 and P2, and universal primers M13F (GTT TTC CCA GTC ACG AC) and M13R (CAG GAA ACA GCT ATG AC).
4. Genomic DNA of *P. syringae* strain that encodes the EOI.
5. PrimeSTAR HS DNA polymerase (CloneTech) or equivalent.
6. Molecular-biology-grade water.
7. Agarose for gel electrophoresis.
8. TBE gel running buffer (10.8 g Tris, 5.5 g boric acid, 4 mL 0.5 M ethylenediaminetetraacetic acid (EDTA), fill to 1 L with dH_2O).
9. Ethidium bromide (10 mg/mL, final concentration 0.5 μg/mL).
10. DNA ladder.
11. DNA loading dye.
12. pCPP5371 (P_{hrp}-GW-Cya destination vector) plasmid DNA [13] (*see* **Notes 1** and **2**).
13. pENTR/SD/D-TOPO kit from Thermo Fisher Scientific or equivalent.
14. *P. syringae* pv. tomato DC3000.
15. LR Clonase II from Thermo Fisher Scientific.
16. Competent *E. coli* cloning strain such as DH5α or TOP10.
17. Antibiotic stocks of kanamycin (50 mg/mL) and gentamicin (10 mg/mL), dissolved in dH_2O and filter sterilized.
18. KB (King's medium B) liquid and agar solidified medium (20 g Bacto peptone, 0.4 g $MgSO_4{\cdot}7H_2O$, glycerol, 1.5 g K_2HPO_4 (*see* **Note 3**); fill to 1 L with dH_2O; add 18 g agar for solidified media, autoclave).
19. LB (Luria-Bertani) liquid and agar solidified medium (10 g of tryptone, 5 g yeast extract, 10 g NaCl, 1 mL 1 M NaOH, fill to 1 L with dH_2O. Add 15 g agar for solidified media, autoclave).
20. Kit for plasmid minipreparation.
21. Primers to confirm final Cya reporter plasmid P3 (F primer, TGA GCA TGC TAC CGA GTA ACG CAG CT) and P4 (R primer, AGT GGT ACC GAT ATC GAA TTC TTA GCT GT).
22. 300 mM sucrose, filter sterilized.
23. 1 mm gap electroporation cuvettes.

24. Cell electroporator.
25. 14 mL disposable culture tubes.
26. 2× NEB OneTaq.

2.2 Plant Inoculation

1. Plants to inoculate (e.g., *Nicotiana benthamiana, Arabidopsis*, tobacco) (*see* **Note 4**): Enough plants to conduct three infiltrations per test strain.
2. *P. syringae* pv. tomato DC3000 pCPP5388 (pCPP5371::P_{hrp}-avrPto-Cya) secretion positive control.
3. *P. syringae* pv. tomato DC3000 T3SS (−) strain such as CUCPB5113 (Δ*hrcQ$_b$-U*::SpR) pCPP5388 (pCPP5371::P_{hrp}-avrPto-Cya) secretion negative control [14, 15].
4. Long wooden inoculation dowels.
5. 10 mM $MgCl_2$ (*see* **Note 5**).
6. 1 mL needleless syringes.
7. Dissecting needle.
8. Spectrophotometer.
9. Kimwipe or paper towel.
10. Large soft-tip black Sharpie marker (*see* **Note 6**).
11. 2.2 mL round-bottom microcentrifuge tubes.
12. 4 mm diameter disposable biopsy punches or cork borers.
13. Liquid nitrogen and dewar.
14. 4.5 mm copper BBs.

2.3 cAMP ELISA and Bradford Assay

1. cAMP ELISA kit (Enzo).
2. 12 × 75 mm Pyrex culture tubes or similar.
3. Multichannel or electronic repeater pipette.
4. Microplate reader that can read at 405 and 595 nm.
5. 0.1 M HCl (*see* **Note 7**).
6. Kimwipes.
7. Vacuum aspirator.
8. Orbital shaker.
9. 96-well plates.
10. dH_2O.
11. Bradford reagent.
12. Bovine serum albumin (BSA) standard.
13. Ethanol vapor “bubble breaker.” A squirt bottle modified with its feed tube cut to 5 cm. filled one-quarter with 96% ethanol.

2.4 Data Analysis

Basic data analysis software package.

3 Methods

3.1 Construction of a Cya Fusion Reporter Construct with Effector of Interest and Introduction into P. syringae DC3000 (See Note 8)

1. Design gene-specific primers to amplify the coding sequence of your EOI in order to clone into the Gateway entry vector pENTR/SD/D-TOPO. The forward primer, P1, must contain a 5'CACC followed by the ATG start codon, and the reverse primer, P2, must exclude the stop codon. This will ensure directional cloning into the entry vector.
2. Use a high-fidelity DNA polymerase enzyme that generates blunt-ended products to amplify your EOI from *Pseudomonas* or the appropriate host. Set up the polymerase chain reaction (PCR) as described in what follows.

   10 μL 5× PrimeSTAR buffer with Mg^{++}.

   4 μL 2.5 mM dNTP mix.

   1 μL 10 μM forward primer (P1).

   1 μL 10 μM reverse primer (P2).

   100 ng genomic DNA.

   0.5 μL Taq DNA polymerase at 2.5 units/μL.

   Sterile molecular-biology-grade water to a final volume of 50 μL. Add contents to a PCR tube and mix well with a pipette.
3. Run reaction in a thermocycler with the following conditions.

   95 °C for 5 min.

   98 °C for 10 s.

   55 °C (or appropriate annealing temperature) for 15 s.

   72 °C 1 min/kb.

   Repeat cycle steps 2–4 for 30 cycles.

   Final extension 72 °C for 5 min.

   Hold at 12 °C.
4. Perform agarose gel electrophoresis using a 1% gel in TBE running buffer, 0.5 μg/mL ethidium bromide, to confirm product amplification. A strong, single band at the expected size must be observed. Alternatively, gel-purify the PCR desired product using the adequate kit.
5. Perform TOPO cloning reaction to introduce the PCR product into pENTR/SD/D-TOPO.

   4 μL fresh PCR product.

   1 μL salt solution.

   Sterile water to 5 μL.

   1 μL TOPO vector.

   Mix components in a tube and incubate at room temperature for 5–30 min.

6. Transform 2 μL of the cloning reaction into chemically competent *E. coli* cells and select for transformants on LB with 50 μg/mL kanamycin. Screen colonies obtained on the selection plate for presence of desired clone, using colony PCR with gene-specific primers.
7. For the colony PCR, prepare a PCR strip or plate by pipetting 9 μL/well of sterile water. Pick individual colonies with a toothpick and mix it into the water, taking care to not pick too many cells. The water should not be turbid to the eye. Boil the cells at 99 °C for 5 min in a thermocycler and cool to room temperature. Give the tubes or plate a quick spin to bring down the contents of the tube. Add 11 μL of enzyme mix per sample, prepared as follows: 10 μL 2× NEB OneTaq, 0.5 μL 10 μM primer 1 (P1), 0.5 μL 10 μM of primer 2 (P2). Enzyme mix can be prepared as a master mix based on the total number of samples. Run the samples in a thermocycler with the following conditions:

   94 °C for 30 s.

   30 cycles of:

   - 94 °C for 15–30 s.
   - 45–68 °C for 15–30 s.
   - 68 °C for 1 min/kb.

   Final extension 68 °C for 5 min.

   Hold at 12 °C.
8. Prepare purified plasmid DNA of selected clone by miniprep and sequence confirm the EOI gene with M13F and M13R primers (*see* **Note 9**).
9. Perform an LR (Clonase Gateway recombination reaction) Gateway reaction to introduce the EOI into the pCPP5371 Cya destination vector. This will generate a fusion protein of the EOI and the Cya at its C-terminus.

   100–300 ng pENTR/SD/D-TOPO::EOI, entry clone.

   300 ng pCPP5371, P_{hrp}-GW-Cya destination vector.

   4 μL 5× LR Clonase II reaction buffer.

   TE buffer, pH 8.0 to final 16 μL.

Add all components to a microcentrifuge tube, mix well, and incubate at room temperature for 1–2 h. Add 2 μL proteinase K to each reaction and incubate for 10 min at 37 °C.

10. Transform 2 μL of the cloning reaction into chemically competent *E. coli* cells, and select for transformants on LB with 10 μg/mL of gentamicin.
11. Screen colonies obtained on the selection plate for the presence of the desired clone using colony PCR with primers P3 and P4. The expected product should be the size of the EOI (bp) + 1.4 kb.

12. Prepare purified plasmid DNA by miniprep of the selected clone and sequence the EOI with gene-specific primers P1 and P2.
13. Introduce Cya reporter plasmid pCPP5371::EOI into *P. syringae* DC3000 by electroporation. Grow DC3000 from frozen stocks on KB agar medium at 30 °C for 1–2 days. From a freshly grown plate, inoculate a 5 mL culture in liquid KB medium and grow overnight at 30 °C.
14. Harvest cells by centrifugation at 3500 relative centrifugal force (RCF) for 5 min at room temperature. Wash cells twice with 5 mL 300 mM sucrose at room temperature. Resuspend in a final volume of 100 μL 300 mM sucrose. Add 100–200 ng pCPP5371::EOI and mix gently with a pipette.
15. Take a 1 mm electroporation cuvette, and add the mixture of DC3000 cells and DNA. Electroporate the cells at room temperature at 1.8 kV, 25 μF, 200 Ω. Electroporate an aliquot with no plasmid DNA as a negative control.
16. Add 1 mL liquid KB, transfer the cells to a 14 mL culture tube, and recover for 2 h at 30 °C with shaking at 250 rpm. Plate 50 and 150 μL each on separate KB agar plates with 10 μg/mL gentamicin. Store the remainder of the electroporation mix at 4 °C in case it is required to replate higher cell volumes later.
17. Incubate plates for 3–4 days at 30 °C until well-spaced single colonies are obtained.
18. Streak single colonies onto KB agar plates with 10 μg/mL gentamicin and prepare frozen 15% glycerol stocks of the DC3000 containing pCPP5371::EOI.

3.2 Plant Inoculation (See Notes 10 and 11)

1. Bring plants from the greenhouse or chamber 1 or 2 days prior to plant inoculation and keep them on the laboratory bench. We have used *Nicotiana benthamiana*, tobacco, and *Arabidopsis* for inoculation of the Cya reporter strains. *N. benthamiana* plants should be 4 to 6 weeks old, tobacco 5 to 8 weeks old, and *Arabidopsis* 4 to 5 weeks old.
2. Start fresh plates for each DC3000 strain, pCPP5371::EOI, pCPP5388 (positive control), and T3SS- pCPP5388 (negative control) on KB agar with 10 μg/mL gentamicin and grow at 30 °C for 1–2 days.
3. From the primary plates use a wooden dowel to scoop up an entire isolated colony, resuspend in 150–200 μL liquid KB, and mix well by vortexing. Spot the entire volume onto a fresh plate of KB agar with 10 μg/mL gentamicin. Swirl the plate around to spread the bacteria on the entire surface of the plate. Refresh all the strains similarly. Leave the plates open in the laminar flow hood for about 5–10 min to dry and then incubate at 30 °C overnight (up to 18 h).

4. Use an inoculating loop or pipette tip to harvest a pea-sized scoop of bacterial cells from the plate and resuspend in 5 mL 10 mM $MgCl_2$.
5. Use a spectrophotometer to measure the optical density at 600 nm (OD_{600}) of the cell suspension. Prepare a 20-fold dilution of the resuspended cells by mixing 950 μL 10 mM $MgCl_2$ with 50 μL culture. Measure the OD_{600} of the dilution and calculate the OD of the culture by multiplying the observed OD value by 20.
6. Adjust the resuspended cells to an OD_{600} of 0.05 with 10 mM $MgCl_2$ to a final volume of 10 mL for plant inoculation. Use the following formula for the adjustment:

   Volume of culture = $(0.05/(\text{observed } OD_{600} \times 20))^{*}10$

   Make up to 10 mL with 10 mM $MgCl_2$.

   In our spectrophotometer, this OD_{600} value corresponds to 5×10^7 CFU/mL. The exact correlation between OD_{600} and CFU/mL will need to be optimized since different spectrophotometers may vary. The important point to remember is that plants should be inoculated with a culture adjusted to 5×10^7 CFU/mL.
7. At least three infiltrations should be done for each strain to create biological replicates (EOI, positive and negative controls). In *N. benthamiana*and tobacco, the different strains can be inoculated on the same leaf as long as the inoculation zones do not overlap. For *N. benthamiana* and tobacco, choose well-expanded leaves, which are the fourth or fifth from the apex. For *Arabidopsis*, choose leaves from the center triplet in the whorl and inoculate two or three whole leaves per sample.
8. Using a 1 mL needleless syringe, inoculate plant leaves with the different strains. Make sure to include the positive and negative controls. A needle may be used to gently prick the leaf and then inoculate the suspension. Pat dry the inoculation zone with a Kimwipe or paper towel. For *N. benthamiana* and tobacco, outline the inoculation zone with a wide-tip black Sharpie to delineate the area. For *Arabidopsis*, mark the petioles of inoculated leaves with the black Sharpie.
9. Leave the plants on the laboratory bench for 6 h (*see* **Note 12**).
10. Take 2.2 mL round-bottom microcentrifuge tubes and add two copper BBs to each tube. Label the tubes with the name of the different strains and the replicate numbers (e.g., EOI_1, EOI_2, EOI_3, and so on). Have these tubes and a dewar of liquid nitrogen ready at the time of tissue harvesting.
11. After 6 h, use a biopsy punch or cork borer to excise ~1 cm^2 leaf tissue from each infiltration zone. Eight leaf discs collected with a 4 mm biopsy punch have a combined total area of

approximately 1 cm^2 of leaf tissue. Working quickly, push a long wooden dowel through the cork borer or biopsy punch to eject the collected leaf discs into pre-labeled tubes. Close the tubes and flash freeze in liquid nitrogen as soon as possible.

3.3 Direct cAMP Assay

Direct cAMP ELISA kit is used to quantitatively measure cAMP in the leaf samples. See the manufacturer's instructions for additional details on conducting the assay and data analysis.

1. Bring all cAMP ELISA buffers to room temperature before use (*see* **Note 13**).
2. Vortex the frozen tubes rapidly so the copper BBs grind the tissue into a fine powder. Proceed to perform the assay. Ground tissue can also be frozen at −80 °C until use (*see* **Note 14**).
3. Add 300 μL 0.1 M HCl to the frozen tissue. Vortex vigorously to mix. Tissue will turn brown in a few minutes. This is normal.
4. Centrifuge at ≥12,000 relative centrifugal force (RCF) for 10 min. Resuspend the pellet by vortexing and physical agitation and recentrifuge at ≥12,000 RCF for 10 min to create a tight pellet. Transfer the supernatant to a fresh microcentrifuge tube, taking care to avoid any leaf debris (*see* **Note 15**).
5. Extracted supernatants of each sample need to be diluted so that cAMP concentrations are within the range of the cAMP standard curve. This must be empirically determined for each EOI vector/host plant/bacterial concentration combination. We have used 10-fold to 300-fold dilutions in different experiments. More concentrated samples may be required to measure lower levels of translocation. The 50-fold dilution described in what follows has been used successfully in *Arabidopsis*.
6. Using a multichannel pipette and a 96-well plate, set up the following dilution series:

   Mix 50 μL extract supernatant with 200 μL 0.1 M HCl (1:5).

   Mix 30 μL 1:5 diluted sample with 270 μL 0.1 M HCl (1:10) (1:50 final).
7. Prepare cAMP standards. Diluted standards should be used within 60 min of preparation. Glass or polypropylene, but not polystyrene, tubes may be used with the standards.
8. Label five 12 × 75 mm Pyrex culture tubes #S1– #S5.
9. Pipet 900 μL of the 0.1 M HCl into tube #S1.
10. Pipet 750 μL of the 0.1 M HCl into tubes #S2– #S5.
11. Add 100 μL of the 2000 pmol/mL standard stock into tube #S1. Vortex vigorously.
12. Add 250 μL of tube #S1 to tube #S2 and vortex vigorously.

13. Repeat **step 12** for tubes #S2 to #S5 to prepare standard curve. Standard concentrations for tubes #S1 to #S5 are as follows: 200 pmol/mL, 50 pmol/mL, 12.5 pmol/mL, 3.13 pmol/mL, and 0.78 pmol/mL.
14. Set up ELISA plate. Every well is set up in duplicate. Sixteen wells will be used to generate assay standards (*see* **Note 16**). The positive and negative control DC3000 strains will require 12 wells (2 strains with technical duplicates of 3 biological replicates). Your EOI samples will require six wells apiece (technical duplicates of three biological replicates).

The following steps can be performed with a multichannel pipette.

15. Pipet 50 μL of neutralizing reagent into each well *except the blank wells.*
16. Pipet 100 μL of the 0.1 M HCl into the B_0 (0 pmol/mL standard) wells.
17. Add 150 μL of 0.1 M HCl to the nonspecific binding (NSB) wells.
18. Pipet 100 μL of standards #S1 to #S5 to the bottom of the appropriate wells.
19. Pipet 100 μL of the samples to the bottom of the appropriate wells.
20. Pipet 50 μL of the blue conjugate into each well except the blank wells.
21. Pipet 50 μL of the yellow antibody into each well except the NSB and blank wells.

The blank wells should be empty, the NSB wells should be blue, and all other wells should be green.

22. Seal the plate with sticker provided in kit and secure to an orbital shaker with tape. Shake at approximately 100–200 rpm for 2 h at room temperature.
23. During incubation, prepare the wash buffer and determine protein concentrations using a Bradford assay (*see* Subheading 3.4).
24. Wash buffer prep: Calculate the amount of wash buffer needed: #wells*1.2 mL/well = total volume wash buffer. Prepare a 1:20 dilution of wash buffer concentrate in dH_2O.
25. Wash the plate: Empty contents of the wells (into the sink) and add 400 μL wash buffer to every well.
26. Repeat **step 25** twice more for a total of three washes.
27. After the final wash, empty the wells and firmly tap plate on Kimwipe to remove any remaining wash buffer.
28. Use a vacuum aspirator to remove the last traces of wash buffer.
29. Add 200 μL of the substrate solution to each well.

30. Incubate for 1 h at room temperature in the dark without shaking. The wells with lower cAMP concentrations will become more yellow.
31. Pipet 50 μL of the stop solution into each well.
32. Measure absorbance at 405 nm.

3.4 Bradford Assay

All standards and samples should be run in duplicate. Sixteen wells will be used for the standards. BSA concentration in the prediluted standard kit: 125, 250, 500, 750, 1000, 1500, and 2000 μg/mL. Use water for the 0 μg/mL "standard."

These steps can be performed with a multichannel pipette.

1. In a 96-well plate add 99 μL water to each standard well and 80 μL water to each sample well.
2. Add 1 μL of the standards to the appropriate wells (1:100 dilution).
3. Add 20 μL of the 1:5 diluted samples to the appropriate wells (1:25 final dilution).
4. Add 100 μL Bradford dye to each well, pipetting up and down to mix.
5. Wells with higher protein concentrations should turn bluer.
6. Use a bubble breaker to remove any bubbles in the wells.
7. Incubate 10 min at room temperature.
8. Measure absorbance at 595 nm.

3.5 Data Analysis

The principle behind the data analysis is as follows. (1) Generate standard curves of pmol cAMP/mL and protein μg/mL concentration. (2) Use the standard curves to calculate the pmol cAMP/mL and protein μg/mL of your samples. (3) Divide the sample cAMP value by the sample protein value to get pmol cAMP/μg protein. This analysis can be conducted using Microsoft Excel.

Use the mean value of technical duplicate wells for all calculations.

cAMP concentration calculation from 405 nm absorbance values.

1. Subtract the blank well value from the NSB, B_0, standard, and sample values.
2. Subtract the blank-adjusted NSB value from the blank-adjusted B_0, standard, and sample values. These are the Net OD values.
3. Use the Net OD values to calculate the percentage of bound cAMP for the standards. $(S1/B_0)*100$, $(S2/B_0)*100$....
4. Plot the percentage of bound (y) for the standards (linear scale) vs. the pmol/mL values (x) of the standards (log scale). The cAMP standard concentrations for S1–S5 are as follows:

200 pmol/mL, 50 pmol/mL, 12.5 pmol/mL, 3.125 pmol/mL, and 0.781 pmol/mL.

5. Fit a logarithmic curve to the data points to determine $y = m*\ln(x) + b$.
6. Calculate the cAMP pmol/mL for each sample by solving for *x*.

   x[sample cAMP pmol/mL] = e^((*y*[sample percent bound] − *b*)/*m*).
7. Multiply the sample cAMP pmol/mL by the dilution factor (in this protocol, 50) to get the cAMP pmol/mL of the undiluted samples.

Protein concentration calculation from 595 nm absorbance values.

8. Plot the protein μg/mL (*y*) for the standards (linear scale) vs. the A595 values (*x*) of the standards (linear scale). Protein standard concentrations are 100-fold dilutions of the BSA standards P1–P5 = 1.25 μg/mL, 2.5 μg/mL, 5 μg/mL, 7.5 μg/mL, 10 μg/mL, 15 μg/mL, and 20 μg/mL.
9. Fit a linear curve to the data points to determine $y = m * x + b$.
10. Calculate sample protein μg/mL by solving for *y*:

    y[sample protein μg/mL] = (*m**(*x*[A595])) + *b*.
11. Multiply the sample cAMP pmol/mL by the dilution factor of 25 to get the protein μg/mL of the undiluted sample.
    Calculate sample pmol cAMP/μg protein
12. Divide the sample cAMP pmol/mL by the sample protein μg/mL. The mL units cancel, resulting in pmol cAMP/μg protein of the undiluted sample.
13. Determine the mean and standard deviation for the pmol cAMP/μg protein values of the three biological repeats of each sample. The T3SS- control strain should have a calculated pmol cAMP/μg protein value of less than 1. The positive control strain will vary based on the host plant and experiment but will likely be in the 10s to 100s range of pmol cAMP/μg protein (*see* **Note 17**).

4 Notes

1. pCPP5371 carries the *ccdB* toxin gene in the Gateway cassette and must be maintained in a CcdB-resistant *E. coli* strain such as DB3.1.
2. The *hrp* promoter is the *avrPto* promoter *hrp* box region.
3. For LM and KB liquid media do not add K_2HPO_4 prior to autoclaving or it will precipitate. Mix up a 100× phosphate stock, 75 g K_2HPO_4, in 500 mL dH_2O, and filter sterilize. Add to media after it cools.

4. *Arabidopsis* Col-0 is grown in an environmental growth chamber at 23 °C constant temperature with a 14 hours of light/10 hours of dark cycle and ~100 μmol light. Plants are covered with humidity domes that are cracked open 2 cm and benefit from growth with very limited airflow and 3–4 cm between each plant. *N. benthamiana* and tobacco can be chamber grown at 28 °C during the day and 23 °C at night on a 12 L/12D cycle and ~200 μmol light or grown under greenhouse conditions with 26 °C during the day and 22 °C at night, 16 L/8D conditions. *Nicotiana* are fertilized every 2 weeks with 1 g/L of Peter's 20:20:20 water-soluble fertilizer.
5. Depending on how the *Arabidopsis* are raised, they may wilt temporarily when infiltrated with 10 mM $MgCl_2$. We have used 0.25 mM $MgCl_2$ regularly as a substitute, which does not cause wilting.
6. Fine-point Sharpies should not be used as they will tear the leaves.
7. Depending on your dilution scheme, the kit may not provide sufficient 0.1 M HCl. You may mix your own to use in place of what is provided in the kit.
8. The methods described here explain in detail the generation of a reporter plasmid containing the EOI fused at the C-terminus with the Cya reporter gene. Expression of the EOI-Cya fusion protein is driven by a *hrp* promoter, which is induced in plants.
9. Some long EOI genes may require additional sequencing to obtain full coverage.
10. The effector AvrPto, encoded by *P. syringae* pv. tomato DC3000, serves as a positive control for the assay because it has been shown to be translocated to high levels, while a T3SS mutant of DC3000, incapable of translocation, is used as the negative control for the assay [10].
11. The assay is flexible and can be performed on different plants. We have used *Nicotiana benthamiana*, *Arabidopsis*, and tobacco plants for the assay. Other researchers have used tomato plants using a similar protocol.
12. We sample at 6 h, as this time point precedes any dramatic changes in bacterial population. Collection at later time points may complicate analyses owing to an increased bacterial replication in some samples based on host compatibility or induction of the hypersensitive response (HR) in incompatible hosts. We would not recommend collecting samples from leaf tissue that has undergone pathogen-induced cell death (disease or HR).
13. In multiplate versions of the kit, the blue conjugate should be aliquoted.
14. The leaf discs frozen along with the copper BBs can be ground to a powder by vigorous vortexing. It helps to first open the lid

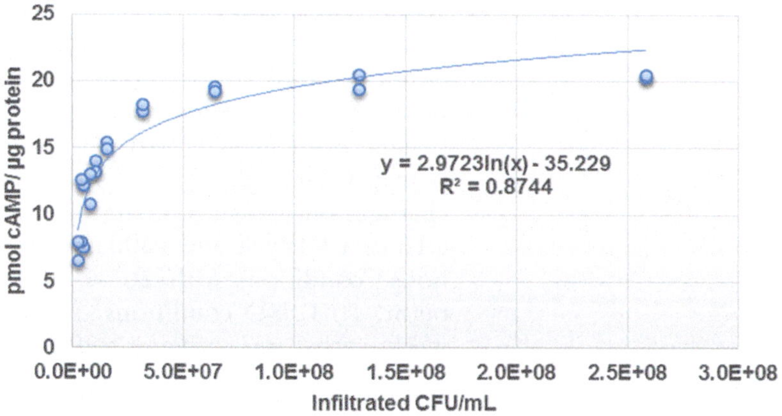

Fig. 1 Nonlinear relationship between inoculum levels and cAMP accumulation in Cya reporter-based T3SS translocation assays. Suspensions of *Pto* DC3000 transformed with *avrPto1-cya*-expressing plasmid pCPP5312 were infiltrated into the panels of two tobacco leaves at concentrations ranging from around 2.6 × 10^8 to 2.7 × 10^6 CFU/mL. Two 1 cm diameter leaf discs were harvested 6 h postinoculation per infiltration area to determine soluble pmol cAMP/µg protein. The mean soluble protein concentrations of 88 processed discs per leaf were used to calculate pmol cAMP/µg protein

of the microcentrifuge tube for a few seconds to release pressure due to liquid nitrogen. Following this, vortex at full speed for 15–20 s until the tissue is ground up. If large pieces of leaf still remain, simply drop the tube back into liquid nitrogen and proceed with other tubes, and complete the grinding after a few minutes. This will prevent tissue thawing. Once the tissue is ground up, add 0.1 M HCl and leave on the bench until all the samples are processed. Proceed to the remaining steps once all the samples have been resuspended in HCl.

15. HCl extracted supernatant can be refrozen for reanalysis, although the measured cAMP values will decrease.
16. We do not run the Total Activity (TA) standard described in the manufacturer's protocol. The TA value is not used in later calculations, and its omission allows all cAMP standards to be run in two columns, simplifying plate setup design.
17. Special consideration should be taken when interpreting translocation values quantitatively. The pmol cAMP/µg protein values increase logarithmically with increasing bacterial concentrations. *See* Fig. 1, which is recalculated and plotted from data collected and published in [16].

Acknowledgments

The authors would like to thank Dr. Lisa Schechter, Dr. Hai Li Wei, Dr. Sebastien Cunnac, Dr. Alan Collmer, and Dr. Sheng Yang He for significant contributions to the development and refinement of the procedure described in this chapter. This work was supported by National Science Foundation grant IOS-1025642 and the Gordon and Betty Moore Foundation (GBMF3037).

References

1. Cunnac S, Chakravarthy S, Kvitko BH, Russell AB, Martin GB, Collmer A (2011) Genetic disassembly and combinatorial reassembly identify a minimal functional repertoire of type III effectors in *Pseudomonas syringae*. Proc Natl Acad Sci U S A 108:2975–2980
2. Fouts DE, Badel JL, Ramos AR, Rapp RA, Collmer A (2003) A pseudomonas syringae pv. tomato DC3000 Hrp (Type III secretion) deletion mutant expressing the Hrp system of bean pathogen *P. syringae* pv. syringae 61 retains normal host specificity for tomato. Mol Plant-Microbe Interact 16:43–52
3. Alfano JR, Collmer A (2004) Type III secretion system effector proteins: double agents in bacterial disease and plant defense. Annu Rev Phytopathol 42:385–414
4. Wei CF, Kvitko BH, Shimizu R, Crabill E, Alfano JR, Lin NC, Martin GB, Huang HC, Collmer A (2007) A *Pseudomonas syringae* pv. tomato DC3000 mutant lacking the type III effector HopQ1-1 is able to cause disease in the model plant *Nicotiana benthamiana*. Plant J 51:32–46
5. Schechter LM, Vencato M, Jordan KL, Schneider SE, Schneider DJ, Collmer A (2006) Multiple approaches to a complete inventory of *Pseudomonas syringae* pv. tomato DC3000 type III secretion system effector proteins. Mol Plant-Microbe Interact 19:1180–1192
6. Schechter LM, Roberts KA, Jamir Y, Alfano JR, Collmer A (2004) *Pseudomonas syringae* type III secretion system targeting signals and novel effectors studied with a Cya translocation reporter. J Bacteriol 186:543–555
7. Garcia JT, Ferracci F, Jackson MW, Joseph SS, Pattis I, Plano LR, Fischer W, Plano GV (2006) Measurement of effector protein injection by type III and type IV secretion systems by using a 13-residue phosphorylatable glycogen synthase kinase tag. Infect Immun 74:5645–5657
8. Sory MP, Cornelis GR (1994) Translocation of a hybrid YopE-adenylate cyclase from *Yersinia enterocolitica* into HeLa cells. Mol Microbiol 14:583–594
9. den Dulk-Ras A, Vergunst AC, Hooykaas PJ (2014) Cre reporter assay for translocation (CRAfT): a tool for the study of protein translocation into host cells. Methods Mol Biol 1197: 103–121
10. Schechter LM, Valenta JC, Schneider DJ, Collmer A, Sakk E (2012) Functional and computational analysis of amino acid patterns predictive type III secretion system substrates in *Pseudomonas syringae*. PLoS One 7:e36038
11. Crabill E, Joe A, Block A, van Rooyen JM, Alfano JR (2010) Plant immunity directly or indirectly restricts the injection of type III effectors by the *Pseudomonas syringae* type III secretion system. Plant Physiol 154:233–244
12. Wei HL, Chakravarthy S, Worley JN, Collmer A (2013) Consequences of flagellin export through the type III secretion system of *Pseudomonas syringae* reveal a major difference in the innate immune systems of mammals and the model plant *Nicotiana benthamiana*. Cell Microbiol 15:601–618
13. Oh HS, Kvitko BH, Morello JE, Collmer A (2007) *Pseudomonas syringae* lytic transglycosylases coregulated with the type III secretion system contribute to the translocation of effector proteins into plant cells. J Bacteriol 189:8277–8289
14. Badel JL, Shimizu R, Oh HS, Collmer A (2006) A *Pseudomonas syringae* pv. Tomato avrE1/hopM1 mutant is severely reduced in growth and lesion formation in tomato. Mol Plant-Microbe Interact 19:99–111
15. Lam HN, Chakravarthy S, Wei HL, BuiNguyen H, Stodghill PV, Collmer A, Swingle BM, Cartinhour SW (2014) Global analysis of the HrpL regulon in the plant pathogen *Pseudomonas syringae* pv. tomato DC3000 reveals new regulon members with diverse functions. PLoS One 9:e106115
16. Kvitko BH, Ramos AR, Morello JE, Oh HS, Collmer A (2007) Identification of harpins in *Pseudomonas syringae* pv. tomato DC3000, which are functionally similar to HrpK1 in promoting translocation of type III secretion system effectors. J Bacteriol 189:8059–8072

Chapter 34

Monitoring Effector Translocation using the TEM-1 Beta-Lactamase Reporter System

Julie Allombert, Anne Vianney, and Xavier Charpentier

Abstract

Among the bacterial secretion systems, the Type III, IV, and VI secretion systems enable bacteria to secrete proteins directly into a target cell. This specific form of secretion, referred to as *translocation*, is essential for a number of pathogens to alter or kill targeted cells. The translocated proteins, called *effector proteins*, can directly interfere with the normal processes of the targeted cells, preventing elimination of pathogens and promoting their multiplication. The function of effector proteins varies greatly depending on the considered pathogen and the targeted cell. In addition, there is often no magic bullet, and the number of effector proteins can range from a handful to hundreds, with, for instance, a substrate of over 300 effector proteins of the Icm/Dot Type IV secretion system in the human pathogen *Legionella pneumophila*. Identifying, detecting, and monitoring the translocation of each of the effector proteins represents an active field of research and is key to understanding the bacterial molecular weaponry. Translational fusion of an effector with a reporter protein of known activity remains the best method to monitor effector translocation. The development of a fluorescent substrate for the TEM-1 beta-lactamase has turned this antibiotic-resistant protein into a highly versatile reporter system for investigating protein transfer events associated with microbial infection of host cells. Here we describe a simple protocol to assay the translocation of an effector protein by the Icm/Dot system of the human pathogen *Legionella pneumophila*.

Key words Effector protein, Type IV secretion system, β-lactamase fusion, CCF4, Fluorescence, *Legionella pneumophila*

1 Introduction

Protein delivery from pathogen to host represents a major and unifying theme in microbial pathogenesis. From the pathogen's perspective, sending its own proteins to the target host cell represents an efficient strategy for interfering with the host cellular functions, preventing exposure to the host defense mechanism or even subverting the cells for its benefit. There are, however, barriers preventing the diffusion of proteins out of the pathogenic cell and blocking the import of proteins to the host cell side. For instance, in Gram-negative bacteria, a protein would have to get across three

Laure Journet and Eric Cascales (eds.), *Bacterial Protein Secretion Systems: Methods and Protocols*, Methods in Molecular Biology, vol. 1615, DOI 10.1007/978-1-4939-7033-9_34,

membranes; the inner and outer membranes of the bacteria and the host cytoplasmic membrane.

Remarkably, bacteria have evolved several multisubunit molecular machines to achieve the feat of *translocating* a protein from the pathogen cytoplasm directly to the cytoplasm or the target cell.

The process is often referred to as *translocation*, and the translocated proteins, whose assumed function is to have an effect on the cell functions, are called *effectors*. The multisubunit molecular machines capable of translocating effectors include the Type III, IV, and VI systems of Gram-negative bacteria and the Type VII system of Mycobacteria [1]. The set of translocated effector molecules tends to be unique to each pathogen and reflects the unique needs and specific niches of each bacterial species. A major challenge is to identify the substrates of these systems and track their translocation in the host cell.

Several methods have been reported to specifically detect this fraction of a bacterial protein that found its way to the host cell. The first of them is the popular CyaA system involving translational fusions of effectors with the calmodulin-dependent catalytic domain of the *Bordetella pertusis* toxin CyaA [2]. This enzyme converts cellular adenosine triphosphate in cyclic adenosine monophosphate (cAMP) in the presence of the eukaryotic protein calmodulin. Levels of cAMP production can be subsequently quantified. A less popular but clever method involves a translational fusion with the phosphorylable Elk peptide fused to the nuclear localization signal (NLS) from the large T antigen of SV40 [3]. The NLS sequence directs the fusion protein to the cell nucleus where the Elk tag is phosphorylated and can be detected with phosphospecific Elk-peptide antibodies. Another method is based on fractionation with digitonin that solubilise the eukaryotic plasma membrane but not the prokaryotic membranes [4]. As for the Cya and Elk-tag systems, these assays require disrupting the eukaryotic cell and then performing an analysis. The beta-lactamase translocation assay was developed to overcome these limitations and analyze translocation in living cells [5]. The beta-lactamase translocation assay (*see* Fig. 1) takes advantage of the fluorescent substrate CCF2-AM (or CCF4-AM) initially developed for the detection of TEM-1 beta-lactamase activity within eukaryotic cells [6]. The substrate consists of a coumarin and a fluorescein fluorophore connected by a beta-lactam ring. Because of the fluorophores' spatial proximity, the fluorescence energy coming from the excitation of the coumarin moiety is entirely transferred to the fluorescein moiety, resulting in the emission of a green fluorescence. Enzymatic cleavage of the beta-lactam ring by the TEM-1 beta-lactamase frees the coumarin moiety, which, under excitation, now emits a blue fluorescence. This shift in fluorescence can be directly observed in the infected host cell with an epifluorescence microscope or quantified with a spectrofluorometer. Translocation in the eukaryotic host cytoplasm of an effector-TEM-1 fusion

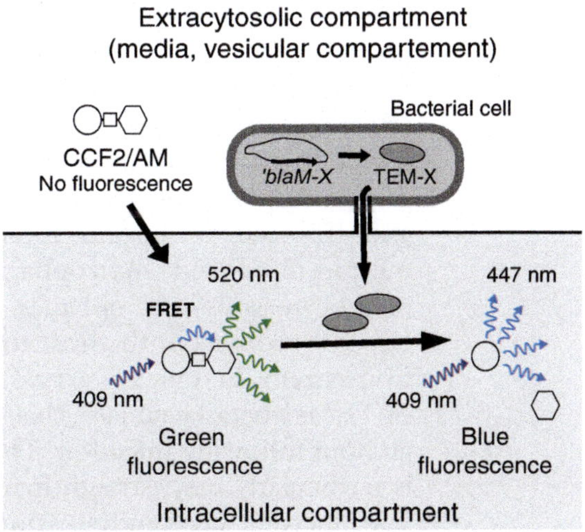

Fig. 1 Schematic representation of TEM-1 reporter system to assess effector protein translocation in live eukaryotic cells. Upon passive entry into eukaryotic cell, the nonfluorescent esterified CCF2/AM (or CCF4/AM) substrate is rapidly converted by cellular esterases into a charged and fluorescent CCF2. Excitation of the coumarin moiety (*circle*) at 409 nm results in fluorescence energy transfer (FRET) to the fluorescein moiety (*hexagon*), which emits a green fluorescence signal at 520 nm. Injection of an effector fused to TEM-1 into a CCF2-loaded cell induces catalytic cleavage of a CCF2 beta-lactam ring (*square*), disrupting FRET. This produces an easily detectable and measurable change in fluorescence from green to blue emission

protein triggers a change in fluorescence of the host cell, which makes the analysis of effector translocation rapid, easy, and reliable. Because fewer than 100 molecules of TEM-1 can be readily detected within a cell [6], the system was sensitive enough to detect the translocation of a weakly produced fusion. Of note, the TEM-1 enzyme is naturally secreted in the periplasmic space by the Sec pathway. Thus, to use TEM-1 as a reporter of the ability of a fused protein to drive it to another secretion system, it is necessary to use a version deleted of its N-terminal secretion signal. Owing to the properties of a secreted protein, TEM-1 can efficiently unfold and refold and is highly permissive of protein fusion. This likely makes it compatible with secretion by most secretion systems, and numerous studies have used it to demonstrate the translocation of effector proteins by the Type III, IV, and VI systems (for a review, *see* [7]). The system has also been successfully used to detect proteins secreted by the protozoan parasite *Toxoplasma gondii* in its host [8]. More than just a convenient way to demonstrate the translocation of an effector, the beta-lactamase translocation reporter system can be used to monitor the kinetic parameters of the translocation process [9, 10]. The assay can be miniaturized to a 384-well format and is compatible with a high-throughput screening to identify small molecules that could inhibit translocation and prevent infection [10, 11]. Translocated effector proteins can also identify cells targeted by a pathogen in an infected host [12, 13].

Here we provide a protocol for testing the translocation of effector proteins that are the substrate of the Icm/Dot Type IV secretion system of the human pathogen *Legionella pneumophila*. *L. pneumophila* infects alveolar macrophages in the human lungs, and this cellular infection can be recapitulated in vitro using monocyte-derived macrophages (THP-1, U937 cells). Upon phagocytosis by macrophages, *L. pneumophila* delivers numerous effector proteins in the host through its Icm/Dot Type IV secretion system. Ectopically expressed as fusions with the mature form of TEM-1 beta-lactamase, their translocation is detectable less than an hour following infection. The beta-lactamase translocation assay is particularly easy, straightforward, and quick. It requires only a few pipetting steps and no sample processing. Typically, the results of the assay are obtained about 3 h post infection. The protocol provided here could easily be adapted to other pathogens, secretion systems, and cellular infection models.

2 Materials

2.1 Bacterial Strains, Beta-Lactamase Constructs, Host Cells

1. *L. pneumophila* strain (Paris, Lens, Philadelphia-1).
2. Plasmids for expression of beta-lactamase fusion proteins: The plasmid pXDC61 and its derivative used in this protocol are available from the nonprofit plasmid repository addgene (addgene.org, plasmids #21841, #21842, #21843, #21844) (*see* **Note 1** and Fig. 2).
3. U937 cell line ATCC number: CRL-1593.2™.

2.2 *Legionella pneumophila* Media and Bacterial Growth

1. ACES-buffered Yeast Extract (AYE) medium: For 1 L dissolve 12 g yeast extract and 10 g N-(2-Acentamido)-2-aminoethanesulfonic acid (ACES), adjust pH to 6.9 with 1 M KOH. Add 10 mL cysteine 40 g/L and 10 mL iron pyrophosphate 30 g/L. Fill volume to 1 L with distilled water and filter sterilize.
2. Charcoal yeast extract (CYE) plates: For 1 L dissolve 10 g yeast extract and 10 g ACES, adjust pH to 6.9 with 1 M KOH, add 15 g agar and 2 g activated charcoal, and autoclave. Add 10 mL filter sterilized cysteine 40 g/L and 10 mL filter-sterilized ferric nitrate 25 g/L. When appropriate add 5 μg/mL chloramphenicol and 1 mM isopropyl β-D-1-thiogalactopyranoside (IPTG) (*see* **Note 2**).
3. Disposable 13 mL polypropylene snap-cap tubes, sterile.
4. 1.5 mL microcentrifuge tubes, sterile.
5. 30 °C incubator.
6. Orbital shaker, 30 °C.
7. Spectrophotometer and cuvettes.

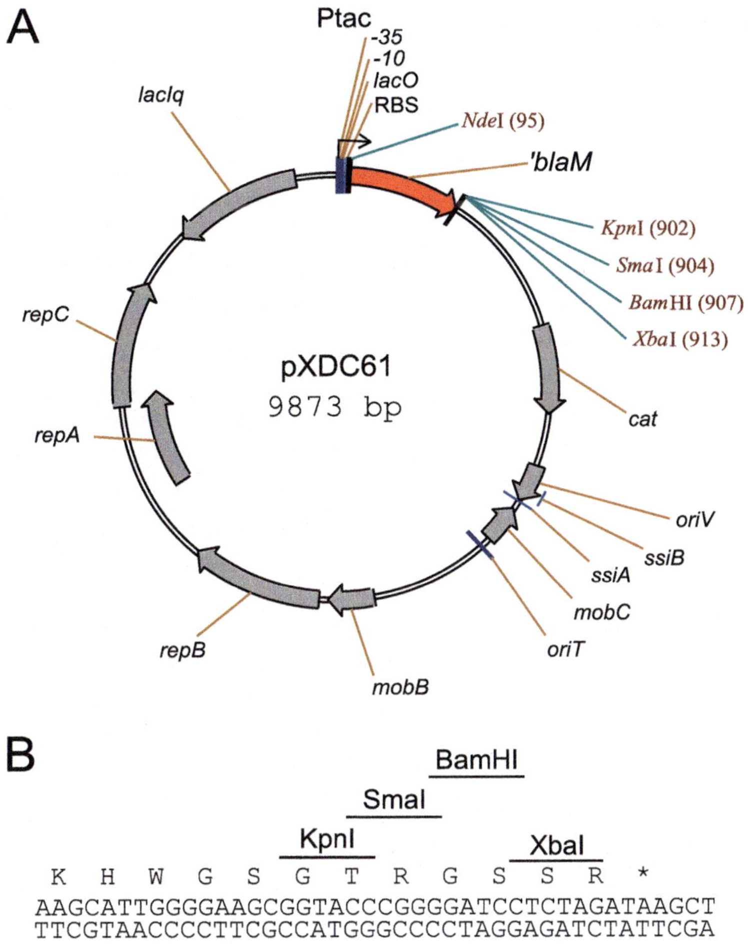

Fig. 2 Map of plasmid pXDC61 to express beta-lactamase TEM-effector fusion proteins. (**a**) The pXDC61 plasmid is a mobilizable (*oriT*) plasmid derived from the broad host range plasmid RSF1010. The plasmid confers chloramphenicol resistance. The '*blaM* gene encodes the mature form of the TEM-1 beta-lactamase deprived of its N-terminal secretion signal and is controlled by an IPTG-inducible promoter. A polylinker (KpnI, SmaI, BamHI, XbaI) is placed at the 3′ end of the '*blaM* gene. (**b**) Detail of polylinker for in-frame cloning of effector genes

2.3 Cell Culture and Differentiation

1. RPMI medium supplemented with 10% fetal bovine serum (FBS) and glutamine (i.e., RPMI 1640 GlutaMAXTM, Gibco). When appropriate add 5 μg/mL chloramphenicol and 1 mM IPTG.
2. Phorbol 12-myristate 13-acetate (PMA): 0.1 M.
3. Culture flask, 25 cm^2, sterile.
4. Disposable 15 mL polypropylene snap-cap tubes, sterile.
5. 96-well black polystyrene microplates with clear bottom, sterile.
6. CO_2 incubator, 37 °C.
7. Malassez counting chamber.

2.4 Translocation Assays

1. LiveBLAzer-FRET B/G loading kit (Invitrogen). This kit includes the CCF4/AM substrate (*see* **Note 3**).
2. Probenecid stock solution: 0.1 M. Dissolve 1.25 g probenecid (Sigma) in 22 mL 0.4 M NaOH by vigorous agitation. Add 22 mL 100 mM phosphate buffer, pH 8.0, and stir to dissolve the precipitate that could form. Check pH and, if necessary, adjust it to 8.0 with 1 M NaOH (if pH < 8) or HCl (if pH > 8). Distribute in 1 mL aliquot and store at −20 °C.
3. RPMI medium.
4. Fluorescence microplate reader equipped with a dual monochromator (e.g., Tecan Infinite M200) or with an excitation filter at 405 nm and emission filters at 460 nm (blue fluorescence) and 530 nm (green fluorescence). Determine whether the plate reader reads the plate from top or bottom.
5. Inverted fluorescence microscope equipped with a beta-lactamase filter set (Chroma Set # 41031; Excitation filter: HQ405/20× (405 ± 10); dichroic mirror: 425 DCXR; emission filter: HQ435LP (435 long-pass)). Alternatively a 4′,6′-diamidino-2-phenylindole (DAPI) filter set (340 to 380 nm excitation and 425 nm long-pass emission) may be used to observe the blue fluorescence, while the green fluorescence may be observed with a green fluorescent protein (GFP)/fluorescein filter set.

3 Methods

3.1 Growth of Infecting *Legionella pneumophila* Strains

The infecting strain should be grown under conditions that have been previously determined to result in a successful infection. These conditions may vary depending on the strain and species used but should include chloramphenicol to maintain the plasmid and IPTG to induce expression of the tested effector fusion proteins.

1. Streak *L. pneumophila* strains carrying pXDC61-derived plasmids from a frozen stock to CYE plates supplemented with chloramphenicol and then incubated at 30 °C for 5 days.
2. With a sterile loop, scrape off a few colonies and transfer to a 1.5 mL microcentrifuge tube containing 1 mL sterile ultrapure water. Resuspend bacteria by repeated pipetting.
3. Measure the optical density at 600 nm (OD_{600}) of a tenfold diluted bacterial suspension.
4. In a sterile 13 mL tube, inoculate 2 mL AYE supplemented with chloramphenicol and IPTG with appropriate volume of previous bacterial suspension to reach a starting OD_{600} of 0.3. Incubate at 30 °C in orbital shaker for 3 days (*see* **Note 4**).
5. Validate the beta-lactamase fusions production by western blot (*see* **Note 5**).

3.2 Maintenance and Differentiation of U937 Target Cells

U937 cells are monocytes that grow as a suspension and should be maintained at a cell density of between 1.10^5 and 2.10^6 viable cells/mL.

1. Seed a 25 cm^2 culture flask with U937 cells from a frozen stock or from a previous culture flask in 10 mL RPMI medium supplemented with glutamine and FBS. The culture flask is incubated at 37 °C in a CO_2 incubator for 5 days.
2. Determine the cellular concentration of the U937 cell culture with a Malassez counting chamber.
3. Transfer 1.10^7 cells to a sterile 15 mL conical tube and centrifuge 5 min at 880 g.
4. Discard the supernatant and gently resuspend the pellet of cells in 10 mL RPMI supplemented with glutamine and FBS (preheated to 37°C) by slow repeated pipetting. This gives a cell suspension of 1.10^6 cells/mL. Add 0.5 μL PMA.
5. Distribute 100 μL of cell suspension per well of a 96-well microplate (10^5 cells/well). Leave three wells without U937 cells (medium alone). They will be used for the blank fluorescence measurement.
6. Incubate at 37 °C in CO_2 incubator for 3 days. Following this incubation, the previously spherical and nonadherent cells should now be differentiated into macrophage-like cells that adhere to the bottom of the well and display a spread-out morphology.

3.3 Detection of Effector Translocation Using a Fluorescence Plate Reader

1. Grow *L. pneumophila* strains as described in Subheading 3.1.
2. Measure OD_{600} of liquid cultures of *L. pneumophila* strains. Adjust to $OD_{600} = 1$ with sterile ultrapure water. This gives a bacterial suspension at 10^9 bacteria/mL.
3. Add 200 μL of the resulting suspension to 800 μL RPMI supplemented with glutamine, FBS, chloramphenicol, and IPTG. Incubate these bacterial suspensions (2.10^8 bacteria/mL) for 2 h at 37 °C in the CO_2 incubator.
4. Add 10 μL of the bacterial suspension in wells of the 96-well microplate containing differentiated U937 cells (Subheading 3.2). This gives a multiplicity of infection (ratio bacteria to differentiated cell) of 20 (*see* **Note 6**). The bacterial suspension of each tested *L. pneumophila* strain is added to three different wells. Centrifuge the microplate for 10 min at 600 × *g* (*see* **Note 7**). Incubate at 37 °C in a CO_2 incubator for 1 h.
5. During the incubation time prepare the CCF4-AM loading solution of the LiveBLAzer-FRET B/G loading kit. Determined the number of tested wells, *n*. Mix *n* × 0.12 μL CCF4-AM 6X solution with *n* × 1.08 μL solution B. Vortex for 10 s. Add *n* × 15.8 μL solution C and *n* × 3 μL probenecid

0.1 M (*see* **Note 8**). Vortex for 10 s. This loading solution should be kept away from strong light and is stable for 4 h at room temperature.

6. Add 20 μL of the loading solution to each of the tested wells of the 96-well microplate, including to the 3 wells that contain media alone (*see* **step 5** in Subheading 3.2). Incubate for 2 h in the dark at room temperature.
7. If you have access to a fluorescence plate reader with bottom read capabilities, go directly to **step 9**. If the microplate reader is only equipped with a fluorescence top reading module, an extra step is needed because the fluorescence signals are quenched by the red solution C used in the loading CCF4 solution.
8. Following the 2 h CCF4 loading in the dark (**step 6**), discard delicately the liquid contained in each well of the 96-well microplate, including the 3 wells without cells. Replace with 50 μL RPMI at room temperature and without FBS (dispensing medium with FBS tends to create bubbles that may interfere with fluorescence measurements).
9. Put the plate (lid on) in a fluorescence plate reader to start the measurements. Measure successively the blue fluorescence (Ex. 405 nm, Em. 460 nm) and green fluorescence (Ex. 405 nm, Em. 530 nm). Both measurements should be performed on blank wells (medium alone) in addition to the tested wells.
10. After collecting the raw data, perform a blank subtraction on each fluorescence read. To evaluate the secretion efficiency of the beta-lactamase fusions, divide the blank-subtracted blue fluorescence signal by the blank-subtracted green fluorescence signal (*see* **Note 9**). An example of expected results is shown in Fig. 3.

3.4 Visualization of Effector Translocation Using Fluorescence Microscope

Following fluorescence quantifications, the infected cells may also be observed under a microscope to assess the percentage of translocation-positive (blue) and translocation-negative cells (green).

1. Follow the protocol of Subheading 3.3 until **step 8**.
2. Place the plate on an inverted microscope equipped with a 40× or 60× objective.
3. Observe cells with the beta-lactamase filter set (Ex: 405 ± 10; dichroic mirror: 425; Em: 435 long-pass). Using this filter set, both green and blue cells can be visualized simultaneously.
4. Alternatively, blue cells can be visualized with a (DAPI) filter set (340–380 nm excitation and 425 nm long-pass emission). Green cells can be visualized with the filter set commonly used for the visualization of GFP. Overexposure of the cells with this filter set may bleach the fluorescein moiety of CCF2

A

	Raw data		Blank-substracted data		
	Blue fluorescence	Green fluorescence	Blue fluorescence	Green fluorescence	Ratio blue/green
Blank	4243	5956			
Uninfected cells	6491	26945	2248	20989	0.11
	7889	27669	3646	21713	0.17
	6993	25331	2750	19375	0.14
WT/FabI	7918	28346	3675	22390	0.16
	8314	30343	4071	24387	0.17
	6835	23625	2592	17669	0.15
WT/LepA	33543	11799	29300	5843	5.01
	34327	12111	30084	6155	4.89
	36127	12802	31884	6846	4.66
Δ*dotA*/LepA	7719	29373	3476	23417	0.15
	8317	31114	4074	25158	0.16
	7581	24710	3338	18754	0.18

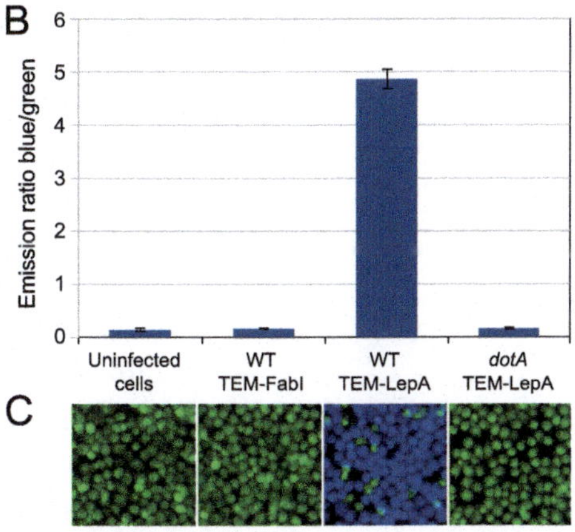

Fig. 3 Typical results of a translocation assay. (**a**) Raw data, blank-subtracted fluorescence signals (relative fluorescence unit, RFU), and fluorescence ratio (emission 460 nm/530 nm) obtained for secretion by wild-type (WT) or Δ*dotA L. pneumophila* strain Lens of the known Dot/Icm effector LepA and the nonsecreted cytoplasmic protein FabI. Secretion assays were done in triplicate for each strain. Fluorescence measurements were made with a Tecan M200 Infinite microplate reader equipped with a monochromator and a fluorescence top reading module. The measurement program includes fluorescence readings with an excitation wavelength of 405 nm, emission wavelengths of 460 and 530 nm, and a gain set at 135. (**b**) Graphical representation of mean emission ratio for LepA and FabI and corresponding standard deviations. Depending on the gain set for each of the two fluorescence measurements, this ratio can change significantly. Data acquired under different gain or in different plate reader can be normalized by setting the 460/530 ratio of the uninfected cells to 1. (**c**) Fluorescence images of a typical translocation assay. Upon excitation at 405 nm, two images were captured at 460 and 530 nm and merged

(or CCF4), which could result in the observation of blue fluorescence even in the absence of effector translocation.

5. If the images of the blue and green cells are acquired separately, the two images should be merged. Typical images are shown in Fig. 3.

4 Notes

1. Beta-lactamase fusion plasmids are derived from the pXDC61 plasmid (Fig. 2) [14]. They are introduced into the *L. pneumophila* by electroporation. These plasmids are constructed by cloning the coding sequence of the candidate effector gene in frame with the '*blaM* gene encoding the mature form of the TEM-1 beta-lactamase deprived of its N-terminal secretion signal. The candidate gene is cloned downstream of the '*blaM* gene in order to leave intact the potential C-terminal secretion signal of the candidate protein. The appropriate polylinker is shown in Fig. 2b. If the nature of the secretion signal is unknown, it is advisable to generate and test both fusion proteins at the C- or the N-terminus of the TEM-1 beta-lactamase.

An NdeI site is available at the start codon of '*blaM*. The expression of these gene fusions is controlled by an IPTG-inducible promoter.

2. Culture conditions must be optimized depending on the used *Legionella* species and strains in order to obtain bacteria in a virulent state (stationary phase). Here, we use CYE agar plates and AYE liquid medium, but bacteria grown on buffered CYE agar plates and LGM (Legionella growth medium) are equally infective.
3. According to the supplier, "CCF2-AM and CCF4-AM differ by two carbons in the bridge linking the coumarin moiety to the lactam ring. Both are in the membrane-permeable, esterified forms, and can be used for assays in intact cells. CCF4-AM has better solubility properties (soluble for >24 h) than CCF2-AM and is thus best suited for screening applications. In addition, CCF4-AM has slightly better FRET and thus slightly lower background than CCF2-AM." In our experience, CCF2/AM and CCF4/AM perform equally well. We have not found significant differences between the two compounds.
4. *L. pneumophila* grown on CYE agar plates may exhibit a heterogeneous population with a large part of filamentous bacteria. Therefore, bacteria are grown in liquid cultures before host cell infection in order to work with a homogeneous and more infective population.
5. It is advisable to assess the correct production and stability of the beta-lactamase fusions. This can be done using conventional western blot techniques, which will not be described here. We recommend using the beta-lactamase monoclonal antibody clone 8A5.A10, which is available from a variety of suppliers.
6. The response of the beta-lactamase reporter should be determined as a function of multiplicity of infection (MOI). For *L. pneumophila* and phagocytic cells, between an MOI of 1 and 10 the beta-lactamase system seems to behave linearly, and above 25 bacteria/cell, the system appears to saturate [10]. Therefore, we use an MOI of 20 as a compromise between sensitivity and linearity. Care should be taken to ensure that the MOI is not too high, for instance, the TEM-FabI negative control should not produce blue fluorescence. This situation occurs in *L. pneumophila* when the MOI is over 50.
7. Motile *L. pneumophila* can make contact with host cells without centrifugation. However, this experiment is based on the fluorescence signals of a host cell monolayer at a specific time point. Thus, the infection must be synchronized by centrifugation in order to work with a host cell monolayer that is homogeneously infected.

8. The addition of probenecid in the CCF4 loading solution is required to inhibit organic anion transporter [15] and facilitates the loading of CCF4 by inhibiting efflux from cells.
9. Expected results are shown in Fig. 3. For Dot/Icm effector proteins, you should see an increase of the blue fluorescence and a decrease of the green fluorescence in comparison to the nonsecreted protein (FabI) or in comparison with the TEM-effector fusion in the Δ*dotA* mutant impaired for its Icm/Dot Type IV secretion system.

References

1. Costa TRD et al (2015) Secretion systems in Gram-negative bacteria: structural and mechanistic insights. Nat Rev Microbiol 13(6):343–359
2. Sory MP, Cornelis GR (1994) Translocation of a hybrid YopE-adenylate cyclase from Yersinia enterocolitica into HeLa cells. Mol Microbiol 14(3):583–594
3. Day JB, Ferracci F, Plano GV (2003) Translocation of YopE and YopN into eukaryotic cells by Yersinia pestis yopN, tyeA, sycN, yscB and lcrG deletion mutants measured using a phosphorylatable peptide tag and phosphospecific antibodies. Mol Microbiol 47(3):807–823
4. Lee VT, Anderson DM, Schneewind O (1998) Targeting of Yersinia Yop proteins into the cytosol of HeLa cells: one-step translocation of YopE across bacterial and eukaryotic membranes is dependent on SycE chaperone. Mol Microbiol 28(3):593–601
5. Charpentier X, Oswald E (2004) Identification of the secretion and translocation domain of the enteropathogenic and enterohemorrhagic Escherichia coli effector Cif, using TEM-1 beta-lactamase as a new fluorescence-based reporter. J Bacteriol 186(16):5486–5495
6. Zlokarnik G et al (1998) Quantitation of transcription and clonal selection of single living cells with beta-lactamase as reporter. Science 279(5347):84–88
7. Pechous RD, Goldman WE (2015) Illuminating targets of bacterial secretion. PLoS Pathog 11(8):e1004981
8. Lodoen MB, Gerke C, Boothroyd JC (2010) A highly sensitive FRET-based approach reveals secretion of the actin-binding protein toxofilin during Toxoplasma gondii infection. Cell Microbiol 12(1):55–66
9. Mills E, Baruch K, Charpentier X, Kobi S, Rosenshine I (2008) Real-time analysis of effector translocation by the type III secretion system of enteropathogenic Escherichia coli. Cell Host Microbe 3(2):104–113
10. Charpentier X et al (2009) Chemical genetics reveals bacterial and host cell functions critical for type IV effector translocation by Legionella pneumophila. PLoS Pathog 5(7):e1000501
11. Harmon DE, Davis AJ, Castillo C, Mecsas J (2010) Identification and characterization of small-molecule inhibitors of Yop translocation in Yersinia pseudotuberculosis. Antimicrob Agents Chemother 54(8):3241–3254
12. Marketon MM, DePaolo RW, DeBord KL, Jabri B, Schneewind O (2005) Plague bacteria target immune cells during infection. Science 309(5741):1739–1741
13. Geddes K, Cruz F, Heffron F (2007) Analysis of cells targeted by Salmonella Type III secretion in vivo. PLoS Pathog 3(12):e196
14. de Felipe KS et al (2008) Legionella eukaryotic-like type IV substrates interfere with organelle trafficking. PLoS Pathog 4(8):e1000117
15. Steinberg TH, Newman AS, Swanson JA, Silverstein SC (1987) Macrophages possess probenecid-inhibitable organic anion transporters that remove fluorescent dyes from the cytoplasmic matrix. J Cell Biol 105(6 Pt 1):2695–2702

Chapter 35

Effector Translocation Assay: Differential Solubilization

Irina S. Franco, Sara V. Pais, Nuno Charro, and Luís Jaime Mota

Abstract

The identification of effector proteins delivered into mammalian host cells by bacterial pathogens possessing syringelike nanomachines is an important step toward understanding the mechanisms underlying the virulence of these pathogens. In this chapter, we describe a method based on mammalian tissue culture infection models where incubation with a nonionic detergent (Triton X-100) enables solubilization of host cell membranes but not of bacterial membranes. This allows the isolation of a Triton-soluble fraction lacking bacteria but enriched in proteins present in the host cell cytoplasm and plasma membrane. Using appropriate controls, this fraction can be probed by immunoblotting for the presence of bacterial effector proteins delivered into host cells.

Key words Bacterial protein secretion system, Type III secretion, Effector, Translocation, Detergent solubilization, SDS-PAGE, Immunoblotting

1 Introduction

Gram-negative bacteria possess different macromolecular structures, known as type III, type IV, or type VI secretion systems, for the delivery of effector proteins directly from the bacterial cytoplasm into eukaryotic or prokaryotic host cells [1]. This protein delivery or injection process is normally described as *effector translocation*. Demonstrating that a particular bacterial effector protein is injected into mammalian host cells during infection is not a trivial task, as effectors are often delivered in minute amounts and can be short-lived within the host cell. Various assays have been developed to monitor effector translocation, for example, using *Bordetella pertussis* calmodulin-dependent adenylate cyclase [2] or mature TEM-1 β-lactamase [3] reporter assays (described in Chapters 33 and 34). Here we describe a method for assessing effector translocation into mammalian cells by differential solubilization. It consists in the infection of tissue culture cells by a bacterial pathogen, followed by lysis of the infected mammalian cells using a detergent

Laure Journet and Eric Cascales (eds.), *Bacterial Protein Secretion Systems: Methods and Protocols*, Methods in Molecular Biology, vol. 1615, DOI 10.1007/978-1-4939-7033-9_35, © Springer Science+Business Media LLC 2017

that does not affect the integrity of bacterial membranes. The nonionic detergents Triton X-100 [4] and digitonin [5] have been widely used for this purpose, based on the inability of Triton X-100 to solubilize outer membranes of Gram-negative bacteria (although it can solubilize inner membranes) [6–8], and on the specificity of digitonin for cholesterol-rich membranes [9]. Subsequent high-speed centrifugation allows the separation of detergent-soluble (supernatant) and insoluble components (pellet) of the lysate, where the supernatant comprises cytoplasmic and plasma membrane components (including delivered effector proteins) and the pellet retains unbroken bacteria and nuclei that remained intact. Analysis of these fractions by immunoblotting makes it possible to confirm the presence of an effector protein of interest in the supernatant fraction, which is taken as evidence of effector translocation. A critical control involves probing for a bacterial protein that is not delivered into host cells. This ensures that during experimental manipulation there was no contamination of the supernatant fraction with cytosolic bacterial proteins.

Differential solubilization can be applied to monitor effector translocation by different Gram-negative bacteria and types of host cells. Furthermore, if effector-specific antibodies are available, differential solubilization can be used to monitor the translocation of endogenously expressed and nonmodified effector proteins. This is in contrast to other effector translocation assays that require the modification of the gene encoding the effector to produce a protein with an epitope tag or fused to a reporter protein, usually expressed from a plasmid and often from an exogenous promoter. The differential solubilization procedure using Triton X-100 is illustrated by the two protocols detailed in what follows, consisting in the monitoring of type III secretion (T3SS)-mediated translocation (1) of the *Yersinia enterocolitica* effector YopE, expressed from its own promoter using a nonmodified wild-type strain to infect RAW 264.7 murine macrophage-like cells (*see* Fig. 1), and (2) of the *Salmonella enterica* serovar Typhimurium (*S.* Typhimurium) effector SteA with a C-terminal double hemagglutinin epitope tag (SteA-2HA), expressed from its own promoter but encoded in an exogenous low-copy plasmid, during infection of HeLa cells (*see* Fig. 2). While YopE is translocated by extracellular *Yersinia* into host cells [2], SteA can also be translocated into the host cell cytoplasm from intracellular *Salmonella* residing within a membrane-bound vacuole [10], which further illustrates the versatility of the procedure.

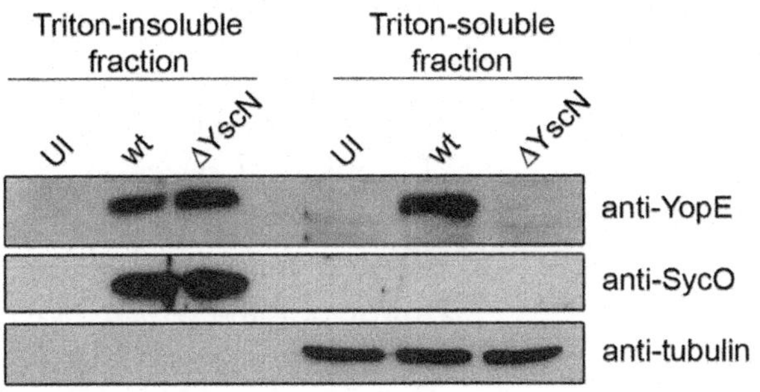

Fig. 1 Effector translocation by *Yersinia enterocolitica* during infection of RAW 264.7 macrophage-like cells. RAW 264.7 cells were infected by wild-type (wt) or T3SS-defective (*yscN* mutant) *Y. enterocolitica* bacterial strains. Triton-soluble and Triton-insoluble fractions from uninfected cells (UI) and from cells infected by wt or *yscN* mutant (ΔYscN) bacteria were prepared as described in Subheadings 2.1 and 3.1. The sample fractions were analyzed by SDS-PAGE and immunoblotting as described in Subheadings 2.2, 2.3, and 3.3. YopE is a *Y. enterocolitica* effector protein; SycO is a *Y. enterocolitica* T3S chaperone [15], used to control for possible significant bacterial cross-contamination of the Triton-soluble fraction (and as a loading control of the Triton-insoluble fraction); tubulin is a host cell protein used as a loading control of the Triton-soluble fraction

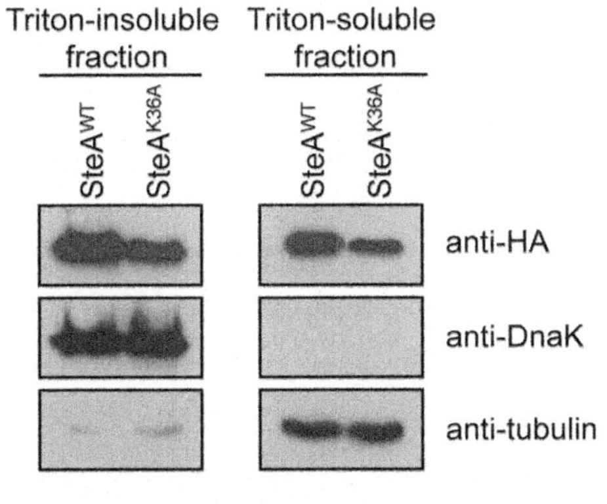

Fig. 2 Effector translocation by *Salmonella enterica* serovar Typhimurium (*S.* Typhimurium) during infection of HeLa cells. HeLa cells were infected by *S.* Typhimurium *steA* mutant bearing a plasmid encoding C-terminal 2 × HA epitope-tagged wild-type SteA (SteA^{WT}-2HA) or mutant SteA with lysine residue 36 replaced by alanine (SteA^{K36A}-2HA). Triton-soluble and Triton-insoluble fractions from cells infected by the two strains were prepared as described in Subheadings 2.1 and 3.2 (*see* **Note 29**). The samples were analyzed by SDS-PAGE and immunoblotting as described in Subheadings 2.2, 2.3, and 3.3. SteA is a *Salmonella* effector protein [10, 14], DnaK is a bacterial molecular chaperone used to control for possible significant bacterial cross-contamination of the Triton-soluble fraction (and as a loading control of the Triton-insoluble fraction), and tubulin is a host cell protein used as a loading control of the Triton-soluble fraction. Note the detection of residual levels of tubulin in the Triton-insoluble fraction

2 Materials (See Note 1)

2.1 Cell Culture, Infection, and Preparation of Cell Extracts

1. Cell lines: HeLa (clone HtTA-1) and RAW 264.7 cells (European Collection of Authenticated Cell Cultures, ECACC).
2. Bacterial strains and plasmids: *Y. enterocolitica* E40 (pYV40) (wild-type) and *Y. enterocolitica* E40 (pMSL41) (yscN$_{\Delta 169-177}$; deficient in YscN ATPase that is essential for the activity of the *Yersinia* T3SS) [11], *S.* Typhimurium *steA* mutant (an isogenic derivative of *S.* Typhimurium strain NCTC 12023 [identical to ATCC 14208s]) [12], carrying low-copy pWSK129-derived plasmids (six to eight copies per cell) [13] expressing C-terminal 2 × HA epitope-tagged wild-type SteA (SteA^{WT}-2HA) or mutant SteA with lysine residue 36 replaced by alanine (SteA^{K36A}-2HA) under the control of the *steA* promoter [10, 14].
3. Dulbecco's Modified Eagle's Medium (DMEM) supplemented with 10% (v/v) heat-inactivated fetal bovine serum (FBS) (DMEM + FBS), stored at 4 °C. Commercial 500 mL bottles of heat-inactivated FBS are stored at −20 °C. The FBS is thawed by incubation at 4 °C for 48 h, followed by preparation of aliquots in 50 mL tubes stored at −20 °C. To prepare DMEM + FBS, the aliquots are thawed in a 37 °C water bath and added to a commercial 500 mL bottle of DMEM. Do not add antibiotics to cell culture medium.
4. Earle's Buffered Salt Solution pH 7.4 (EBSS). Store at room temperature.
5. TrypLE™ Express (Thermo Fisher Scientific). Store at room temperature.
6. Phosphate-buffered saline (PBS 1×): 137 mM NaCl, 2.7 mM KCl, 10 mM Na_2HPO_4, 1.8 mM KH_2PO_4, pH 7.4. Store at room temperature. Prepared by diluting a stock of commercial PBS 10× in double-distilled water (ddH_2O), followed by sterilization by autoclaving.
7. Nalidixic acid 3.5 mg/mL: Dissolve appropriate amount in 0.1 M NaOH and filter (0.22 μm) sterilize. Store at −20 °C. Keep working aliquots at 4 °C.
8. Kanamycin 50 mg/mL: Dissolve appropriate amount in ddH_2O and filter (0.22 μm) sterilize. Store at −20 °C. Keep working aliquots at 4 °C.
9. Lysogeny broth (LB) medium: Dissolve appropriate amount of LB powder in ddH_2O and sterilize by autoclaving. Store at room temperature. Freshly supplemented with kanamycin (to 50 μg/mL) to grow *S.* Typhimurium strains.
10. LB agar: Dissolve appropriate amount of LB powder in ddH_2O, add agar to 1.6% (w/v), and sterilize by autoclaving

(store at room temperature). Allow to cool to 55 °C, and add adequate amounts of nalidixic acid (to 35 μg/mL) or kanamycin (to 50 μg/mL) to grow *Y. enterocolitica* or *S.* Typhimurium strains, respectively. The plates can be stored at 4 °C for up to 2 months.

11. Brain heart infusion (BHI) medium: Dissolve appropriate amount of BHI powder in ddH_2O and sterilize by autoclaving. Store at room temperature. Freshly supplemented with nalidixic acid (to 35 μg/mL) to grow *Y. enterocolitica* strains.
12. Triton X-100, stock solution at 10% (v/v) in PBS 1× (stored at 4 °C): Incubate Triton X-100 at 37 °C for 30 min, within the biological safety cabinet measure an adequate volume of Triton X-100, and add it to appropriate volume of sterile PBS 1× (e.g., 5 mL Triton X-100 to 45 mL PBS 1× in a 50 mL tube), mix well, and incubate 30 min at 37 °C.
13. Gentamicin 10 mg/mL. Store at 4 °C.
14. Protease inhibitor cocktail. Store at −20 °C.
15. CO_2 incubator, microbiology incubators, class II biological safety cabinet, water bath, shaking water bath with adjustable temperature, mini centrifuges.

2.2 Sodium Dodecyl Sulfate Polyacrylamide Gel Electrophoresis

1. 1.5 M Tris–HCl, pH 8.8: Dissolve an appropriate amount of Tris base in ddH_2O, adjust pH to 8.8 with HCl, adjust to desired volume with ddH_2O, and sterilize by autoclaving. Store at room temperature.
2. 1.0 M Tris–HCl, pH 6.8: Dissolve an appropriate amount of Tris base in ddH_2O, adjust pH to 8.8 with HCl, adjust to desired volume with ddH_2O, and sterilize by autoclaving. Store at room temperature.
3. Acrylamide/bis-acrylamide (37.5:1 solution). Store at 4 °C.
4. Sodium dodecyl sulfate (SDS) 20% (w/v): Dissolve an appropriate amount of SDS in ddH_2O. It is not required to sterilize the solution. Store at room temperature.
5. Ammonium persulfate (APS) 10% (w/v). Store at 4 °C (*see* **Note 2**).
6. *N*, *N*, *N*, *N*′-tetramethyl-ethylenediamine (TEMED). Store at 4 °C.
7. 12% SDS polyacrylamide gel electrophoresis (PAGE) gels (for two mini gels): Prepare resolving gel: 6.5 mL H_2O, 4.5 mL acrylamide/bis-acrylamide (37.5:1 solution), 3.8 mL 1.5 M Tris–HCl, pH 8.8, 75 μL 20% (w/v) SDS, 150 μL 10% (w/v) APS, 6 μL TEMED (*see* **Note 3**). After polymerization, prepare stacking gel: 7.34 mL H_2O, 1.25 mL acrylamide/bis-acrylamide (37.5:1 solution), 1.25 mL 1 M Tris–HCl, pH 6.8, 50 μL

SDS 20% (w/v), 100 μL APS 10% (w/v), 10 μL TEMED (*see* **Note 4**).

8. Protein molecular weight marker. Store at −20 °C (*see* **Note 5**).
9. Tris-glycine buffer: 0.025 M Tris, 192 mM glycine, 0.1% (w/v) SDS. Prepare a 10× Tris-glycine stock solution without SDS (0.25 M Tris and 1.92 M glycine, using adequate amount of Tris base, glycine, and ddH_2O). Store at room temperature. This stock solution is used to prepare the Tris-glycine running buffer, using adequate amounts of ddH_2O and 20% (w/v) SDS. Store at room temperature.
10. SDS-PAGE loading buffer 5×: 0.25 M Tris–HCl, pH 6.8, 10% SDS (w/v), 50% (v/v) glycerol, 0.5 M β-mercaptoethanol, 0.5% (w/v) bromophenol blue. Store at −20 °C.
11. SDS-PAGE mini-gel caster and migration apparatus.

2.3 Immunoblotting

1. Transfer buffer: 0.025 M Tris, 192 mM glycine, and 20% (v/v) methanol. Store at room temperature.
2. PBS 10×: 1.37 M NaCl, 0.027 M KCl, 0.1 M Na_2HPO_4, 0.02 M KH_2PO_4. Weigh adequate amounts of each of the reagents, dissolve in ddH_2O, adjust to final volume, and sterilize by autoclaving. Store at room temperature.
3. Washing solution (PBST): PBS 1× containing 0.2% (v/v) Tween 20. Store at room temperature.
4. Blocking solution: PBST containing 4% (w/v) skim milk powder: Dissolve the appropriate amount of skim milk powder in PBST (*see* **Note 6**). Prepare fresh and store at 4 °C for up to 2 days.
5. Stripping buffer: 25 mM glycine, pH 2, 1% (w/v) SDS: Dissolve an adequate amount of glycine in ddH_2O, adjust pH to 2 with HCl, add 20% (w/v) SDS to a final concentration of 1% (w/v), and adjust to desired volume using ddH_2O. Store at room temperature.
6. Nitrocellulose membranes, 0.2 μm pore size (*see* **Note 7**).
7. Ponceau solution: 0.1% (w/v) Ponceau solution in 0.5% (v/v) acetic acid: Dissolve Ponceau S in H_2O and glacial acetic acid.
8. Whatman paper.
9. Autoradiography films.
10. Primary antibodies (all stored at −20 °C): Mouse monoclonal anti-DnaK (clone 8E2/2; Millipore; used at 1:5000); rat monoclonal anti-HA (clone 3F10; Roche; used at 1:1000), mouse monoclonal anti-TEM-1 (QED Bioscience; used at 1:500); mouse monoclonal anti-α-tubulin (clone B-5-1-2; Sigma-Aldrich; used at 1:1000); rabbit polyclonal anti-SycO ([15]; used at 1:500); rabbit polyclonal anti-YopE ([16]; used at 1:1000).

11. Secondary antibodies: Mouse and rabbit horseradish peroxidase (HRP)-conjugated secondary antibodies (used at 1:10,000). Store working aliquots at 4 °C and stocks at −20 °C.
12. Immunodetection kit such as Western Lightning Plus-ECL (Perkin Elmer) or similar reagent.
13. Gel transfer apparatus.
14. Gel imaging apparatus.

3 Methods

3.1 Infection of RAW 264.7 Cells by Y. enterocolitica and Preparation of Triton-Soluble and Triton-Insoluble Fractions

1. RAW 264.7 cells are maintained in DMEM + FBS (with no antibiotics) at 37 °C in a humidified atmosphere with 5% (v/v) CO_2. The cells are used for up to 15–20 passages. The cells are routinely tested for mycoplasma contamination, using Venor®GeM Advance (Minerva Biolabs GmbH).
2. The day before the infection prepare RAW 264.7 cells and grow *Y. enterocolitica* strains: (1) seed RAW 264.7 cells at a density of 1×10^6 cells per well in six-well tissue culture plates; and (2) grow *Y. enterocolitica* in 5 mL BHI, overnight at 26 °C, with continuous shaking (130 rpm).
3. Dilute the bacterial cultures grown overnight to an optical density at 600 nm (OD_{600}) of 0.2 in fresh BHI and resume growth at 26 °C with continuous shaking (130 rpm) for 2 h (*see* **Note 8**).
4. To induce expression of the *Yersinia* T3SS genes, quickly shift the bacterial cultures to a shaking water bath (130 rpm) at 37 °C and incubate for an additional 30 min (*see* **Note 9**).
5. Centrifuge 1.5 mL of the bacterial culture (17,000 × *g*, 1 min; *see* **Note 10**) and resuspend the bacterial pellet in 1 mL DMEM + FBS and measure the OD_{600}.
6. Calculate the volume of the bacterial suspension that needs to be added to the RAW 264.7 cells to have a multiplicity of infection (MOI) of 50, i.e., 5×10^7 bacteria per well (*see* **Note 11**).
7. Add the calculated volume to the seeded RAW 264.7 cells. Carefully swirl the plates to obtain an even infection.
8. Incubate the infected cells for 3 h at 37 °C in a humidified atmosphere of 5% (v/v) CO_2.
9. After 3 h of infection, replace the medium of the infected cells by DMEM + FBS (previously warmed at 37 °C) containing 50 μg/mL gentamicin to kill extracellular bacteria (50 μg/mL) and incubate for an additional 2 h at 37 °C in a humidified atmosphere of 5% (v/v) CO_2.

10. From this point, all manipulation should be done on ice and using ice-cold solutions.
11. Wash infected cells twice with ice-cold 1× PBS.
12. Add 250 μL 1× PBS containing 0.1% (v/v) Triton X-100 and a protease inhibitor cocktail (*see* **Notes 12** and **13**).
13. Incubate cells for 10 min on ice.
14. To remove cells from the wells, pipet up and down several times (about 15–20 times) and transfer cells to a 1.5 mL tube.
15. Centrifuge samples at 17,000 × *g* for 15 min (*see* **Note 10**) at 4 °C. Remove the top 200 μL of supernatant and repeat this centrifugation step (*see* **Note 14**). Recover the top 100 μL of this second centrifugation step and add 25 μL 5× SDS-PAGE loading buffer (this is the Triton-soluble fraction).
16. Remove all supernatant from the pellet of the first centrifugation and resuspend it in 200 μL 1× SDS-PAGE loading buffer (this is the Triton-insoluble fraction).
17. Incubate samples for 10 min at 95–100 °C.
18. Use immediately 30 μL of the Triton-soluble fraction and 20 μL of the Triton-insoluble fraction for immunoblotting (*see* subsequent discussion) or keep samples at −20 °C or −80 °C until use.

3.2 Infection of HeLa Cells by S. Typhimurium and Preparation of Triton-Soluble and Triton-Insoluble Fractions

1. HeLa cells are maintained in DMEM + FBS (with no antibiotics) at 37 °C in a humidified atmosphere with 5% (v/v) CO_2. The cells are used for up to 15–20 passages. The cells are routinely tested for mycoplasma contamination using Venor®GeM Advance (Minerva Biolabs GmbH).
2. The day before the infection prepare HeLa cells and grow *S.* Typhimurium strains: (1) seed HeLa cells at a density of 2.5×10^5 cells per well in six-well tissue culture plates; and (2) grow *S.* Typhimurium in 5 mL LB overnight at 37 °C, with continuous shaking (130 rpm) (*see* **Note 15**).
3. Dilute 1:33 the bacterial cultures grown overnight in 5 mL fresh LB medium and grow the bacterial culture for 3 h 30 min at 37 °C with continuous shaking (130 rpm) (*see* **Note 16**).
4. 5–10 min before the bacterial incubation has ended (**step 3**), wash once the seeded HeLa cells with previously warmed EBSS and incubate for 15–20 min at 37 °C in a humidified atmosphere of 5% (v/v) CO_2.
5. Measure OD_{600} of the bacterial culture.
6. Dilute the bacterial culture in 5 mL EBSS (previously warmed at 37 °C) to have a MOI of 100 (i.e., to 1.25×10^6 bacteria/mL) when adding 2 mL of this suspension to the seeded HeLa cells (*see* **Note 17**).

7. Remove EBSS and add 2 mL bacterial suspension to monolayer of HeLa cells. This corresponds to the beginning of the infection (time zero).
8. Incubate the infected cells for 15 min at 37 °C in a humidified atmosphere of 5% (v/v) CO_2.
9. Wash the infected cells three times with DMEM + FBS (previously warmed at 37 °C) containing 100 μg/mL gentamicin (added fresh from the gentamicin 10 mg/mL stock solution just before the washing steps).
10. Incubate the infected cells in DMEM + FBS containing 100 μg/mL gentamicin for 1 h at 37 °C in a humidified atmosphere of 5% (v/v) CO_2.
11. Replace the medium of the infected cells by DMEM + FBS (previously warmed at 37 °C) containing 16 μg/mL gentamicin (added fresh from the gentamicin 10 mg/mL stock solution just before the washing steps).
12. Incubate the infected cells for a total of 14 h of infection, using as reference the time zero of infection (*see* **step 7** of Subheading 3.2) (*see* **Note 18**).
13. Wash HeLa cells with 1× PBS.
14. Add 250 μL TrypLE Express to the HeLa cell monolayer and incubate the cells for 5 min at 37 °C.
15. Add 1 mL DMEM + FBS and collect the cells into a 1.5 mL tube by extensively pipetting up and down (15–20 times).
16. Centrifuge at 17,000 × *g* for 1 min (*see* **Note 10**), discard supernatant, and wash cells in 1 mL ice-cold 1× PBS.
17. Repeat centrifugation and washing step (**step 16**) (*see* **Note 19**).
18. From this point, all manipulation should be done on ice and using ice-cold solutions.
19. Resuspend the pellet in 100 μL ice-cold 1× PBS containing 0.1% (v/v) Triton X-100 and a protease inhibitor cocktail.
20. Incubate for 10 min on ice with occasional homogenization.
21. Centrifuge cell lysates at 18,620 × *g* for 15 min at 4 °C (*see* **Note 10**) to separate the Triton-soluble from Triton-insoluble fraction as described in **steps 15–17** of Subheading 3.1.
22. Centrifuge samples at 18,620 × *g* for 15 min (*see* **Note 10**) at 4 °C. Remove the top 80 μL of supernatant and repeat this centrifugation step (*see* **Note 14**). Recover the top 40 μL of this second centrifugation step and add 10 μL 5× SDS-PAGE loading buffer (this is the Triton-soluble fraction).
23. Remove all supernatant from the pellet of the first centrifugation and resuspend it in 100 μL 1× SDS-PAGE loading buffer (this is the Triton-insoluble fraction).

24. Incubate samples for 10 min at 95–100 °C.
25. Use immediately 30 μL of the Triton-soluble fraction and 20 μL of the Triton-insoluble fraction for immunoblotting (*see* following discussion) or keep samples at −20 °C or −80 °C until use.

3.3 SDS-PAGE and Immunoblotting

1. Load samples on separate wells of a 12% SDS-PAGE (*see* **Notes 3**, **20**, and **21**).
2. Run the SDS-PAGE for 70 min at 150 V (*see* **Note 22**).
3. Process the SDS-PAGE for transfer into nitrocellulose membranes (*see* **Note 23**).
4. After transfer, evaluate the efficiency of protein transfer by staining the membrane(s) with a 0.1% (w/v) Ponceau solution: Sink the membrane in a few milliliters of the Ponceau solution and incubate with gentle shaking for 1–5 min, destain with distilled H_2O until protein bands are visible, and use a pen or a pencil to label the bands of the molecular weight marks and the position of lanes. If appropriate, cut the blotting membrane in strips to be detected against specific primary antibodies.
5. Using a flat-bottom incubation vessel (e.g., a Petri dish), incubate the membrane(s) in blocking solution for at least 1 h at room temperature, with gentle rocking (*see* **Note 24**).
6. Dilute primary antibody in the blocking solution and incubate for at least 1 h at room temperature, with gentle rocking (*see* **Note 25**).
7. Remove the primary antibody solution and store it at −20 °C (*see* **Note 26**).
8. Add an excess volume of PBST and rinse the membrane(s) by gentle swirling of the immunoblotting incubation vessel. Discard the PBST solution.
9. Add an excess volume of PBST and incubate the membrane(s) for 10 min at room temperature, with gentle rocking. Repeat twice (*see* **Note 27**).
10. Discard the PBST solution and incubate the membrane(s) for 1 h with appropriate HRP-conjugated secondary antibodies diluted in blocking solution.
11. Discard the secondary antibody solution and wash the membrane as indicated earlier in **steps 8** and **9**.
12. Perform immunoblot detection using ECL detection system and acquire the final image using an imaging system or by exposure to ECL autoradiography films followed by processing in a dark room using photography developer and fixer solutions (*see* **Note 28**).

13. If a membrane must be reprobed with other primary antibodies, wash the membrane in PBST (as described earlier in **steps 8** and **9**) and incubate it in an excess volume of stripping solution for 20 min at room temperature, with gentle rocking.
14. Wash with PBST (as described earlier in **steps 8** and **9**).
15. Proceed with the immunoblotting procedure, restarting from the earlier **step 5**.

4 Notes

1. Prepare all solutions using ddH_2O, unless otherwise indicated, and analytical-grade reagents. Prepare all reagents at room temperature and store them at the indicated temperatures. Follow regulations and guidelines for the manipulation and disposal of chemicals, mammalian cell cultures, and biosafety level class II organisms. All solutions and materials used for the manipulation of mammalian cell cultures must be sterile and manipulated only within a biological safety cabinet, and those used for bacterial cultures must also be sterile and manipulated by aseptic techniques.
2. The recommended procedure in classical molecular biology laboratory textbooks (e.g., Sambrook et al., *Molecular Cloning: A Laboratory Manual*) is that the 10% (w/v) APS solution should be prepared fresh. We normally prepare a 10 mL stock solution of 10% (w/v) APS that we store at 4 °C and use within several weeks with no noticeable effect in the performance of the SDS-PAGE.
3. The description is for 12% SDS-PAGE, but the concentration of the resolving gel can be adjusted according to the molecular mass of the proteins being analyzed by recalculating the volumes of acrylamide/bis-acrylamide and ddH_2O.
4. To facilitate the visualization of the wells, we normally add ~50 μL of a 2% (w/v) solution of Orange G (Sigma-Aldrich) to the stacking gel.
5. It is convenient to follow running of the SDS-PAGE and to label the bands in the nitrocellulose membrane with colored protein markers (*see* **step 4** in Subheading 3.3), but other types of markers can be used.
6. Other blocking agents (e.g., bovine serum albumin (BSA) or fish gelatin) can be used, but skim milk powder works very well.
7. Polyvinylidene difluoride (PVDF) membranes can also be used, but they need to be soaked in methanol for 15–30 s prior to use in immunoblotting.

8. The bacterial growth and infection conditions described are for *Y. enterocolitica* and need to be adapted to each particular bacterial species.
9. It is critical that the shift be done quickly and that a water bath be used to incubate the bacterial cultures at 37 °C.
10. Basically top speed in a microcentrifuge for 1 min; this is the *g* force at maximum rotations per minute in the microcentrifuge we normally use for this.
11. To accurately calculate the MOI for infection, it is necessary to establish the relation between the colony forming units (CFU)/mL of a culture grown in liquid medium and its corresponding OD_{600}. For *Y. enterocolitica* cultures we consider that an OD_{600} of 1 corresponds to 5×10^8 CFU/mL.
12. Optimal conditions to detect effector translocation for each particular experiment might have to be determined empirically, such as by the duration of the infection or the solubilization agent used. In our hands, 0.1% (w/v) Triton-X100 works well for monitoring effector translocation by *Y. enterocolitica* or *S.* Typhimurium into tissue culture cells. Other commonly used detergents in this type of assay include 0.2% (w/v) saponin [17] and 0.02% (w/v) digitonin [5, 18].
13. The infected cells can also be recovered as described for the *S.* Typhimurium infection of HeLa cells (**steps 13–20** in Subheading 3.2).
14. This is a critical step, and disturbing the pellet when recovering the supernatant must be avoided. The second centrifugation step described is designed to circumvent this problem.
15. The bacterial growth and infection conditions described are for *S.* Typhimurium and need to be adapted to each particular bacterial species.
16. These are incubation conditions that induce expression of the genes encoding the *Salmonella* pathogenicity island-1-encoded T3SS (SPI-1 T3SS) whose effectors promote invasion of HeLa cells. If infecting macrophages, the overnight culture of *S.* Typhimurium can be opsonized (or not) and used directly in macrophage infection [19].
17. *See* **Note 11**. For *S.* Typhimurium cultures we consider that an OD_{600} of 1 corresponds to 1×10^9 CFU/mL. Make sure to mix extremely well (vortex and invert the tube 10–15 times) the bacterial suspension.
18. These infection conditions are to monitor the translocation of *S.* Typhimurium effector proteins by the SPI-2 T3SS, as SteA in this protocol [10, 14], which is induced within the vacuole where *Salmonella* resides within host cells. Translocation of SPI-2 effector proteins normally can be detected 6–8 h after

bacterial inoculation of the tissue culture cells. Incubating the HeLa cells with *S.* Typhimurium for 14 h is convenient because the infection can be done late in the afternoon and the samples collected early in the morning.

19. This procedure makes it possible to obtain a pellet of infected cells and to concentrate the protein extract by adjusting the volumes of 1× PBS containing 0.1% (w/v) Triton X-100 used to lyse the infected cells.
20. Several controls should be used to rule out possible cross-contamination of the obtained fractions. To rule out the possibility of bacterial lysis and consequent release of bacterial components into the Triton-soluble fraction, the blots should be probed with an antibody against a bacterial nontranslocated protein (*see* Figs. 1 and 2). Additionally, the presence of a host cell cytosolic protein in the Triton-soluble fraction (e.g., tubulin) can be confirmed (*see* Figs. 1 and 2). Demonstration of translocation of an effector by a particular secretion system is normally done by using a secretion-defective strain (*see* Fig. 1).
21. If the proteins to be detected have significantly different molecular weights, the blotting membrane can be cut in strips to incubate each one separately with the appropriate primary antibody. If the bands of the proteins are expected to be too proximal, the membrane must be stripped and reprobed.
22. The running conditions indicated can be adjusted accordingly, depending on the molecular mass of the proteins that need to be analyzed, and the running time of the SDS-PAGE can be controlled visually based on the migration of the bromophenol blue dye or of the prestained protein markers.
23. Semi-dry or wet electroblotting apparatus can be used. Wet electroblotting transfer is known to facilitate transfer of proteins with a molecular mass >100 kDa.
24. The blocking step can also be done overnight at 4 °C or after blocking for 1 h at room temperature; the membranes can be kept at 4 °C overnight or for 2 or 3 days. Use an excess volume of blocking to ensure that the membranes are fully covered at all times.
25. The incubation with the primary antibody can also be done overnight at 4 °C. Exact conditions should be optimized for each specific antibody. The volume of the antibody solution used must ensure that the membranes are fully covered at all times, but the volume should be minimized because antibodies are usually expensive or scarce. For a small membrane, 5 mL of antibody solution in a 90 mm Petri dish is usually enough.
26. The antibodies diluted in blocking solution can normally be reused at least five or six times. If a decrease in performance is

noticed, the procedure can be repeated with a fresh dilution of the antibody.

27. If appropriate, the membranes can be left for longer in PBST (at least 1–2 h).
28. Other detection reagents or image acquisition systems can be used.
29. The use of an uninfected control is recommended (*see* Fig. 1) but can be dispensable if the primary antibody used is known not to originate a background signal in immunoblotting (as was the case with the anti-HA antibody used in the experiment illustrated in Fig. 2).

Acknowledgments

This work was supported by the Unidade de Ciências Biomoleculares Aplicadas—UCIBIO, which is financed by national funds from Fundação para a Ciência e a Tecnologia (FCT) (UID/Multi/04378/2013) and cofinanced by the ERDF under the PT2020 Partnership Agreement (POCI-01-0145-FEDER-007728) and by FCT research grants PTDC/BIA-MIC/2821/2012 and PTDC/BIA-MIC/116780/2010. Irina Franco is recipient of apostdoctoral fellowship (SFRH/BPD/102378/2014) from FCT and Sara V. Pais holds a fellowship (PD/BD/52210/2013) within the scope of the PhD program Molecular Biosciences (PD/00133/2012) funded by FCT.

References

1. Costa TR, Felisberto-Rodrigues C, Meir A, Prevost MS, Redzej A, Trokter M, Waksman G (2015) Secretion systems in Gram-negative bacteria: structural and mechanistic insights. Nat Rev Microbiol 13:343–359
2. Sory MP, Cornelis GR (1994) Translocation of a hybrid YopE-adenylate cyclase from *Yersinia enterocolitica* into HeLa cells. Mol Microbiol 14:583–594
3. Charpentier X, Oswald E (2004) Identification of the secretion and translocation domain of the enteropathogenic and enterohemorrhagic *Escherichia coli* effector Cif, using TEM-1 beta-lactamase as a new fluorescence-based reporter. J Bacteriol 186:5486–5495
4. Collazo CM, Galan JE (1997) The invasion-associated type III system of *Salmonella typhimurium* directs the translocation of Sip proteins into the host cell. Mol Microbiol 24:747–756
5. Lee VT, Anderson DM, Schneewind O (1998) Targeting of *Yersinia* Yop proteins into the cytosol of HeLa cells: one-step translocation of YopE across bacterial and eukaryotic membranes is dependent on SycE chaperone. Mol Microbiol 28:593–601
6. Schnaitman CA (1971) Effect of ethylenediaminetetraacetic acid, Triton X-100, and lysozyme on the morphology and chemical composition of isolate cell walls of *Escherichia coli*. J Bacteriol 108:553–563
7. Schnaitman CA (1971) Solubilization of the cytoplasmic membrane of *Escherichia coli* by Triton X-100. J Bacteriol 108:545–552
8. Birdsell DC, Cota-Robles EH (1968) Lysis of spheroplasts of *Escherichia coli* by a non-ionic detergent. Biochem Biophys Res Commun 31:438–446
9. Esparis-Ogando A, Zurzolo C, Rodriguez-Boulan E (1994) Permeabilization of MDCK

cells with cholesterol binding agents: dependence on substratum and confluency. Am J Phys 267:C166–C176

10. Domingues L, Holden DW, Mota LJ (2014) The *Salmonella* effector SteA contributes to the control of membrane dynamics of *Salmonella*-containing vacuoles. Infect Immun 82:2923–2934
11. Sory MP, Boland A, Lambermont I, Cornelis GR (1995) Identification of the YopE and YopH domains required for secretion and internalization into the cytosol of macrophages, using the *cyaA* gene fusion approach. Proc Natl Acad Sci U S A 92:11998–12002
12. Figueira R, Watson KG, Holden DW, Helaine S (2013) Identification of *Salmonella* pathogenicity island-2 type III secretion system effectors involved in intramacrophage replication of *S. enterica* serovar Typhimurium: implications for rational vaccine design. MBio 4:e00065
13. Wang RF, Kushner SR (1991) Construction of versatile low-copy-number vectors for cloning, sequencing and gene expression in *Escherichia coli*. Gene 100:195–199
14. Domingues L, Ismail A, Charro N, Rodriguez-Escudero I, Holden DW, Molina M, Cid VJ, Mota LJ (2016) The *Salmonella* effector SteA binds phosphatidylinositol 4-phosphate for subcellular targeting within host cells. Cell Microbiol 18:949–969
15. Letzelter M, Sorg I, Mota LJ, Meyer S, Stalder J, Feldman M, Kuhn M, Callebaut I, Cornelis GR (2006) The discovery of SycO highlights a new function for type III secretion effector chaperones. EMBO J 25:3223–3233
16. Diepold A, Amstutz M, Abel S, Sorg I, Jenal U, Cornelis GR (2010) Deciphering the assembly of the *Yersinia* type III secretion injectisome. EMBO J 29:1928–1940
17. VanRheenen SM, Luo ZQ, O'Connor T, Isberg RR (2006) Members of a *Legionella pneumophila* family of proteins with ExoU (phospholipase A) active sites are translocated to target cells. Infect Immun 74:3597–3606
18. Denecker G, Totemeyer S, Mota LJ, Troisfontaines P, Lambermont I, Youta C, Stainier I, Ackermann M, Cornelis GR (2002) Effect of low- and high-virulence *Yersinia enterocolitica* strains on the inflammatory response of human umbilical vein endothelial cells. Infect Immun 70:3510–3520
19. Drecktrah D, Knodler LA, Ireland R, Steele-Mortimer O (2006) The mechanism of *Salmonella* entry determines the vacuolar environment and intracellular gene expression. Traffic 7:39–51

Chapter 36

Quantitative Determination of Anti-bacterial Activity During Bacterial Co-culture

Juliana Alcoforado Diniz, Birte Hollmann, and Sarah J. Coulthurst

Abstract

Anti-bacterial activity assays are an important tool in the assessment of the ability of one bacterium to kill or inhibit the growth of another, for example during the study of the Type VI secretion system (T6SS) and the anti-bacterial toxins it secretes. The method we describe here can detect the ability of a bacterial strain to kill or inhibit other bacterial cells in a contact-dependent manner when co-cultured on an agar surface. It is particularly useful since it enumerates the recovery of viable target cells and thus enables quantification of the anti-bacterial activity. We provide a detailed description of how to measure the T6SS-dependent anti-bacterial activity of a bacterium such as *Serratia marcescens* against a competitor prokaryotic organism, *Escherichia coli*, and also describe possible variations in the method to allow adaptation to other attacker and target organisms.

Key words Gram-negative bacteria, Protein secretion system, Type VI secretion system, Co-culture assay, Anti-bacterial activity, Bacterial competitive fitness, Toxin/immunity pair

1 Introduction

To gain a fitness advantage in mixed microbial communities, many bacteria have developed the ability to kill competitor prokaryotic cells. Protein secretion systems are an important weapon in this war, particularly the Type VI secretion system (T6SS), which can be utilised to kill both closely and distantly related competitors efficiently [1]. This versatile nanomachinery [2], which in some cases can also be used against eukaryotic targets, is widespread in Gram-negative bacteria. In the last few years, work by different groups has resulted in the identification of many new anti-bacterial toxins, also called effectors, delivered directly into target bacterial cells by the T6SS. These include a variety of enzymes able to disrupt the bacterial cell wall, cell membrane and nucleic acid. Together with these toxins, the secreting organism possesses cognate immunity proteins that are responsible for specific neutralisation of the cognate effector to provide self-resistance [1, 3, 4].

Laure Journet and Eric Cascales (eds.), *Bacterial Protein Secretion Systems: Methods and Protocols*, Methods in Molecular Biology, vol. 1615, DOI 10.1007/978-1-4939-7033-9_36, © Springer Science+Business Media LLC 2017

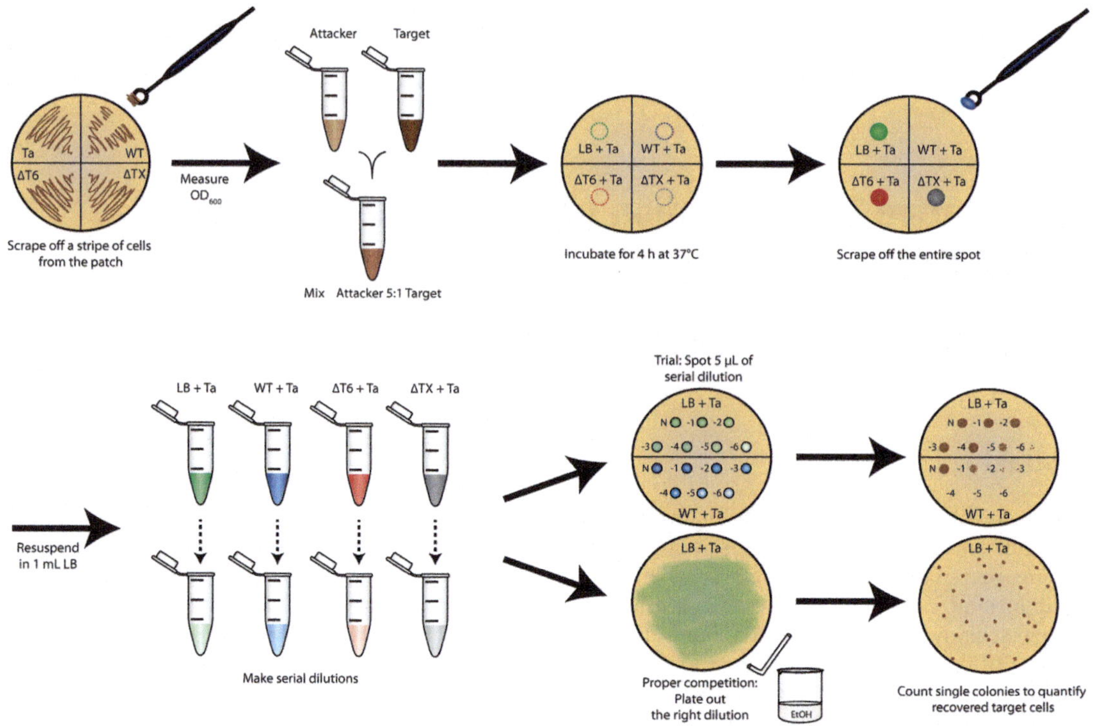

Fig. 1 Schematic overview of anti-bacterial activity assay. Following overnight growth on a solid agar surface, the target strain (Ta) and different attacker strains (wild type, WT; T6SS inactive mutant, ΔT6; a mutant lacking a toxin of interest, ΔTX; and a no-attacker, medium-only control, LB) are mixed in a ratio of 5 attackers to 1 target and spotted onto a solid agar surface. After a defined incubation period at the required temperature, the co-culture spots (LB + target, *green*; wild type + target, *blue*; ΔT6SS + target, *red*; and Δtoxin + target, *grey*) are scraped off, the cells resuspended and serial dilutions prepared. For an initial trial, these dilutions from neat (N) to 10^{-6} (−6) are spotted on an agar plate supplemented with antibiotic selective for growth of the target only. After incubation, estimation of target recovery from the trial plate is used to determine the dilution which will provide a few tens of single colonies per plate in the actual experiment. For the actual experiment, an appropriate volume of the correct dilution of the co-culture is spread on a selective plate with a glass spreader and the colonies are counted following overnight incubation; replicate experiments provide fully quantitative data. *See* text for full details

An important tool in identifying new toxin/immunity pairs and also monitoring the level and impact of the T6SS-dependent anti-bacterial activity of a particular strain is a co-culture-based anti-bacterial activity assay, such as that described here. This assay is performed on the solid surface of an agar plate, since T6SS-mediated anti-bacterial activity requires intimate cell–cell contact to permit the puncturing device of the machinery [5] to physically interact with the target cell [6, 7].

In brief, the accessible method described here involves co-culture of an attacker and a target strain of bacteria on the surface of an agar plate for a defined time, followed by the use of antibiotic selection to kill the attacker strain and allow the recovery and enumeration of viable target cells. An overall schematic depiction of this method is given in Fig. 1. To determine the number of surviving target cells following exposure to the T6SS-wielding attacker, a standard serial dilution-based viable count of target cells is performed. This method is quantitative and reliable, with an extended dynamic range (target cell recovery from 10^1 to $>10^{10}$ colony-forming units (cfu) can be quantified owing to the serial dilution approach). The alternatives to this assay use colorimetric or fluorescent reporters to distinguish target cells within a co-culture [7–9]. These approaches have the advantage of convenience but have disadvantages in that they reduce the dynamic range of the output or have potential issues in separating or distinguishing attacker and target cells during quantification.

In *Serratia marcescens*, the technique described here has been successfully implemented to demonstrate the existence and impact of T6SS-mediated anti-bacterial activity and to identify new T6SS-dependent anti-bacterial toxins [10–13]. This technique, or minor variations of it, has also been used to measure T6SS-mediated anti-bacterial activity in other organisms, including *Vibrio cholerae*, *Pseudomonas aeruginosa* and *Agrobacterium tumefaciens* [14–16]. Overall, this assay can be an important tool in demonstrating T6SS-dependent activity against competitor organisms, characterising the functionality of mutants in the T6SS and confirming the identification of new toxin/immunity pairs. It could also be applied to other inter-bacterial competitive strategies beyond the T6SS. Here we describe in detail how to perform an anti-bacterial activity assay of *S. marcescens* (attacker) against *Escherichia coli* (target) as an example, along with ways to adapt the assay to other systems of interest.

2 Materials

1. Liquid LB medium: 10 g tryptone, 10 g NaCl, 5 g yeast extract, 1000 mL deionised water. Mix, adjust pH to 7.5 and autoclave at 121 °C for 20 min.
2. LB agar: 10 g tryptone, 10 g NaCl, 5 g yeast extract, 1000 mL deionised water, 12 g Select agar. Mix, adjust pH to 7.5 and autoclave at 121 °C for 20 min.
3. Antibiotic: Dissolve 10 mg streptomycin sulfate in 1 mL deionised water and pass through a syringe filter with pore size 0.2 μm to sterilise. Aliquot and store at −20 °C.
4. LB agar plates: Following autoclaving, bring the molten agar to 55 °C. Under sterile conditions, dispense 20 mL into each 90 mm single-vented plastic Petri dish, then allow to cool and set.
5. LB agar plates plus antibiotic: Prepare LB agar plates as just described, except this time add streptomycin to a final concentration of 100 μg/mL (1/100 dilution) while the molten agar is at 55 °C and mix well before pouring the plates.

6. Sterile disposable inoculation loops 10 μL.
7. Bent glass rod and ethanol for spreading cell suspensions on agar plates.
8. Optional: Tally counter or pen-style colony counter for counting colonies.
9. 30 and 37 °C static laboratory incubator.
10. Laminar flow cabinet (if unavailable, alternative plate-drying methods can be used).

3 Methods

Carry out all procedures at room temperature and under sterile conditions.

1. Streak out the required target and attacker strains from freezer stocks and grow to single colonies according to their normal requirements. In this example, the target strain is a streptomycin-resistant strain of *E. coli* K12, such as strain MC4100 [17] (*see* **Note 1**). The attacker strains are selected according to the experiment; they should include, at a minimum, the wild type (e.g. *S. marcescens* Db10) and a T6SS inactive mutant strain (lacking one of the core components), plus as many other mutants as required.
2. Using a sterile inoculating loop, inoculate a single colony from each of the strains onto an LB agar plate as a 'patch' (Fig. 1) and incubate overnight at the optimal growth temperature for that strain. Up to five strains can be patched together on the same plate, taking care to keep them apart.
3. Dry the LB plates to be used for the co-culture spots the next day; for this, keep them open for 2 h in a laminar flow cabinet and then store at room temperature overnight.
4. Using a sterile disposable inoculating loop, scrape off a stripe of cells from each of the overnight patches described in **step 2** and resuspend in 0.5 mL of sterile LB in a sterile 1.5 mL microcentrifuge tube (agitate the loop, remove and then vortex the tube).
5. Measure the optical density (OD) of the resuspended cells for each of the target and attacker strains and normalise to an OD_{600} of 0.5 in a final volume of 100 μL of sterile LB medium (e.g. if $OD_{600} = 2.5$, then add 20 μL of the culture to 80 μL of medium).
6. Mix together normalised attacker and target cells at a ratio of 5 attacker: 1 target (e.g., 50 μL attacker +10 μL target) (*see* **Note 2**). Do this for each attacker and also include 5 LB: 1 target as a no-attacker control.

7. Spot 25 μL of each mixture onto an LB agar plate (all the spots from one replicate, i.e. one co-culture spot for each attacker, go on the same plate). When the co-culture is being performed at 37 °C (*see* **Note 3**), pre-warm this plate.
8. Wait 5 min to allow the spots to dry and immediately put plates in the incubator at 37 °C for 4 h (*see* **Note 4**).
9. Following the incubation period, scrape off each spot using a sterile disposable loop and resuspend all the cells on the loop in 1 mL sterile LB; mix thoroughly on a vortex for approximately 30 s or until the pellet is completely resuspended.
10. Prepare serial tenfold dilutions of each resuspension from neat to 10^{-6} using sterile LB (e.g. 90 μL LB plus 10 μL preceding dilution). Make sure to change pipette tips and to vortex 5 s between each dilution step. Then put diluted samples onto antibiotic-containing selective medium to enumerate the surviving target cells, in one of two formats described in **steps 11** and **12**.
11. Trial: When testing a particular attacker/target combination for the first time, it is advisable to perform a trial experiment to determine the correct dilution to spread on a selective plate to yield a few tens of single colonies. For this, perform a full serial tenfold dilution of the resuspended cells, from neat to 10^{-6}, and spot 5 μL of each dilution onto an LB + streptomycin plate, as shown in Fig. 1. Incubate at 37 °C (or the target organism's preferred growth temperature) overnight or until single colonies have grown.
12. Proper experiment: Prepare a suitable dilution of the resuspended cells for each attacker/target pair (based on the trial) and spread 50 μL or 100 μL onto LB + streptomycin plates. We recommend using a bent glass rod, sterilised by dipping in ethanol and removal using a flame, to spread the cell suspension evenly over the agar surface. Incubate at 37 °C (or the target organism's preferred growth temperature) overnight or until single colonies have grown. It is important to obtain well-spaced single colonies; if this does not happen, adjust the volume or dilution spread on the plate.
13. Count the colonies on the enumeration plates, using a tally or pen-style colony counter if preferred. Calculate the number of viable target cells recovered, expressed as cfu per co-culture spot. *See* also **Notes 5** and **6**.

    Example (Fig. 2):

    Target with LB only: 30 colonies in 50 μL of 10^{-6} dilution → $30 \times (1000/50) \times 10^{6} = 6 \times 10^{8}$ cells/spot.

    Target with wild-type *S. marcescens*: 11 colonies in 100 μL of 10^{-2} dilution → $11 \times (1000/100) \times 10^{2} = 1.1 \times 10^{4}$ cells/spot.

    Target with T6SS mutant: 45 colonies in 100 μL of 10^{-6} dilution → $45 \times (1000/100) \times 10^{6} = 4.5 \times 10^{8}$ cells/spot.

	R1	R2	R3	R4	Mean	SEM
LB	6.0E+08				6.0E+08	
WT	1.1E+04				1.1E+04	
ΔT6SS	4.5E+08				4.5E+08	

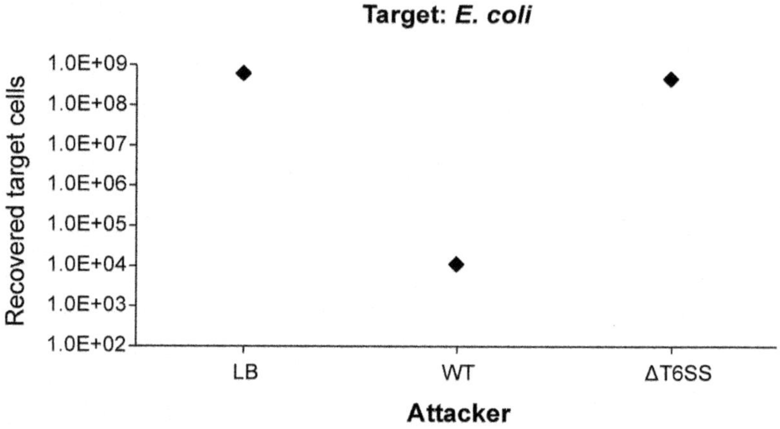

Fig. 2 Graphical representation of data generated by anti-bacterial activity assay. *Top*: Table showing number of colony-forming units per co-culture spot resulting from a typical assay (from the example in the main text). Here just the first replicate (R1) is represented so the mean corresponds to a single replicate and there is no standard deviation (SEM); however, in the proper experiment at least four replicates should be performed and the mean ± SEM presented. *Bottom*: Graph generated using Microsoft Excel with data provided in table. The *y*-axis shows the number of recovered target cells, presented as the number of colony-forming units per co-culture spot, and the *x*-axis shows the attacker strains: no-attacker control (LB), wild type (WT) and an inactive T6SS mutant (ΔT6SS)

4 Notes

1. Streptomycin-resistant strains: Target selection does not have to utilise streptomycin; however, an antibiotic to which the attacker is fully sensitive and the target is fully resistant is required. This can be achieved using an intrinsic resistance of the target strain or, alternatively, a chromosomally encoded, stable resistance determinant can be introduced using standard genetic methods.
2. Co-culture ratios: The initial ratio of attacker:target in the co-culture spot can be varied. In our experience, 5:1 or 1:1 normally gives the best outcome. However, especially if the

two organisms are mismatched in terms of growth rate or the killing effect is very small, more extreme ratios can work better.

3. Incubation temperatures: The temperature at which co-culture spots are incubated can be varied and optimised to suit the attacker/target combination. The attacker strain *S. marcescens* can grow at 37 °C or 30 °C; in this case, the choice is based on the target strain.
4. Incubation times: The incubation time for the co-culture of attacker and target can also be varied.
5. Number of replicates: To obtain quantitative data, at least four replicates are required. For convenience, two replicates per day, obtained an hour apart, is ideal. Start the replicates from fresh patches of cells.
6. If desired, the recovery of the attacker following co-culture can also be determined simultaneously, for example if the final attacker:target ratio is of interest. In this case, the cells recovered from the co-culture are enumerated in parallel on plates containing a second antibiotic, one to which the target is sensitive and the attacker is resistant.

Acknowledgements

This work was supported by Coordenação de Aperfeiçoamento de Pessoal de Nível Superior (CAPES, Ph.D. studentship to JAD) and the Wellcome Trust (Senior Fellowship to SJC).

References

1. Alcoforado Diniz J, Liu YC, Coulthurst SJ (2015) Molecular weaponry: diverse effectors delivered by the Type VI secretion system. Cell Microbiol 17:1742–1751
2. Cianfanelli FR, Monlezun L, Coulthurst SJ (2016) Aim, load, fire: the Type VI secretion system, a bacterial nanoweapon. Trends Microbiol 24:51–62
3. Durand E, Cambillau C, Cascales E et al (2014) VgrG, Tae, Tle, and beyond: the versatile arsenal of Type VI secretion effectors. Trends Microbiol 22:498–507
4. Russell AB, Peterson SB, Mougous JD (2014) Type VI secretion system effectors: poisons with a purpose. Nat Rev Microbiol 12:137–148
5. Shneider MM, Buth SA, Ho BT et al (2013) PAAR-repeat proteins sharpen and diversify the type VI secretion system spike. Nature 500: 350–353
6. Russell AB, Hood RD, Bui NK et al (2011) Type VI secretion delivers bacteriolytic effectors to target cells. Nature 475:343–347
7. Schwarz S, West TE, Boyer F et al (2010) Burkholderia Type VI secretion systems have distinct roles in eukaryotic and bacterial cell interactions. PLoS Pathog 6:e1001068
8. Gueguen E, Cascales E (2013) Promoter swapping unveils the role of *the Citrobacter rodentium* CTS1 type VI secretion system in interbacterial competition. Appl Environ Microbiol 79:32–38
9. Hachani A, Lossi NS, Filloux A (2013) A visual assay to monitor T6SS-mediated bacterial competition. J Vis Exp 20:50103

10. Alcoforado Diniz J, Coulthurst SJ (2015) Intraspecies competition in *Serratia marcescens* is mediated by Type VI-secreted Rhs effectors and a conserved effector-associated accessory protein. J Bacteriol 197:2350–2360
11. English G, Trunk K, Rao VA et al (2012) New secreted toxins and immunity proteins encoded within the Type VI secretion system gene cluster of *Serratia marcescens*. Mol Microbiol 86: 921–936
12. Fritsch MJ, Trunk K, Diniz JA et al (2013) Proteomic identification of novel secreted antibacterial toxins of the *Serratia marcescens* Type VI secretion system. Mol Cell Proteomics 12:2735–2749
13. Murdoch SL, Trunk K, English G et al (2011) The opportunistic pathogen *Serratia marcescens* utilizes Type VI secretion to target bacterial competitors. J Bacteriol 193:6057–6069
14. Hood RD, Singh P, Hsu F et al (2010) A type VI secretion system of *Pseudomonas aeruginosa* targets a toxin to bacteria. Cell Host Microbe 7:25–37
15. Ma LS, Hachani A, Lin JS et al (2014) *Agrobacterium tumefaciens* deploys a superfamily of type VI secretion DNase effectors as weapons for interbacterial competition in planta. Cell Host Microbe 16:94–104
16. Macintyre DL, Miyata ST, Kitaoka M et al (2010) The *Vibrio cholerae* type VI secretion system displays antimicrobial properties. Proc Natl Acad Sci U S A 107:19520–19524
17. Casadaban MJ, Cohen SN (1979) Lactose genes fused to exogenous promoters in one step using a Mu-lac bacteriophage: in vivo probe for transcriptional control sequences. Proc Natl Acad Sci U S A 76: 4530–4533

Index

Laure Journet and Eric Cascales (eds.), *Bacterial Protein Secretion Systems: Methods and Protocols*, Methods in Molecular Biology, vol. 1615, DOI 10.1007/978-1-4939-7033-9, © Springer Science+Business Media LLC 2017

T

U

W

X

Y

Printed by Printforce, the Netherlands